地方重塑
LOCAL REMODELING

国际公共艺术奖案例解读1

总策划 汪大伟　主编 金江波

上海大学出版社

目录
CONTENTS

序
FOREWORD

这本册子是对来自全球的 142 个以"地方重塑"为主旨的公共艺术案例的解读。

在全球化背景下，人类文化多样性问题日益受到关注。文化生态的健康与否既是衡量一个地区文化潜力的重要指标，也是保护地域文化遗产的关键要素。"地方重塑"不仅是局部的、实时性的地方再造，也是整体的、历时性的地区文化生态工程。重视推进文化生态研究，把文化置于生态之中，用历史视角解读地域文化变迁，研究文化演变与这一地区其他生态系统如自然、经济、社会等的关系。这些研究既是了解一个地区的重要资源，也为艺术家制定地区可持续发展方略提供借鉴和启示。

虽然世界不同地区存在不同的文化差异性，但公共艺术是文化认同的载体，一方面以艺术的形式表达群体的文化本质，另一方面公众又以艺术的方式参与其中并形成文化认同。因此，在"地方重塑"的过程中，公共艺术对地区的介入，就文化学意义而言，正是一种认同与再造的结合，"地方重塑"的理念也在这种认同中具有了普遍意义。

"地方重造"将艺术引入大众日常生活，引导大众重新认识、界定自己的生存环境，赋予地方建设以更大的开放性。加入艺术元素的社区环境生出一种双向机制：日常生活既因为艺术的到来而获得新的层级结构和发展潜质，导致新的文化价值思考，同时，生活中的许多元素又对艺术的存在方式及发生逻辑产生冲撞，从而产生出更加具有社会属性和公众意义的艺术样式，在交叉中使艺术类型更加多元化并产生新的动力。

我相信，"地方重造"将是公共艺术的永恒主题。

上海大学美术学院院长、教授
汪大伟
2014 年 10 月 31 日

编者语
EDITOR COMMENTS

公共艺术日益成为国家文化战略的重要组成部分。公共艺术概念的倡导与传播，是走向成熟和文明社会的现实反映与发展需求。当今的公共艺术不仅仅是美化环境与改善空间的美学方式，更重要是对于区域建设、解决现实问题和重塑地方精神的一种运作机制。

我们所提倡的"地方重塑"的理念是希望从不同角度来关注和诠释城市生活与地域文化。关注空间环境，关注人文、历史脉络与公众日常生活，体现出公共艺术对于重塑城市文明及重塑城乡文化生态的意义。公共艺术普遍存在于广大人民群众活动、休闲、观赏的公共领域，因此人文关怀精神、对人的尊重和体贴成为最基本的出发点。一件有价值的公共艺术作品能够重塑空间面貌，更重要的是能提升特定空间中人的精神风貌。这些作品的成功之处在于将公共空间和文化心理及价值观念的阐释紧密融合，营造新的人文情怀，将公众的表达诉求与公共领域文化功能统筹起来，体现人的主体意识与主体精神，从而实现公共文化的价值理想。

围绕着"地方重塑"的价值观与理念发展的需求，我们构建了一个覆盖来自全球六大洲各地区的公共艺术研究员人才网络，他们为我们带来了从 2006 年 1 月 1 日到 2011 年 9 月 30 日期间的件 142 公共艺术作品。这些提名的作品在不同层面上具有标杆性，作品充分体现以艺术家的创造力为核心、突出"地方塑造"为主题概念的公共艺术项目。

提名作品包括壁画、雕塑、社区改造、空间转换、艺术活动等多种形式，艺术家通过公共艺术的方式表达出对城市发展的特定思考。公共艺术语境下的壁画和雕塑既不再是以装饰功能为主的"环境壁画"或"城市雕塑"，也不是以表现艺术家个体语言和艺术风格为主的载体，而是具有明确社会价值和文化理想的艺术形态。这些公共艺术作品从不同层面演绎了当代文化的深刻命题，体现出艺术家、艺术作品与城市环境之间的文脉关系。公共艺术的方式表达了公众对公共领域文明的特定思考，不同区域之间由于千差万别的自然、人文等因素，因此也为创造出不同主题和模式的"地方重塑"样本提供了无限可能。我们希望这些案例能给我们带来新的文化视野与国际经验，更好地为当下的中国社会需要什么样好的公共艺术项目而服务。

上海大学美术学院院长、国际公共艺术奖发起人之一汪大伟教授曾这样归纳道："我们借国际公共艺术奖的活动机会，构建了一个来自全球的六大洲七大地区公共艺术研究员组成的研究网络，他们代表了这个领域中全球的最高水准，而我们能够参与组织、构建这样的人才智库网络，这将为我们国家的公共艺术学科研究提供更高的学术平台。"

我们汇编这本"地方重塑"公共艺术案例解读册子，希望能将国际化研究的成果共享给业界。我们明白"解读"不是"呈现"，本册汇编的内容有专家的评述，亦有业界精英的访谈，更有博士生们独到的阅读体会。我们汇编成综合的文本，希望从不同的视线和角度来体察、解析这些来自不同文化与制度下的公共艺术案例，发现和挖掘它们背后的故事。让它们如镜子般折射出我们自己的艺术生态与运作机制，提示我们进一步思考需要什么样"好的公共艺术"来到我们的生活中。

这本《地方重塑——国际公共艺术奖案例解读1》册子，是我们围绕着每两年举办的"国际公共艺术奖"活动的学术成果梳理。这也将是我们坚持公共艺术学科发展长期计划的重要部分。作为首届案例汇编的解读册子，我们希望既提供给业界以解读的线索和启示，也更希望这些"案例"留给业界更多的解读空间。

上海公共艺术协同创新中心执行主任
上海大学美术学院副教授、硕导
金江波
2014 年 10 月 20 日

专家谈公共艺术

EXPERT VIEWS

路易斯·比格斯
Lewis Biggs

这么多年来，艺术给社会带来的变化和艺术基础发展是有目共睹的。作为一名策展人，我一直在关注这个主题。这也是为什么多年我以来一直在推动公共艺术的发展情况，并关注如何使美术从博物馆和造型艺术中解放出来，走进城市生活和公共空间的原因。我希望我能为中国的艺术基础发展也做点贡献。

这次评奖和论坛是一次非常国际和专业的学术活动，我喜欢了解不同背景下公共艺术的状况。国际评奖和论坛是个挑战，因为有许多迷人的文化差异。事实上，如何定义公共艺术以及如何看待公共艺术非常重要，因为公共艺术必须同公共空间相连。这种文化差异深深吸引了我。

路易斯·比格斯（Lewis Biggs）简介：
国际公共艺术奖评审委员会主席，国际公共艺术奖（IAPA）评委会主席，国际公共艺术协会主席。 曾任英国文化协会视觉艺术部展览馆员、圣保罗双年展英国专员、泰特美术馆馆长、利物浦双年展创始人和利物浦双年展主席等。

汪大伟
Wang DaWei

公共艺术是空间中的艺术，有的人把公共艺术定义为大众服务的理念，也有的把公共艺术看作是一种运作机制。我认为公共艺术是用艺术的语言和方式去参与和解决公共问题。

从 1998 年起，我就提出以学科建设的思路来发展公共艺术。至今，虽然已为城市建设做出了不少贡献，但我们认识到，公共艺术要在学科建设和学术研究方面取得长足的发展，还需要一个国际对话的渠道。

大家逐渐认识到公共艺术是提升人们生活幸福指数的有机组成，这样的认识过程和发展形势，意味着我们需要一个平台，进行广泛的国际交流，并向公共艺术发展迅速的西方国家取经。这是发起国际公共艺术奖的原因之一。

在学科建设过程中，我们尚处于起步阶段，对公共艺术的理解以及为社会服务的方式方法的认识方面都不够全面。但是中国的城市化进程与城镇建设发展迅猛，愈发感到公共艺术在其中作用的重要性。因此，与国际合作，通过交流对话，结合中国的国情，逐步探索和构建公共艺术的评价体系，与国外同行平等合作，互相促进。

党的"十八大"明确提出加快城镇化发展，如何在城镇化过程中提升人的生活质量和幸福指数，就成了一个值得关注的问题。这不仅是投入硬件，更重要的是文化建设。而艺术恰恰是最为人民群众喜闻乐见的文化建设方式。艺术可以释放心灵，可以缓解矛盾，让并不富裕的生活充满幸福与满足，由此可见公共艺术在城乡发展过程中的重要性。设立这个奖项，可以开拓思路，通过国际合作来引进、参与，甚至帮助解决中国城镇化过程中存在的一些问题。

总而言之，是鉴于学科的需求，建立一个国际合作的评价体系的需求，以及中国自身发展的要求，我们设立了"国际公共艺术奖"这样一个平台。这次是以研究作为基础，进行推荐评选的评奖机制。与其说是评奖，倒不如说是全球公共艺术研究网络的建设，也是我们共同构建评价体系的过程。

"国际公共艺术奖"应该聚集全世界公共艺术最优秀的创作人才和策划人到举办地，为当地解决建设问题。这样，今后谁要解决问题就要把申请这奖的举办权。是以解决举办地实际问题的方式来扩大公共艺术奖的影响力。

汪大伟（Wang DaWei）简介：
上海大学美术学院院长、教授、博导，《公共艺术》杂志主编。 现任中国美术家协会理事、中国美术家协会平面设计艺委会副主任、上海市文联委员、上海创意设计工作者协会主席、上海美术家协会副主席、上海公共艺术协同创新中心主任、教育部高等院校艺术设计专业指导委员会委员、 教育部艺术硕士专业指导委员会委员、上海地铁建设环境艺术委员会委员、上海双年展艺术委员会委员、上海艺术博览会艺术委员会委员。

杰克·贝克尔
Jack Becker

"中国的公共艺术是将艺术融合于空间，并且和城市建设等有着莫大的关联。"就此，他诚恳地提出了六个方面的建议，包括派遣有才华的艺术家游学海外、与欧美进行交流合作、从小即开展公共艺术教育、支持更多对话和交流、给予公共艺术家充分的时间。有理由相信，这些建议对中国公共艺术的发展将有所助益。

杰克·贝克尔 (Jack Becker) 简介：
美国，艺术家，"公共艺术预测"执行董事、《公共艺术评论 "杂志发行人
Artist, Executive Director Forecast Public Art, Publisher of Public Art Review Magazine

蕾哈娜·戴文波特
Rhana Devenport

我觉得对于公共艺术，有三个问题需要探讨：第一，为什么要做公共艺术？第二，谁创造了公共艺术以及为什么？第三，谁会关心公共艺术？这三个问题好像太生硬了，但是的确值得我们思考。我们必须要考虑整个公共艺术创立的动机是什么以及跟观众是如何联系起来的。

蕾哈娜·戴文波特（Rhana Devenport）简介：
新西兰奥克兰国立美术馆馆长
Director，AUCKLAND ART GALLERY TOI O TĀMAKI

长谷川祐子
Yuko Hasegawa

对于传统来说，公共艺术会包括环境设计、建筑，往往是在现有景观的基础上补充一些文化价值。出于一些象征性意义或者是出于增加本地的旅游价值的考虑，人们希望使用一些公共艺术。但我认为，今后公共艺术会更多地增强文化和公众之间的关系。它不应该依靠单一的艺术形式，应该增强地域性，而且能够从本地居民那儿取得精华，能够增强本地的文化和历史背景。很重要的是，公共艺术能够帮助本地发现更多的可能性，能够释放本地居民更多的潜力。我想这是对于公共艺术的展望。

当然在这个过程当中，会给本地社区带来很多经济利益。比如说，丰富他们的文化，增加本地的旅游价值，我想除此之外会有更多的价值，比如说让人们发挥更多的想象力。策展人也好，这些项目的协调者也好，他们发挥着非常重要的作用。我觉得我们这次公共艺术论坛，应该能够触发我们去思考，艺术怎么样能够充分利用人际资源，能够使我们的公共艺术成为人和社区之间以及文化之间的桥梁。

我想摆脱欧洲的视角，希望让本地人能够描绘自己的文化蓝图。因此我们可以看到，它会赋予本地更多的文化氛围。在这样的思维的指引下，沙迦双年展上出现了诸如 EMESTO NETO、SANAA、SuperFlex 等既融合了伊斯兰风格，又契合公共艺术理念的优秀作品。

我觉得我们要关注当地人，也要关注一些本地人，他们会有很多的潜力，他们可以用自己的想法，用自己的想象力去创造公共空间的各种可能性，这是我自己的一种体会。也是我在沙迦这个双年展工作当中的一些体会。

长谷川祐子（Yuko Hasegawa）简介：
日本东京当代艺术馆（MOT）总策展人
Chief Curator of Museum of Contemporary Art, Tokyo (MOT)

保科丰巳
Hoshina Toyomi

今天，日本的当代艺术与社会以及日常生活的关系愈加密切，尤其是 2011
年 3 月的日本东部大地震以来，以自然环境和人类自身为中心的"社区设计"
被作为公共艺术的新主题，以艺术的力量作为推进"造乡运动"、"地域
振兴"等事业的重要环节的策划越来越多。相对于以东京为中心的当代艺术，
这样的"艺术工程"正在日渐显示出向地方展开的可能性。

保科丰巳 Hoshina Toyomi 简介：
东京艺术大学美术学部学部长，教授
Professor，Dean Faculty of Fine Art of Tokyo University of the Arts

曹意强
Cao YiQiang

我认为应该从三个角度讨论公共交通的艺术：一是艺术怎样融入整体的地铁建设，而不是作为一种装饰，如何使艺术成为地铁建造包括功能各方面的一个有机组成部分。二是在受限的建筑空间里面，怎样利用艺术，让艺术再重新融入到建筑里面去，使它构成一个整体。三是从理论方面，就是刘悦笛研究员，主要谈生活美学，里面涉及公共艺术的价值问题。非常有意思。

我一直在想，我们考虑艺术的时候，往往把艺术当成一种丰富公共环境的东西，其实这种考虑在某种程度上是没有真正体现艺术更深层的价值。我认为艺术在给我们提供美的过程中，其实在不断塑造我们的一种敏感性，而且对于我们的思想、思维以及对于我们的科学创造包括我们对于生活问题的处理，提供的都不光是一种想象力，而是一种非常高级的技术手段。公共艺术不是一种添加物，而是解决我们的技术问题，解决我们的能源、环保、安全以及生活问题的一种智慧，同时也是一种创造性手段。这给我非常深的体会。也就是说，我们说的公共艺术，其实是艺术与科学、与我们生活的一个有机结合体。我认为，公共艺术用中文解释，就是人类整体环境的有机艺术。

曹意强（Cao YiQiang）简介：
国务院艺术学科评议组召集人，中国美术学院艺术人文学院院长、教授
Convener,Aartistic Disciplines Consultative of the State Council
Professor, Dean, College of Arts and Humanities of China Academy of Art

方晓风
Fang XiaoFeng

我关注的核心点是公共性。我觉得因为公共性，尤其在中国我觉得还是讨论得不够充分的一个概念。公共性意味着不同人群对公共事务的平等介入，由这一认识出发，艺术的目的、手段以及艺术所应作出的姿态都必须有相应的调整。对于普通公众而言，姿态可能是最易感知的内容，姿态本身也应被视为艺术家创作的一部分并得到充分的讨论。而艺术究竟应该以何种姿态介入生活，是个正在被逐步认识和探索的问题。

现代艺术应更进一步去精英化。艺术家，如果能够去掉这种权利影子，艺术才能够得到更彻底的尊重。另外一个问题，就是参与，一方面是有利益引导的参与，另一方面我们是想激发对公共事务或者公共话题讨论的参与。所以，我个人对公共艺术的理解是艺术在这个地方只是手段，而不是目的，目的反而是公共性，所以公共艺术我的理解是，以艺术为手段的，对公共性的一种追求。

方晓风 Fang XiaoFeng 简介：
清华大学美术学院《装饰》杂志主编、教授
Professor, Editor, "ZhuangShi" by the Academy of Arts&Design ,Tsinghua University

翁剑青
Weng JianQing

于我而言，艺术与生活之间最密切的关系，就在于它的实践性，包括公共艺术在内的当代艺术实践，需要对于现实问题和需求做出积极的介入。对于 21 世纪的中国公共艺术而言，其社会和文化的使命，主要在于积极参与城镇一体化进程中的公共空间、公共生活场所的建构与改造。创造性的介入和干预城镇社区的日常生活形态和观念形态的传承与变革。而不仅仅是政治文化和商业资本的一种传播工具，或仅仅是纯粹的视觉美学的一种张扬。所以我说，公共艺术的创作和表现方式，显然应该立足于其地域的文脉、地域的特性和地域性的价值的揭示与发挥，从而使之成为振兴和再造地方社会的文化方式，和富有生机的艺术途径。"

翁剑青（Weng JianQing）简介：
北京大学艺术学院教授
Professor, School of Arts, Peking University

杭间
Hang Jian

荷兰哲学家海因茨说，美学来源于城市生活，后者是前者的土壤。然而，这其中的内容，有的是会变成一个形式的躯壳。有的虽然有思想，但是有可能这个思想可能是一个假象。另外不可否认，在所谓的一些城市的公共艺术的过程当中，有很多是由商业策略推动的。所以我个人认为，尤其要警惕创意产业这四个字，在创意产业的口号下，公共艺术的娱乐化生产会变得令人怀疑，因为这里面我很想知道，它的内在的资本、它的文化的价值以及这个文化的消费之间，究竟是一个什么样的关系。

其次，随着新一代的成长，公共性的本质蜕变为娱乐知识的服务，公共生活的民主精神被不断地肢解，这样的状况在东亚已经变得比较明显。那么在这样的一个过程当中，我想公共艺术如何在保持刚才有的先生发言当中提到的，公共性的本质也是值得思考的。

并且，我认为这种在美学城市的发展过程当中，公共艺术所出现的城市化、躯壳性以及商业性影响的这样一些因素，如果在专制制度下出现就更有危险性。专制制度一个非常重要的特点就是不是经过充分吸收大多数的意见而产生的决定，往往通过权力交换、利益交换而来实现某一个所谓的艺术样式或者是公共行为的建成。在这样的一种情形下，可能会遭受到非常大的困难。

有一种解决办法是回到公共艺术家的状态。所谓的公共艺术家，我指的是那些城市的游荡者，对于他们来说，城市如果变成这些艺术家的最好的隐居地，能够让他们自由的游荡或者闲逛。这样的艺术可能产生某一种超越一般价值之上的结果。当然，同时这种艺术原本状态的产生必须跟民众、社区，跟民众的意愿，跟社区需求产生一种非常密切的结合。

杭间（Hang Jian）简介：
中国美术学院 教授、中国包豪斯研究院院长、中国美术学院美术馆馆长
Professor, China Academy of Art

顾骏
Gu Jun

我认为公共性不是艺术家所认为的公共性，而是民众实际生活中体现的公共性。所以在这一点上来说，艺术家是我们许多城市公共行动的发起者和动员者，我非常欣赏这句话，但是我又反过来觉得，作为发起者和动员者，公共艺术家能否成功，可能同他的技术水平、艺术表现手法和他个人的能力有关。但是根本上可能还要取决于公共艺术家的理念与公民实际需求的吻合程度，以及两者之间的良性互动。

我们在四川美院的设计里看到了对农业的尊重、其实这也是对环境的尊重，对我们曾经生活状态的尊重。并且，四川美院保留的农村、农业和农民的原生态，恰恰要同今天农民的公共需求联系起来考虑。如今，中国农村已不再是人人渴望进城的阶段了，相反，现在城里人越来越希望到农村来玩玩，特别是最近一段时间，中国城市里面遭遇大规模的雾霾，所以去农村洗洗肺也成为一种时尚，更重要的是，越来越多的农民经过城市生活和农村生活的比较之后，他们开始发现农村生活是更人性化的生活。因而，农民对于农村和城市两种生活的不同方式的选择是今天四川美院保留农村、保留农民、保留农业的原生态的更深刻的理念。所以我想有许多时候，艺术家的理念很重要，没有理念，在某种程度上就没有公共艺术，但是仅仅只有艺术家个人的理念，却不能得到公众的认可，那么我可以承认他是艺术家，很难承认他是公共艺术家。

顾骏 Gu Jun 简介:
上海大学社会学院教授
Professor, School of Sociology and Political Science, Shanghai University

黄英彦
Huang YingYan

我们坚持赋予艺术以专业独立性，通过长期对艺术机构和大型艺术活动提供无偿资助和营运支持的方式，为中国当代艺术在本土的良性发展机制和华侨城自身的文化品质与公共艺术环境建设，树立了受到海内外称誉的"华侨城模式"。在我看来，公共性就是艺术应该跟公众发生关系。让公众对艺术能够触手可及，成为我们所在城市的高品质的文化生活的窗口。

黄英彦（Huang YingYan）简介：
上海天祥华侨城投资有限公司公共关系总监
PR Director, OCT(Shanghai) Land Co.,Ltd

孙珊
Sun Shan

上海公共空间主要存在四个问题：第一是公共空间的数量和面积不足，服务半径过大。第二是尺度不够宜人，有广场并不代表有人愿意在里面停留。设施的配置和活动以及人性化的设施，都是考虑不足的。第三是上海有很多历史建筑，但历史建筑的元素挖掘不够，风貌展现不够，很多历史建筑还在隐藏，而且被保护和利用得也不是非常好。第四是上海空间中的艺术设计品质还有待提升。

我们也学习了一些国际上好的做法，譬如巴塞罗那、纽约、里昂等城市的做法。目前，我们在做的第一件事情，就是充分利用空间。第二件事情，我觉得可以做的就是对历史古建和古树名目进行一个全方位梳理，纳入到前面讲的一些公共活动和串联的体系里面，并且把它纳入到整个的管理机制里面。第三是以街区为单位，全面推进整个城市的更新。

我认为在"十八大"提出的新型城镇化的概念里面，不同的城市是有不同的路径的，我们上海的新型城镇化更加应该注重内涵式的发展，进一步聚焦发展质量，全面提升城市品质，在空间上大有作为，最后通过空间和艺术的发展，建设魅力上海，提升整个城市的硬实力和软实力。

孙珊（Sun Shan）简介：
上海市城市规划设计研究院副院长、教授
Professor, Vice Dean, Shanghai Urban Planning Institute

王受之
Wang ShouZhi

自改革开放二三十年以来，中国的城市发展很快，民众开始有了公民意识，有了公民意识，公共艺术开始走向一条路，就会说话了。中国的公共艺术和城市的关系，即当中国城市里面的居民逐步产生了比较强烈的公众意识的情况下，中国的公共艺术会走向一条应该走的道路。目前，中国的城市化越来越快。这一届的大会，提出一个口号叫"城镇化"，我觉得这是中国一个伟大的进步，不说"大都会化"。有了这个发展，中国的公共艺术其实就可以扮演一个更加活跃的角色，在很多地方创造很多人能够生存的空间。另外，公共艺术在中国城市里面扮演的角色，不完全是一个艺术欣赏的对象，应该是给城市居民带来一个欣赏艺术的空间。

而我们所讲的公共艺术，有几种，其中一种是公共拥有的艺术。现在可以说所有艺术品都是公共拥有，因为国家就是我们的代表，我们假设这么说。但是另外一种是公共参与的艺术品。第三个就是让政府对于艺术品有一定的重视，对于公共艺术能够有一定的投入，重视这个是一个城市必须的东西。最后一个就是提高市民的审美能力，让市民终于有一天能够把公共艺术变成公众参与的艺术，公众创作的艺术。

王受之（Wang ShouZhi）简介：
汕头大学长江艺术与设计学院副院长、副教授
Professor, Dean, Yangtze Art and Design College, Shantou University

杨奇瑞
Yang QiRui

关于公共艺术的文化探讨与学术研究，在中国已经历时多年，观点多元而丰富。而城市化进程，在中国呈现出"剧变"的特征。 城市化进程都具有城市更新、历史街区改造这样一个特点。所以我的作品都是关于历史街区改造的。在这个作品的思考当中，我觉得作为我们这代人做艺术，总是想在作品当中有一点内涵和思想性。因此要尊重艺术与城市空间、艺术与大众、艺术与本土文化的关系以及艺术家的独立精神，并且始终反映"源于生活，取之于生活，还给生活，还给百姓"的公共艺术理念。

杨奇瑞（Yang QiRui）简介：
杨奇瑞中国美术学院公共艺术学院院长、教授
Professor, Dean, College of Arts and Humannities of China Academy of Art

孙振华
Sun ZhenHua

公共艺术的本体问题有三种可能：一是一个公共艺术的发起者或者一个组织者或者一个创作者他给公众创造了一个公共艺术作品，还有一种方式和公众一起完成了一件公共艺术作品。最后还有一种方式就是让公众自己完成了一个公共艺术的作品，或者说实践的一个过程，我觉得这个实际上是不一样的。

说到参与的问题，其实参与也有不同的参与方式。包括欣赏，能够得到或者欣赏到公共艺术这是一种参与，能够创造公共艺术，这也是一种参与。最后还有一种，你是否能够参与到公共艺术的决策，赋予公众民主权利。其实在很多时候，人们不是先想好了什么是公共艺术以后才去做的，而是做的过程中将是一个长期的定义公共艺术的过程，这个过程以后再来回顾的话肯定是充满趣味的。

孙振华（Sun ZhenHua）简介：
深圳雕塑院院长、教授
Professor, Dean of ShenZhen Sculpture School

毛竹晨
Mao ZhuChen

后世博的公共艺术就我个人而言，更希望看到我所经历的，或者说上海人和世界所有游客共同经历的记忆，这个城市的梦想，它的文脉，以某种形式浓缩被提炼，在这个地方继续传承下去，可能若干年之后，大家回到这个地方，通过某一件作品，某一个公共艺术的形式，依然能对这个时代、这个事件感兴趣，我想我要看到的可能就是这个。

毛竹晨（Mao ZhuChen）简介:
上海世博发展集团创意策划部总经理
General Manager, Creative Planning Fivision of Expo Development Group

丁乙
Ding Yi

上海"世博源"公共艺术规划方案的主题概念是汇、源、聚、韵。整体公共艺术设置方面有两个比较重要的思考，一个是固态、动态和互动的，一个是永久性和临时性和虚拟性的。同时，在整个作品布局时，从景观和整体形态都做了相应的考虑。包括从艺术品互相之间的编排、连接方面做了很多思考。对作品能够反映某种地域特征以及中国的传统，包括视觉的角度、心理层面等都做了一定研究，对一些公共艺术的点位和人流之间的关系做出分析。此外，在整个实施方案中，围绕"六点一线"进行公共艺术和世博轴之间关系的定位。通过六个阳光谷的改造，在整个最上层的平面上进行公共艺术的概念设计。六个阳光谷的改造归纳了六个词：寻找城市源头、记录城市印迹、延续城市文脉、关注当代生活、展现文化新潮、浓缩生活体验。定位为"一个是水的谷，一个是自然谷，一个是乐谷，一个是亭谷，一个是影像谷"。另外，在二楼全长1045米南北贯通世博园平台上面设计一条以观光车为线路的丰富的线，这个线又对应了下面的商业主题：乐活、时尚、潮流、品位。可以说，这个项目既是通过文化传播的方式增添公众购物新体验，也是对公共艺术的概念、形式、功能和意义的综合实验案例的探索。

丁乙（Ding Yi）简介：
上海复旦视觉艺术学院美术学院副院长、教授
Professor, Vice Dean, Academy of Fine Arts of Shanghai Institiute of Visual Art, Fudan University

王中
Wang Zhong

我认为公共艺术在中国更强调艺术营造空间。而艺术营造空间的背后我们更应该强调艺术激活空间。公共艺术应该是植入公众土壤中的一个种子，它应该诱发文化生长，它应该让文化具有生长性，这是最重要的。 而城市公共艺术的建设，是一种精神投射下的社会行为，艺术和美不是唯一的目标，而是一种态度、一种眼光、一种体验，甚至是一种生活方式。让城市文化从日常生活中彰显出来，并与当下生活发生关联，将艺术植入城市肌体，激活城市公共空间，使艺术成为植入城市公共生活肥沃土壤中的"种子"，诱发文化的"生长"，延伸喜悦、激发创意，让艺术成为城市的精神佳肴，令城市焕发生机和活力，并为城市带来富有创新价值的文化积累。

王中（Wang Zhong）简介：
中国美术学院城市设计学院副院长、教授
Professor , Vice Dean, School of City Design of Central Academy of Fine Arts

艾米·巴拉克
Ami Barack

艺术作品存在于公共场所，这会让大家有种归属感和参与感。在博物馆或美术馆这类场所中，艺术品的存在是理所当然的，公众如果公开质疑就显得不合情理了，但是在你的居住地、购物的超市或常去坐坐的广场，人们就有权发表自己的看法。

公共艺术越来越流行，比以往任何时候都来得重要，也将变得越来越好，艺术作品是雕塑还是设施？是用于氛围已定的现有社区，还是氛围仍会受到委员会影响的新社区？是长期的还是临时的？如果是临时的，那后续呢？社区是如何参与的？艺术家是如何参与的？任何人如果考虑到公共艺术委员会，特别是试图引领或启动环境转变的委员会，都应当慎重，不应期望太多。在真正重要的层面——在我们与居住、工作、购物的场所的接触中——我们预期会见到大量的公共艺术作品。这些艺术品应该是赏心悦目的，看到它们，我们会因此心情愉快，或许为之着迷，当然也会产生好奇。这样的日常接触不应该变成无趣的例行公事，而且我们对这类物品和图像所做出的反应，背后必然有深层原因。

艾米·巴拉克（Ami Barack）简介：
法国独立策展人
Independent Curator

比尔·菲特斯吉本斯
Bill Fitxgibbons

如今，指定场所艺术在国际上越来越流行，这种艺术使公众感受并培育公众的持续场所感。它使公众和私人场所更活跃，同时是定型的环境再次焕发生机，让看到这些艺术作品的人们得到激励，感到振奋。

比尔·菲特斯吉本斯（Bill Fitxgibbons）简介：
美国独立艺术家
Independent Artist

刘悦笛
Liu YueDi

我认为目前公共艺术的问题不仅仅是艺术的问题，公共艺术的本质和公众性是相关的，公共艺术可以回到生活美学加以看待。一方面公共艺术是按照现代艺术原则与美学理论构建而来的矛盾修饰法，并逐渐拓展到当代艺术的疆域之中，而今是否要走出这种现代主义模式，回到生活美学，来重新思考公共艺术的公共定位。我认为公共艺术应该更强调参与者参与到公共艺术所营造的情境当中，这样可以把公共艺术分为三种：第一种是前卫艺术，第二种是环境空间艺术，第三种就是民众生活艺术。

刘悦笛（Liu YueDi）简介：
中国社会科学院哲学研究所副研究员
Associate Professor, Insititute of Philosophy of Chinese Academy of Social Sciences

潘力
Pan Li

我认为以地方重塑为主题的公共艺术的概念，绝对不是外来的艺术家对当地的一种施舍、一种外来力量的符号性的添加，而是把它作为一种激励机制，去激活当地自身的动力，只有这样，以地方重塑为核心的公共艺术才能真正有一个持续的发展空间，也才能真正能够造福当地，美丽乡村建设是美丽中国建设一个最重要的环节，其核心就是激发人们对生活的追求，而不是在于艺术家个人对公共艺术本体的表现。受中国国情影响，它决定了中国的公共艺术面貌和国外的公共艺术面貌是必然不同的。

而从长期讲，要获得什么样的成功，就必须要知道通过推出公共艺术要达到什么样的目标。比如说，我们希望公共艺术能够增加游客量，或者希望能够迎来各个年龄层次的人士，或者说我们希望能加强荣誉感等。公共艺术能让每个乡村都变得不同，就必须要侧重于这个地区或这个乡村的独特历史和文化，对这个地区或者城镇的品牌起到促进作用。让来自外部的艺术家能融入到日常的工作中，随着他们进一步了解这个乡村的独特之处，才能够真正有独特的体验。因此，首先要突破常规和传统，也就是对公共艺术定义的传统；第二需要不同媒介创造公共艺术，不仅仅是视觉艺术。我们需要有不同的资源来创造这个公共艺术。同时也必须有专业人士和协调员在创作过程中发挥作用。另外，还要有一些教育性的项目，让本地的居民对这种艺术形式有所了解。而谈到地方重塑和公共艺术之间的关系，我们必须看到，我们要创造的是有意义的社会空间，这是核心。我觉得公共艺术就是关爱，我们要关爱我们所生活的这个世界的健康，我相信我们需要把艺术家的才能跟社区真正的需要结合起来。

潘力（Pan Li）简介：
上海大学美术学院教授
Professor, College of Fine Art, Shanghai University

施明强
Shi MingQiang

我认为发展关键是重塑农村的文化生态，让城市居民到乡村寻找精神家园，让农民激活文化自觉和发展信心，为新农村置入内生动力，进而催生由文化意义、经济产业带动农民致富。由此看来美丽乡村建设本身就是一个大的公共艺术课题。于是，我们把公共艺术和地域文化、现代农业、生活体验、课外素质教育、民间文化、生态文明、社团组织、地方特色节会等结合起来，使得乡村重塑成为一种模式。此外，公共艺术组织介入美丽乡村建设，需要加强与社会科学研究机构、政府部门的双向沟通，以期达成统一思想和合作的架构，形成互惠互利、互动发展的良性的体制；第二是制度层面问题，如何通过跨界研讨以及政府官员培训等方式，扩大公共艺术的普及宣传和对社会的影响力，进而增加与政府间的交流合作，使公共艺术的社会介入和解决公共问题职能纳入政府的社会制度层面。第三是农民主体意识和文化自觉的问题。农民的素质如何提升，还有文化自觉的培养，是农村持续健康发展的一个重要问题。如何加强对美丽乡村的跟踪考察，着力培养智慧农民，完成农民教育培训体系，值得特别关注。

施明强（Shi MingQiang）简介：
浙江省玉环县龙溪镇党委书记
Party Secretary, LongXi Town,ZheJiang Province

塔库巴·艾肯
Ta-coumba Aiken

我觉得艺术家和学生都要学会和各方合作。公共艺术更是如此，一定要跟相关的比如政府机构合作。要知道他们有什么要求，他们有什么想法，而且合作的一个好处，就是合作总是能够让我们碰撞出很多的新想法，有很多新主意。

塔库巴·艾肯（Ta-coumba Aiken）简介：
美国，艺术家
Sculptor

德瑞克·查理斯
Derrick Cherries

我认为公共艺术能够鼓励公众的参与和体验，而且能丰富他们的体验。从这几天接触到的公共艺术作品中，我看到作品都体现了某种身份和某种目的。我们需要能够识别在这些作品背后的驱动力和愿望。此外，我看到了全面的知识的重要性，要应用于不同的情形。在一个环境中的知识，在另外一个环境中可能就是缺乏的。所以说，我们需要其他知识的补充。此外，需要帮助培养我们受众的专业知识。公共艺术的观众，艺术的专业知识可能并不多，这就需要我们去帮助他们，让他们知道公共艺术是什么样的内容，他们会看到什么样的公共艺术作品。我这里没有一个一成不变的模式，但是确实我们需要去积极地做起来。

德瑞克·查理斯（Derrick Cherries）简介：
新西兰，奥克兰大学 Elam 美术学院院长
Head of Elam School of Fine Arts, University of Auckland.

荣念曾
Rong NianZeng

我觉得艺术的公共性，艺术在不同时代与环境的关系，只要引发到公共的讨论与关注的话，它就显示公共性。今天我们将公共放在艺术前面，可能就是对艺术的一种批评，对艺术现在在经济、政治强势下的一个批评，同时也是对艺术缺乏教育的一个批评。

艺术留存到今天，必然有它的公共性，今天的艺术能流传到明天，也一定有它的公共性。我们关注的是"流传"这两个字，关注这两个字本身的机制和运营以及背后的政策与价值观，它是不是真的能走在社会公共理念发展的前端，能否发展成为拓展、维护社会公共空间的力量。我想这里关心的都是公共艺术的机制、营运、政策以及价值观，了解这些客观的情况，再去审视艺术创作与当下环境公共的关系，自然就会发展出相对的评估准则，这样才能协助我们更深入的审视论述艺术创作本身。我想大家不会反对说公共艺术应该是可以影响公共空间的艺术，同时也是应该用来探讨目前跟未来空间的艺术。

当然，公共是一个非常政治的名词。公共空间在发展中国家能否独立于第一部门和第二部门就是政府和企业，还是有挑战的。它的严肃性、辩证性，都是可以拓展和强化公共空间。我个人觉得这个论谈本身就是一个公共空间最好的案例。我们其实是靠这里每个朋友，怎么让这个论坛有严肃性、辩证性，然后让公共空间跟公共艺术开拓新的论述。更实际一点，是通过公共和艺术的讨论，让我们的文化发展有更健康、更前瞻的机制和营运。

荣念曾（Rong NianZeng）简介：
香港当代文化中心主席
Director of HongKong Insititue of Contemporary Culture Center

徐明松
Xu MingSong

轨道交通的公共艺术论坛给我两点启示：一点是对公共艺术的认知中不断触及人与自然、人与环境的关系。人与自然的关系中，存在着很多需要解决的城市问题，这也是公共艺术所要面对的一个重要发展方向。另外一点是我们对公共艺术在地铁公共空间的发展，实际看到了它的历史文脉，或者每个国家的地铁发展过程中都有自身城市发展的特点。

还有一点，我们现在很多地铁公共艺术的创作手段和创作方法是多元化的，具有发展的前景。实际上，我们在伦敦地铁近些年的项目实施过程中，可以看到概念艺术、当代艺术在地铁空间的呈现。为我们地铁艺术的多元化或者是国际化、多样性，提供了一个很好的参照。

徐明松（Xu MingSong）简介：
上海书画出版社副总编辑
Deputy Chief Editor , Shanghai Fine Arts Publishing House

"2013 国际公共艺术奖"的创办与评选

访上海大学美术学院院长 汪大伟教授

汪大伟教授是"国际公共艺术奖"发起人之一，现任上海大学美术学院院长、《公共艺术》杂志主编。他最早提出以学科建设的思路来发展公共艺术，致力于推动中国公共艺术的发展，近年来策划、参与过多项公共艺术项目，如上海轨道交通总体设计、上海世博会主题设计等。访谈中，汪大伟阐述了"国际公共艺术奖"评奖活动的创办初衷、评选过程以及评奖机制，并介绍了国际公共艺术论坛的主题与内容。他指出，举办"国际公共艺术奖"评奖活动的最重要的目的是，聚集全世界最优秀的公共艺术创作和策划人才到举办地，以解决当地的实际问题。

邱家和 / 采访

邱家和：为什么要发起国际公共艺术奖？

汪大伟：从 1998 年起，我就提出以学科建设的思路来发展公共艺术。至今，虽然已为城市建设做出了不少贡献，但我们认识到，公共艺术要在学科建设和学术研究方面取得长足的发展，还需要一个国际对话的渠道。

首先，从世界范围来看，公共艺术虽然备受关注，可是公共艺术被作为话题提出来还是在近 25 年到 30 年的时间里。我国对公共艺术的认识更只是从环境装饰开始，把它看成城市建设的一部分。今天才逐渐认识到公共艺术是提升人们生活幸福指数的有机组成。这样的认识过程和发展形势，意味着我们需要一个平台，进行广泛的国际交流，并向公共艺术发展迅速的西方国家取经。这是发起国际公共艺术奖的原因之一。

其次，在学科建设过程中，我们尚处于起步阶段，对公共艺术的理解以及为社会服务的方式方法的认识方面都不够全面。但是中国的城市化进程与城镇建设发展迅猛，愈发感到公共艺术在其中作用的重要性。因此，与国际合作，通过交流对话，结合中国的国情，逐步探索和构建公共艺术的评价体系，与国外同行平等合作，互相促进。

第三，出于我国对公共艺术的切实需要。"十八大"明确提出加快城镇化发展，如何在城镇化过程中提升人的生活质量和幸福指数，就成了一个值得关注的问题。这不仅是投入硬件，更重要的是文化建设。而艺术恰恰是最为人民群众喜闻乐见的文化建设方式。艺术可以释放心灵，可以缓解矛盾，让并不富裕的生活充满幸福与满足，由此可见公共艺术在城乡发展过程中的重要性。设立这个奖项，可以开拓思路，通过国际合作来引进、参与，甚至帮助解决中国城镇化过程中存在的一些问题。

总而言之，是鉴于学科的需求，建立一个国际合作的评价体系的需求，以及中国自身发展的要求，我们设立了"国际公共艺术奖"这样一个平台。

当然，奖项只是一个说话的由头。因为一谈到"奖"就要有一个标准，就要产生一个评奖机制。所以，这个奖究其实质是渗透和引领的体现，学科建设也好，文化推广也好，某种程度上就是构建一个学科体系。

我们曾对公共艺术的社会影响力、公共艺术能起到什么样的引领作用以及什么是好的公共艺术的标准等进行热烈的讨论，这就是我们共同构建评价体系的过程。这次是以研究作为基础，进行推荐评选的评奖机制。与其说是评奖，倒不如说是全球公共艺术研究网络的建设。那么，如何建立公共艺术奖的影响力？我在和其他国家的几位评委谈话时说到，奥斯卡凭借着家喻户晓的明星和传播快速的电影赢得了权威和影响力，而我们这样一个白手起家的奖项将如何获得影响力？我认为"国际公共艺术奖"应该聚集全世界公共艺术最优秀的创作人才和策划人到举办地，为当地解决建设问题。这样，今后谁要解决问题就要申请这奖的举办权。是以解决举办地实际问题的方式来扩大公共艺术奖的影响力。譬如，这次我们做了中国的三个实际案例：一是后世博的公共空间利用，二是每天人口流通量600万人次上海轨道交通公共艺术策划，三是乡镇公共艺术。这三个实践案例聚集了一批世界顶尖的艺术家和策展人，以期能切切实实地为这三个区域带去帮助，即使只是发展思路上的启发。我们就是用这样的方式来拓展影响力。

"国际公共艺术奖"的参选作品不是由艺术家个人申报，而是先在全球挑选15位研究员，由他们对本大区里的公共艺术作品进行筛选推荐，同时写明推荐理由。经过这样一个评选体系，初步推选出了141个案例。在此基础上，我们以"地方塑造"为主题，考量这些通过初选的案例对主题的把握度。（值得一提的是，"地方塑造"是以公共艺术的各种方式对当地产生影响力，在这个过程中，涌现出许多新奇巧妙的工艺、材料、创意和方式方法。）经过研究评选出26个提名案例，在这26个中再聚焦6个。这就是评选的一般过程。我们的评委是来自全球各地顶尖的策展人和评论家，他们与上述过程一起，构成了一个完整的评选机制。并且，这样的网络结构可以成为惯例，延续到今后的国际公共艺术奖中。

邱家和：为什么选择美国《公共艺术评论》杂志作为共同发起方？

汪大伟：《公共艺术评论》是美国唯一评论公共艺术的杂志，创办于1979年。这本杂志对美国乃至西方社会都很有影响力。同时，我们也是中国唯一的公共艺术杂志，与这样有影响力的同行建立战略合作，共同来做这个奖，是希望通过评奖来推广公共艺术。

邱家和：为什么选择路易斯·比格斯担任评委会主席？根据什么原则挑选评委成员？

汪大伟：发起国际公共艺术奖以及和路易斯合作，完全是机遇使然。路易斯是利物浦双年展的创始人，举办过十届双年展。他在打造利物浦双年展这个品牌的过程中，始终关注如何让造型艺术从博物馆解放出来，走进城市、生活和公共空间。这就是公共艺术，使当代艺术和美术馆划清界限。他一直希望公共艺术能进入社区，他觉得这对西方来说是最难做到的事。当路易斯得知我们在上海的曹阳新村做了一个公共艺术项目时，非常感兴趣，派了一个观察员跟随我们三个月。开幕的时候他亲自来看，他说那次活动让他很有触动。于是，我们就有了共同话语，一直愉快合作。

我主动向他提出，是否可以共同发起一个国际公共艺术奖来引导和推广公共艺术，他说这与他的想法以及他所从事的工作不谋而合——即让公共艺术关注到人，不是关注物质空间。我们真是一拍即合。在讨论主题"地方重塑"时，我们的意见更是出奇的一致。我和他都认为公共艺术是对一个地方的改变，更重要的是影响人的精神风貌和归属感的提升，具有一种引导性，因此商定用"地方重塑"来命名。

我们共有四个合作伙伴，路易斯和约翰·麦科马克都是在国际平台上操大牌的人，有很好的人脉。约翰原是新西兰亚洲文化基金会的主席、新西兰国家美术馆馆长；杰克是《公共艺术评论》的创始人，积聚了全球公共艺术的信息。加上我们学科发展的急需，更重要的是中国的城市化发展受到全球公共艺术研究的关注。因此，我们一拍即合。这样，四个人分别代表了不同国家和不同的社会身份。我是学校的老师，路易斯是策展人，约翰代表基金会，杰克是媒体人。这样很有趣的四个人组合在一起，就发起了评选国际公共艺术奖的活动。

邱家和：你如何看待这次评选的结果？

汪大伟：我认为，评选结果与我们所主张的主题是吻合的，但肯定会遗漏一些同样优秀的作品，因为评委的文化背景不一样，来自巴西、土耳其、北美、欧洲、日本、中国，文化差异使他们对"地方重塑"的理解不一致。最初，我们采取的办法是第一轮投票之后只讨论有票提到的案例，在讨论过程中就可以看到明显的差异。例如，土耳其评委提出的几个方案都没有被采纳，她提出异议，再复审再投票，还是没被选上。土耳其评委很不高兴，但是只能服从规则。

六个获奖案例是大家评选出来的，代表了六位评委比较集中的意见，符合主题，有普适的意义。我们认为，只有不断完善，才能逐渐构建起一个趋向完好的评价体系。譬如，在这个过程中，我们还邀请国内学者为这 26 个案例做点评，渗透着评委的观点和智慧，很有指导意义。

Q: 为什么要举办国际公共艺术论坛？从评奖到论坛是如何转变的？

汪大伟：当时，我们谈到评奖结果出来后的宣传方式，例如像奥斯卡颁奖一样走红地毯，请电视媒体做宣传，找国际公关公司定点推广等，但这些方案都被一一否定。

首先是资金的原因，上述方式所需要的大笔钱财是我们所担负不起的。其次，我坚守这个奖项的学术性。本身就是学校策划这次颁奖，那么宣传也要符合这个身份。同时，我的目的是学科建设与学术研究。于是，我提出以国际论坛的方式来替代普通的传播。其实，论坛的方式是一种在学术界的传播，这样既能起到宣传的作用，又有利于公共艺术研究的深入。

这次论坛的嘉宾来自五湖四海，而且都是研究人员。这样一来可以起到信息沟通的作用，二来可以做成一个公共艺术研究的平台，用时间和质量来扩大影响力。论坛的方式正契合其严谨的学术身份，又有上述多种好处，何乐不为？

邱家和：这个论坛为什么要选择三个公共艺术项目来论证，这三个项目是怎么选出来的？

汪大伟：我选这三个项目，是为了帮助论坛增加渗透力。前面已经讲到，结合举办地寻找实际问题，联手国际顶尖艺术家和策展人来解决问题，是扩大论坛影响力的好办法。

所以，当时我提出这个建议后，便得到他们三位的认可。我们酝酿这个奖项是在 2011 年，世博会刚结束，后世博园区遇到了公共空间如何再利用的问题，所以我们主动找了世博集团，与他们一拍即合。他们也希望我们能够为世博轴的六个阳光谷提供公共艺术的改进方案。

邱家和：这个方案实际是谁在做？主题是什么？

汪大伟：在国内我们选择了丁乙的团队，还有中央美院的王中——他在国内有许多很有影响的公共艺术作品。国际上则有两个，英美各一家；另外，对于 800 万到 1000 万人流量的上海轨道交通网的公共艺术策划，我们则请国际上一些有成功经验的策展人和艺术家来做方案，我校也有一个团队在做地铁公共艺术规划，这就形成了一个对接交流；此外，我们还在浙江省玉环县设立了一个分论坛，进行美丽乡村建设的公共艺术策划，并且希望今后在那里发展一个公共艺术创作基地。

希望通过聚焦这三个实际案例，能够为中国公共艺术的地方重塑概念探出一条路子，也让论坛为中国建设贡献一份力量。

邱家和：请你谈谈论坛的组成构架和内容。

汪大伟：关于论坛形式，我们改变以往分两个部分的惯例，这次是分成三个板块。

第一个板块是六个获奖案例的介绍，第二个板块是思想交锋，每人做 10 分钟的观点陈述。陈述人有社会学家、理论学者、媒体人、政府官员、企业家和艺术家，以不同身份阐述自己的观点，请四位国际评论员点评观点。然后，与会人员可以对话、碰撞、讨论和辩论。第三个板块是解决实际问题及三个实际项目论证。

邱家和：你刚才讲到评奖和论坛将会长期举办，今后大概是什么方向？

汪大伟：我们做这个奖是长期打算，两年一届。第二轮研究成果的征集已经开始。

我认为这个奖首先要有专业性。我和路易斯商议，第一届评奖和论坛，我们一边做，一边制定出标准。今后谁要申办这个奖，这些规则和标准必须做到。所以这次工作量非常大，从 VI 设计，整个流程设计、评奖程序设计，全是一张白纸，重头做起。但把这些都制定好了以后，我将出一本规范手册，这本手册就是其他地方今后申办这个评奖活动奖的程序规则。

随后就可以凭借在专业圈里的影响力去影响政府。如何让政府来参与这件事？那就是帮助政府来做实事。此外，我希望由各地的大学和政府联手举办。并且，主办方一定要由学校来承担，因为学校是知识聚集与学术研究的所在地。

邱家和：上海大学近年来在公共艺术方面做过哪些事情？

汪大伟：上海大学美术学院一直将公共艺术和城市建设结合在一起。可以说，上海的城市雕塑百分之五十以上出自我们学院的设计系和雕塑系。

从 20 年前开始，我们从事上海轨道交通公共艺术设计。从做壁画开始，后来转入到环境空间设计，到后来整个轨道网络的视觉形象设计和整个轨道交通网络的导视系统的规范标准设计。这两项设计已成为企业的规范性文件，即轨道交通在造新线和车站时必须拿这两本文件作为投标文件。轨道交通 7 号线、8 号线、9 号线、人民广场换乘枢纽、地铁下沉式广场等大型公共空间以及徐家汇的换乘通道等，都是我校美院的设计。

美院还参与了上海世博会的主题策划、主题演绎的全过程。中国馆、城市文明馆以及演艺中心最早的策划书的撰写。另外，我们直接参与了城市足迹馆的主题策划和世博会博物馆的总体设计。国家级层面上场馆的展示总体设计由三个院校独立承担：中央美院、中国美院和上大美院。此外，世博中心 14 万平方米的公共艺术设计，大型壁画与会议室的大厅装饰摆设、雕塑、壁画全是我院独立承担。世博会之后，宝山国际民间艺术博览馆从策划到设计实施也都是美院承担。

作为上海大学美术学院，我们必须关注城市公共文化生活，为这个城市做贡献。如果能够把公共艺术做强、做好，那么上大美院就不愧为这个都市的美术院校。我们也一直把发展公共艺术作为建设"都市美院"的核心。而公共艺术也恰是国画、油画、版画、建筑和其他各门类设计的综合。也就是说，公共艺术不仅是对接社会服务的平台，更是把各个专业汇聚在一起的公共学科。譬如，国画可以作为一门学科存在，但是也可以去这个平台上实现与社会服务的对接。我当然不反对一个院校把自己关在象牙塔里面，发展自我陶醉的小众艺术，相反，我们亦是很看重这一块，虽说是个人原创的自留地，但也是鼓励原创，滋养艺术发展的土壤。学校有对社会发展应负的责任和义务，因此提出公共艺术作为"都市美院"发展的方向。

公共艺术学科体系可以分为三个层次：第一个是公共艺术技术实验中心，是国家级的实验教学中心，全国在艺术类只有18个点，上大美院就是其中一个。我们的本科专业既有国画、油画、版画、雕塑纯艺术和建筑设计、建筑规划工科、视觉传达、环境设计、数字媒体设计等设计类，还有美术史论、艺术管理、理论类等组成。公共艺术技术实验中心使我们的学生从熟悉材料、工艺到通过动手激发创造力的一个公共艺术实验教学平台，也是公共艺术研究基地；第二是公共艺术创作中心，是上海市人文重点学科，主要是公共艺术理论建设与公共艺术的创作研究，这个创作中心是面向老师的一个平台；还有就是公共艺术创意中心，是上海市第一批九个高校知识服务平台，这一层次是将各专对接社会，用知识服务促进产业转型和城市发展。

希望国际公共艺术奖能够拉动我们学术水平的提升，能更好地和各个专业有机渗透。就像雕塑未必都是公共艺术，国画、油画怎么走入公共文化生活，这都是我们要解决的课题。

有的人认为，公共艺术是空间中的艺术，有的人把公共艺术定义为大众服务的理念，也有把公共艺术看作是一种运作机制。我认为公共艺术是用艺术的语言和方式去解决公共问题。

当代艺术最重要的两个标志：一是干预生活，二是走出象牙塔、走向社会，和传统观念决裂。西方当代艺术以个人行为和个人主义极度膨胀来对抗社会，以批判性来抗衡社会。而在中国，我们更需要解决问题而不是简单的批判，而是重在建设。因此在这种意义上可以说，公共艺术是中国当代艺术的方向，也是符合国情应该提倡艺术形式。

当我院完成曹杨新村的公共艺术项目后，我的体会是，公共艺术是能解决公共问题的，比如说公众的沟通。在美国也提社区重建，是因为发现他们的制度出了问题，极度的个人化，使得人和人的沟通不畅。他们希望能在社区通过文化建设来打通不同种族不同文化的隔阂。

这次评奖中有个美国的经典案例"在我死之前"，非常精彩。艺术家在一个废弃的房子上刷了一个黑墙，重复标了一个填充题：在我死之前，我想要做什么？男女老少、不同文化背景的人都去填空，大家由此产生心灵沟通。后来通过网络复制，整个西半球都在参与。然而，中国人肯定认为这不是好的公共艺术，因为中国文化是忌讳讲"死"的，这就是文化背景的不同。

（录音整理：李恩、徐蓉蓉、傅梦婷，文稿已经本人确认。）

我为什么这么做？ ——公共艺术之我见

"Why did you do it?"
Some personal reflections on public art

路易斯·比格斯（Lewis Biggs）曾任英国泰特美术馆长和利物浦双年展艺术总监，为利物浦双年展为利物浦成为"欧洲文化之都"做出重要贡献。本文阐述艺术作品走出美术馆进入公共空间的重要意义，描述利物浦双年展中的公共艺术与民众以及政府的互动，依据具体案例探讨公共艺术与城市发展的关系。

Lewis Biggs/ 文

One day in 2010, when I still lived in Liverpool, I took a visitor to see Antony Gormley's sculpture Another Place at Crosby Beach, where the River Mersey leaves the Port of Liverpool and heads for the Atlantic Ocean. It had been my idea, and the achievement of my colleagues, to put Gormley's sculpture on this beach when I was Director of Liverpool Biennial. My visitor was an art dealer: her gallery shows Antony Gormley's work, and she also takes care of some business for my wife, who is an artist, a painter of still life in many forms. We enjoyed the strong breeze from the sea while we walked around the sculpture - art and nature together. She was delighted by the experience, and I told her how we had worked for many years with great difficulties to complete the permanent placing of the artwork. When we left the beach she said: 'But Lewis, why did you do this?' My answer was: 'Because it's beautiful'.

Of course her simple question contained many other unspoken ones, and my simple response hid a great many other answers. I shall come back to some of those other questions and answers in a moment. But what she said next was, I felt, very important. She said that she had noticed that many very good artists she knew who had 'conquered the market' then looked for a new challenge: they wanted also to reach a public that would never buy art, and to make an impact on society in ways that had nothing to do with the art gallery or museum.

After I had been Director of Tate Liverpool for ten years, I began to feel constrained by the ways in which art is able to gather meaning within the museum and gallery system. Within this system, art is always appropriated by its two most powerful validators: the art market and the history of art. Some museum directors, when they reach this point, challenge themselves with building programmes.

But I had already been responsible for completing the building of Tate Liverpool as well as having been on the management committee for the building of London's Tate Modern, and I wanted something different. Like the artists who are no longer excited by the market, I wanted to see if art could reach a public unable to buy art, and unmoved by the history of art. And if art could reach this public, it would be exciting to see what effects it might have, on individuals and on society.

My first Liverpool Biennial was in 2002 (I had become Artistic and Executive Director in 2000) and although much of the art was placed inside museums and galleries, I signalled my intentions for the Biennial by commissioning a number of outdoor 'pavilions' around the city, on the streets and in disused buildings. I hope that the five editions of Liverpool Biennial for which I was responsible established a reputation for an imaginative engagement between art and urban environment (our slogan was 'engaging art, people and place'). Another way in which I took the Biennial in this same direction was by working as a commissioning agency for art in public space (outside the time-frame of the Biennial festival). In this way, Liverpool Biennial became one of a handful of similar events around the world that focus on place making through art (the best known are Skulptur Projekt Münster in Germany, Echigo-Tsumari in Japan, SITE Santa-Fé in New Mexico).

So my visitor's question 'why did I do it?' contained some others: why, when my house did not have a view of the beach and the sculpture – and in fact I lived some miles away; why, when my children were too old to play on this beach, and I had no dog to take for walks there; why, when I had no business interest in the locality that could benefit from the sculpture's presence; why, when most of my peers in the art world, whose admiration I must surely want, would never bother to visit this beach; why, when the difficulties, time and effort in undertaking this work far outweighed the financial rewards that my organisation received?

And when I answered all these questions by saying 'I did it because it's beautiful' I meant to suggest at least four other answers through the word 'beauty': the beauty of place making; the beauty of fun; the beauty of regeneration and the beauty of hope. I should explain more.

When I was a child, my blind grandfather lived next door, and I used to lead him by the hand to go walking. We would walk to the woods, and I would tell him what I could see. One thing I noticed was the way that nature is always decomposing but always repairing itself. When I began to live in a city, I could see that the constant change – the decay and reconstruction of buildings – is the same process. But the dependency between my grandfather and I made me very conscious of how he responded to his senses: that his state of being was very much influenced by the sensory qualities of his surroundings. So I became aware that the quality of our immediate physical environment has a huge influence on the possibility of human health, in body and mind; and I became aware of my own need to make beautiful places. (As a teenager I wanted to be an architect). I believe now that some social illnesses are directly caused by a bad physical environment. This is common sense, not rocket science! But why are some of our cities built like prisons, as if to punish their inhabitants for being alive?

'Beauty' differs for different people and in different cultures, but the word is useful even so: something stimulating or comfortable, refreshing or challenging, desirable or thought-provoking, emotionally exhausting (like a dance) or health-inducing (like a good meal). A beautiful place is one that is loved by the people who use it; and they will look after it because they appreciate it (this is true even of wilderness). One aim of place making must be to create a relationship between the place and the people who use it (whether residents or visitors) so that the place gives pleasure to the users and in return the users look after the place. This is just what most people do in their own homes. The beauty – and the relationship – may be mainly sensual, or it may be more intellectual, to do with the history of the place or its metaphoric possibilities; or all of these. And of course these kinds of relationships are exactly the same as those that we can have with works of art. So there is an intrinsic connection between making art and making places, a complete continuity of intention, as is recognised in the history of gardens or in Bernini's urban planning in Rome.

Liverpool Biennial's Urbanism09 programme grew out of the impulse to improve the quality of the environment of a canal-side neighbourhood in north Liverpool. One of the key contributions was Janta Manta 2009 by Raumlabor Berlin. The sculpture was based on an Indian astrological instrument and monument: a way to read the stars and so traditionally a way also to gain knowledge of the future. It was the centrepiece of a range of other initiatives, some created by artists, others by local environmental groups or schools, that together brought the quayside of the canal back into use. Once the canal was in use again, it became an asset rather than being perceived as a threat to security or sanitation.

Another example of Liverpool Biennial's place making was Dancing Queen 2006 by Peter Johanssen, a Swedish artist whose cheeky contribution to the city of the Beatles was to set up a museum to the Swedish pop group Abba. This flat-pack Swedish house was set up on an area of grass in front of the most celebrated three buildings in Liverpool. This area of grass had formerly been a ferry and bus terminus, and despite the grass being laid down, the relationship of the open green space to the iconic buildings alongside was uneasy: people did not know how to make use of it. The absurdly loud red house, blaring Abba's song Dancing Queen, helped people understand that the imposing surroundings did not have to stop them enjoying themselves on the grass, and now the place is much better used as a picnic spot and part of a river front promenade.

One aspect of the beauty of place making is to create a sense of fun: fun produces energy and can induce creativity. Turning the Place Over 2007, by Richard Wilson, occupies a building opposite the entrance to a metro station used by thousands of office workers every day. It introduced a sense of fun to a derelict and depressing street in the city of Liverpool, where property development had been halted for many years.

Arbores Laetae 2008 by DSRNY is equally light hearted in spirit, and draws attention to a triangle of 'waste' ground at a 'gateway' road junction where traffic turns off the inner ring road onto one of the axis routes into the city. DSRNY decided to celebrate this 'turning' by planting 17 trees and making three of them turn – at different speeds. In the Linnean system, trees have latin names, so these Hornbeam trees have been given the name 'joyful trees'.

The beauty of regeneration is present in nature's ability to renew itself every year, and is most often celebrated in annual festivals any society that is close to nature. Renewal in the city follows a different rhythm, but is also sometimes recognised, for instance in the ceremony when a roof is completed on a new building, or through the cyclical redecoration of people's homes. So in writing of the different kinds of beauty that have motivated me to be involved with art and place making, I must not overlook the fact that the main reason that Antony Gormley's Another Place is on Crosby Beach is because a visionary government official asked me to put an artwork there that would 'increase footfall' to the neighbourhood, and so improve the economy. He should be justly proud that his initiative has been such an incredible success.

Renewal in nature and in the urban environment are different in that in the cities we somehow expect regeneration to be better than what was there before rather than simply a new version of what was there before. Humanity appears to be addicted to development, improvement, growth and so the expectation of regeneration through place making is that it will involve an improvement of the environment along with social and economic relations. Two commissions by Liverpool Biennial that exemplified such an improvement were Rockscape 2008 by Atelier BowWow, architects based in Tokyo, and Rotunda Folly by Gross Max landscape architects based in Edinburgh.

Rockscape was created in an empty lot in Liverpool's city centre where a house had been destroyed by bombing in 1942. Atelier Bowwow's method was to construct an outdoor theatre on this site, and in the process construct visitors to the site as the audience and passers-by as the actors. In fact, they created the conditions for 'street theatre' whether or not the participants were aware of their respective roles. Certainly they made a sad site more beautiful and stimulated the imaginative possibilities of the people of Liverpool.

The brief to Gross Max from the Rotunda community college was to re-purpose the waste ground in front of the college by creating something that would give the students and staff something to be proud of, something that would increase the visibility of the college in the locality. The response was to design a strip garden to be maintained by the local residents, and a 'Folly' that would provide a focus for events organised by the college. The garden and Folly have become local landmarks, and both are well loved and well maintained. Although they were built with a temporary licence in 2008, the ensemble has become 'permanent'.

The attitude of hope has a beauty that is hard to define, but it is real enough: we tend to like optimists and dislike pessimists. Clearly hope is about the future, but usually it also entails a belief that the future will be different from − and somehow better than - the present. Maybe this is nearly the same as our delight in novelty and fashion, a recognition that any change is stimulating and therefore to be desired. So regeneration and hope go together as beautiful ingredients of place making. The recognition of potential for change, growth and development − such as we see in children and young adults - leads to a positive relationship to the future.

Liverpool Biennial worked with a group of former coal miners in the town of St Helen's, near Liverpool, to commission an artist to transform a slag heap (hill of waste material from the mine) into a monument to the colliery on that site. After a period of study about the possibilities for commissioning different kinds of art, the miners chose to work with Jaume Plensa, an artist from Barcelona. His first proposal was considered by the miners to be too much of a memorial to the past: reflection on their situation had changed their view. Now they wanted a 'memorial to the future', which was unknown but would certainly not be coal mining. So Plensa produced a second proposal, an elongated head of a young girl, and this was accepted as appropriate by the miners. The completed commission Dream 2009 by Jaume Plensa has become the symbol of St. Helen's and is thought of with great pride by the local residents.

In 1998, while I was still Director of Tate Liverpool, I co-curated with Robert Hopper an exhibition called Artranspennine98, involving many new commissions in public space at 30 sites from coast to coast across the north of England. SuperLambanana 2008 by Taro Chiezo was commissioned as a temporary intervention in a heritage area of the city, and was intended by the artist to refer to the 'sunrise industry' of genetic engineering as a part of Liverpool's future. It is still in the city 14 years later, so well loved that it has generated every kind of merchandise from keyrings to coffee mugs and T-shirts.

I myself was uncertain as to whether it would be well-thought of by the people of Liverpool, but within a few months something happened that made me more certain of a good outcome. I was called for a meeting to the office of the recently appointed Chief Executive of Liverpool City Council (the local government executive). On the oak panelling behind his desk, dominating the vision of any visitor, there were hung three huge framed photographs of SuperLambanana. It was clear that our CEO not only felt some identification with the energy of this yellow monster, but that his hopes for the future were somehow captured by this weird and humorous sculpture.

Within months, he had launched Liverpool's successful bid to be nominated European Capital of Culture 2008, generally understood to be the critical event in Liverpool's regeneration over the past 15 years. During 2008 the city organised that every school make their own small scale replica of SuperLambanana, and these were then arranged in procession around the city centre. I like to think that maybe it was SuperLambanana that was responsible for suggesting the bid to our CEO – perhaps even for its success. But I'm also convinced that it's best for patrons, politicians and developers, as well as curators, to focus on commissioning good art, and let the consequences look after themselves.

案例
CASE STUDY

欧洲地区提名作品
EUROPE NOMINATED CASES

艺术提法里提
Art Tifariti

艺术家：格戴姆·伊兹克团队（国际艺术团队）
地点：阿拉伯撒哈拉西部被解放的领土上
推荐人：吉乌希·切科拉
Artist：Gdeim Izik collective（International Art Collective）
Location：Tifariti Western Sahara Liberated Territories
Researcher：Giusy Checola

作品描述

2007 年正值持续 40 年的撒哈拉西部战争结束，艺术团队开始组织一年一度的公共艺术活动，目的是宣布撒哈拉西部人民对他们的土地、文化、根源和他们的自由的拥有权，关注"地方可见度"、"地方承诺"、"地方希望"和"地方认同"。其共同点在于，他们的选择和创作目的都是为了帮助撒哈拉西部人民，让世界看到他们和西方世界的冲突，并努力争取实现联合国三十多年前做出的承诺：完全实现自治。

第一期"艺术提法里提"在当地民众的参与下取得了巨大成功，将提法里提与世界各地联系起来。艺术家们以不同的方式来表现撒哈拉西部人民和社会的现状，如阿尔及利亚艺术家团队在医院的墙上绘制的《废墟的重生》，该地区曾在战争期间被炸毁，一直未得到修建；另一部分作品则强调战争造成的本质性影响。作品中一些是永久性的，其余是临时性的，在博物馆室外展示，它们均已成为提法里提文化遗产的一部分。

该项目展示了撒哈拉西部的冲突并提高了社会意识，使这个地区被重新认识和认同，当地人民也积极投入参与实地创作，有关撒哈拉西部的信息被传播到世界各地。艺术家对于实地创作不断投入是源于为撒哈拉的艺术家创造了新的设施和福利，他们每年都积极参与各种艺术项目，不断参与公共空间项目并与国际艺术家开展合作。

解读

格戴姆·伊兹克团队的公共艺术作品安置于阿拉伯撒哈拉西部被解放的领土上，构成作品的材料主要由当地提供。废弃的汽油桶、子弹壳，甚至因炮击而残留在地面的弹坑都可以作为构成作品的主要成分。作品具有鲜明的反战和追求和平的艺术倾向，强调撒哈拉西部人民对于安定、和平、自由的强烈渴望。这种朴素的表现形式造就一种浪漫、梦幻的感觉，物象的极度反差令人感到恍如隔世。艺术家们希望通过作品，让事实穿越时空的阻隔，为世人翻开那段因战乱而尘封的历史记忆——对外宣布撒哈拉西部人民对他们的土地、文化、根源和自由的拥有权，关注"地方可见度"、"地方承诺"、"地方希望"和"地方认同"，呼吁世界的正义力量关注与推动当地的和平解放事业。（马熙迓）

Artwork Description

The war of Western Sahara, which lasted 40 years, came to an end in 2007. The Gdeim Izik art collective has organized a public art event every year since, whose goal is to proclaim the rights of the Western Sahara's people to their land, their culture, their roots and their freedom. In particular, the event highlights the values of "visibility of place", "commitment to place", "hope in place", and "identity of place". All the projects are selected and implemented on the basis of their potential to help the people of Western Sahara communicate the nature of their discord with the Western world, and to realize their long sought-after self-determination (a goal articulated by the United Nations more than thirty years ago).

Since its first edition Art Tifariti has achieved great success, connecting Tifariti to the rest of the world through the efforts of all the local participants. The human and social situation of Western Sahara is communicated in different ways by the participating artists. Rebirth of the Ruin, for instance, is a mural created in 2008 by the Argelian Artists' Collective (Argel). Painted on the walls of the partially bombed Tifariti hospital, the artwork drew attention to the substantial lasting impact of the war. The artworks include both permanent and temporary installations. They are all exhibited outside a museum context, and have become part of the cultural heritage of Tifariti.

The project has generated greater awareness of the conflict in Western Sahara while bringing a stronger sense of identity to the region through local participation in acts of creativity. In connection with this project information about Western Sahara has been broadcast throughout the world. Thanks in part to the provision of new facilities and continued support each year, local artists continue to create new artworks, participate in public space projects and cooperate with international artists.

Artwork Excellence

The excellence of the project is confirmed because it answers its aims: "to show the conflict of the Western Sahara to the world through art and raise awareness to the society." As a result, more information has been spread by the mass media about this event any other event in the Western Sahara. The individual art projects are created in strong connection with site and context, involving local Saharawis as active participants in place making. In the western world, the festival has raised awareness and generated political debate and support. The constant commitment by the artists to placemaking has resulted in new facilities and benefits for the Saharawi artists, who are eager every year to participate in the different art projects. They develop their own ideas, evolving continuously through contact with public space projects and international artists.

你我之间
Between You & Me

艺术家：马丁、英格·瑞比克（荷兰）
地点：荷兰蒂尔堡
推荐人：汉娜·皮尔斯
Artist: Martin & Inge Riebeek（Netherlands）
Location: Tilburg, Netherlands
Researcher : Hannah Pierce

作品描述

坐落于荷兰一所新建的舞蹈音乐学校的庭院里，该作品由一大型 LED 显示屏安装在信号杆上面，显示屏上方安装有可移动摄像头。

广场人行道地面上埋设有动态感应 LED 灯，当行人进入中央灯光圈内，LED 灯颜色会随之发生变化。摄像头会聚焦行人面部四秒钟然后向上移动继续从上方拍摄记录来访者。

形象会后继显示于 5m×3m 的大屏幕，描绘出参与者在庭院的公共空间中更为广阔的环境。

面部识别软件的运用意味着大屏幕上的图像总是集中在参与者的面部。很短一段时间之后，观者将会看到自己的面部动态慢动显示在屏幕上大概 20 秒钟左右，与此同时系统将会由此筛选一条标语在广场另一边的一个液晶显示仪上出现。

在广场上由过路行人参与生成的影像和文字信息调节着整个公共空间的气氛。任何时候一个人或者一群人进入这个互动的感光地带，会随之生成不同的文本和颜色。通过拍摄和投射影像，该公共空间被转换成一种舞台甚至是表演空间。这样一来，任何人参与所选择的站立位置都被放大，他们以此可以从远距离更好地观察自己，同时更细腻地体验自己和周围人的关系以及周围的空间。公共空间在此跃升为舞台感的背景式空间。被企业和资本买卖大范围使用的广告用大型户外液晶显示屏占据着公共空间，观者被动接受与他们自身无法控制的信息，疏离了他们对公共空间的感官主动权。

瑞比克的装置作品警惕这样一种不平衡，通过完整呈现观看者本人的面貌来作为视频终端的视觉输出。"你我之间"这件作品提出这样一个问题：你置身于公共空间之中的自我空间是什么形式的？其他人的空间又是从哪里开始的？在公共空间中和他人的接触是无法避免的，正是这种人与人相互之间的关系造就了城市。当这件互动装置作品没有被激活的状态下以数码平台的角色播放各种影像艺术作品。

解读

作品通过人与多媒体互动的形式，为公众提供一个展现自我的互动公共空间，人们在走进作品享受多媒体回馈的同时，潜移默化地成为了整个作品展示的一部分，并带动更多的人们加入其中。在互动公共空间中，人们无可避免地发生接触，互相探寻彼此空间的起始与结束，引发人们注意到人与人之间的一种无法由自身控制而触发的信息交换，这正是当代城市生活中人与人关系的鲜明缩影。现代城市生活的繁忙，令人与外界的交往多了一份冷淡，众多的朋友仿佛都是无缘的过客。艺术家马丁·英格瑞比克通过多媒体介入的形式，促使任何人在任何时候都可以参与到公共交往中，为人们开启了一扇心灵沟通的大门。与以往商业用途的电子多媒体设备投放不同，"你我之间"更是一场纯粹的艺术感官盛宴，是一场无言的个人与群体的聚会。（马熙达）

Artwork Description

Between You & Me is located in a public courtyard outside Factorium, a new school of dance and music in Tilburg, the Netherlands. The work consists of a large LED screen mounted on a mast, which in turn is connected to a movable camera on a fixed track above it. Motion-sensitive LED light fixtures were embedded into the pavement around the square beneath so that these lights change colour when a person enters the central light circle. The camera then films a close up of individual standing in the light circle for four seconds and moves upwards on the track continuing to document the visitor from above. This image is subsequently shown on the 5x3m panel screen, depicting the participant in the wider context of the courtyard. Facial recognition software means that the image on the screen is always centralised around the participants face. A short while later, the viewer sees himself or herself in slow motion, broadcast onto the big screen for 20 seconds, along with a system-generated slogan that is also displayed on a second panel mounted on another wall across the square to the participant's left.

The activities that take place on the square, and the visitors, colors, images and texts on the LED screens, influence one another and set the atmosphere for the public space. At any time when an individual, or group of people, enter the interactive light circle, other texts and colours appear. Through the act of filming and projecting images, the function of this public space shifts to a stage or performance site. In consequence, anyone who chooses to stand on the spot temporarily becomes the focal point of a greater whole, and is able to observe not only themselves from a distance but their relationship to other people and the space the are inhabiting. The surroundings suddenly shift function to that of a stage or site of performance.

With the vast majority of large exterior screens used for advertising, and corporations and big business able to occupy a space that the viewer has no influence over, there is a sense of disconnection between the public and their sense of ownership or representation in public space. The Reibeek's installation recognised this imbalance and aimed to address it through making the viewer an integral part of the screen's output. Between You & Me addresses the question of what your own space within the public space is and where that of others begins. Contact with others is unavoidable in public space and it is these interrelationships that make a city what it is. When the artwork is not activated, the big screen functions as a digital platform for video art.

Artwork Excellence

Between You & Me has been recognized both regionally and nationally in the Netherlands for excellence in public art commissioning. In 2011 the work was awarded the Brabantse Straatkunstprijs, a publically voted regional street art award for which over 150 works were nominated. Video is often a significant component in Martin & Ige Riebeek's public works, Between You & Me demonstrates a further complexity in public new media art that required strong collaboration with a leading audio-visual team to make the delivery of the interactive work possible. The placemaking qualities of the work are the result of five years of consideration in the design and development of the installation, which stimulates a high level of engagement with viewers. By handing over the content of the screen to the audience, the viewers are instilled with a sense of ownership of the public square. The installation attracts a huge number of people to the courtyard who would have previously only walked by, enhancing the sociability of the area and culturally enriching Tilburg by facilitating additional video art presentations.

黄上的蓝
Blues on Yellow

艺术家：丹尼尔·布伦（法国）
地点：比利时布鲁塞尔
推荐人：汉娜·皮尔斯
Artist: Daniel Buren(France)
Location: Brussels,Belgium
Researcher: Hannah Pierce

作品描述

"黄上的蓝"是由艺术家丹尼尔布瑞根据比利时布鲁塞尔司法部的环境制作的一件大型装置作品。这是一个连接布鲁塞尔上城区和下城区的一个交通要道，之前一直被忽略的城市重要的公共空间，尽管在地理位置上靠近中央火车站和各种文化机构。

该装置作品由 89 块分布在广场两侧的固定在黄色旗杆上的蓝色旗帜组成，这件作品被借此用来象征司法部在更新和发展司法决策上的新步伐。

黄色旗杆在广场上以网格坐标的形式被安置，正方形等距排开。旗杆高度从 7.5 米到 12.2 米之间不等，这使得每个旗杆顶部都是同样的海拔高度，因为地面本身海拔高度有稍许出入。旗杆上每一面丝质旗帜都为蓝底色的三角形图案。这些三角形以 8.7 厘米的宽度和竖条纹蓝白相间形成交替，是艺术家个人的惯用符号。旗帜被固定成这种方式意在防止大风情况下旗面被缠绕在旗杆上。每个旗杆顶部装有 LED 灯确保它们在夜晚还能依稀可见，同时在广场和皇家大道上发挥出路灯的功能。

皇家大道从司法部前一直通到该作品的中间，这样一来黄蓝相间的作品就能同时被广场和大道另一段同时看到。场地本身的开阔性使得艺术家运用了很多旗帜的想法，这种强烈的色彩对比能够从视觉上吸引广场两面方向人群的注意力。处于地势较低的观众可以在大道立交桥下面从作品一端穿行到另一端，而马路行人和公路上使用其他交通工具的人可以从马路两边俯瞰作品的非常醒目的设计。

由旗杆组成的方格阵形占据了整个广场，旗帜迎风飘扬，给它们身后的建筑物蒙上了神秘的面纱。旗帜轻盈的材质使得整件作品总是不断处于动态之中，随风鼓动的蓝色旗帜上由三角形打破其视觉完整性，而不同蓝色的挑选更是强化了这一视觉深度。大面积大胆的蓝黄交织经常被视为都市丛林，点亮了整个广场，从整个城市环境和建筑物中张扬而出。

解读

"黄上的蓝"是由艺术家丹尼尔布瑞根据比利时布鲁塞尔司法部的环境制作的一件大型装置作品。该装置作品由 89 块分布在广场两侧的固定在黄色旗杆上的蓝色旗帜组成，这件作品被借用此来象征司法部在更新和发展司法决策上的新步伐。

场地本身的开阔性使得艺术家运用了很多旗帜的想法，这种强烈的色彩对比能够从视觉上吸引广场两面方向人群的注意力。处于地势较低的观众以及马路行人都可以在不同的角度欣赏到作品的魅力。

蓝色的旗帜在空中摇摆着，给人一种轻盈、灵动的感觉，同时，也给它们身后的建筑物蒙上了神秘的面纱。大面积大胆的蓝黄交织经常被视为都市丛林，点亮了整个广场，从整个城市环境和建筑物中张扬而出。（冯正龙）

Artwork Description

Blues on Yellow is a large-scale installation by Daniel Buren situated in the Place de la Justice in Brussels, Belgium. This crossing point connects the upper and lower parts of Brussels and was previously overlooked as potential public space in the city, despite its close proximity to central train stations and cultural institutions. Consisting of 89 blue banners mounted onto yellow masts spread across the two sides of the Place, the work was commissioned to signify a new step in the regeneration and redevelopment programme of the Place de la Justice.

The yellow masts are laid out as though on a grid, in equidistant squares across the Place. These masts vary in height between 7.5m and 12.2m, which allows their elevation to reach the same height since the ground level across the site is not even. Each mast bears a patterned silkscreen banner in one of five shades of blue broken up by large triangles. These triangles alternate between the blue of the flag and vertical stripes of white 8.7cm wide, a trademark of the artist. The banners are mounted in such a way that prevents the flag element from wrapping around the mast in the event of strong winds. An LED light is installed at the top of each mast, which makes the work both visible in the evening and contributes to the functional street lighting of the square and the Boulevard de l' Empereur.

The boulevard passes over the place de la justice and through the centre of the artwork, as a result Blues on Yellow is designed to be viewed both from both below in the square and from above on the upper street level. The split nature of the site informed the artist' s decision to propose the multiple flagged masts, as the strong colours of the banners successfully draw together the two sides of the place visually. Lower level viewers are able to pass under the Boulevard overpass from one set of masts to the other, while pedestrians and road users on the upper level are able to view the work on either side simultaneously due to its height and striking design.

The grid of masts occupies the entire Place, and the banners, on catching the breeze, mask the facades of the buildings behind them. The lightweight material of the banners mean the work is constantly in motion. The variations of blue, with the interjecting triangles mix visually when moving in the wind, which couldn' t have as much depth had only one blue been chosen. The mass of blue and the bold yellow masts, often described as an urban jungle, embellish the square and detract from the city environment and surrounding buildings.

Artwork Excellence

Blues on Yellow was inaugurated within the context of a larger redevelopment project of the Place de la Justice called Chemins de la Ville. Deputy mayor Henry Simons identified the potential of a public art commission to improve the sectors image. La Région de Bruxelles-Capitale commissioned the work upon the request of the Municipality of Brussels. On the advice of the Comité d'Art Urbain, Daniel Buren was commissioned in 2004 due his extensive experience of installing public artworks that create new spaces within existing environments. Architect Bruno Corbisier was contracted as the Project Manager to oversee the projects execution. The total cost of the work was a little over 1 million Euro. This was to include maintenance of the work for 6 years, a service contract was put in place to replace the banners every six months over this period. A further 2.4 milion Euro was spent on roadworks around the site. Of this total budget, 2 million Euro was financed by the Région de Bruxelles-Capitale.

柱廊
Collonade

艺术家：乐都园艺工作室（德国）
地点：德国莱比锡
推荐人：薇拉·托尔曼
Artist: Atelier le balto（Germany）
Location: Kleinliebenau, Germany
Researcher: Vera Tollmann

作品描述

作品描述克莱里伯诺骑士封地教堂坐落在古老的王者大道上，历史悠久，最早的记载能追溯到 1309 年。中世纪时期，沿途的商人曾在这座供奉商人守护神圣尼古拉斯的教堂里停留休息。教堂内部巴洛克式的风格可以追溯到 1787 年，当时的风貌几乎完好地保留到了今天。直到 20 世纪 70 年代，人们仍可以上教堂做礼拜。2001 年，这所教堂被划为施科伊迪茨城所有，2005 年又象征性地以 1 欧元卖给了一位名叫亨里克·莫拉斯卡的老师。同年，文化和朝圣者联合会成立，旨在修复并保护该神圣的文物，以供后世瞻仰。每年平均有至少 50 位朝圣者在去西班牙圣地亚哥德孔波斯特拉的路上停留于此。

文化和朝圣者联合会成员最初打算请艺术家在教堂里做艺术创作，但鉴于对 18 世纪巴洛克风格的尊重，负责此事的当代艺术博物馆建议保留教堂的原有风貌，转而聘请位于德国柏林的园艺建筑工作室"乐都"在教堂外围设计一个花园。

乐都的设计成果"柱廊"由以下元素构成：首先，一条1米宽、抬离地面25厘米的木板路环绕在教堂外部。走在这条木道上能够一览教堂独特的建筑风格，周边景色也尽收眼底。另外，西北角建有一个平台，设有座位，可作为音乐会或其他活动场所。通道两旁还设计有达40厘米深的沟渠，种有柳树。这与当地村落的景色相呼应，很多小路旁都有类似的沟渠。

新建的平台和添加的植物让围绕其中的教堂呈现给人一种浑然一体的视觉感，凸显了村子的特色。除此之外，由于文化和清教徒联合会经常在这里举行活动，教堂入口处铺有高质量的石板路，十分耐用。教堂的南面有一片草坪，其间点缀着颜色各异、品种繁多的植物。"柱廊"围绕在这座朝圣教堂的周围，为教堂重构了空间，淡化了其地理位置不佳的缺点（该教堂紧邻路边），让朝圣者和去教堂做礼拜的人享受到更怡人的空间。设计名"Lustgang"是比较老的一个园艺术语，意思是穿梭于花园之中的一种美好体验。通道结构的独特设计考虑到了立体感，不同的角度能欣赏到各异的景观，加强了视觉享受。

解读

克莱里伯诺骑士封地教堂建筑年代久远，古朴的教堂建筑见证了世间千百年的兴衰荣辱。但随着时代更迭，教堂现有的基础设施已无法满足现代人的需求，亟待改善。艺术团队乐都园艺尊重当地文化和朝圣者联合会的请求，旨在保护教堂巴洛克风格原貌的前提下，对教堂进行一系列改造工程，并在外围增设各式公共基础设施，力求与当地景色融为一体。改造工程并没有影响当地居民、教徒的正常朝圣活动，同时为人们提供了现代化的便利，使教堂的独特风貌得以令后世瞻仰。这种逐级优化的改造形式，最终达到了一种最优化的理想状态——不破坏区域传统文化的前提下更好的进行现代化改造，同时促进区域特色景观的发展，实现被开发区域自然、文化、经济、社会的整体协调发展。（马熙逵）

Artwork Description

The Rittergutskirche Kleinliebenau is a historical monument situated on the old Via Regia. The church was mentioned for the first time in 1309. In the Middle Ages it offered shelter to the traders travelling along that road. It has been dedicated to St. Nicholas, the patron of merchants. The baroque interior, which is today almost entirely preserved, dates back to 1787. During the 1970s the church was still in use for services. In 2001 it became property of the town of Schkeuditz and in 2005 the historical building was sold to the teacher Henrik Mroska for the symbolic amount of one Euro. That same year, the Kultur- und Pilgerverein Kleinliebenau e. V. (Cultural and Pilgrim Association Kleinliebenau) was founded with the mission to restore and preserve the sacral historical place for the generations to come. Each year an average number of 50 pilgrims (sometimes even more) make a stop off in Kleinliebenau on their way to Santiago de Compostela in Spain.

Members of the Association initially wanted to invite artists to make artistic interventions in the church. But paying tribute to the baroque interior of the 18th century, the mediator for the commission, Galerie für Zeitgenössische Kunst Leipzig (Museum of Contemporary Art), suggested leaving it intact and instead bringing in the Berlin-based landscape architectural studio atelier le balto and commissioning them to design the garden around the building of the church.

Atelier le balto' s Collonade (Lustgang) is made up of the following elements: A raised wooden walkway, 100 cm wide and about 25 cm above ground, runs around the outside of the church. As you walk along it, it provides an excellent view of the unusual structural features of the church and the characteristics of the surrounding landscape. This walkway opens up towards the graveyard. At its northwestern corner, the platform is built up to provide seating accommodation for concerts and events. The pathway is flanked by hollows, up to 40 cm deep, planted with willows. These reflect the characteristic motif of the ditches that are found along the roadside in many parts of the village.

The optical framing of the church by the platform and the plants accentuates the special character of the village with its church. The entrance area is paved with robust high-quality stonework, to withstand frequent usage during the popular events held by the cultural and pilgrims' association. To the south of the church, a colourful meadow with a large variety of plant species is planned. The Collonade project starts off at the area surrounding the pilgrim church, (re)creating space for the church. The rather unfavourable position of the site (the church is situated close to a road) is thus softened, creating far more pleasant surroundings for the gatherings of pilgrims and churchgoers. The title Lustgang refers to an old term taken from landscape art, meaning a pleasant means of experience, like walking through a garden. The path is intentionally structured in order to generate and stimulate perception at various levels.

Artwork Excellence

Collonade (Lustgang) was commissioned in the framework of the European program Neue Auftraggeber/Nouveaux Commanditaires. Neue Auftraggeber is an operational model addressing a diverse range of participants. The artwork does not solely originate from the needs and thoughts of its individual creator, but also from the needs and thoughts of society, represented by people who agree to fully assume roles as decision makers like the artists. The project in Kleinliebenau is a part of the life of the community because members of the community participated in the process. The artwork Collonade by atelier le balto fits very well the needs and interests of its commissioners. It is an inexpensive construction and a modern answer to romantic collonades. Church visitors can use it for recreation. Thus, the artwork perfectly integrates into the public space and resists becoming l'art pour l'art.

梦
Dream

艺术家：约姆·普兰萨（西班牙）
地点：英国默西赛德郡圣海伦斯
推荐人：汉娜·皮尔斯
Artist: Jaume Plensa（Spain）
Location: St. Helens, Merseyside UK
Researcher: Hannah Pierce

作品描述

"梦"是一尊地标性雕塑，坐落于以前的萨顿庄园煤矿旧址。完成于2009年5月，它标志着圣海伦斯长期的愿望得以实现了，为煤矿旧址留下标记，煤矿自从1991年起关闭了，作品是向该地区采煤历史的致敬。这件作品采用了一位年轻女孩的头部形象，她的眼睛闭着，看上去像在做梦一样的状态中。她在底座上休息，底座上刻着铭文"梦想萨顿庄园"，引用小型，圆形的木块，是每个矿工携带着，作为鉴定的一种手段。

"梦"是艺术家与前任矿工以及更广泛的当地社区成员进行对话，随后做出的简洁的回应，这些人希望能有一件艺术品指引着更光明的未来，给后代们创造一种沉思的空间。该作品受到了圣·海伦斯以前座右铭的启发：ex terra lucem（拉丁原文），即"来自土地的光"。这座20米高的巨型雕塑矗立于高出海平面100米的位置，特别地突出，可以俯瞰全英国最繁忙的高速公路之一。雕塑的面板是用预制的混凝土和西班牙白云石制作而成。这使得雕塑外观非常白皙，几乎会发光，根据当天的天气和时间而改变。

面向南方的位置意味着它的轮廓和阴影随着太阳由东到西的移动而变换，每当夜晚时，雕塑就被照亮了。

艺术品的设计及规模和旧址的自然结合在一起——该旧址曾经遭受过累累破坏——提出了相当的技术挑战，将这个雄心勃勃的概念转化成一座 20 米高的雕塑。由于其尺寸和重量，最终配置由 54 个单独的面板构成了雕塑的头部，雕塑的基座有 36 块面板，每一块都有艺术上的要求，和两倍的连接系统用来把面板连接在一起，为了整体效果而制作了一个集成的整体结构。雕塑的大型三维造型是一些高度专业化的分包商共同合作的产物：奇塔姆山建筑有限公司作为主承包商，考代克制作了 90 个单独的模具，埃文斯混凝土铸造和组装了 90 块混凝土面板，ICP 协助进行了安装。

当地社区参与了创造这件艺术品方方面面的工作。此外，前任矿工提名了这个旧址，发展简单化，选择艺术家，激励概念，更广泛的当地社区得到通知，通过各种不同的方式参与这个项目。当地居民起到了协调的作用，无论在正式的计划过程实施中，还是实施期间，他们都与建议计划书保持紧密的联系。串联着主要的"梦"委派任务，由海伦娜伙伴关系代表圣海伦斯议会"大艺术小艺术"发起一些列公共艺术项目，进一步促进全体社区参与计划。

解读

作品"梦"的初始构想，是为填补当地因煤矿关闭后而产生的经济空白，向该地区曾经的采煤历史致敬。作品的设计基于曾遭受破坏的旧址之上，并旨在与自然的结合，因此在建造过程遇到了相当的技术挑战。在作品落成后的时间里，"梦"已超出了它本身的替代和纪念作用，悠远、梦幻的年轻女孩形象，带给人们无限遐想。它的朴实无华，传递着如诗般的缠绵，宁谧祥和中留给人们对未来磨蚀不尽的痴梦一场。梦的邂逅是道不完的岁月如梭，有时来不及细细珍藏，转瞬即逝，却在消散中留下雕琢般的记忆。"梦"在某种程度上成为凝聚观者的一种介质，吸引着无数的善男信女。各种梦想在此盛开，续写着今世的铮铮誓言，为人们对美好未来 种下了"梦"的种子。（马熙逵）

Artwork Description

Dream is a landmark sculpture located on the site of the former Sutton Manor Colliery. Completed in May 2009, it marks the fulfillment of a long held ambition of St. Helens to mark the colliery site since its closure in 1991 and to pay tribute to the area's coal-mining history. The work takes the form of the head of a young girl with eyes closed, seemingly in a dream-like state. She is resting on a plinth bearing the inscription "Dream Sutton Manor" referencing the small, circular "tally" each miner carried as a means of identification. Dream is the artist's response both to the brief and to subsequent conversations with the ex-miners and members of the wider local community who wanted a work that would look to a brighter future and create a contemplative space for future generations. The concept draws inspiration from the former motto of St. Helens: ex terra lucem "from the earth comes light."

The 20-metre tall sculpture stands at just over 100 metres above sea level and prominently overlooks one of the UK's busiest motorways. The panels were fabricated in pre-cast concrete and Spanish dolomite.

This lends a very white, almost luminescent finish to the sculpture with its appearance altering according to the weather and time of day. The south-facing position of the piece means that its contours and shading transform as the sun moves across the face from east to west, while at night the sculpture is illuminated.

The scale and design of the artwork, together with the nature of the site—a former spoil heap—presented considerable technical challenges to being able to translate the ambitious concept into a 20-metre high sculpture. Due to its size and weight, the eventual configuration was made up of 54 individual panels for the head element of the sculpture and a further 36 for the plinth, each one determined both by artistic requirements and the two-fold connection system used to hold the panels together and in effect create an integrated monolithic structure. The large three-dimensional form is the result of a collaborative effort of a number of highly specialised sub-contractors: Cheetham Hill Construction Ltd. as lead contractor, Cordek to manufacture the 90 individual moulds, Evans Concrete to cast and assemble the 90 concrete panels, and ICP to help install them.

The local community was involved in all aspects of developing the artwork. In addition to the ex-miners' role in nominating the site, developing the brief, selecting the artist, and inspiring the concept, the wider local community was informed and involved in a variety of different ways.

Local people performed a consultative role, and were kept closely informed of proposals both before and during the formal planning process. In tandem to the main Dream commission, a series of public art initiatives were devised and delivered by Helena Partnerships on behalf of St. Helens Council "Big Art's Little Art" to further the overall community engagement programme.

Artwork Excellence

Through the host of awards that Dream has won for its tourism, technical, planning and community engagement merits—as well as for its artistic excellence—the work is recognized on both a local and national level. In the two years since its launch Dream has established itself as an icon, appearing in place marketing and inward investment literature, annual reports and private sector advertising, gaining worldwide coverage for St. Helens.

The placemaking quality of the work is apparent in a number of ways. Locals have a new leisure resource that draws many new visitors to the site and region. Dream attracts approximately 60,000 to 70,000 people each year. Since its completion in 2009, the site of the artwork has been continually enhanced, broadening its substantial impact on regeneration. This has included the upgrading of new visitor facilities such as paths, seating and the provision of digital visitor interpretation materials. The artwork is now the catalyst for the creation of a new 300 hectare regional forest park and outdoor pursuits destination.

Dream represents a best-practice model for commissioning large-scale public art that could be applied and adapted elsewhere. The excellence of the project is evident in the quality of the process and partnerships underlying its successful development and delivery, in particular the involvement of the local community in all of these aspects.

戴荆冠的耶稣像
Ecce Homo

艺术家：马克•瓦林格（英国）
地点：英国伦敦
推荐人：汉娜•皮尔斯
Artist: Mark Wallinger （UK）
Location: London, UK
Researcher: Hannah Pierce

作品描述

马克•瓦林格创作的"戴荆冠的耶稣像"是皇家艺术学会委派的三件作品中的第一件，目标是把目光焦点重新聚集在伦敦特拉法加广场上那空空的基座上。白色大理石树脂制作的真人大小的男子塑像站在基座的边上，表现了耶稣基督被本丢•彼拉多推到群众面前接受审判的那一刻。本作品的英文原名的意思是"你们看这个人！"这是彼拉多把基督推给暴徒们时说的一句话。这尊塑像表现了一位非常人性化的基督，胡须刮得很干净，低着头，目光垂向地面，强调了他那一刻的脆弱无助。这座基督像的双手被绑在背后，头上戴着用一圈圈铁丝网绕成的镀金的荆棘冠。这个塑像的铸造过程借用了真人的肖像，而采用雕刻或其他方法是达不到如此逼真的效果的。结果，这尊塑像在周边宏大的伦敦市中心加上四周矗立的历史上杰出军人塑像的包围下显得渺小了。

特拉法加广场是伦敦市中心的著名地标，也是主要的旅游景点。这是一个具有重要历史价值的地点，其中的纪念碑和雕塑有各自的遗产分类。站在伦敦特拉法加广场的东北角，第四个基座建于 1841 年，作为威廉四世的巨型骑马塑像的基座，结果他在遗嘱里没有留下足够的钱用来支付这尊雕像。"戴荆冠的耶稣像"于 1999 年 7 月安装在此基座上，之前的 158 年里基座都一直空着。自特拉法加广场自 1800 年代早期建成后就一直被视为全国的民主和抗议中心。在广场上经常举行政治、宗教和各类问题的集会和示威活动。

瓦林格的作品占据了千禧年的景点，内容涉及基督的诞生但作品本身却不是基督教的。将著名宗教人物展现在伦敦中心地点，这一做法是对多元文化的英国的宗教庆祝活动所隐含的歉疚状态的质疑。作为一件很公开的作品，它在基督徒和非基督徒中都引起了很大的反响。瓦林格并不是想要歌颂基督的诞生，而是通过对基督作为一名政治犯形象的描绘来表现基督献身的那一刻。瓦林格这一作品中表现的这个瞬间是很少在宗教雕塑中出现过的，他把一个现代版的耶稣基督放到众人的面前接受审判，而反过来又把一件苦差事交给了公众，让他们作出选择是否要谴责他。

解读

基督教会认为耶稣是上帝的儿子和救世主，耶稣的言行和教导形成基督教教义的基础，众人向耶稣祈求解脱自己的人生磨难。同时因为古代人加之于耶稣身上的种种超脱自然的神迹，耶稣带给人一种身处苦难的救世主形象。随着现代科学的兴起，人们的思想已彻底改变。宗教观也不例外，教徒更愿意以理性和经验为基础去认识宗教传统和信仰价值。艺术家马克·瓦林格展现给世人一个普通人形象的耶稣，头戴荆冠，身着简单的裹布，体格健壮而面无表情，与以往宗教色彩的耶稣形象形成对比。"戴荆冠的耶稣像"并不意味着对宗教信仰的异端和言不由衷，而是借助艺术的表现形式，理性的梳理出作品隐含的政治、宗教问题，将质疑与思考留给观众，由观众自发的去挖掘作品真正的内涵。（马熙迤）

Artwork Description

Mark Wallinger's Ecce Homo was the first in a series of three commissions by the RSA that aimed to refocus attention on the empty plinth in London's Trafalgar Square. The life-size sculpture of a man in white marble resin stood, perched at edge of the plinth portrayed Christ at the moment he was presented to the crowds by Pontius Pilate for judgment. "Ecce homo," meaning "behold the man," was stated by Pilate as he handed Christ over to the mob. The figure depicted a very human Christ, clean-shaven with a bowed head and eyes cast towards the ground, emphasizing his vulnerability. The sculpture stood with hands bound behind his back, adorned with a gold plated crown of thorns made rings of barbed wire. The process of body casting to create the sculpture lent the effigy a human likeness impossible to achieve through carving or any other method. In consequence, the figure appeared dwarfed by the scale of its formidable London environs and the military heroes that surrounded it in the square.

Trafalgar Square is a significant landmark in central London and a major tourist attraction. It is a site of significant historic value and its monuments and statues also have individual heritage classifications. Standing in the northeast corner of London's Trafalgar Square, the Fourth Plinth was built in 1841 as a pedestal for an equestrian statue of William IV, who did not leave enough money in his will to pay for it. Ecce Homo was installed on the site in July 1999 at which point the plinth had previously remained unoccupied for 158 years. Since its construction in the early 1800s, Trafalgar Square has been regarded as a centre of national democracy and protest. Rallies and demonstrations are frequently held in the Square on a range of political, religious and general issues.

Wallinger's work occupied the site over the Millennium, and addressed the birth of Christ without being a Christian work. The presentation of a prominent religious figure in a central London location questioned the apologetic nature of religious celebration in multicultural Britain and as a very open work resonated with Christians and non-Christians alike. Wallinger did not seek to glorify Christ's birth and instead chose to represent the day of his death by portraying Christ as a political prisoner. Wallinger's work identified a moment rarely represented in religious sculpture, one that portrayed a contemporary Christ as someone placed for judgment in front of a public who in turn was tasked to choose whether or not to condemn him.

Artwork Excellence

Commissioning a high-quality public art work in a prominent location in the centre of London that had previously lain unused for over 150 years represented a significant step in re-purposing highly visible spaces in metropolitan areas. Mark Wallinger was able to present his work to the broadest possible audience, and through this initiated an ongoing dialogue about the relationship between contemporary public art commissioning and public space. Following the enormous public interest generated by Wallinger's commission and the two art works that followed it, the decision was made not to install one of the three pieces permanently but to continue an ongoing series of temporary works of art commissioned from leading national and international artists. The Fourth Plinth programme has achieved international status and remains a focus for press and public attention, stimulating discourse about public art at an international level.

石头们
Folk Stones

艺术家：马克·沃林格（英国）
地点：英国福克斯通
推荐人：汉娜·皮尔斯
Artist: Mark Wallinger（UK）
Location: Folkestone,UK
Researcher: Hannah Pierce

作品描述

马克·沃林格的这件作品是由 2008 年在英国肯特郡东海岸举办第一届福克斯通三年展时永久展示的一件户外作品。"石头们"由 19240 块作了记号的鹅卵石铺撒成边长为 9 米的正方形，地点在福克斯通镇的俯瞰英吉利海峡的丽斯长廊——维多利亚式悬崖顶散步区。原初作品方案只是为三年展创作的暂时性作品，后来由于受到当地人和游客的欢迎便被永久保留下来。

在草坪中，鹅卵石被平铺在正方形水泥底上，这些石子来源于悬崖下面的海岸边的石子。每个石子上由手绘数字记号从 1 到 19240，19240 是 1916 年 7 月 1 日索姆河战役打响的第一天牺牲的英国士兵的人数。福克斯通港是第一次世界大战法国和佛兰德斯战役士兵们出发登船的海港。福克斯通港在此意义上是战争的纪念地，每一块鹅卵石都代表着一位牺牲的士兵。艺术家沃林格同时也揭示出用一个数字对等一条生命的薄弱之处以及通过统计数据来体现个人生命价值这一举动的无力之感。

晴朗的天气里从丽斯长廊可以依稀看到法国的海岸线。这是作品选址该处的原因，艺术家希望观者在观看这件作品的时候可以同时看到法国。福克斯通三年展当时不得不充分地与安理会协商以获得在该地点制作艺术作品的批准。在作品制作过程中当地一些居民并不理解这一行为，在解释了作品意图和选址背后的意义之后被当地社区热情接受了。为了答谢福克斯通当地居民对"石头们"这件作品的接受和响应更广泛的观众，经过艺术家同意，该作品被永久性保留在原处。福克斯通三年展经济上不足以支持该作品时由安东尼瑞诺兹画廊出资使得该作品得以在丽兹长廊保留下来。

作品完成之后的四年来，该地点作为当代公共艺术同时作为战争纪念地被频繁光顾，吸引着国内外的观众前来拜访。每当战争纪念日，当地社区会组织悼念活动来反思每一块鹅卵石所象征的生命。该处被明确标记出是一件艺术作品，同时还注有简短的作品说明。

由于作品原初计划只是暂时性选址，所以作品的大小尺寸被明确标出以维护作品的长期性。风吹日晒使得其中很多石头上的数字日渐模糊，必须被不断重新描清楚。为了确保每一个数字都是准确无重复的，最近对每一块石头的位置和数字都做了详细的检查以确保它们被规整在正方形内。

解读

作品"石头们"灵感来自在法国和弗兰德斯战场上浴血奋战的英军士兵，被编号的石头们分别代表着索姆河战役第一天牺牲的每一名英军将士。艺术家马克·沃林格希望通过公共艺术的形式，使观者铭记，在战争面前，个人的生命显得如此薄弱，人权、自由、尊严等人性的真理荡然无存，战火使无数的家庭承受着失去亲人和家园的痛苦。战争的爆发归根结底还是因人类的丑恶造成的。在这场磨难中，人类已不受命运的掌控，宝贵的生命被漠视，错误的弥补被不惜一切代价的复仇所替代，人类遗失的不再是真理，而是对于生命的责任。无论是敌人还是朋友，他们的生命同样需要我们的重视。作品《石头们》唤醒了人们尘封已久的内心，它带给我们不仅是对战争的反思，还有对生命和幸福生活的珍视。 （马熙逵）

Artwork Description

Mark Wallinger's Folk Stones is a permanent work commissioned by the first edition of Folkstone Triennial, an art event inaugurated in 2008 on the East Kent coast. Folk Stones consists of 19,240 numbered stones set into a 9m square on The Leas, the town's Victorian cliff-top promenade overlooking the English Channel. Initially proposed as a temporary work for the duration of Triennial, the work was so well received by residents and visitors that is has been acquired for the site on a permanent basis.

The stones are laid out in a large square concrete base set into the grass and are near identical to those found on the shore below. The stones are individually hand-painted with a unique number between 1 and 19,240. The precise number of beach pebbles laid out on the site is the exact number of British soldiers who lost their lives on 1 July 1916, the first day of the Battle of the Somme. Folkstone's harbour was the embarkation point for millions of soldiers going to fight on the battlefields of France and Flanders in the First World War. Folk Stones is in this sense a memorial to the lost, each stone corresponding to a soldier killed. However, Wallinger is also addressing the failure of trying to equate a human life to a number, the impossibility of adequately representing an individual through a figure or statistic.

From The Leas, on a clear day, the French coast can easily be seen on the horizon. This very much informed the selection of the site, as the artist felt it was important to be able to see France in the distance when viewing the work. Folkestone Triennial had to negotiate extensively with the council to gain permission to install a temporary work on the site. During the installation period some resistance was met from residents, however on explaining the inspiration and local significance of the work it was widely supported by the local community. In response to the positive reception of Folk Stones by the inhabitants of Folkestone and the wider audience, the decision was made, with the artist, to acquire the work for Folkestone on a permanent basis. This was fully supported by Wallinger and the Anthony Reynolds Gallery who, when Folkestone Triennial was unable to raise the capital funds to purchase the work, donated Folk Stones to remain on The Leas.

In the fours years since its installation, Folk Stones has become both a visitor attraction, drawing national and international crowds as a contemporary art work and a memorial, with large numbers of the local community gathering at the work on Armistice Day to reflect on the loss of lives that each stone represented. The work has been clearly signposted as an artwork, and a brief description is given outlining the events the work refers to.

As the work was only intended to be installed temporarily, measures have since been put in place to maintain the work in the long term. Due to its prolonged exposure to the elements, some of the numbering became increasingly less visible and required repainting. To ensure that no number is ever duplicated, a full survey of the stones was recently undertaken to map out the location of each numbered stone within the square.

Artwork Excellence

The excellence of Folk Stones can be recognized in the decision of both Folkestone Triennial and the artist to ensure that the work was installed on a permanent basis on The Leas. This action was informed by the positive responses of the local residents who identified with the personal significance of the project. Mark Wallinger was so keen to support this demand to keep the work in Fokestone that he donated the work to the Roger De Haan Charitable Trust.

Since 2008 the work has established its placemaking value in two ways. As a contemporary artwork it is one of the town's most popular visitor destinations, drawing a national and international audience. It has also established itself as a site of remembrance and memorials for residents and visitors alike, who now gather at the artwork each Armistice Day to reflect on its meaning. Folk Stones has been used widely in press coverage of Folkestone Triennial, and this work is one of many that are contributing to a new legacy of cultural excellence in Folkestone.

第四基座
Fourth Plinth

艺术家：托马斯·舒特（德国）、马克·奎恩（英国）、
凯塔琳娜·弗里奇（德国）、因卡·索尼巴尔（英国）
地点：英国伦敦特拉法加广场
推荐人：汉娜·皮尔斯
Artist：Thomas Schutte(Germany), Marc Quinn(UK),
Katharina Fritsch(Germany), Yinka Shonibare(UK)
Location：Trafalgar Square, London,UK
Researcher：Hannah Pierce

作品描述

特拉法加广场是英国最著名的广场，正中央是 19 世纪特拉法加海战中英军统帅纳尔逊的塑像。广场外围四周还有四个石像基座，其中三个已站上了历史名人，只有广场西北角的第四个基座一直空着。第四基座原为威廉四世骑马雕像的基座，因资金缺乏而空置 150 年。1998 年，英国皇家艺术学会决定向艺术家征集作品，后来伦敦市长专门成立了"第四基座委员会"，使"第四基座"在最近 13 年里被用来轮流展示非永久性的当代雕塑作品。

第一件作品是马克·渥林格的大理石雕塑"瞧！这个人"。唯一没有展示雕塑物体的作品是葛姆雷的"一个和另一个"，这个活生生的纪念碑从 2009 年 7 月至 10 月，让每一位参与者在基座上在一小时之内按自己的意愿来表现自己。2011 年，展出了因卡·索尼巴尔的雕塑作品 "瓶中的纳尔逊战舰" ，这是第一件和特拉法加广场的历史有直接关联的作品。2012 年 2 月，一座 4 米多高骑着摇晃木马的男孩的青铜雕塑成了这里的新主人，既是向不远处的纳尔逊雕像致意，也令人重新思考战争的意义。

"第四基座委员会" 倾向于轻逸、幽默的方案，意在为伦敦营造丰富的当代视觉文化，更重要的是为了让大众参与到对于公共空间内的当代艺术作品的讨论中来。这些讨论由它所处城市中心的重要位置以及一系列活动而得到激发。

解读

特拉法加广场西北角的第四基座因资金的缺乏而长期空置， "第四基座委员会" 希望艺术家们在此轮流展示非永久性的当代艺术作品，来填补现有的空缺。 "第四基座" 作为当代艺术作品的展示媒介，在功能实现与节约成本的平衡中取得了极大的成功。它不仅通过艺术家的个人创造来丰富当代视觉文化，引发大众对历史与现实、过去与未来的思考，而且弥补了历史的缺憾，使艺术的精彩从此循环。公共艺术不仅要在空间中争取最具优势的艺术表现，而且要经受得起时间的拷问。世间没有一成不变的标准，却有着一样的解决方案，适合大众喜好的作品才更具蓬勃的生命力。《第四基座》最大的智慧在于它的求新与永恒，不同的时代、不同的社会需求，会迸发出它更加巨大的艺术潜能。（马熙逵）

Artwork Description

Trafalgar Square is the most famous city square in Britain. A statue of Lord Nelson, the British commander in the 1805 naval battle of Trafalgar, stands in the center. Four stone plinths are situated around the periphery, statues depicting other historic figures mounted on three of them. Only the fourth plinth near the northwest corner of the square stood bare. This fourth plinth was built as a pedestal for an equestrian statue of William IV, but with insufficient funds left in his will to pay for the statue's completion the plinth remained unoccupied for 150 years. A scheme was initiated in 1998 by the Royal Society for the Encouragement of the Arts (RSA) to commission sculptural works for the site. Later, the mayor of London established a special "fourth plinth board ", which over the past 13 years has guided the exhibition of a series of temporary contemporary sculptures.

The first was Mark Wallinger's Ecce Homo. The only commission that did not present a sculptural object was Gormley's living monument One & Other. Exhibited from July through October, 2009, almost thirty-five thousand people applied to participate. Yinka Shonibare's sculpture Nelson's Ship in a Bottle was exhibited during 2011, the first commission to relate directly to the history of Trafalgar Square. Rocking Horse Boy, a four-meter high bronze sculpture was erected in February 2012. It is a tribute to the nearby statue of Nelson and also an incitement to rethink the meaning of war.

"The fourth plinth board" tends to select works that have a light, humorous side which enrich London's contemporary visual culture. The foremost goal is to involve the public in discussion by means of contemporary artworks in a public space. By virtue of the high profile site and the series of activities which are run in conjunction with the exhibits, the program has stimulated many lively discussions.

Artwork Excellence

The aim of the Fourth Plinth is to place the highest quality contemporary art at the centre of the London, building upon the city's rich contemporary visual culture. After the initial success and high level of public support shown for the first three commissions initiated by the RSA, the Mayor of London recognised the excellence and value of the programme and took the project on as the responsibility of the Greater London Authority to continue it. The rolling commissions continue to receive abundant national and international press attention and are recognised as part of a world-class art commissioning programme.

A key element of the Fourth Plinth is to involve the public in debate about contemporary art in public spaces and this is stimulated through its prominent location in the centre of the city and its supporting programme of exhibitions, conferences and talks, including a community and education programme. The selection process itself is opened to the public to respond to, with 17,000 people commenting on last year's shortlisted proposals at the exhibition, submissions and via the Fourth Plinth website following their unveiling. While public opinion is often divided about individual commissions, the programme itself has received a huge amount of positive support from the public and visitors to the capital.

佛朗哥喷泉
Franco Fountain

艺术家：费尔南多·桑切斯·卡斯蒂略（西班牙）
地点：西班牙卡尔达斯
推荐人：薇拉·托尔曼
Artist: Fernando Sánchez Castillo（Spain）
Location: Caldas, Galicia Spain
Researcher: Vera Tollmann

作品描述

"佛朗哥喷泉"是根据西班牙历史和民主主义而形成的概念项目。佛朗哥的半身像安置在步行区，毗邻卡尔达斯省温泉小镇上最具象征性且最受欢迎的地标之一——乌米亚河畔的卡巴莱拉公园。

在他的作品中，艺术家费尔南多·桑切斯·卡斯蒂略的目的是调侃过去的英雄人物，诸如弗朗西斯科·佛朗哥，尽可能地引起观者的反应。卡斯蒂略安装了一种技术装置使水柱从这位独裁者的嘴里喷出。这样的想法是想看看人们是否敢喝来自领导者嘴里的水，或者说是过去的历史是否仍然沉重地压在人们心头以致能够引起纷争。这件公共艺术品尽管第一眼看似简单，却对西班牙意义非凡并且激发了一场重要争论。从某种程度上讲这是一场测试，测试人们是如何联系过去的？佛朗哥的形象还有多强大？

艺术家称前统治者的雕塑遍地都是，仿佛一切都应该归功于他。"我设计的这个喷泉，使从不规则的喷射口中饮水成为可能。这也可能建立了一种与雕塑间一反常态的新联系。"

佛朗哥以铁腕政策统治了西班牙近四十年之久。卡斯蒂略在具有讽刺意味的不同场景中使用了他的形象，大多涉及水这个要素。在一件作品中他用了佛朗哥的两条眉毛，据他称是从一个人手里买来的，此人于1975年11月打造了这位总司令的死亡面具。人们能够通过使用放大镜在一个样品袋中看见这对眉毛。此外，这对眉毛正以一个不公开的价格出售。2008年，卡斯蒂略设计了以四位世界领导人和独裁者为一个系列的青铜喷泉作为索恩斯比克地区的公共艺术项目，名为《吐水的领导人》。该作品位于池塘边，是传统的巴洛克式喷泉。水柱间断地从喷射口中喷出。例如，斯大林朝路易十四吐唾沫。

卡斯蒂略的作品"上和下"（2006—2008），当有人投一枚硬币时独裁者弗朗哥才会直立，犹如雄伟的骑马雕塑。这算得上是流于指尖的个人崇拜，因而每个人都略感内疚。

然而，在西班牙像弗朗哥这样的人物比在其他国家更能得到强烈认可。因此，这座暗示为不可侵犯的佛朗哥喷泉遭到了油漆的喷射，喷上的字迹是"法西斯主义"。

费尔南多·桑切斯·卡斯蒂略以雕塑、表演和视频为媒介塑造了众多集权、压迫却令人印象深刻的人物的不朽作品，涉及了西班牙的历史。

解读

作品"佛朗哥喷泉"试图通过调侃的形式，展现西班牙历史上的统治者弗朗西斯科·佛朗哥。作品形式尽管看似简单，普通的水泥方柱上，一尊人物铜像从嘴里喷射出不规则的水柱，却引发了一场西班牙民众对英雄人物历史功过的激烈纷争，从侧面反映出人们如今对以往历史的评价态度。艺术家费尔南多·桑切斯·卡斯蒂略有意采用这种讽刺和调侃的艺术表现形式，带有明显的个人政治倾向和观点。艺术家对于西班牙国内遍地的统治者雕像表示厌烦，认为这是历史遗留的没有任何实际意义的个人崇拜，英雄人物的个人功过评判应该交给本国最广大的人民群众。统治者雕像不规则的喷水，使人们接近作品取得饮水成为可能，建立了一种人与雕塑间的反常态的关系，更具政治讽刺意味。（马熙逵）

Artwork Description

Franco Fountain is a conceptual project based on the history of Spain and democracy. The bust-fountain of Franco was installed in a pedestrian area right next to one of the most emblematic and visited areas of the spa town of Caldas: the park-garden and carballeira, bordering the river Umia.

In his work, artist Fernando Sánchez Castillo's intention is to play with heroes of the past, such as Francisco Franco, in a way that possibly provokes a viewer's reaction. Castillo installed a technical mechanism so that a stream of water emerges from the dictator's mouth. The idea was to see if people dared to drink from the mouth of the leader, or whether history still weighed heavily from the past that divided Spain. This public artwork, which at first glance looked simple, was conceptually very important for Spain, and stirred up the population in an important debate. In a way, it was a test: How do people relate to the past? How strong can Franco's image still be?

The artist says former dictator sculptures were everywhere, as if allthings were thanks to him. "I designed a water fountain which it is possible to drink from its irregular jet. It was possible to establish a perverted new relation with the sculptures. "

Franco ruled Spain with an iron fist for four decades. Castillo has used his figure in different ironic settings, mostly involving the element water.

For one piece he used two of Franco's eyebrows, which he claims to have bought from the man who made the generalissimo' s death mask in November 1975. They could be seen in a sample bag from behind a magnifying glass, and were on sale for an undisclosed sum. In 2008, he designed a bronze fountain with a constellation of four different world leaders and dictators for a public art project in Sonsbeek called Spitting Leaders. Sited in a pond, the work was in the tradition of Baroque fountains. Water projected at intervals from their mouths; Stalin spat at Louis XIV, for example.

In Castillo's work Up and Down (2006-2008), dictator Franco arises as the imposing equestrian statue, but only if someone puts in a coin. It was a cult of personality at your fingertips, so everyone felt a little guilty.

However, in Spain the likeness of Franco is more strongly recognized than in other countries. Therefore it is no wonder that the suggestively intimate Franco Fountain was attacked with spray paint. "Fascismo" the paint read.

Fernando Sánchez Castillo creates monuments as collective figures of power, repression and memory, with references to the history of Spain in the mediums of sculpture, performance and video.

Artwork Excellence

More than 36 years after heading for the military dictatorship, the figure General Francisco Franco is still provoking the public. It's because of the artist Fernando Sánchez Castillo, who has used the iconic dictator in different fountain-sculptures; this one is his most "intimate" one. The fountain was spray-painted and had to be removed after a couple of days. One equally important aspect of this project is the professional curation of Kaldarte, a ten-year campaign and art project defending the freedom of art expression in public spaces.

"免费屋"和"明天的市场"
Freehouse, Markt van Morgen

艺术家：珍妮・凡・赫斯维克（荷兰）
地点：荷兰鹿特丹
推荐人：薇拉・托尔曼
Artist: Jeanne van Heeswijk（Netherlands）
Location: Rotterdam, Netherlands
Researcher: Vera Tollmann

作品描述

"免费屋"是一个 Jeanne van Heeswijk 和迷宫公司合作为鹿特丹城市
（Kosmopolis Rotterdam）创作的原创项目。这个概念来自一个中世纪
和巴洛克的称之为"外人"空间模型，即那些没有进入主流政治和社会
生活的场所。在与之相关的"明天的市场"（Markt van Morgen）项目
中，主创人员重塑了 Afrikaanderwijk 市场——在鹿特丹南部的一所破旧
的市场。在那里居住许多移民并有很高的失业率。荷兰人以商人闻名，在
荷兰很多令人愉快的城市广场是由市场组成的。然而今天，严格的规定，
越来越少的供应和销售量的下滑已导致市场遭受冲击，Afrikaandermarkt
的也不例外。经过制造商，商人和教育机构进行广泛的研究和采访后，
Van Heeswijk 和合作开发人 Dennis Kaspori 及项目领导人 Annet van
Otterloo 通过艺术形式将该区域的发展可行性以可视化的方式呈现在公众

眼前。他们使用 1:1 的比例，将其未来生活的状态 "活生生" 的展现出来。这个团体与艺术家、设计师、当地的企业家以及其他工作组共同制定了新的市场计划，恢复其成为一个文化生产活动丰沛和充满社会活力的地方。这个市集由当地政府组织，艺术团体和市场摊贩合作。"免费屋" 积极倡导新的 "管理方案"。该方案由一系列的介入措施组成，运行过程中逐渐制定新的规章和测试新的组织形式，通过为期两年的合作，两年当地官方的审议，"免费屋" 成为了新 Afrikaanderwijk 市场的官方合作伙伴。在与设计师和艺术家合作中，Van Heeswijk 为当地的文化生产设置了三个方式。在 "适合自己" 这个主题中，服装及衣着配件由当地女性组成的缝纫和手工艺团体生产出来，与此同时 MWFH 是一个时尚产品的收集中心，它与 "免费屋" 缝纫工作室进行合作。在 "幸运米财富烹饪"（Lucky Mi Fortune Cooking）这个主题中，移动厨房和点心实验室这两个概念被创作了出来。根据新设计的跨文化的食谱，食客可以品尝到鹿特丹南部口味。这些举措已经壮大为 "邻居工作室"（Wijkatelier）和南部邻居的厨房（Wijkkeuken van Zuid）两个工作室。通过这些方式 Afrikaandermarkt 市场内部产生交易。摊贩可以购买布料和食品等等并在自己的摊位售卖如服装，图案，配饰，餐具和食谱等。他们也可以在鹿特丹的其他地方做这些买卖。现在 "明天的市场" 已经开始在鹿特丹闻名。

今天，这个市场有一个由市场摊贩组成的督导小组；一个市场摊贩代表联盟；"免费屋" 和当地政府一起计划对市场的管理。Van Heeswijk 负责的《免费屋》艺术家和组织部分。该项目目前由专家和当地人组成的团队合作运行。而此团队是为目前的状况特别设立的，必要时可以改变其配置。

"免费屋" 也启发了新的举措，如烘培店苏茜的季节蛋糕，或快速环保的运输服务以及环保运输服务。

解读

"免费屋" 和 "明天的市场" 是一个专注于促进公众交流的艺术计划，在某种程度上是对未来生活形式的积极探索。该计划构建了一个适合大众日常交往的公共空间，引导人们在户外空间扮演更积极的角色，同时促使当地经济、文化的良性发展，这正是当代城市公共艺术的重要职责与创作导向之一。"免费屋" 和 "明天的市场" 为公众建立了一个开放性的创造机制，由政府组织、艺术团体和市场摊贩合作，提出 "适合自己" 的计划主题，人们可以在这里生产、交易、游玩，体验一种更具活力的文化生产生活。这不仅积极推动了公众间的情感交流，人们在参与的过程中学习和分享精巧的技艺，并且有效地匹配了他们的特长，在合作与交流中创造出引领市场的新产品。（马熙逵）

Artwork Description

Freehouse is an ongoing project by Jeanne van Heeswijk and The Maze Corporation in cooperation with Kosmopolis Rotterdam. The concept derives from a medieval and Baroque model of a space dedicated to "outsiders" without access to mainstream political and social life.

In the related Market of Tomorrow (Markt van Morgen) project, the group focused on reshaping the Afrikaanderwijk market, a rundown market located in a neighborhood south of Rotterdam with a high unemployment rate and many immigrants. The Dutch are traders, and many of the most pleasant city squares in the Netherlands are market squares. Today tighter regulations, decreased supplies and declining sales have caused markets to suffer, and the Afrikaandermarkt is no exception. After conducting extensive research and interviews with manufacturers, businesspeople and educational institutions, Van Heeswijk—together with co-developer Dennis Kaspori and project leader Annet van Otterloo—visualized opportunities for the area. The plan was presented in 1:1 scale, as a living model of its own future. The group worked with artists, designers, local entrepreneurs, and others to develop a new program for the market, restoring it as a place of cultural production and community vitality.

The market is centrally organized by the local government. In cooperation with market stallholders, Freehouse actively advocated a management program. This program was set up through a two-year series of interventions, development of regulations and tests of new organizational forms. After two years of bureaucratic deliberations, Freehouse became an official partner in the new Afrikaanderwijk market.

In collaboration with designers and artists, Van Heeswijk set up three pathways for local cultural production. In "Suit It Yourself," clothes and accessories are being developed with women from local sewing and craft groups, while MWFH is a fashion collection produced in collaboration with the Freehouse sewing studios. And at "Lucky Mi Fortune Cooking," a mobile kitchen and snack lab that devises new intercultural recipes, diners can sample the flavors of south Rotterdam. These initiatives have grown into the Wijkatelier (Neighborhood Studio) and the Wijkkeuken van Zuid (Neighborhood Kitchen from South). These initiatives conduct a trade with the Afrikaandermarkt, buying fabrics and foodstuffs there and selling items such as clothing, patterns, accessories, dishes and recipes on their own stalls, as well as at other locations in Rotterdam and elsewhere. The Market of Tomorrow has become well known in Rotterdam.

Today there is a market steering group made out of market stallholders, market union representatives, Freehouse, and local government who together plan the management of the market. Van Heeswijk is responsible for the artistic and organizational part of Freehouse. The project is now run in cooperation with a project team of experts and locals. The team has been specifically put together for this occasion and its setup can vary when necessary.

Freehouse has also inspired new initiatives, like the baked-goods shop Suzy's Season Cake, or Fast Flex Feijenoord, an environmentally-friendly transport service.

Artwork Excellence

Van Heeswijk offers an interesting alternative for change in southern Rotterdam's Afrikaanderwijk neighbourhood in the form of the Freehouse project, which has helped the district to grow stronger culturally and economically.

Van Heeswijk's starting point for her urban projects is her observation that city dwellers feel less and less involved in the processes and design of public space. Therefore she employs dialogue and confrontation to provoke a new engagement. She initiates and organizes processes in public space that result every time in products that suit the place in question and the people who live, work and play there. Freehouse is a site specific project that encourages a new cultural entrepreneurship. There, Van Heeswijk has succeeded in bringing together local businesspeople, youth, artists and designers to exchange knowledge, experience and ideas. The project has set up several working co-ops that produce new products and services. They work towards self-sustainability and they are managed and organized by the locals themselves.

黄金
Gold

艺术家：赫苏斯·帕洛米诺（西班牙）
地点：西班牙韦尔瓦塞拉艺术中心
推荐人：吉乌希·切克拉
Artist: Jesús Palomino（Spain）
Location: Location：SIERRA Art Center, Huelva, Spain
Researcher: Giusy Checola

作品描述

2012 年 7 月，艺术家赫苏斯·帕洛米诺在西班牙塞拉艺术中心安装了一个直径 8 厘米的金环，金环安装在位于艺术中心山坡坡面上一个横置人造水坑的内部空间中。金环位于"caña"（西班牙语，"甘蔗"之意）的主要导管之内，这个传统方法在阿拉塞纳的塞拉是用于从山坡上挖出平行隧道以收集干净的水源。这个金环就埋藏在这其中，而观众是看不到的。

从形式上看，"黄金"这个作品是一个特定地点的装置，这个装置包括了制造和安置一个金环到一个洞穴的井里。所有从山上流下来的水通过这个贵金属分流到池塘中，作为塞拉艺术中心及其地产的花园灌溉和水消耗之用。

这个行为除了充满诗意之外，从概念上一方面重新定义了一个地方，这标志着被称为村民的群体存在着，但其确切存在已经被忽略了的一个空间；另一方面，这个项目反映了对大自然元素的一种变态反应，而这现在已经成为了一种标志。这个地方频繁地遭受旱灾，真正的财富是水，而不是黄金。这是一个简单而且非常重要的元素。

这个项目是为塞拉艺术中心的开幕而接受其馆长和总监、艺术家鲁本·巴罗索的个人委托而进行的。由于塞拉艺术中心已经着眼于环境保护项目，因而最终概念是很容易定义的，接下来就看大自然和想象力了。这个项目只花费了非常少的金钱来完成，而这部分钱是艺术中心通过人们捐赠和赞助获得的。这些生活在桑塔·安娜村子里善良和慷慨的人们对这个项目起到了最根本的作用。

因为这个塞拉艺术中心所在地是属于市政厅拥有的公共财产，所以"黄金"这个作品也已被视为是公众的财产。公众可以在未来几年中免费使用这个地方和这些资源。

这个艺术行为是一种永久性的介入。正如赫苏斯·帕洛米诺的每一个装置艺术作品，也正如他一向说的"设计提供了一种对诸如人权、生态、文化对话和民主批评上的审美和道德上的意见，正如当面对各种现实的情况时需要紧急反应、解决办法、最好是社会变革时所需要的创造性行动"。

解读

正如赫苏斯·帕洛米诺的每一个装置艺术作品，也正如他一向说的"设计提供了一种对诸如人权、生态、文化对话和民主批评上的审美和道德上的意见，正如当面对各种现实的情况时需要紧急反应、解决办法、最好是社会变革时所需要的创造性行动"。

"黄金"这件公共艺术装置作品，以艺术永久性地介入生活与环境，体现环境保护的理念，不仅创意新颖，而且重视公众的积极参与。从项目本身来说，观众会疑惑不解，如果对创作背景有所了解，这件装置作品的教育意义就已显现。通过视乎变态的艺术手法，用"黄金"珍贵难得作比，彰显对大自然的关怀，呼吁人们珍惜那"取之不尽，用之不竭"的水资源。

由于生活在桑塔·安娜村子里善良慷慨的人们对这个项目起到了最根本的作用，因此"黄金"这件作品也已被视为是公众的财产。公众可以在未来几年中免费使用这个地方和这些资源。（冯正龙）

Artwork Description

In July 2010, artist Jesús Palomino installed a golden ring 8 cm in diameter in the inner space of a horizontal water pit excavated on a hillside at SIERRA Art Center. The golden ring is located at the main pipe of the "caña," a traditional method in the Sierra of Aracena of collecting clean water by excavating a horizontal tunnel on the hillside. The golden ring is buried and not visible for the audience.

Formally, GOLD is a site-specific installation that consists of the manufacturing and placement of a gold ring into the well of a cave. All the water that flows from the mountain, and then is distributed to the ponds for garden irrigation and water consumption at the SIERRA Art Center and premises, goes through this precious metal.

Apart from being a poetic action, conceptually this work is on the one hand a resignification of the place, marking a space whose existence was known by the villagers but whose exact location had been ignored. On the other hand, the project reflects a kind of metamorphosis of the elements, which have now become signs. In this area that suffers frequent draughts, the real treasure is water instead of gold. It is a simple but extremely important element.

The project was commissioned personally to the artist by Rubén Barroso, curator and director of SIERRA Art Center, for its opening. Because SIERRA Art Center is already oriented to environmental projects, it was easy to define the final concept. The natural site and the imagination did the rest. The work was produced with a very little money provided by the center after an attempt to obtain more funds from other grants and sponsors. The good will and generosity of the people who live in the village of Santa Ana la Real played a fundamental role.

GOLD has been conceived as work of communal property, since the whole area where SIERRA Art Center is located is a public property owned by the City Hall. The public can use the place and its resources for free during the coming years.

This action is a permanent intervention. Like each installation by Jesús Palomino, it was, according to Palomino, "designed to provide an aesthetic and ethical comment on issues such as human rights, ecology, cultural dialogue and democratic criticism... as creative actions (finally, no more than imaginative proposals) in the face of real current situations that demand urgent reflection, resolution, and hopefully, transformation."

Artwork Excellence

The excellence of the project could be related to the "wave of transformation" unexpectedly provoked by the fact of the burial of a golden ring at a natural site in relation to water.

When I first visited the place, the whole area was abandoned and presented a poor environmental situation after several years of derelict and poor maintenance. Once I pointed the abandoned place of the horiozontal water pit, a real "wave of transformation" started. The City Hall sent a crew of 10 workers to clean up the rural paths aorund the area. The entrance of the water pit was as well cleaned and arranged to provide a better access to the public. Indeed, the location was recovered as a public land mark for the village of Santa Ana la Real. The place was transformed into a new place of meaning and encounter by the action of the buried golden ring. The very function of the pit was again underlined and resignified. The stories of abandoned treasures buried by the Arabs at the ancient times were resounding and echoing the traditional oral stories told at the Sierra de Aracena. The gold gave back life to the natural site by the hand of an artistic action of resignification and transformation.

(Jesús Palomino, March 2012, interview with the artist)

历史和故事实验室
History and Stories Lab

艺术家：马西莫 • 巴托里尼（意大利）
地点：意大利都灵
推荐人：吉乌希 • 切克拉
Artist: Massimo Bartolini（Italy）
Location: Turin , Italy
Researcher: Giusy Checola

作品描述

该地点历史和故事实验室的基本原则是叙事，无论是单独的个人还是在小团队里，每个学生、教师和参观者都能够找到可供叙述自己故事的一片空间。这就是都灵重建工程发起的位于米拉非欧力北部的项目之一《都市 2》，那是一片 20 世纪 30 年代围绕规模巨大的菲亚特汽车厂建造起来的、最早被都市化的郊区。

实验室发展成为新赞助计划的一部分，授权普通的欧洲人通过一位指定的策展人 / 中间人（现在可以通过这个地址 a.titolo.）都可以委派作品。项目的出资人是一群米拉非欧力北部的老师和学生。昂塞尔美迪教堂，原先用作礼拜堂如今却废弃的建筑，其历史可以追溯到该城市最早的移民，这座教堂的位置差不多就在学校的操场当中。虽然那些学校儿童不得进入，但他们经常在废弃的教室那里作猜测。

学校的老师一直在和学生一起工作纪录和保存与这一地区有关的故事和历

史，他们把这座废弃的教堂确定为继续进行此类作品并与更广泛的公众面进行接触的潜在的"实验室"。教师们通过 a.titolo. 委派马西莫·巴托里尼参与项目合作。他们继续在项目的成型中起着积极的作用，巴托里尼的作品和艺术哲学来实现自我教育。

完成后的实验室由两个主要空间组成。以前教堂的元素，包括艺术品和祭坛的隔栏被保存了，而在中殿，巴托里尼通过增加长凳和空书架，目的是它成为供人沉思默想的空间。"我想让这块地方空出来，提供给孩子和所有人体验空荡荡的感觉，尤其是在城市里。"第二个空间包括功能化的书架，投影空间，实验室桌子和一个独特的、用来放备案柜的平面地板。

该概念、计划和作品的创作开始于 2002—2007 年，项目通过艺术家、一群教师、a.titolo. 中间人、"都市 2"项目的技术员和建筑师、建造这个项目的工匠们之间的对话来展开。2004 年，前教堂私人业主把这个建筑提供给都灵市和公众免费使用 25 年。

"历史和故事实验室"在目前的春季实验期间由学校使用，在特殊场合如邻里节日和艺术类活动时，包括举办其他艺术家的特定地点的展览时对公众开放。

解读

在形式与功能的背后，是一份万物平等的"关怀"——构建事物的一切原材料，一切工艺，无所谓贵贱和高低，乃至于没有废物和珍宝。作品"历史和故事实验室"将废弃已久的老教堂，改造成供教师为孩子们传授知识的神圣殿堂，扩展了教堂原有的对公众的传播和吸引功能，巧妙的构建了一个极具实用价值的公共艺术空间。在教堂内部的空间设计中，艺术家马西莫·巴托里尼又赋予了它新的视觉语言，简洁、几何式的设计风格与极具风土气息的巴洛克风格相融合，使走进它的人们不仅能感受到的历史沉积的沧桑，同时为每个学生、教师和参观者提供了完美的沉思空间。"历史和故事实验室"不仅是一场简单的建筑改造工程，更是对于教堂命运的升华，它不仅成为历史的见证，并且将长久的与生活凝合。（马熙逵）

Artwork Description

The principle function of the History and Stories Lab of the site is narration; whether alone or in small groups, each student, teacher, and visitor finds a space for his or her story. It is one of several projects launched by Turin's redevelopment program, Urban 2, in Mirafiori Nord, a suburb that first urbanized around the massive Fiat factory built at its heart in the 1930s.

The lab was developed as part of the New Patrons program, which empowers everyday Europeans to commission artworks through a designated curator/mediator—in this case, the Turin-based collaborative, a.titolo.

The patrons for the project were a group of Mirafiori Nord teachers and their students. Anselmetti Chapel, an abandoned house of worship that dated to the original settlement of the city, sat literally within the school grounds. Although off-limits to the school children, they often speculated on the deserted chapel.

Teachers at the school had been working with students on recording and preserving neighborhood stories and history and they identified the disused chapel as a potential "laboratory" for such work to continue and to intersect with the broader public. The teachers, through a.titolo, commissioned Massimo Bartolini to collaborate on the project. They continued to take an active role in shaping the project and educated themselves on Bartolini's work and artistic philosophy.

The finished lab consists of two main spaces. Elements of the former chapel were preserved, including artworks and the sanctuary rail, while in the nave, Bartolini added benches and empty bookshelves, intending a space for contemplation. "I wanted to leave this place empty, to offer the children and all people the possibility to experience emptiness, especially in a city." A second space includes functional bookshelves, projection spaces, lab tables, and a unique, dimensional floor made from archival filing cabinets.

The conception, planning and creation of the work took place between 2002 and 2007, through meetings between the artist, the group of teachers, the a.titolo mediators, Urban 2 technicians and architects, and artisans building the project. In 2004, the private owners of the former chapel offered use of the building to the City of Turin and the public for free for twenty-five years.

The Laboratorio di Storia e Storie (History and Stories Lab) is currently used by the schools during the spring laboratories, and it's open to audiences on special occasions, such as the neighborhood festivals and arts-related activities, including site-specific exhibitions of other artists.

Artwork Excellence

Laboratorio di Storia e Storie (History and Stories Lab) is primarily a long term dialectical space; its functions are open and they do not end with the educational project from which it was generated. To the requests of the New Patrons commissioners, Massimo Bartolini has replied with a flexible and multifunctional framework. The work is a place that can assume different shapes, depending on who will use it and the nature of activities that will be carried out.

Through this project it was possible to serve the needs of a school through a work of art; create a unique educational space for learning and research; conceive a space for one school available to all schools of the city; save a building of great symbolic and sentimental value for the residents of the neighborhood, as well as of great importance for the history of the area; carry out restoration and consolidation of the building, otherwise very difficult to achieve; open it to public access and uses; empower citizens in the design and redesign of common space; give new contemporary forms and functions to historical architecture while respecting its original function as a religious building.

祭坛
Ingeborg Bachmann Altar

艺术家：托马斯·赫希霍恩（瑞士）
地点：德国柏林
推荐人：薇拉·托尔曼
Artist: Thomas Hirschhorn（Swiss）
Location: Berlin, Germany
Researcher: Vera Tollmann

作品描述

托马斯·赫希霍恩的作品"祭坛"是由各种五颜六色的动物，装着新鲜或枯萎的鲜花的花瓶，装着图书的盒子，电视，图片和手写诗歌和情书等共同组成的装置。这个装置被摆放在柏林亚历山大广场 U-Bahn 站的两条线路中间——这个地方通常由小混混或无家可归的人所占据。在这里托马斯·赫希霍恩创建了一个新的空间，一个"高文化"的空间，其代表人物奥地利作家 Ingeborg Bachmann 在此受到人们的纪念和尊敬。

"祭坛"看起来像流行偶像戴安娜王妃的悲惨死亡死后的场景——图片、卡片和鲜花堆积如山。其结果是一种感人的世俗宗教祭奠，用平庸的大众消费产品传达对死者的追忆。但"祭坛"却肯定是颠覆性的：因为像 Ingeborg Bachmann 这样的人永远无法达到的流行明星的地位，或被大众如此崇拜。

托马斯·赫希霍恩用这个临时性的装置作品，仿佛创作了一个街头纪念碑——自发的集中悼念点。而死者可能就是在这个地方故去。托马斯·赫希霍恩说：我很感兴趣，个人与个人的连接，各种事物的发散性混合，发自内心的，爱的消息给死者的礼物背后没有审美的意图。但却是强有力的，纯粹的力量。奥地利作家 Ingeborg Bachmann 是我喜爱的作家。我喜欢她的文字，我喜欢她的生活，所以我说，'我爱'。"他补充说："她也曾在这个世界上存在生活工作。"

因为这个装置作品，人们开始阅读 Bachmann's 微妙的，哲学的诗。开始引用她诗歌及书里的文字。

Altars 很喜欢这样的作品及作品引发的思索可以在任何地方发生，可以是不同的城市，路边，广场或者是任何房子的墙上。正由于作品的脆弱性，这样的作品才更有力。不论是当地的居民还是警察都不不敢去搬动它。

托马斯·赫希霍恩已经为四位已故艺术家作家在分别四个城市创建了临时街头祭坛，或者称之为"反英雄"、"反纪念碑"坛。他们分别是：Piet Mondrian, Otto Freundlich, Raymond Carver and Ingeborg Bachmann。对于每一个"祭坛"，他都用低档的材料，并引用了相同的美学概念。性格各异的人都可以被广泛的划分为"高文化"人群，可以被持续当做街道的突出物。就这一点来说，托马斯·赫希霍恩认为，祭坛是个人化的艺术化得宣言。我想要告诉你我的英雄是谁。这些祭坛是告诉大家我所爱的某人死了，我要纪念追忆他。能证明一个人喜爱另一个人并与之相联系是重要的。英雄不能改变，但祭坛的位置可以改变。神坛可能已在其他城市，国家。它们可能是在不同地点的街道上，路边或街角。凭借改变自己在人们"道德线"的位置，这些纪念的场所可以成为国际化的场所。这是我感兴趣的。

托马斯·赫希霍恩的作品曾属于系列的展览项目："U2 的亚历山大广场"。从 1990 年到 2007 年，由柏林的新美术学会 (NGBK) 举办。在东德时期，几次主题为"世界上的一部分"海报比赛，在亚历山大广场地铁站举办。

解读

艺术家托马斯·赫希霍恩试图通过生活中最简单、平常的材料，构建极具表现力的公共艺术作品。作品摒弃与生活疏远的方法，依据多种介质的自然真实来反映艺术的真实，并且通常都安排在一些特殊的场地以产生不一样的互动效果。"祭坛"这一作品不寻求将观众作为作品的一部分，而是让观众作为纯粹的观看者，通过对作品的欣赏而产生思维的启迪。作品的装置手法趋向于在"混乱"中探求美的存在，不理会秩序的审美，通过一系列的"混乱"组合来表达、反映介质，在某一个角度或者某一个方面向观众传达作者对社会矛盾的洞察和观念。作品"祭坛"带给观众的艺术传达是多种艺术介质的叠加，艺术家只是悟道者的领路人，不同的观众对作品的感悟亦会存在变量。（马熙逵）

Artwork Description

Thomas Hirschhorn' s Ingeborg Bachmann Altar was a colorful installation of stuffed animals, vases with fresh or wilted flowers, boxes of books, a television, pictures and notes with handwritten poems and love letters. By locating the work between two lines at U-Bahn station Berlin Alexanderplatz, a place usually occupied by punks or homeless people, Hirschhorn created a new space where "high culture" and its representative, Austrian writer Ingeborg Bachmann, was celebrated and revered.

Ingeborg Bachmann Altar looked like the accumulations of pictures, messages and flowers people assemble when pop idols die, such as was the case with the tragic death of Princess Diana. The result was a touching quasi-religious kitsch, a posthumous worship with banal products of mass consumption. But the Ingeborg Bachmann Altar was certainly subversive: people like Bachmann could never reach the status of pop stars or become an object to reach the mass worship.

With the temporary installation, Hirschhorn gave the impression of a street monument, a spontaneous collection of condolences for a person who might have died on this spot. "This volatile mix of things, love messages and gifts to the deceased followed no aesthetic intent. I'm interested but the personal connection, which is expressed in it. It comes from the heart. She is pure energy," says Thomas Hirschhorn.
"The Austrian writer Ingeborg Bachmann is somebody I love. I like her texts and I like her life—so I say love. She was a person who lived and worked in resistance to the world," he added.

Because of the work, people started reading Bachmann's subtle, philosophical poems, quotations that were included in the work, along with her books themselves.

Altars like this can arise anywhere: in different cities, on the roadside, in squares, on house walls. Because of their fragility, they have power. Neither pedestrians nor police officers would have dared to remove it.

Thomas Hirschhorn has dedicated temporary street altars, "anti-heroic" "anti-monuments," to four deceased artists and writers, set up in different cities: Piet Mondrian, Otto Freundlich, Raymond Carver and Ingeborg Bachmann. For each "altar" he used low-grade materials and referenced the same aesthetic. All the personalities can be broadly classified as "high culture" and consequently might be considered out of place on the street. On this point Hirschhorn comments, "An altar is a personal, artistic statement. I want to fix my heroes. The altars attempt to memorialize a person who is dead and who was loved by someone else. It is important to testify to one's love, one's attachment. Heroes can't change, but the altar's location can change. The altars could have been made in other cities, countries. They could be in different locations—on a street, a side passage, in a corner. These local sites of memory become universal sites, by virtue of their location. That is what interests me."

Thomas Hirschhorn's work was part of the exhibition project series "U2 Alexanderplatz," organised by Neue Gesellschaft für Bildende Kunst (NGBK), Berlin from 1990 to 2007. During GDR times, several poster competitions on the subject "piece of the world" took place at the underground station Alexanderplatz.

Artwork Excellence

Ingeborg Bachmann Altar was installed in an unusual urban situation—in a busy underground station in Berlin. It had an unexpected style—the artwork did not look like an art piece in the first place because of its similarity to spontaneous altars. The piece looked like a temporary monument assembled by many people, like a community piece. It became an open artwork because people added objects or took some away. It constantly evolved because of the public's involvement (people knew or not know it was an artwork). Hirschhorn used a quotidian form, thus he created a new kind of memorial that was transformative, informative and accessible. In a place where people usually rush through, he very convincingly made people stop.

伊索拉艺术中心
Isola Art Center

艺术家：伯特·泰斯（卢森堡）
地点：意大利米兰
推荐人：吉乌希·切克拉
Artist: Bert Theis (Luxembourg)
Location: Milan - Isola quarter, Italy
Researcher: Giusy Checola

作品描述

自从 2001 年起，米兰的伊索拉当地居民参与捍卫他们唯一的公共空间——废弃的工厂"工匠棒"和附近的公园，目前正预备把该地区改建成一个几百万美元的豪华住宅用地和高档商品零售场所。伊索拉艺术中心同年开工了。

从开创之日起，伊索拉艺术中心成为了一个开放式，极具活力的实验性平台，融汇了国际当代艺术，新兴艺术和理论研究，能够满足居民成分主要为混合的工人阶级居住区的需求与渴望，把居住区转型则是首要任务。项目把当地的一场围绕都市发展的斗争选为作品地点，对居民的活动及其选择采取了明确的支持立场。它需要把地点特异性的概念延展开，发展成斗争特异性的新概念。这个地点应该由居住和工作在此地的居民来决定。如果这些人，比如居住在伊索拉地区的居民，被组织和动员起来，斗争特异性艺术的状况就产生了。

在这些人中间,选择参与和支持伊索拉艺术中心项目的有评论家和策展人,比如侯瀚如,马克·斯科特尼,罗贝托·平托和瓦西夫·科尔屯;还有音乐家和艺术家,如斯蒂夫·皮科洛,阿德里安·帕西,阿尔伯托·嘎如蒂,恩佐·乌巴卡,爱娃·马里萨尔蒂,卢卡·维托尼,洛里斯·切契尼,卢卡·潘克拉奇,马西莫·巴托里尼,马力欧·埃洛,塔尼亚·布鲁格拉,斯特芬诺·阿里恩提,以及其他许多人。

"伊索拉艺术中心"是一个分散的组织,在不同地点开展工作,包括一个在"工匠棒"的驻场空间。这是一个没有预算的项目,运作完全靠精力、热情和团结。其组织结构不是自上而下的:没有主管、没有策展人决定项目计划。也不是一个艺术家运营的组织。相反的,这个中心具有机动灵活性、敞开式、根茎结构,没有事先确定的高下等级。参加一些项目的嘉宾策展人和艺术家(车臣紧急双年展,由伊芙琳·朱阿诺策展;女性使改变(音译),艺术家来自于中国广东,由玛蒂娜·科佩尔-杨展;艺术社区——500多位艺术家肖像,伍尔夫冈·特拉格尔创作。)还有其他一些项目,比如《野餐》,伊索拉艺术中心邀请了都灵、佩斯卡拉,皮亚琴察的艺术空间参与设计,然后在米兰进行展览,开放场地,提供服务,比如通信和设置援助。这些来自外面的支持对参与项目的艺术家和策展人与中心提供了额外的帮助(这些人中,有马克·斯科特尼,罗贝托·平托,阿德里安·帕西以及多人)。其他项目由一组艺术家参与,在中心的保护下自主运作。

这个中心工作的一个关键方面是周边居民的参与。居民通过多种方式参与活动:帮助完成了一些作品(比如托马斯·萨拉切诺的《航空太阳能博物馆》);当地家庭接待国际艺术家住在自己家里;参与活动、表演和实验室活动;建议在邻近场所做新的项目。这类参与行动从一开始就是自发的。这类活动的重心自始至终一直是保护公共空间,不让它们被私有化。他们一已贯之的努力现在有了成果:市议会新的管理部门正在委托伊索拉艺术中心管理小部分地区,它带有居住区中的一片绿色区域和一座艺术展馆,被称为伊索拉青椒。预定在 2012 年春季开放。

解读

"伊索拉艺术中心"是一个分散的组织,工作地点不定,项目没有固定的资金支持,运作完全靠成员们的精力、热情和团结。这个中心具有机动灵活性、敞开式的结构。自中心开放后,便具备公共实验性功能,融合国际当代艺术与新兴艺术理论研究,能够满足居住于当地社区的多数工人阶级居民的需求与渴望。"伊索拉艺术中心"旨在对于都市发展而引发的生存焦虑提出质疑,引导居民积极参与各项艺术计划,使参与到活动的公众明确保护公共空间的重要性,防止公共空间被恶意私有化,力求每位居民都能享有艺术计划带来的文化与思想革新。时间无涯,生命有期,也正为了生活的美好,才会引发大众对于公共空间权利的维护,而这众多的烦恼都会随着人性的克制而得到永远的解脱。(马熙逵)

Artwork Description

Since 2001, the inhabitants of Isola neighborhood in Milan have engaged in the defence of their only public space, the abandoned factory "Stecca degli artigiani" (Artisans' Stick) and the nearby park, which are currently slated to be turned into a multi-million-dollar complex of luxury housing and upscale retail establishments. Isola Art Center began in the same year.

Since its outset, Isola Art Center's goal has been the creation of an open and dynamic experimentation platform, combining international-level contemporary art, emerging art, and theoretical research, with the needs and desires of a residents in a mixed, working-class neighborhood— with the overarching mission of neighborhood's transformation. The choice of working within a local urban struggle, taking a clear stance in favour of the citizens' movement and the alternatives it fights for, requires the extension of the concept of "site-specificity" to the new notion of "fight-specificity." A site is determined by the people living and working in it. If those people—as in the Isola neighborhood—are organized and mobilizing, the conditions arise for a "fight-specific" art.

Among the people who choose to participate and support the aims of Isola Art Center project are critics and curators like Hou Hanru, Marco Scotini, Roberto Pinto, and Vasif Kortun; and musicians and artists like Steve Piccolo, Adrian Paci, Alberto Garutti, Enzo Umbaca, Eva Marisaldi, Luca Vitone, Loris Cecchini, Luca Pancrazzi, Massimo Bartolini, Mario Airò, Tania Bruguera, Stefano Arienti, and many others.

Isola Art Center has functioned as a decentralized organization working out of various locations, including a squatted space in Stecca degli artigiani. It is a "no-budget" project, functioning only through energy, enthusiasm, and solidarity. Its organizational structure is anything but top-down: There is no director or curator deciding the program and projects. Neither is it an artist-run organization.

Instead, the center is characterized by its flexible, open, rhizomatic structure, lacking a pre-established hierarchy. Some of its projects include guest curators and artists (for example, Chechnya Emergency Biennial, curated by Evelyne Jouanno; Women Shi Gaibian, with artists from Canton, China, curated by Martina Koeppel-Yang; The Art Community, with over 500 artist portraits by Wolfgang Träger.) In other projects, like Picnics, Isola Art Center has invited art spaces from Turin, Pescara, and Piacenza to design and wholly realize shows in Milan, opening its venues, offering services such as communication and set-up assistance. Such "external" contributions have proved complementary to the projects of artists and curators more consistently involved with the center (among them, Marco Scotini, Roberto Pinto, Adrian Paci, and many others). Other projects consist of contributions of groups of artists autonomously operating under the center's umbrella.

A key aspect of the center's work is participation from neighborhood residents. Inhabitants participate in many ways: helping in the realization of some works (such as Tomas Saraceno's museo aero solar); hosting international artists with local families; participating in actions, performances, and labs; proposing new projects in neighborhood venues. Such participation has been, from its outset, spontaneous.

At the center of these activities there was and there is, always, the protection of the public space of the quarter from privatization. The continuity of its efforts is now giving results: the new administration of the City Council is going to entrust Isola Art Center to manage a small lot, with a green area for the neighborhood and an art pavilion, called Isola Pepe Verde. It is slated to open in spring 2012.

Artwork Excellence

IIsola Art Center has sustained a 10-plus year commitment to the Isola neighborhood through its flexible approach to programming and organization, as well as an unflagging alignment with the desires and struggles of the neighborhood. As such it is a model for such practices and serves as an alternative to top-down public art approaches. In terms of placemaking, the center—through numerous locations and an ever-shifting cast of participants—has brought artistic programming into its unique territory while engaging and supporting residents in their ongoing process of claiming public space. A culmination, or next chapter, of this effort will be inagurated with Isola Pepe Verde, a green- and art-space, which the City Council has appointed Isola Art Center to manage.

梅·韦斯特
Mae West

艺术家：芮塔·麦布莱德（美国）
地点：德国慕尼黑
推荐人：薇拉·托尔曼
Artist: Rita McBride（USA）
Location: Munich, Germany
Researcher: Vera Tollmann

作品描述

"梅·韦斯特"是 21 世纪更新版的埃菲尔铁塔。作品位于圆形的艾弗纳广场和环绕慕尼黑市内地区的米特拉莱环路，艺术家从这些圈和旋转运动处得到启发，形成作品开始的主要几何形状。雕塑虽然高，但恰如其分的网格结构使它显得不那么突兀，而具有最大限度的渗透性、通透性及开阔性。汽车从广场下方地道直接穿过，而电车穿越雕塑。作品所具备的潜质能使迄今为止受交通和多样环境主导的广场转变成为象征城市生活活力的地方。

直到高达 15 米的第一根环形大梁处，所有的钢管都裹着碳纤维增强塑料，这种材料主要用于造船。直径从 275 毫米锥形向上到顶端的 225 毫米高于大梁的管子则全是碳纤维增强塑料制成的。每根在上方的管子为 40 米长、重约 550 千克（为相似钢管的六分之一重）。整个高处的碳结构就地组接，并连接到预置的钢结构基础上。

慕尼黑的艾弗纳广场是"米特拉莱东环路"隧道项目的一部分。一个城市内的市政建设项目中支持艺术的资金比率为2%，此城市隧道建设项目隶属于"慕尼黑艺术与建设项目(QUIVID)"，在艺术项目的预算范围内，但由于该项目规模浩大，艺术方面的预算被限制到0.4%。

"建设和公共区域中的艺术"机构负责项目的咨询工作，选择艾弗纳广场为作品放置点的原因是因为它处于伊萨尔河河边，因而成了东部高楼大厦、西南和北部公寓楼房和英国公园之间重要空隙的象征。

艺术委任机构和当地城市地区理事会用了5年时间才最终决定在此地放置艺术作品。最终选定芮塔·麦布莱德的"梅·韦斯特"方案，意在为这个交通中心带来令人意外的新地标，并使混杂的周遭两层阳台房屋和高楼变得更为协调（"梅·韦斯特"高52米，而Hypo大楼为114米，威斯汀大酒店为75米）。芮塔·麦布莱德受委任后，由她和CGB Carbon GroßbauteileGmbH组成的"梅·韦斯特"工作团队成立了

据麦布莱德说，雕塑似人体腰部的曲线让她联想到梅·韦斯特，她是好莱坞女演员、舞者、编剧及20世纪30年代的性感偶像，因此采用了"梅·韦斯特"作为作品的名称

解读

"梅·韦斯特"位于德国慕尼黑市区内，现已成为当地具有地标性的雕塑作品。雕塑外形酷似人体腰部的曲线，这使作者联想到20世纪30美国著名女星梅·韦斯特，因此采用了"梅·韦斯特"作为作品的名称。"梅·韦斯特"占地面积庞大，但开放性的设计丝毫没有使其所占土地闲置，汽车从广场下方地道直接穿过，而电车在地面穿行雕塑。人们可以乘坐多种交通工具，全视角的去欣赏它。"梅·韦斯特"由32根碳纤维增强塑料杆交叠构成，直线的交叉以及自然的曲度的配合，形成了多变的几何图形，这与作品整体趋于稳定的形态形成对比。"梅·韦斯特"既保持了理性的严谨和科技感的冷漠无情，又通过多样的内部结构阐述一种繁杂琐碎的结构美，具有强烈的艺术个性化，赋予了形式独立的价值和审美。（马熙逵）

Artwork Description

Mae West is a 21st-century update of the Eiffel Tower. The sculpture's principal geometric form, which the artist developed beginning with a circle and a rotating movement, refers to the round Effnerplatz where the work is sited and to the entire Mittlerer Ring road encircling Munich's inner city districts. Despite its height, the sculpture is unobtrusive thanks to the sophisticated grid-like structure of the stick construction allowing a maximum of permeability, transparency and openness. While car traffic is directed through a tunnel below the plaza, the tramway will pass through the sculpture. The sculpture has the potential to turn this plaza so far dominated by traffic and its heterogeneous surroundings into a place that symbolizes the vitality of urban life.

Up to the first circular girder at a height of 15 meters, the sticks consist of steel tubes coated with carbon fiber-reinforced plastic (CFRP), which is mainly used in boat building. Above this girder, the tubes are made entirely of CFRP, tapering towards the upper end from a diameter of 275mm to 225mm. Each tube in the upper part is 40m long and weighs roughly 550 kg (one sixth of a comparable steel tube). The whole upper carbon construction was assembled on site and then connected to the pre-installed steel foundation.

Munich's Effnerplatz is part of the tunnel project "Mittlerer Ring Ost." Within the framework of Munich's art and construction program (QUIVID) and a city regulation allocating 2% of building costs in municipal construction projects to art, this urban tunnel construction project was equipped with a budget for art projects. But due to the fact that this was a very large-scale project, the arts budget was limited to 0.4%.

The Commission on Art in Construction and in Public Places, responsible for project consulting, chose Effnerplatz as the location for the future work of art because of its topographic location at the Isar River rim, symbolizing an important hiatus between the high-rise buildings in the east, the apartment buildings in the south-west and north of Munich, and the landscape park Englischer Garten.

It took five years for the art commission and the local City District Council to decide on an artwork for the site. It ultimately selected Rita McBride proposal for Mae West, which was intended to provide this traffic hub with an unexpected landmark and to "harmonize" between the heterogeneous surrounding buildings consisting of two-storied terraced houses and high-rise buildings (Mae West is 52 meters high, while the Hypo building is 114 meters high the Westin Grand hotel is 75 meters high). After McBride was commissioned, the Mae West Working Group, comprised of the artist Rita McBride and CGB Carbon Großbauteile GmbH, was founded.

According to McBride, the slender sculpture's waist-like silhouette inspired her to choose Mae West as a title. Mae West was a Hollywood actress, dancer, script writer and the sex symbol of the thirties.

Artwork Excellence

The sculpture Mae West is based on a highly elaborate geometry. Thus, this sculpture has a long-distance effect (for those approaching Effnerplatz by car, for example), as well as an immediate on-site effect. It provides a strong visual experience that changes the perception of the surrounding urban architecture.

纳粹政权下的同性恋受害者纪念碑
Memorial to Homosexuals Persecuted
Under Nazism

艺术家：埃尔默格林（挪威）、德拉格赛特（丹麦）
地点：德国柏林提尔加藤公园
推荐人：薇拉·托尔曼
Artists：Michael Elmgreen (Norway) & Ingar
Dragset (Denmark)
Location：Tiergarten, Berlin, Germany
Researcher：Vera Tollmann

作品描述

这是德国第一座此类纪念碑，联邦议院各党派经过数十年的争论后才得以建成。但在揭幕仪式之后依然时常遭到破坏。自 1934 年开始，纳粹政权开始了对同性恋者的迫害，许多同性恋者被关进集中营，被害者人数至今无法准确统计。第三帝国崩溃后，歧视和迫害同性恋的法律还延续到 1969 年，直到 2002 年才被联合政府废除。历史学者和政界人物经常忽略了同性恋者也是纳粹大屠杀的受害者。

纪念碑的形状和颜色借鉴了彼得·埃森曼的欧洲犹太屠杀纪念碑，位于提尔加藤公园附近的一个较隐秘的空间，这里在夜晚成为同性恋者相聚的场所。参观者透过灰色立方体正面的一扇窗可以看到一部两个男人正在接吻的短片。立方体内的电影两年更换一次，由一系列不同的艺术家创作，每部影片都表现男同志或女同志恋爱的场景。

根据作者的诠释，一座纪念碑应该具有活动的有机体的特质，有着动态变化，而不是静态的最终陈述。两位艺术家根据纪念碑周围的环境密切地调整他们的美学构思，用雕塑的形式塑造了这座纪念碑，在大屠杀纪念碑对面，他们选择使用贴切的正式语言。纪念碑寻求多种方式与大家亲密交流，旨在面对公众个人层面上，而不是构成一种一般的场景。

解读

20 世纪初，世界医学界否定了同性恋性取向与道德相关的观念。认识到同性恋是人性的一种自然流露，并非内心的扭曲，应尊重他们的个性化情感的发展。在欧洲，受基督教的影响，绝大部分国家都认为同性恋违反神灵的意志，将同性恋认为是一种精神疾病。当人们开始为遭受纳粹屠杀的犹太人建立纪念碑时却遗忘了同遭迫害的同性恋者，作品旨在唤起人们对同性恋者的正确对待和认同。"纳粹政权下的同性恋受害者纪念碑"整体为灰色立方体，透过立方体正面的小窗可观摩到一系列同志恋爱的场景。艺术家们强调这座纪念碑除了静态的陈述，还应具备动态的变化，纪念碑周围的环境和群体交流同样是构建其美学架构的组成部分。所以这里除了纪念沉寂的过往以外，还为同性恋群体提供了一个亲密交流的公共空间。（马熙达）

Artwork Description

This monument is the first national monument of its kind, constructed after decades of debate and frequently vandalized since its dedication. Homosexuals were persecuted by the Nazi regime beginning in 1934. Many were arrested and brought to concentration camps, but the number of victims remains unknown to this day. After the Third Reich collapsed, the gay community continued to be persecuted by the same law until 1969. It was not until 2002 that they were pardoned by the coalition government. History scholars and politicians often ignore that homosexuals were also victims of the Nazi Holocaust.

In its shape and color the Memorial is reminiscent of Peter Eisenman's Memorial for the Murdered Jews of Europe, although it stands in a more intimate situation in Tiergarten Park, a long-established meeting place for gays in the evenings. On the front side of the grey cuboid is a window through which visitors can see a short film of two kissing men. Every two years, the film inside the memorial will be changed. Created by different artists, each film offers an interpretation of a scene of gay or lesbian love.

According to the artists, a monument should possess an active quality like an organism; it should be an expression of dynamic change rather than a statement of static finality. The two artists developed their aesthetic idea with close attention to the surrounding environment. Located near the Memorial to the Murdered Jews of Europe, the shape of the sculpture adheres to conventional forms of the monument. But rather than constituting a general conception, the monument seeks to communicate with the public on a personal level in relation to a variety of senses.

Artwork Excellence

According to Elmgreen & Dragset's interpretation, a monument should have the character of a living organism subject to dynamic change rather than a static and final statement. The two artists have closely adapted their aesthetic conceptualisation to the monument's immediate surroundings. For their sculptural rendition of the memorial, they have chosen to appropriate the formal language of the Holocaust Memorial directly opposite. In Elmgreen & Dragset's version, however, the cubic sculptural shape of Eisenman' s work acquires an additional layer. The memorial seeks in various ways to exchange the monumental with the intimate, aiming to confront the public on a personal level rather than constitute a general spectacle.

穆尔塔土里．提佩尔
Multatuli Tippel

艺术家：提昂（荷兰）
地点：荷兰哈德威克
推荐人；薇拉•托尔曼
Artist: Tiong Ang（Netherlands）
Location: Harderwijk, The Netherlands
Researcher: Vera Tollmann

作品描述

"穆尔塔土里•提佩尔"是艺术家提昂创作的一件以研究为基础、有着明确地点的装置作品，在荷兰城市哈德威克举办的"庇护所 07：都市空间中的公共艺术"展览上展出。

提昂选择了以前殖民地后院粮库里的门房作为他这件干预作品的实施地点。这幢大型的驻军大楼曾是 19 世纪时荷兰帝国主义者为荷兰西印度群岛（如今的印度尼西亚）招募志愿军的征兵处。作为"陌生人"军团，荷兰殖民军吸引了欧洲和其他地区成千上万的年轻人前来应征。为接待新兵，哈德威克市内很大一块区域变成了红灯区，一连串的酒吧和妓院开张营业。

艺术家放置了穆尔塔土里•提佩尔的大型肖像画，他是一位伟大的 19 世纪荷兰作家，是最早公开批评荷兰殖民政策的人之一。这幅肖像画就被放置在大楼最早的入口处的窗户后面的等候室内。一件拼贴影像作品摆放于另一扇一扇窗旁，内容由两部荷兰故事片的片段构成（1975 年的《娼妇凯蒂》和 1976 年的《马克思·哈弗拉尔》），将 19 世纪殖民地士兵和妓女的图像重建，和肖像画并列在一起作为对比展示。提昂的作品连接了旧时资产阶级逃避现实的两个文学世界：一个是水性杨花的著名女士《娼妇凯蒂》所代表的现实世界，由尼尔·多夫摄制；另一个是殖民地现状的轮廓图像，由《马克思·哈弗拉尔》作者穆尔塔土里勾勒。红绿色的管状电灯增强了装置的矛盾性质。

提昂的作品融合了两种概念手法：基于流程的媒体艺术和单独的图画，提出了知觉的问题、集体记忆、社会和文化的隔阂和脱位，所有的内容都嵌入矛盾和环境这个瞬息万变的领域。使用媒介将他的主题具体化——无论是绘画，录像装置，社会项目还是环境——提昂提出了关于在这个道德和文化杂交的世界里我们如何表达自我形象的概念的问题。他怀疑视觉形象和故事讲述的真实性和短暂性。

1961 年，提昂生于印度尼西亚的泗水，他在荷兰长大，现在居住在阿姆斯特丹。他的作品立足于自己的生平背景，他是生于印尼的华人后裔，后来移民去了荷兰。他的作品意在寻找一种跨文化的碰撞，他的作品在国际展览上被多次展出过，包括"2001 年威尼斯双年展"。提昂最近才去了印度尼西亚，这之前一直没有回去过，也不会讲当地的语言。提到印度尼西亚的殖民地历史艺术家有着痛苦的个人隔阂，而对荷兰的那段殖民记忆基本已经模糊。

解读

"穆尔塔土里·提佩尔"具有很强的艺术家个人风格，作品通过构建形式具有矛盾性的视觉空间，连接了旧时资产阶级逃避现实的两个文学世界。艺术家提昂的作品采用两种形式的表现手法，一种是建立在多媒体基础上的视觉艺术和纯粹的图画，提出了"知觉的问题"、"集体记忆"、"社会和文化的隔阂和脱位"，艺术作品的表现内容都融入到社会实际问题和矛盾的解决中。作品无论是绘画艺术还是录像装置，社会问题还是历史与环境的更迭，艺术家提昂提出在由文化与道德共同支配的社会中，公民应更清楚地了解自我形象的问题。由于视觉形象和故事讲述的真实性难以保证，并且不能保证其长远性，作品采取了某种形式极端的表现手法以最大限度上确保作品主题的印象深刻。（马熙逵）

Artwork Description

Multatuli Tippel was a research-based and site-specific installation work by Tiong Ang exhibited at "Shelter07: Public Art in Urban Space " in the Dutch city of Harderwijk.

As a location for his intervention, Tiong Ang chose the former lodge of the duty officer of Koloniaal Werfdepot (the colonial yard depot). This large garrison building is where military volunteers were recruited for Holland's imperialist endeavors in the former Dutch Indies (current

Indonesia) in the 19th century. As a "strangers' legion," The Dutch colonial army attracted thousands of young men from all over Europe and beyond. To accommodate the recruits, a large part of the inner city of Harderwijk transformed into a red light district, with a string of bars and brothels.

Ang placed Multatuli Tippel, a large painted portrait of Multatuli, a great 19th C. Dutch writer who was among the first to openly criticized the Dutch colonial policies, in the waiting room behind the windows of the original entrance to the building. A collage video accompanies the painting at the other window. It consists of excerpts of two Dutch feature films (Keetje Tippel, 1975 and Max Havelaar, 1976), juxtaposing reconstructed images of colonial soldiers and prostitutes in the 19th century. Tiong Ang's work connects two former literary worlds of bourgeois escapism: the reality of Keetje Tippel, a famous woman of easy virtue, recorded by Neel Doff; and the contours of colonial reality, sketched by Multatuli in Max Havelaar. Green and red tube lights enhance the ambivalent nature of the installation.

Tiong Ang's work blends the conceptual approach of process-based media art and the solitary, pictorial practice of painting to address issues of perception, collective memory, social and cultural estrangement and dislocation, all embedded in an ever-changing field of contradictions and contexts. Using media specific to his topic—whether painting, video installation, social projects or environments—Ang poses questions about how we negotiate notions of self image in an ethically and culturally hybridized world. He questions authenticity and impermanence of visual imagery and storytelling. Born in 1961 in Surabaya, Indonesia, Tiong Ang grew up in the Netherlands and now resides in Amsterdam. His works, based on his own biography as an Indonesian-born Chinese who migrated to The Netherlands, engage in the search of intercultural encounters and have been shown in many international exhibitions, including the 2001 Venice Biennale. Tiong Ang has not returned to Indonesia until quite recently and does not speak the language. This personal estrangement painfully and urgently correlates with the Dutch amnesia in general when referring to Indonesia as colonial history.

Artwork Excellence

Multatuli Tippel is a very strong example of a context-responsive, research-based intervention in public space. It is a work that explores the medium specific conditions of its space and its historical subject in an exemplary way.

我的城市
My City

艺术家：多位艺术家
地点：土耳其多个城市
推荐人：里奥·谭
Artist: Multiple artists
Location: Multiple cities, Turkey
Researcher: Leon Tan

作品描述

"我的城市"是一件大型艺术项目，由英国文化协会和合办组织共同发起。该项目的目的是促进土耳其和欧洲之间的文化理解。该项目由欧盟和欧盟土耳其民间社会对话机构：文化桥梁项目和英国文化协会共同出资赞助。该项目涉及向在土耳其的欧洲艺术家委派公共艺术任务，向驻欧洲的土耳其艺术家委派驻场项目，同时也开展一个教育项目。

以下报告着重介绍为"我的城市"而委派的，在土耳其的恰纳卡莱、伊斯坦布尔、科尼亚、马尔丁、特拉布宗开展的公共艺术项目，由艺术家马克·瓦林格，安德里亚斯·佛噶拉西，琼安娜·拉吉克弗斯卡，克雷蒙斯·冯·魏德迈和米娜·亨里克森进行创作。下面是对五个公共项目的简短介绍。

"失忆电影院"，位于恰纳卡莱，马克·瓦林格创作。

"失忆电影院"，由马克·瓦林格创作，他受到港口城市恰纳卡莱的委派，地点选择在一座清真寺旁边。使用了回收的锈迹斑斑的金属，盒子一样形状的电影院影射了那些频繁出入港口的船运集装箱。电影院放映了电影《尤利西斯》，进入内部的参观者能看到时间滞后 24 小时的录像片，放映着水和半岛的画面。它把过去与现在并列呈现，甚至带有一种本体论的陈述意味，表示由于现在和过去要永远共存下去，因而现在并不见得胜于过去。

"全景画（正确的观点）"，位于伊斯坦布尔，由安德里亚斯·佛噶拉西创作。

全景画在 18—19 世纪发展并且得以推广，它将观众置身于一个圆形（360度）的画面中，因而营造出一种如同亲临一座城市或景观中的幻觉。安德里亚斯·佛噶拉西的全景画却将观众置身于"伊斯坦布尔的文字联想上"，通过"舱门"或开窗户的全景结构，间歇地提供外部景观。这些风景经过了战略性的选择，把关注关键地点锁定在：卡德柯依，亚洲分部；贝西克塔斯，位于博斯普鲁斯海峡；马斯拉克，商业区；艾敏厄努，位于加拉塔大桥。通过文本和视觉片段的组合，佛噶拉西没有向观众展现统一完整的伊斯坦布尔，而是展现了伊斯坦布尔的许多城市、历史和联想。

"瓦尔特·本雅明在科尼亚"，位于科尼亚，琼安娜·拉吉克弗斯卡创作。

琼安娜·拉吉克弗斯卡在科尼亚的项目选择了巧合的一年：1923 年现代土耳其共和国成立了，同年瓦尔特·本雅明的《翻译者的任务》出版。拉吉克弗斯卡的作品由语言和物质之间关系的分析构成，涉及将本雅明的文章翻译成奥斯曼土耳其语。为了欣赏这个行为，人们必须意识到 1928 年穆斯塔法·凯末尔颁布法令以后，土耳其语原本受到的阿拉伯和波斯语的影响被清除了。《瓦尔特·本雅明在科尼亚》出版发行了 500 本，这个版本是由德语版本翻译成奥斯曼和现代土耳其语。它被存放在科尼亚的优素福·阿迦图书馆（这是鲁米陵寝的延伸），还拍一部名为《去图书馆漫步》的电影，叙述了本雅明去科尼亚的一次虚构的旅程。拉吉克弗斯卡同时在市政厅附近创作了一件公共雕塑，大理石上铭刻着本雅明的文章（德语、奥斯曼和现代土耳其语），雕塑经过流水的冲刷。

"太阳电影院"，位于马尔丁，克雷蒙斯·冯·魏德迈创作。

"太阳电影院"是一座位于马尔丁的露天电影院，与伊斯坦布尔建筑师一起合作而建，由当地电影院协会运营。艺术家的想法是建造一座面向大众，展示电影院正宗原理的电影院。创作的成果是一座屏幕、一座露天剧场和一个投影基地。黎明时，太阳光投射在屏幕面板上，能够表演皮影戏。黄昏时，太阳光通过后方的镜面板，光线反射到南面。不用说，这块屏幕也能够放映电影。冯·魏德迈的项目参考了阿拉伯古老的光线研究，它与绘画中的透视科学同时产生。与此同时，它与黑盒子一起削弱了移动图像的常见关联，以及录像创作中无关紧要的过程，还有当今智能手机使得后期制作变成可能，它把电影院放置在一个"位于城市与美索不达米亚平原之间"的历史性纪念景观中。

"喜欢说真话的人"，位于特拉布宗，米娜·亨里克森创作

米娜·亨里克森的项目以青年人为本，涉及工作坊、城市观光、超越公认的智慧和特拉布宗观念的协同思考，作为表达年轻人观点的一种手段。

它由三股线索构成，《喜欢说真话的人》的标题得名于一支当地乐队 Üç Hürel 的歌曲"心之歌，我不能超越"。首先，亨里克森安装了一座霓虹灯作品，位于雅儒兹·赛里木大街的一座墙上，作品名为"讲述真正的爱情""责备不公"。其次是出版了一本城市指南书，由工作坊制作，里面的文字与摄影抒发了参与者对自己和对城市的感受。第三，由亨里克森组织或者说提供一个平台，让来自项目的参与者组成了一个音乐组合在特拉布宗的主广场上进行开幕之夜的演出。除了霓虹灯装置，亨里克森的作品被认为是一种参与性的项目，她的作品挑战了对特拉布宗的普遍观念来和定义。

解读

"我的城市"实际上是一场融合多项艺术计划的大型艺术项目，由欧洲多国文化组织参与。该项目依托民间组织的交流活动，促进土耳其和欧洲之间的文化理解，同时也开展一系列的文化交流项目。在"我的城市"系列艺术项目中，表现形式具有多样性，没有严格的艺术命题，关于城市的历史、联想和疑问都可以成为展示的主题。艺术家们注重对土耳其文化的传承与保护，对于历史性的文化与政治展现具有原味的陈述性。与此同时，部分参与性项目具有明显的艺术家个人思考，抒发个人对于城市发展的感受，以及对于传统观念和定义提出挑战。该计划倡导过去与现在和谐共存，在发展中不断融入传统的优良文化与思想，没有绝对的现代化革新措施，是一场倡导性的艺术推进计划。（马熙逵）

Artwork Description

My City was a large-scale art project by the British Council and partner organizations. Its objective was to fostering cultural understanding between Turkey and Europe. Funded by the European Union within the framework of EU-Turkey Civil Society Dialogue: Cultural Bridges Programme and the British Council, the project involved public commissions for European artists in Turkey, residencies for Turkish artists in Europe, as well as an educational program. This report focuses on the public art works commissioned for My City in Çanakkale, İstanbul, Konya, Mardin and Trabzon (Turkey), by the artists Mark Wallinger, Andreas Fogarasi, Joanna Rajkowska, Clemens von Wedemeyer and Minna Henriksson. What follows are short descriptions of the five public projects.

Sinema Amnesia, Çanakkale, by Mark Wallinger

Sinema (Cinema) Amnesia by Mark Wallinger was commissioned for the harbor city of Çanakkale and sited next to a mosque. Recycled from rusty metal, the box-cinema references the containers and ships that frequent the harbor. While the cinema advertizes the film Ulysses, visitors to the interior are confronted with a 24-hour time-lagged video-view of the water and the peninsula. What is presented then is a juxtaposition of the past with the present, perhaps even an ontological statement, that the present does not succeed the past so much as it perpetually co-exists with it.

Panorama (The Right of View), Istanbul, by Andreas Fogarasi

Developed and popularized in the 18th and 19th centuries, panoramas positioned viewers in the middle of a circular (360-degree) image, producing an illusory effect of being immersed in a city or landscape. Andreas Fogarasi's panorama, however, immersed audiences within "associative texts about Istanbul," intermittently providing views onto the outside through "hatches" or windows cut out of the panorama structure. These views were strategically chosen to focus on key sites: Kadıköy, the Asian quarter; Beşiktaş, on the Bosphorus; Maslak, the business district; and Eminönü, at the Galata Bridge. Through the combination of textual and visual fragments, Fogarasi confronted audiences not with a unified Istanbul, but rather an Istanbul of many cities, histories and associations.

Walter Benjamin in Konya, Konya, by Joanna Rajkowska

Joanna Rajkowska's project in Konya picked up on a "coincidental" year: 1923 is when the modern Turkish Republic was founded and Walter Benjamin' s The Task of the Translator was published. Rajkaowska's work consisted of an analysis of the relationship between language and materiality and involved the translation of Benjamin's text into Ottoman Turkish. To appreciate this act, one must be aware that the Turkish language was purged of its Arabic and Persian influences following a decree by Mustafa Kemal Atatürk in 1928. Walter Benjamin in Konja was published in an edition of 500 copies, with the German text, and translations into Ottoman and modern Turkish. It was deposited in the Yusuf Ağa Library in Konya (an extension of Rumi' s mausoleum), and accompanied by a film entitled The Stroll to the Library, portraying a fictional journey by Benjamin to Konya. Rajkowska also created a public sculpture close to the City Hall, in which marble etched with Benjamin's texts (in German, Ottoman and modern Turkish) are washed over with water.

Sun Cinema, Mardin, by Clemens von Wedemeyer

As the title suggests, Sun Cinema by Clemens von Wedemeyer is an open-air cinema in Mardin, created in cooperation with Istanbul architects, and run by the local cinema association. The artist's idea was to construct a cinema that was both widely accessible and would expose the "formal principles of cinema." The resulting creation comprises a screen, amphitheater, and a projector base. At dawn, the sunlight captured on the screen face enables the performance of shadow plays. At dusk, sunlight is reflected to the south by mirror panels on the rear of the screen. Needless to say, the screen face is also capable of displaying films. von Wedemeyer's project refers to ancient studies of light in Arabia, taking place simultaneously with the emerging science of perspective in painting. At the same time, it erodes the common association of the moving image with the black box, as well as with the "immaterial" processes of video production and post-production made possible by smartphones of today, precisely by setting the cinema in nothing less than a historical-monumental landscape "between the city and the Mesopotamian plains."

The One Who Loves Speaks the Truth, Trabzon, by Minna Henriksson

Minna Henriksson's project was a youth oriented one, involving workshops, city tours, and collaborative thinking beyond the accepted wisdoms and perceptions of Trabzon, as a means of portraying the perspective of its young people. It consisted of three strands and took its title The One Who Loves to Speak the Truth from the song Ben Geçerim Gönül Geçmez by Üç Hürel (a local rock band). First, Henriksson installed a neon light work located on a wall facing Yayuz Selim Boulevard entitled Aşık olan doğru söyler. Haksızlığa sitem eyler. Second, a city guidebook was produced out of the workshops, incorporating text and photography by participants on their feelings about both themselves and their city. Third, Henriksson organized, or rather provided a platform for, a music group comprised of participants from the project to perform on the opening night in Trabzon's main square. Besides the neon installation, Henriksson's may be considered a participatory project, which worked to challenge prevailing representations and definitions of Trabzon.

Artwork Excellence

The artistic excellence of this project lies in its ambitious scope, coordination, and the specific interventions of each of the five artists involved in different sites in Turkey. The logistics alone would have proven formidable, especially in hosting the work of international artists in Turkey and arranging for the execution of the projects, not to mention the coordination of Turkish artists in Europe, and the running of a parallel education program. Taken as a whole, My City is placemaking at the scale of cities and intercity relations, and serves as an example for future projects of similar scale. The projects themselves are described in brief, but extensive documentation may be accessed at the British Council's website.

夜曲
Nocturne

艺术家：纳彦·库尔卡尼（英国）
地点：英国盖茨黑德
推荐人：汉娜·皮尔斯
Artist: Nayan Kulkarni, (UK)
Location: Newcastle upon Tyne, Gateshead, UK
Researcher: Hannah Pierce

作品描述

"夜曲"是一件光电装置作品，永久性地安置在伊丽莎白二号地铁桥中，该地铁横穿英国处于纽卡斯尔到盖茨黑德泰恩河峡谷段。大桥全长 360 米，其中有 168 米是横架水泥桥墩之上。承载地铁的桥面位于泰恩河河面上方 25 米的高空中。2007 年前，整座大桥颜色单一而且没有灯光照明，而它东面的其他四座桥都有独特的照明系统。在每十年一次翻新重漆大桥的情况下，有给大桥更新照明的打算。

纳彦库尔卡尼被邀请给大桥设计一个新的上色方案和互动的照明装置。桥身被粉刷成蓝色，白色和棕色的作为陪衬，为建筑照明塑造背景并突出建筑结构形式本身。白色部分为钢铁构架，勾画出整体桥身的长度并为彩色灯光做好底色。白色作为底色能够使投射的其他颜色更加清晰可见，并且节省耗电。白天两种配色的灯光跟随被观察的角度和日光的渐变而发生变化，晚上，LED 灯展示一系列的色彩组合会根据两种因素发生变化：泰恩河河水潮汐的变化还有"地铁夜曲"网站上网民上传的图片。

由潮汐所引发的灯光模式会随着潮涨潮落而微妙改变。当潮水高涨，河水水位会上升，白色光线将会到达最强的亮度。同样的潮涨潮落的速率会相应地和灯光变化的快慢发生呼应，当河水渐渐退潮的时候，灯光色彩会慢慢地移动，当河水涨潮的时候变化速度会稍稍加快。

公众被鼓励积极向"地铁夜曲"网站发送图片，以参与大桥在接下来 15 年中的灯光的变化。每天系统将挑选一张上传的图片，该图片特有的色彩组合被数字化之后传给泰恩河峡谷。事先设置好的程序将保证不会出现重复的色彩组合，也就意味着任何一组变幻闪耀过大桥桥身的色彩组合都是独一无二的。整个照明系统紧紧镶嵌在桥身上，节能的 LED 灯在变幻出 1650 万种颜色和 20 亿种不同色彩组合的同时保持对能源最小的浪费。

解读

作品"夜曲"本质上是对伊丽莎白二号地铁桥的亮化工程，装置于桥上的 LED 灯展示出一系列的互动色彩组合，使大桥的各个部分呈现出如潮汐般的灯光变化，这使年代久远的大桥焕发了新的青春。按照美的规律去改造现有的事物，通过艺术去唤醒公众的审美，这是公共艺术的目的之一。艺术家纳彦·库尔卡尼巧妙地运用模拟自然潮汐变化的灯光，在宁静的夜晚，为人们呈现了一个崭新的梦幻空间。这种浪漫效果的实现离不开艺术家对现代表现材料的发掘与应用，合理采用新科技时代的产物表达艺术内涵已成为现代公共艺术的一种必然趋势。当轻轨车悄然穿行大桥时，人们享受着雾里看花的朦胧，同时陶醉于一个没有声音但却充满动感音乐的世界，多么美好。（马熙逵）

Artwork Description

Nocturne is a permanent light and colour installation built into the fabric of the Queen Elizabeth II Metro Bridge, which carries the Metro train line across the Tyne Gorge between Newcastle and Gateshead in the UK. The full length of the bridge is 360 metres and it spans 168 metres between the concrete piers. It carries the Metros 25 metres above the River Tyne. Prior to 2007, the QEII Bridge was painted a single colour and unlit, while the four bridges to its east all had unique lighting schemes. Consideration had previously been given to the possibility of lighting the bridge in conjunction with its repainting, which takes place on a 10-year cycle.

Nayan Kulkarni was commissioned to design a new paint scheme and an interactive light artwork for the bridge. The bridge is painted a mid-tone blue, with white and dark brown accents, creating a background for the architectural lighting and emphasising the form of the structure. The white areas are within the steel trusses, which establish a clear pattern across the length of the bridge and provide the surface onto which the colours can be projected. Projecting onto white allows for more subtle colours to be visible and for the lighting to operate on a much lower power. By day the two-tone colour pattern on the bridge changes its aspect according to the angle from which it is viewed and the time of day. At night, the LED light units display a pattern of colours that are determined by two sources: The tidal movements of the River Tyne and photographs supplied by members of the public through Nocturne's dedicated website, www.metronocturne.com.

The lighting patterns determined by the tidal algorithms create a subtle pattern of colours that change with the rise and fall of the tides below. When the tide is in, and the water levels within the river are high, the luminosity of the white light is at its most intense. Similarly, the ebb and flow of the tide influences the speed at which the colour messages travel: As the tide meets its ebb, the messages gradually move more slowly across the bridge, and as it flows their speed gradually increases.

The public is encouraged to influence the colours that the bridge will produce over the next 15 years by sending images to the Nocturne website. Each day a donated photograph is digitalised into its own distinctive colours to generate a message that passes forward and back across the Tyne Gorge. Pre-programmed barcodes mean that the lights never repeat the same combination of colours twice, meaning that each band of colour that flows through the structure of the bridge—animating and illuminating it against the night sky—is entirely unique. The lighting is contained within the bridge and its low energy LED lights cause a minimum of light pollution whilst creating a spectrum of 16.5 million colours, giving the possibility of more than 2 billion combinations across the whole of the bridge. The control interface system that programmes the light sequencing of Nocturne is shown in image_i005.

Please note: Due to the theft of the mains power supply cables to the Bridge, the artwork is currently not on, and consequently the website is also temporarily disabled. Nexus are working to solve this by finding an alternative route for the line of copper.

Artwork Excellence

Nocturne's achievements as a public art commission has been recognised nationally. In 2007, it won a Civic Trust Award, which recognises projects that make an outstanding contribution to the quality and appearance of their environment. Technologically it is a challenging work due to both its large scale, close proximity to functioning train lines and the complexity of the lighting controls. The close collaboration between the artist and programming designer has resulted in the delivery of an intricate system that is both responsive to the installation's changing environment and the contributions of the public through the dedicated website. The Nocturne illuminations have transformed a previously overlooked space into a place that reinforces a unique sense of identity for Newcastle and Gateshead. In the five years since its launch, Nocturne has established itself as a major landmark of the North East of England and is recognised as contributing to the region's cultural development and regeneration.

诺沙拉场
Northala Fields

艺术家：彼得 • 芬克（美国）
地点：英国伦敦
推荐人：薇拉 • 托尔曼
Artist: Peter Fink（USA）
Location: London,UK
Researcher: Vera Tollmann

作品描述

"诺沙拉场"项目坐落于伦敦，规模非常大，具有纪念意义。该项目受到了伦敦自治镇伊林的委派，占地面积超过 5 公顷。新公园的发展项目由 FoRM 协会的艺术家彼得 • 芬克和建筑师伊格 • 马可共同设计。项目的创立没有使用纳税人的钱。

四个巨大的土堆，其中一个具有独特的螺旋通道，耸立于天际线上，构成了这个地点的中心。这些定型的土堆使用了伦敦西部大规模拆除工程产生的 65 000 辆货车的废弃材料。于是诺沙拉场成了温布利塔最后的安息处。通过接收这么多废弃材料，该项目赢得的资金用于建造这座耗资 1120 万美元的公园。

使用废弃材料构造这个地貌的好处不仅是财政上的，它还体现在美学上。这个充满活力的设计的核心是那几个大型土堆——最高的有 26 米——不仅为公园挡住来自附近 A40 高速公路的噪声，更是把公园改建成了地貌艺术中的一件不朽之作，而且其规模是欧洲同类作品中最大的。提升这块地方的休闲功能是芬克和马可设计时优先考虑的重点。建成后的这件公共艺术作品包括几处全开放式的钓鱼塘、两座儿童游乐场、一块保留沼泽地、一座航模池、几条自行车道、一些开放式游乐区域和四座土堆。这几处新建的湖泊是伦敦第一处，也是唯一的一处具有都市渔猎功能的场所。新地貌上还建成了一座户外娱乐性圆形剧场，为当地社区和学校提供了开展活动的场所。各种自行车和行人通道连接着周围的自行车和公共交通网络。

这个新地貌的建成为众多的地点和发展项目提供了解决方案。这个项目赢得的资金帮助该地区获得了一系列娱乐和生态方面的利益，而没有增加纳税人的任何负担。

这个项目还为处理伦敦西部建筑项目如温布利体育场、白色城市和希斯罗机场 5 号航站楼等的建筑垃圾方面作出了重大贡献。为了给野生生物制造自然栖息地，除了种植一系列草甸植物外，还增加了林地种植面积。新的水道将给水生和湿地植物提供机会，目前这里还没有这些品种。

"诺沙拉场"赢得了多个奖项，比如"LI 设计奖"和英国景观研究所"超过 5 公顷"项目的最佳设计奖。2009 年，诺沙拉场入围了"世界建筑奖"。艺术家彼得·芬克和建筑师伊格·马可是 FoRM 协会的合作伙伴。FoRM 协会是最具广泛意义的城市规划机构，他们合作开展工作，把生态学融入都市设计和景观建筑环境设计、总体规划、建筑、品牌、照明、艺术、媒体和工程中。

解读

"诺沙拉场"在设计布局的时候除了追求空间的美观和新奇之外，还有机地融入绿色理念。在财政支出合理的情况下，解决建筑废料的掩埋问题，同时在此基础上构建了一个绿色的人工符号，这符合空间规划的合理性和科学性。忙碌于城市心脏的人群，年华像浮云流水，灯红酒绿模糊了整个世界。人们每每怀念的正是那乡间呢喃的甘泉，还有满眼的新绿。设计师彼得·芬克在为公众提供一个大型公共交往的艺术空间之余，营造了一个供人休憩、玩耍、放松的场所，人们可以在此享受自然的阳光和新鲜泥土的气息，过眼忘却都市的嘈杂。这一切理性的设计，有效地解决了建筑废料堆积而带来的视觉垃圾和土地闲置问题，并且为人们延伸了一条从繁华城市通往绿色天堂的道路。（马熙逵）

Artwork Description

The Northala Fields project located in London is large scale and monumental. The project was commissioned by London Borough of Ealing and covers an area of over 5ha. The new park development was designed by artist Peter Fink and architect Igor Marko of FoRM Associates. It was created at no cost to the tax payer.

Four great mounds, one with a distinctive spiral path, dominate the skyline and form the centerpiece of the site. The regularly shaped mounds were constructed from 65,000 lorry loads of waste material from major deconstruction projects in West London: for one, Northala Fields is the final resting place of the Wembley Towers. By accepting waste material, the project also generated funds for the $11.2 million park.

The benefits of constructing the landform using waste material were not just financial, but also aesthetic. At the heart of the dynamic design are the large mounds—the tallest at 26m—which not only shield the park from the noise of traffic on the adjacent A40, but also transform the park into a monumental piece of land-art, the largest of its kind in Europe.

Enhancing the recreation opportunities of the site was a priority for Fink and Marko. The result is a public art project that incorporates fully accessible fishing ponds, two children's playgrounds, a marshland reserve, a model boating pond, cycle paths, open playing fields, and the four mounds. The newly created lakes function as the first and only urban fishery in London. An outdoor entertainment amphitheatre was built into the new landform, providing a venue for local community and school events. Various cycle and pedestrian paths link into the surrounding cycle and public transport networks.

The creation of the new landforms provided a solution to a number of site and development issues. It generated the necessary funds to enable the range of recreational and ecological benefits being sought on the site, eliminating any costs to the taxpayer. It also made a dramatic contribution to shrinking the ecological footprint of West London construction projects such as Wembley Stadium, White City and Heathrow Terminal 5. In order to create natural habitats for wildlife, additional woodland planting was added, along with a range of meadow plants. The new watercourses will provide opportunities for water and wetland flora and fauna that are not currently present on the site.

Northala Fields won serveral prizes, like the LI Award Design and the UK Landscape Institute price for the best design over 5Ha. In 2009, Northala Fields was a finalist in the World Architecture Award.

The artist Peter Fink and architect Igor Marko are partners in FoRM associates. FoRM associates are urbanists in the widest sense, working collaboratively to fuse urban design and landscape architecture with ecology, environmental design, masterplanning, architecture, branding, lighting, arts, media and engineering.

Artwork Excellence

1. Maximising the volume of material within the new landform and hence potential for income generation

2. Reducing adverse affects arising from the adjacent A40, particularly noise, visual and air pollution

3. Delivering new recreation opportunities not currently available within the open space resource of the area focusing on a six lake urban fishery

4. Establishing new ecological opportunities in response to the creation of new topography

5. Creating a major piece of "land art" forming a gateway icon for West London

巴勒斯坦大使馆
Palestinian Embassy

艺术家：卡米拉·马腾斯、托瑞尔·高克索耶尔（挪威）
地点：挪威奥斯陆
推荐人：汉娜·皮尔斯
Artist: Camilla Martens and Toril Goksøyr(Norway)
Location: Oslo, Norway
Researcher: Hannah Pierce

作品描述

2009年秋天，挪威奥斯陆庆祝开斋节期间，卡米拉·马腾斯和托瑞尔·高克索耶尔展出了作品"巴勒斯坦大使馆"。

高克索耶尔和马腾斯的"巴勒斯坦大使馆"采用了热气球的形式，热气球上印着巴勒斯坦的国旗，气球的一边印着"巴勒斯坦大使馆"的英文，另一边用阿拉伯文印着同样的内容。

"巴勒斯坦大使馆"是展出办成了大使馆的正式开放仪式，由大使，政要，娱乐圈人士发表开幕演说，大使和艺术家们在热气球升空前进行了开幕剪彩。热气球升空于奥斯陆市中心，连续四天在城市上空遨游，由于天气状况不佳，在原定四天飞行日程内，热气球成功地进行了两天的飞行。"巴勒斯坦大使馆"在两个层面上概念化地开展项目。

作为巴勒斯坦自治权的视觉化暗示，这个热气球被作为与受邀参与者对话的一个平台。这些计划安排的对话着重于巴勒斯坦人民的处境，他们有限的民主和外交空间。在那些受邀者中间，都是以色列和巴勒斯坦的代表人员，也有挪威政治家和学术界人士。

该次辩论由挪威奥斯陆和平研究所（PRIO）的历史学家斯坦恩·托纳森发起，谈话通过扬声器直接传到地面，使观众能够实况听到采访和讨论。

在奥斯陆上空的热气球内的这些讨论把话题锁定在巴勒斯坦人民的政治前景，提到了巴以双方于1993年签订了奥斯陆协议。这项协议标志着谈判的突破口，并且希望此举能够解决双方之间的冲突。最初被认为是和平与和解的象征，可是不久奥斯陆协议就被认定是一系列冲突谈判之后的不成功例子。

高克索耶尔和马腾斯展出的"巴勒斯坦大使馆"作为行为艺术，象征着一个新的奥斯陆交流平台供受邀与会者交流，并且审视艺术和政治领域之间的关系。尽管该项目有着乌托邦式的愿望，但要成功让参与各方对迫在眉睫的政治问题进行商讨，事实证明是有困难的。在所有受到邀请的挪威政治家中，只有一位前来参加。众多其他大使和代表也都拒绝了这个提议，而相当多的巴勒斯坦活动家和政治家都愿意参加辩论，导致了对话的不平衡。

每一份被拒绝的邀请虽然从某些方面讲致使该项目未能充分发挥出原定的功能，但它对正在进行的围绕巴勒斯坦人苦难的讨论却起到了促进作用，而这正是此作品的最初的目的。

解读

巴勒斯坦地区因历史上的各种复杂纠葛，使得犹太人和阿拉伯人皆认为该地区是他们的固有领土，并不惜为此诉诸武力。由于巴以冲突的持续，巴勒斯坦的经济蒙受巨大损失，人民生活水平低下。艺术家卡米拉·马腾斯、托瑞尔·高克索耶尔希望通过一种诙谐、幽默的行为艺术形式——空中移动"办公"的热气球"巴勒斯坦大使馆"，为不同文化的人们构建一个信息自由交流平台，唤起人们对巴勒斯坦地区纷争的关注，使更多的人了解巴勒斯坦人民现在的处境，宣扬地区发展应该遵循法制、公正、人权、民主。该项目最终没能邀请到众多的大使和代表来对有关巴勒斯坦的政治问题进行商讨，而众多的巴勒斯坦活动家、政治家纷纷参加了活动，导致了对话的不平衡，但这不影响整个计划最初目的——使更多的人关注到巴勒斯坦人民的苦难生活。（马熙逵）

Artwork Description

In the autumn of 2009, during the celebration of Eid, Camilla Martens and Toril Goksøyr staged the opening of a Palestinian "embassy" in Oslo, Norway.

Goksøyr & Martens Palestinian Embassy took the form of a hot air balloon fashioned in the colours of the Palestinian flag featuring the text 'Palestinian Embassy' on one side, and the same again in Arabic on the other. The opening of the Palestinian Embassy was organised as an official inauguration of an embassy, with an opening speech by the ambassador, political appeals, entertainment and an official cutting of the ribbon by the ambassador and the artists, before the balloon take off. The balloon was launched from the centre of Oslo, making trips over the city for four consecutive days, although due to adverse weather conditions the balloon was only able to make successful journeys on two of its four intended flight days.

Palestinian Embassy was conceptualised to work on two levels. As much as a visual cue to symbolise Palestinian autonomy, the balloon would operate as a platform for dialogue between invited participants while in flight. These programmed conversations focused on key issues related to the Palestinian people's situation and sought to emphasize their limited democratic and diplomatic space. Amongst those invited were representatives of both Israel and Palestine, as well as Norwegian politicians and academics. The conversations were led by Stein D. Tønnesson, a Norwegian historian based at the Peace Research Institute Oslo (PRIO) and were transferred directly to the ground through loudspeaker so the audience could follow the interviews and discussions.

Locating these discussions about the political prospects of the Palestinian people in a hot air balloon in the skies above Oslo made reference to the talks between Palestine and Israel that led to the signing of the Oslo Accords in 1993. This agreement signified a breakthrough in negotiations and was hoped to signify a move towards a solution of the conflict between the two parties. Initially considered to be a symbol of peace and reconciliation, the Oslo Accords were quickly acknowledged to be another in a series of unsuccessful conflict negotiations. The presentation by Goksøyr & Martens of the Palestinian embassy as a performance art work symbolised a new Oslo platform for exchange between the invited participants, while investigating the relationship between the fields of art and politics.

Despite the Utopian aspirations of the project, the reality of successfully engaging such parties in discussions on urgent political issues proved difficult. Of all the Norwegian politicians invited to participate in the project only one agreed to attend. Numerous other ambassadors and representatives also declined the offer to contribute to debates which resulted in the dialogue being somewhat unbalanced as considerably more Palestinian activists and politicians were willing to participate. Each declined invitation, while in part leading to the project being unable to realise its full intended potential, contributed to the process of the work and the ongoing dialogue around the plight of Palestinians, which had been the work's original aim.

Artwork Excellence

Goksøyr & Martens conceptualised Palestinian Embassy as a medium for reflection. As a platform for open dialogue within an art context, rather than a soap box for political critique, the work prompts the viewer to reconsider any preconceived notions they may have held about the conflict between Israel and Palestine.

By taking place above the city, Palestinian Embassy was able to reach the broadest possible audience, prompting debate about difficult and urgent political and religious themes that the incidental viewer may not otherwise has considered. This broad reach was also amplified through extensive local and national press coverage, allowing a secondary audience to investigate the art work and its subject matter, widening the debate platform further again.

Despite the declined invitations and disrupted flight schedules, the project successfully purposed artistic strategies to stimulate a dialogue between a number of academic, politicians and activists that couldn't have been achieved elsewhere.

零点
POINT 0

艺术家："公共底座"：达利博·巴卡、马丁·皮亚塞克、
米凯尔·莫拉维塞克、托马斯·德扎东（斯洛伐克）
地点：斯洛伐克布拉迪斯拉法
推荐人：吉乌希·切克拉
Artist: Public Pedestal: Dalibor Bača, Martin Piaček,
Michal Moravčík and Tomáš Džadoň（Slovakia）
Location: Bratislava, Slovakia
Researcher: Giusy Checola

作品描述

"公共底座"是一个公共艺术小组，最初诞生于"什么雕塑赋予城市？"（2006
年）和请愿"停止在我们城市里的'文化政策'！"（2006年），这些项
目表达了对于布拉迪斯拉法官方公共艺术和空间的批评。年青一代的艺术
家、理论家和艺术消费者抱怨斯洛伐克城市内假充艺术的低劣旅游工艺品
太多。最近，这些情绪导致了2010年反对库里希的抗议活动，反对雕塑家杰·
库里希在布拉迪斯拉法城堡里展出的装置作品"斯瓦托普鲁克国王"。

"公共底座"的目标在于创造一个临时的艺术自治区域——一个崇尚自由
表达的另类空间，这类空间似乎已经在斯洛伐克城市里失踪了。公共底座
起到了如同这类空间策展人的作用，在布拉迪斯拉法以及在其他欧洲城市
寻求确立一片公共展示/公共空间，致力于开展临时性展览，及对记忆现

象和对地点／城市的身份确认进行的研究和思考的可能性。组织者的设想是邀请感兴趣的当地和国外艺术家，通过他们的出席并提出发人深思的观点和批判性意见有可能使现行的公共空间概念得到扩展。通过举办临时艺术项目甚至借助争论的方式，"公共底座"努力在艺术家、文化运营商和公众中开展讨论。

组织者托马斯·德扎东追踪这个国家令人失望的官方艺术的根源，他指责部分是由于缺乏资金，部分是委托机构的缺乏教育。"而另一个原因则是在斯洛伐克缺乏批判性思维，"他说道，"导致的结果是建造的纪念碑和雕塑带有民族主义、排外性，而且完全依附于政治影响力。"

"公共底座"目前的项目是"零点"，致力于把布拉迪斯拉法最有名的公共广场"自由广场"的一部分转型，广场最初以捷克斯洛伐克共产党主席克雷蒙特·哥特瓦尔德是名字命名，1989 年"丝绒革命"以后公园改了名，这块场地以前有一尊哥特瓦尔德的纪念碑（于1991年迁移）如今成了空地。"零点"建议重新使用以前的极权纪念碑，包括从下列几位中挑选：帕维尔·阿尔泰莫，索博尔奇·吉斯帕尔，克里斯托弗·金特拉和伊诺拉·内梅斯。

"该项目的目标是突出著名艺术家创作的高水准作品，"德扎东解释道，"只有当这些作品出现在广场上，我们才算迈出认识公共空间及其价值的第一步。我们想要把这类作品引入公共空间，自从社会主义垮台以后它们至少消失了将近 20 年。将前政权的遗骸作为这些活动的背景具有重要意义。重新使用这些废弃的社会主义空间是"零点"这部作品的意义所在。"

为了把讨论进一步深化，"公共底座"在 2011 年 12 月组织了一场题为"公共空间里艺术"的研讨会，构想了一次对于放在公共空间里的艺术背景下的政治的批判性讨论。这次讨论由米拉·克拉托娃和米凯尔·莫拉维塞克组织，参与者包括格露帕·斯珀麦尼克、安德里亚斯·弗嘎拉西和拉卢卡·沃尼亚。它的目标不是寻找公共艺术的正面例子，它们常常借助宣泄性的争论来帮助社会结构的改进。比较而言，这个观念是暴露出在公共空间里机会主义艺术的局限性，以及为对社会产生影响的艺术作品的构成要素作出定义。这包括社会构建"想象的共同体"的一场争论，以及出于这一目的而创作的艺术品和因此而产生的内容。

解读

"公共底座"是一个公共艺术小组，通过相关公共艺术项目，表达了对于布拉迪斯拉法官方公共艺术和空间的批评。"公共底座"的目标在于创造一个临时的艺术自治区域——一个崇尚自由表达的另类空间，这类空间似乎已经在斯洛伐克城市里失踪了。公共底座起到了如同这类空间策展人的作用，在布拉迪斯拉法以及在其他欧洲城市寻求确立一片公共展示／空间，致力于开展临时性展览，及对记忆现象和对地点／城市的身份确认进行的研究和思考的可能性。组织者的设想是邀请感兴趣的当地和国外艺术家，通过他们的出席并提出发人深思的观点和批判性意见有可能使现行的公共空间概念得到扩展。通过举办临时艺术项目甚至借助争论的方式，"公共底座"努力在艺术家、文化运营商和公众中开展讨论。

"零点"是他们目前的项目，致力于把布拉迪斯拉法最有名的公共广场"自由广场"的一部分转型，突出一些著名艺术家的作品，把他们引入到公共空间中，体现公共空间的作用和价值。

"公共底座"以公共艺术项目为媒介，关注公共空间，研究和思考城市相关的种种问题，用艺术来寻求解决的可能性，在公共艺术发展举步维艰的情况下，却一直在崇尚自由表达的另类空间，这种追求与做法带给我们很多启示与反思。（冯正龙）

Artwork Description

Public Pedestal is a public art group that resulted from the initiative What statue for the city? (2006) and the petition Stop "cultural policy" in our cities! (2006), projects that expressed criticism of official public art and space in Bratislava. The younger generation of artists, theoreticians, and art consumers complain that Slovak cities are overloaded with tourist kitsch masquarading as art. More recently, these sentiments launched Antikulich a 2010 protest against the installation of the sculptor Ján Kulich's "King Svätopluk" at Bratislava's castle.

Public Pedestal aims to create a Temporary Autonomous Zone for art—an alternative space that values freedom of expression, which seems missing in Slovakian cities. Public Pedestal acts as a curator of such spaces, seeking to identify public display/public space in Bratislava, as well as in other European cities, which will be devoted to temporary sculptural interventions and exhibitions, with the possibility to research and reflect upon the phenomenon of memory and identity of the place/the city. The idea is to invite interesting local and foreigner artists whose presence might expand the prevailing notion of public space by offering provoking ideas and critical conceptions. By staging temporary art projects, and even controversy, Public Pedestal seeks to start discussions among artists, cultural operators, and the public.

Organizer Tomáš Džadoň traces the group's roots to a general disappointment in the country's official art, which he blames in part on lack of funding and lack of education on the part of commissioning bodies. "The other is a lack of critical thinking in Slovakia," he says, "and the result of this are the construction of monuments, sculptures which one can say are nationalistic, xenophobic and totally depending on political influence."

Public Pedestal's current project is Point 0, an effort to transform part of Freedom Square, Bratislava's most famous public square. Originally named for Czechoslovakian Communist president Klement Gottwald, the park was renamed in 1989 after the Velvet Revolution, and the space formerly occupied by Gottwald's monument (removed in 1991) remains unoccupied. Point 0 proposes reuse of the former totalitarian monument, including options from Pawel Althamer, Szabolcs KissPál, Krištof Kintera and Ilona Németh .

"The aim of the project is to highlight works of high artistic level by well-known artists," Džadoň explains. "Only when these works appear on square can we make a first step to acknowledge the public space and its values. We want to bring such things to public space, which have been missing for at least 20 years, since socialism broke down. It is important that the remains of former regime serves as a background for these activities. Re-using these abandoned socialist spaces is the real benefit of Point 0."

To move the discussion forward, Public Pedestal organized "Politics of art in public space," a December 2011 seminar conceived as a critical discussion about politics in the background of art in public space. Organized by Mira Keratová and Michal Moravčík, participants included Gruppa Spomenik, Andreas Fogarasi, and Raluca Voinea. Its aim was not to search for positive examples of public art, which often contribute to the development of social structures through cathartic controversy. Rather, the idea was to formulate the limits of the opportunistic potential of art in public space and to define features constituting the artwork with possible social impact. This included a debate about socially constructed "Imagined Communities," for which the public art is produced and which determine its content.

Artwork Excellence

In the project Point 0, Public Pedestal is engaged in launching a discussion about public art and public space in the former Eastern Bloc against the particular history of the region and the socio economic situation on the ground today. The group uses a variety of means to initiate discussion: Public agitation and consciousness raising; conferences and other discussions; and commissioning of works that challenge prevailing notions of public art and the uses of public spaces.

圩区杯
Polder Cup

艺术家：AES+F(俄罗斯)、伊萨·冈兹肯（德国）、
梅德尔·洛佩兹（西班牙）和艾谢·厄尔克曼（土耳其）
地点：荷兰奥拓兰德，维特德的正门
推荐人：富利亚·厄尔德姆奇
Artist: AES+F (Russia), Isa Genzken (Germany), Maider
López, (Spain) and Ayşe Erkmen (Turkey)
Location: Ottoland, The Netherlands, façade of Witte de
With Museum Center for Contemporary Art
Researcher: Fulya Erdemci

作品描述

这是一个现场项目，一场别样的足球锦标赛于 2010 年 9 月 4 日在格拉芙斯特鲁姆市的奥拓兰德圩田中举办。洛佩兹在奥拓兰德圩田重新构建了一片足球场，为参与者们创造了一个全新的平台。这项活动不仅开创了一个不同寻常的形式，更反映了荷兰社会共同长期对抗洪水的历史。她希望通过这个项目转换圩区的功能，超越其历来的农业用途，努力产生互动，建立社会的不同领域之间的关系。

在这个项目中，艺术家通过共同的大众语言——足球，将不同社会群体汇聚在一起，以游戏、体育和社会交流的方式达成公众交流的目的。与典型的足球赛不同，洛佩兹把足球场地格局进行了替换，让运河与足球场相交，并改变了足球比赛的规则，鼓励参赛者发明新的比赛规则。通过参与比赛的足球运动员，以及来自各界的观众，圩区将经历一次转型。这个场地的使用功能将被重新诠释，改变了固有的看待空间环境的方式。

洛佩兹的主要计划是通过多样化的公众行动来重塑公共空间，她认为，通过一个共同的"语言"，个人可以与他人彼此互动，产生联系。有了这样的变化或在公共空间进行干预，将赋予公众表达自我的能力，她作品的主要特点是具有表演性质，基于人与人之间的关系，通过在公共空间制造干预性活动，阐明日常生活的政治。

解读

作品"圩区杯"是一种新式的足球比赛形式，它融合荷兰圩区地貌中的自然障碍，特意规划了带有水沟、草坑的不平整足球场，运动员们不得不考虑如何有效地绕过这些自然屏障，并获取团队比赛的胜利。艺术家们的规划是在土地完成种植任务的间歇期内，将待种的土地转化为暂时的娱乐场地，巧妙地开发了土地的潜在价值，为大众构建了一个公共交流平台，人们可以在此通过娱乐增进互相的认识，并更多地与自然相伴。在比赛的过程中笑料百出的运动员，成为周围观众共同欣赏的"艺术品"，这也正是作品希望达到的目的——通过艺术介入，让人们参与和分享快乐。"圩区杯"带给公众娱乐互动的同时，令人们更多地享受自然的芬芳，让欢乐化作每个人心中的一片清凉。（马熙逵）

Artwork Description

Reflecting on the long history of the Dutch struggle to tame the lowland coastal waters, the artists rebuilt a football field on the polder (coastal infill) of Ottoland, Graafstroom, where a special championship match was held on September 4, 2010. Maider López created a brand new platform for the participants at the event. In the unusual form of the platform and its site-specific milieu López hoped to establish a relationship between different social spheres—between the football game and the site's historic agricultural use—thus transforming perceptions of the polder zone's function.

López's key goal is to rebuild public spaces through involving members of the public in a variety of ways. She maintains that individuals can interact with others simply through a common human language. Through her interventions in public space she empowers people to realize their capacity to change. Her work is mostly performative and based on human relations; she enjoys challenging assumptions and articulating the politics of everyday life.

Artwork Excellence

In this project, the artists brought different social groups together through a common public activity: football. The project achieved the public purpose of communication via games, sports and casual social interaction. Unlike a typical football field, however, the playing field was intersected by canals, requiring revisions to the rules and challenging players to invent new strategies. Through the participation of football players and audience members from all walks of life the polder zone was singularly transformed. Through this experience the function of this site has been reinterpreted, and assumptions about space and environment have been brought into question.

柏林市中心雕塑公园
Sculpture Park Berlin_Center

艺术家：艺术共和国（德国）
地点：德国柏林
推荐人：薇拉·托尔曼
Artist: KUNST re PUBLIK (Germany)
Location: Berlin, Germany
Researcher: Vera Tollmann

作品描述

　　"柏林市中心雕塑公园"是一个非传统的户外展览地，自 2006 年至 2010 年期间，先后由柏林市中心的 62 块空地建成。这片区域曾经是柏林墙前的"死亡地带"。随着 1990 年柏林墙的倒塌，土地的产权和所有权变得归属不明，这片地方一度沦落为"无人看顾"的停车区，那里野狗横行、小径丛生、垃圾成堆。后来，这片土地各自回到了产权所有人的手中。2006 年，马蒂亚斯·因霍夫、菲利普·霍斯、马克·罗曼、哈利·萨克斯和丹尼尔·塞佩尔从不同的土地所有人那里得到了使用许可，创建了柏林市中心雕塑公园，并由他们自办的"艺术共和国"对其进行运营和维护。

　　雕塑公园所用的那 5 英亩土地周围是六至八层高的居民住宅以及办公楼，雕塑公园从建立伊始便用围栏将自身与周遭部分环境隔离。整块地方就是一个被乏味的公寓楼和办公大楼所包围的城市废地，杂草丛生，荒树成林，随处散落着二战前琼楼玉宇的遗迹，"柏林市中心雕塑公园"给这片土地上带来了各式各样精心打造的极具艺术性与文化性的活动，让这片土地焕发出了都市魅力。

131

随后的展览系列以轮流管理委员会方式进行：库存，单元格，宣传炒作，土地改革，还有梦幻仙境。2008 年，在"柏林市中心雕塑公园"举办了第五届柏林双年展。"艺术共和国"就不同项目的需要及法律要求与产权所有人进行了沟通协调。

丹尼尔·塞佩尔想在公园里装喷泉以期使土地和财产增值。不过他未经授权挖掘和使用另一口井中的水使得这个善意的初衷引发出了"到底谁是井水所有人？"的老套矛盾来。艺术家艾蒂安·布郎杰开了一间经济实惠的 2 星单人房酒店，酒店内四面墙都贴有广告牌和公益广告。凯尔特瓦塞尔和科柏林在园内挖了三个点，作为历史和考古基地。在西柏林沿柏林墙的地方曾经有几处观景台，这次的挖掘呈现出了这些观景台的大致形态。菲利普·霍斯在离地面 20 米的高空安装了一个高亮白炽灯。三天里，居住在酒店的客人可以通过一个远程遥控来对灯进行开关。

它位于亚历山大广场的西南面，是柏林市中心区和克罗伊茨贝格区的交界处。柏林参议院在提议城建规划时，如是形容柏林市中心雕塑公园，将它定义为一次"极具关键意义的重建"。所谓城建规划其实是指谨慎的"重建历史性柏林"并拆毁之前有些年代和名气的民主德国建筑。不过，雕塑公园毗邻的中心区和克罗伊茨贝格区却并没有如规划般的高速发展起来。四年里，柏林市中心雕塑公园在发展中带动了外围街区并使更多未经使用的城市空间得到了利用。日益增多的房地产项目慢慢改变了柏林市中心雕塑公园的地盘。2009 年，在新建第一幢楼 15 年后，又迎来了百年来的第二幢新建筑：一幢可居住的商务楼。这幢楼房于 2010 年竣工。

解读

杂草丛生，荒树成林，随处散落着"二战"前琼楼玉宇的遗迹，就是"柏林市中心雕塑公园"以前衰败废弃的景象。如今，谁会想到，以前的"死亡地带"却成了今天房地产项目必争的地盘。

"柏林市中心雕塑公园"是一个非传统的户外展览地，自 2006 年至 2010 年期间，先后由柏林市中心的 62 块空地建成。2006 年，马蒂亚斯·因霍夫、菲利普·霍斯、马克·罗曼、哈利·萨克斯和丹尼尔·塞佩尔从不同的土地所有人那里得到了使用许可，创建了"柏林市中心雕塑公园"，并由他们自办的"艺术共和国"对其进行运营和维护。"柏林市中心雕塑公园"给这片土地上带来了各式各样精心打造的极具艺术性与文化性的活动，系列展览如库存、单元格、宣传炒作、土地改革、梦幻仙境以及第五届柏林双年展等，让这片土地焕发出了都市魅力。（冯正龙）

Artwork Description

Sculpture Park Berlin_Center (Skulpturenpark Berlin-Zentrum) was a non-traditional outdoor exhibition venue from 2006 to 2010 on 62 vacant lots of downtown Berlin real estate. The land was formerly part of the "death strip" within the Berlin Wall. When the Wall came down in 1990, property rights and ownership were unclear, so the the strip of land had been used for years a "wild" parking lot, dog run, short cut and garbage dump. Eventually, the various lots were transferred back to their original owners. In 2006, Matthias Einhoff, Philip Horst, Markus Lohmann, Sachs and Daniel Seiple asked permission of the various landowners to appropriate the lots and their art collective,

KUNSTrePUBLIK, operated and maintained the land as Sculpture Park Berlin_Center.

When the project began, Sculpture Park's 5 hectares—surrounded by six- to eight-story residential and office buildings—were partially fenced. An urban wasteland surrounded by drab apartment and office buildings, overgrown with weeds, bushes and trees, and strewn with the remains of pre-WWII edifices, Sculpture Park Berlin_Center transformed the land into an urban stage, with various spontaneous and orchestrated artistic and cultural activities.

The following exhibition series were realized in collaboration with a rotating curatorial committee: Inventory (Bestandsaufnahme), Grid Cell (Parcella), Speculations (Spekulationen), Land Reform (Landreform) and Wonderland (Wunderland). Sculpture Park Berlin_Center hosted the 5th Berlin Biennial in 2008. Depending on the needs of each project and legal necessities, KUNSTrePUBLIK e.V. negotiated permissions with landowners.

With his installation Fountain, Daniel Seiple mimicked a common strategy used to increase land and property value. However, the artist's unauthorized digging and pirating of another well's water contradicted this benevolence and asked the old question: Who owns the water? The artist Etienne Boulanger ran The Single Room Hotel, an affordable 2-star hotel, contained within four walls of billboards and public advertising. Kaltwasser & Köbberling excavated three sites in the park and literally turned over the site's historic and archeological foundations. The excavations mirrored the form of viewing platforms once erected by West Germany along the Berlin Wall. Philip Horst installed at 20 m height above ground a strong light bulb. For three days, residents could turn it on or off with a remote control switch.

Located to the southwest of Alexanderplatz, where the districts Berlin-Mitte and Kreuzberg meet, Sculpture Park Berlin-Center was an area classified for "critical reconstruction," an expression used by the Berlin Senate for urban planning. Basically it means to carefully "reconstruct" the "historical Berlin" \and to demolish GDR architecture and therefore its visibility and history. But the prominent neighborhoods Mitte and Kreuzberg did not grow together as quickly as planned. So in its four years, Sculpture Park Berlin_Center grew to include peripheral blocks and more unused urban space. Then more and more real estate projects slowly changed the Sculpture Park Berlin_Center terrain. In 2009, fifteen years after the first new building, the second new construction of the last 100 years was erected: a mixed-use office and residential building. The project ended in 2010.

Artwork Excellence

Sculpture Park Berlin_Center subverted and expanded the historical notions of a "sculpture park." The integration of objects into a cultivated park landscape—in the tradition of open-air museums—was replaced by process-oriented activities, made visible and negotiated through social, historical and societal contexts. Invited artists confronted the specific situation of the place and initiated artistic interventions, situations and signs. The historical and current significance of the area was thus subject to continuous reinterpretation and discussion, without affecting its basic character as an open urban space.

定位·去
Situa.to

艺术家：阿·提托洛（意大利）
地点：意大利都灵
推荐人：吉乌希·切克拉
Artist: a.titolo（Italy）
Location: Turin, Italy
Researcher: Giusy Checola

作品描述

"定位·去"是由 a.titolo. 这个策展团队和建筑师兼策展人莫里奇奥·西里为 2010 年都灵担任欧洲青年首都期间相关的"你们—我们的时代机构计划"而构思。"定位·去"是一个研究平台、实验室和永久都市观测站，用来自 30 名 30 岁以下的年轻研究者的观点对一座城市进行探索、提问、描述和想象。

项目的目标是创造公共艺术和都市设计项目，与都市文化的新实践沟通，针对目前发生在城市中的转型创作故事，着重对环境、生活品质、新形式的公民和新一代的愿望的叙述。该项目的组织者给它起名为"城上景观"，是对原来的一档题为"城市的面貌"的奥运会节目的戏谑，因为那档节目的资金被转给了"定位·去"。

通过四次为期一周的会议把这个项目组织了起来，包括工作坊、讲座、非正式会议、反馈会议和单独进行实地调研。其结构的意图是制造一个学习方法论，它分两个步骤实行：第一是研究，第二是发展、制作艺术和跨学科项目。

从不同的国家中公开挑选了 200 多名候选人名单，参与者分享都灵在生活、学习和工作方面的经验。建筑师、艺术家、摄影师、作家、表演家、电影人、设计师、学生和人文类学科的毕业生参与了一个教育项目，一个引导性的实践体验，设计成共同分享，探索他们专业的不同方法。鼓励参与者们与城市进行对话并对城市地区和身份的复杂性提出质疑。

"定位·去"的策展人对项目的选择基于对其可行性，包括财政上的可行性。接着就是资金筹集活动。2011 年，30 个项目中的九个项目进入研究阶段，其覆盖面从由青年人建立的一个青年中心到一个壁画项目。

欧共体的最新研究显示，年轻人的人数在人口总数中呈下降趋势。"定位·去"的策展人开发了一种方式，其目标是对于都市生活更深的意识和公民介入的新形式提供评判工具，以期让年轻人更多地参与都市生活，提出可供选择的一般聚集和娱乐活动的模式，给非物质的和创意性的劳动部门提供新的、不同的观点，这个领域可能激活专业途径，目标在于加强公共场地、日常生活、人际沟通和关系的质量。

这个方法主要涉及情境主义者的理论和实践，打算使艺术项目和不由机构城市复兴计划指导的市内地区的公民积极参与之间的互动成为可能，为城市自治改造进程创造条件。项目旨在为年轻人创造机会主动发展，使他们成为发明者和推动者。

解读

"定位·去"是一个研究平台、实验室和永久都市观测站，用来自 30 名 30 岁以下的年轻研究者的观点对一座城市进行探索、提问、描述和想象。

目标是创造公共艺术和都市设计项目，与都市文化的新实践沟通，针对目前发生在城市中的转型创作故事，着重对环境、生活品质、新形式的公民和新一代的愿望的叙述。该项目的组织者给它起名为"城上景观"，是对原来的一档题为《城市的面貌》的奥运会节目的戏谑，因为那档节目的资金被转给了"定位·去"。

该项目包括工作坊、讲座、非正式会议、反馈会议和单独进行实地调研。其结构的意图是制造一个学习方法论，它分两个步骤实行：第一是研究，第二是发展、制作艺术和跨学科项目。

"定位·去"通过设计多种公共艺术和相关项目，以青年研究者为研究团队，针对城市中的众多问题进行挖掘，探讨城市环境、生活品质、趣闻轶事等，另外，鼓励参与式与体验式互动，让参与者在体验中感受城市，体验生活。"定位·去"与其说是有关公共艺术与都市相关项目的研究平台，不如说它是对公共艺术方法论以及运行机制的体现。（冯正龙）

Artwork Description

Situa.to was conceived by a.titolo, a curatorial collective, and Maurizio Cilli, architect and curator, for Y-Our Time institutional program in conjunction with Turin's term as 2010 European Youth Capital. Situa.to is a research platform, laboratory and permanent urban observatory created to explore, question, describe and imagine a city from the point of view of 30 young researchers under age 30.

The aims of the project are to create public art and urban design projects, to communicate new practices of urban culture, and to produce narratives of the transformations currently taking place in the city, with a special attention to the environment, quality of life, new forms of citizenship, and the desires of the new generations. The organizers of the project dubbed it "Look on the city," a play on the original Olympic program "Look of the City," from which funding was transferred to support Situa.to.

The project is organized through four week-long sessions that include workshops, lectures, informal meetings, feedback sessions, and a phase of field research conducted individually. The intention of its structure is to create a methodology of learning by doing that is divided in two steps—one of research and one of development and production of artistic and interdisciplinary projects.

Selected from a shortlist of over 200 candidates chosen by open call from different nationalities, the participants share the experience of Turin as city in which to live, study, and work. Architects, artists, photographers, writers, performers, filmmakers, designers, students, and graduates in the humanities were involved in an educational program and a guided practical experience designed to share and explore the differing approaches of their disciplines. Participants were encouraged to enter a dialogue with the city and to raise questions about the complexity of urban territories and identities.

The selection of projects by the curators of Situa.to was based on the analysis of their feasibility, including financial feasibility. Fund raising activity followed. In 2011, nine of the thirty projects—ranging from a youth-built youth center to a mural project—began the research phase.

Recent studies of the European Community reveal that the young represent a shrinking percentage of the population. The curators of Situa.to developed a methodological approach which aims to offer critical tools for a greater awareness of urban living and new forms of involved citizenship—to improve young people's involvement in urban life, to propose alternative models to simple and general aggregation and entertainment, and to offer new and different perspectives in the immaterial and creative labour sector as a field where it is possible to activate professional paths aimed at the enhancement of the quality of places, everyday life, communication, and relationships.

This approach mainly refers to the theories and practice of the situationists (Situationist International), and intends to make possible the interaction between art projects and forms of active citizenship in areas of the city that are not directed by institutional programmes for urban regeneration, in order to create incubators of autonomous urban transformation processes. The aim is to give young people opportunities to develop initiatives in which they themselves can be inventors and promoters.

Artwork Excellence

The excellence of this project can be judged against the backdrop of rapidly declining percentage of youth as a demographic group in Europe. Although the population is aging, youth remain the key drivers in the vital work of remaining and reinventing society for the better. The challenge in our current demographic situation is one of providing a platform for young voices to be heard and one of creating laboratories where young practitioners can explore, learn, take risks, and refine their skills.

勇气的象征
Symbol of Courage

艺术家：莱斯·阿弥里（伊拉克）
地点：伊拉克提克里特
推荐人：里奥·谭
Artist: Laith al-Amiri（Iraq）
Location: Tikrit, Iraq
Researcher: Leon Tan

作品描述

2008 年 12 月，当美国总统乔治·布什和伊拉克总理努里·马利基会晤时，记者蒙塔兹·扎伊迪脱下脚上的鞋子向美国总统布什掷去。这一画面的视频就像文化病毒一样迅速传遍了全球媒体和社交网络。对于许多伊拉克人，事实上在阿拉伯世界中，扎伊迪成了一位文化英雄。他的行动正是完美表达了对美国中东政策和制造战争的反感。伊拉克雕塑家莱斯·阿弥里铸起一只 8 英尺 ×10 英尺的巨鞋，向这位"飞鞋英雄"致意。这尊雕塑是由莱斯·阿弥里和提克里特城孤儿院的孤儿一起携手完成的，其安装所在地也正是萨达姆·侯赛因的家乡。这座雕塑也被授予了"勇气的象征"以及"光荣与慷慨之像"等不同的名字。

这个只用了两周就完成的雕塑上刻着蒙塔兹的碑文：禁食直至鲜血沥开锋利剑；沉默直到我们说出真相。雕塑由玻璃和青铜制造而成，并填满了人工灌木。当地孤儿院和儿童组织的员工阿卜杜勒•卡迪尔说道："这座雕像是我们对蒙塔兹•扎伊迪最起码的敬意，因为他的这一掷使伊拉克人民备感欣慰。"然而，其他报道表明伊拉克政府声明次日将把这个公共雕塑移走。据运营这个孤儿院的慈善机构主席，Shahah Daham 说："我立即拆掉并摧毁了这个鞋子雕塑，但是我并没有问原因。"

为了理解在伊拉克向他人投掷鞋的文化意义，我们首先要知道鞋是和"底部"紧密相关的，鞋子是在地面泥土层面上的，向别人投掷鞋子是表示一种侮辱，因为它意味着被投掷者只配得到最大的蔑视。在伊拉克，不仅仅只有扎伊迪扔了鞋，鞋子也曾被投向萨达姆•侯赛因的雕像以及乔治•布什的模拟像。然而，只有扎伊迪成功地抓住了机会把鞋子投向了活生生的总统。

尽管这座雕塑的存在时间很短，莱斯•阿弥里的雕塑可以被视为一个成功的场景干预，因为它激发伊拉克社会中产生了一种普遍的情绪。公共场所是社会群体安全感所在，对坦率言论一定程度上的包容是必需的。而这正是扎伊迪和莱斯•阿弥里揭示出的一个社会问题：伊拉克社会（以及政治）如何才能形成一个畅所欲言的环境？更重要的是，政治意见分歧如何不但能够得到容忍，而且还应受到积极鼓励？

解读

"飞鞋事件"发生的时候，超过一百万名殉道者倒在占领军的子弹之下，而在伊拉克全境，有超过五百万名孤儿嗷嗷待哺，一百万名寡妇无依无靠，还有几十万名伤残者在忍受病痛的折磨，数百万名伊拉克人或无家可归，或流亡海外。"勇气的象征"表面上是一只"飞鞋"的雕塑，其实际意义因为"飞鞋事件"上升为政治反抗的标志。作者同时也希望更多的人能够关注到伊拉克人民的艰难生活。

艺术家莱斯•阿弥里通过再造场景的手法，探讨公共场所是否应该有必要的安全机制，对于各种言论应该有其被包容的必要。艺术家强调意见分歧是人类生活中不可避免的，对立双方必然有着各自不可退让的立场，是否应当建立一个畅所欲言的环境，让意见能够有效沟通，值得人们去思考。（马熙逵）

Artwork Description

When the journalist Muntazar al-Zaidi took off and threw both his shoes at U.S. President George Bush in December 2008, while the latter was meeting his Iraqi counterpart Nouri al-Maliki, a video of the action created a viral meme that made its way rapidly across the world's media and social networks. For many in Iraq, and indeed in the Arab world, al-Zaidi became something of a cultural hero, and his act the perfect expression of revulsion at American middle-east policies and war-making. Laith al-Amiri's monumental sculpture, a depiction of al-Zaidi's famous shoe measuring eight by ten feet was created to commemorate al-Zaidi. Entitled variously Symbol of Courage and Statue of Glory and

Generosity, it was a work co-created by al-Amiri and orphans in Tikrit, the site of its installation (and former hometown of Saddam Hussein).

The construction of the sculpture took a mere two weeks and carried the inscription Muntazar: fasting until the sword breaks its fast with blood; silent until our mouths speak the truth. Manufactured of fiberglass and bronze, it was filled with artificial shrubbery. According to staff of the children's organization and orphanage, Fatin Abdul Qader,
  "This statue is the least expression of our appreciation for Muntazer al-Zaidi, because Iraqi hearts were comforted by his throw." However, other reports indicate that the Iraqi government demanded the removal of the public sculpture the next day. In the words of Shahah Daham, head of the charity running the orphanage, "I did take the shoe down immediately and destroyed it; and I did not ask why."

In order to understand the cultural significance of throwing a show at someone in Iraq, it is important to know that the shoe is associated with all that is "base" ; it is at the level of the dirt of the ground. Throwing the show at someone is an insult as it conveys the impression that the recipient is worth only of the utmost contempt. Shoe throwing in Iraq is not limited to al-Zaidi's intervention: shoes were also thrown at statues of Saddam Hussein and had been previously thrown at an effigy of George Bush. Only al-Zaidi, however, succeeded in seizing the moment to direct his throw at the living president.

Artwork Excellence

Despite its short lived installation for one day in 2009, Laith al-Amiri's sculpture may be considered a successful placemaking intervention, insofar as it tunes into and relays a prevailing mood within the Iraqi community. For a public place to feel safe for a community, a certain degree of tolerance for the practice of parrhesia (speaking candidly) is necessary. This is the social problem brought to the fore by both al-Zaidi and al-Amiri: How can the Iraqi public (and political) sphere become one in which speaking candidly? And more importantly, how can political dissensus not only tolerated but actively encouraged and not repressed?

黑色的云
The Black Cloud

艺术家：希瑟、伊凡·莫里森（英国）
地点：英国布里斯托尔
推荐人：汉娜·皮尔斯
Artist: Heather and Ivan Morison（UK）
Location: Bristol, UK
Researcher: Hannah Pierce

作品描述

"黑色的云"是一座大型的亭子，由艺术家希瑟和伊凡·莫里森设计，与建筑系毕业生萨西·雷丁一起合作，临时安装在布里斯托尔的维多利亚公园里。

艺术家使用从威尔士植物园里运来的木材制造亭子，它的结构受到了世界各地建筑的影响，能够应对各种气候和紧急情况。鉴于如此的预言景象和环境问题，在莫里森的艺术实践中体现出了这些明显的主题。"黑色的云"设计成预想未来沸腾的布里斯托尔，在无情烈日的烘烤下变得干燥。亭子由特殊形状木制三角框架构成 152 个连续的环形结构所组成，螺栓连接在一起制造出了一种强烈弯曲的屋顶，和一个敞开中心，体现了印第安棚屋式的公共空间，有着亚马逊地区雅诺马阿米尔印度安人建造的循环结构。受到了能够承受极端环境条件下的乡土建筑的影响，这些木材经过了日式木炭屋的处理手法，木材经过了烧焦处理，产生了防护罩的作用。

"黑色的云"不仅融合了多种建筑风格，在功能上也是多样化的。这件作品同时也是雕塑装置供人凝视欣赏，也是一系列公共活动的平台。这个作品明确以时间为基础，展示为期四个月，由一系列艺术家带领下开展活动，还有公共讨论，亭子可供每个人使用，如果有人希望在凉亭里举办活动的话，免租借费。莫里森发动了一项活动，包括初始的建设活动，"黑色的云"成型了，举办了类似于抬高谷仓的传统仪式，艺术家带领一群志愿者提高亭子结构，通过公共宴会和音乐活动庆祝亭子的建成，体现了传统芬兰人的聚合和阿米什人抬高谷仓的活动。在亭子里继续开展对话活动，讨论面对经济和气候变化下，人类的命运和未来的选择余地，如何在即将到来的坏年中蓬勃发展。论坛的谈话者有科幻小说作者，未来思想者和环境运动者。

最后《黑狗时代报》与艺术家及布里斯托尔的木偶公司远光灯视觉剧场一起合作，举办了一个冬季狂欢节式的唤醒活动，庆祝"黑色的云"开展的最后几天。其他活动包括一整天的音乐活动和篝火庆祝晚会，还有官方举办的活动，当地社团和居民经常使用这个亭子。

这个建筑结构能够鼓励人们进行表演和娱乐活动，正式开展的有舞蹈课、音乐演出或调解活动，或非正式的，观察过路人、牵狗人和孩子的反应。这个亭子的结构形成了自己临时的社区，范围广泛，涵盖着各类人群，他们受到了项目目标的激励，希望参与当地项目，为环境做出贡献。

解读

作品"黑色的云"建筑结构受到多种风格的影响，它的灵感来自亚马逊流域雅诺马阿米尔人部落的循环结构建筑。"黑色的云"除了作为建筑，还是一件供人观赏的雕塑作品，它具有多样的功能性。艺术家希瑟·伊凡·莫里森希望通过设计，在宽阔的公共区域的中心与周边，构建用于庇护人类远离未知灾难的生活空间，带有一定的科幻色彩和对未来的探索。"黑色的云"还将暂时性的搭建在社区和公园内，为当地的居民和各种团体免费提供开展活动的场地。艺术家试图激起一场激烈的争论——如何在繁荣未来，探寻一种解决方案，抵御我们将要面对的气候、经济的骤变，以及因此而产生的诸多不利因素，唤醒人们对自然和生态保护的重新认识，珍惜我们最美且不能失去的家园。（马熙达）

Artwork Description

The Black Cloud was a large-scale pavilion designed by the artists Heather & Ivan Morison in collaboration with architecture graduate Sash Reading and installed temporarily in Victoria Park, Bristol. Built with timber milled from the artists arboretum in Wales the structure draws influence from architecture across the world which respond to climactic scenarios or emergencies. References to such prophetic visions and environmental concerns are identifiable motifs across the Morison's practice. The Black Cloud was designed in anticipation of a future boiling Bristol, baked dry by a relentless burning sun. The pavilion is made up of successive rings of 152 uniquely shaped timber triangular frames, bolted together to create a strong curving roof with an open centre, reflecting the communal space of the shabono, a circular structure built by the Yanomamo Amer-Indians from the Amazon. Informed by the vernacular architecture built to withstand extreme environmental conditions, the timber underwent a Yakisugi treatment creating a scorched, protective shield.

The Black Cloud is not only hybrid in architectural styles but also in its function. The artwork was simultaneously a sculptural installation to be enjoyed and contemplated and a platform for a range of public events. The work was explicitly time based, activated over the course of four months by a series of artist-led participatory events, public discussions and available for use by anyone, free-of-charge if they desired to hold an event in the shelter. Events instigated by the Morisons included an initial building event, The Shape of Things to Come Barn-raising the Black Cloud, which involved the artists leading a team of volunteers to raise the structure and celebrate its arrival through communal feasting and music, reflecting the traditional Finnish talkoot and Amish barn-raising. The programme continued with a discussion held in the shelter on alternative futures and the fate of humanity in the face of climate and economic change, How to Prosper During the Coming Bad Years. Speakers at the forum included science fiction writers, future thinkers and environmental campaigners. Finally the Black Dog Times involved the artists collaborating with Bristol based puppet company Full Beam Visual Theatre to create a carnival-esque winter wake to celebrate the last days of The Black Cloud and the end of time. Other events included a day-long music event and Bonfire Night celebrations

Between officially programmed events, local community groups and residents regularly used the site. The architecture of the structure readily encouraged performance and play, either formally such as dance classes, musical performances or mediation, or informally, as was observed through watching the response of passers-by, dog-walkers and children. The structure formed its own temporary community made up of a broad and varied range of individuals, who were galvanised by the project aims and wanted to contribute to the local environment.

Artwork Excellence

The success of The Black Cloud was due to the relationships formed with the project's delivery partners. The collaboration of the artists with the architect and structural team resulted in an innovative design. The success of the events programme was similarly dependant on relationships with project partners who supported the artist-led events and the engagement of the local community who organized their own events in The Black Cloud. These partnerships not only fulfilled one of the main aims of the structure as an event platform but also brought in audiences who otherwise might not have experienced The Black Cloud.

The Black Cloud has inspired new thinking in relation to public art, such as tackling pre-conceived ideas that work should be a permanent fixture within our public spaces. Such projects challenge and initiate change in the way in which curators, urban planners, local authorities and other artists are working in public space. The project is a best practice model of combining visual experience with embedded participation while retaining a critical framework to situate the project. The value and place-making quality of the Black Cloud has been recognised regionally and nationally. Bristol City Council had hoped to purchase the work and install it in another park, acknowledging its strong local endorsement, but were unable to raise the capital funds. Since then the work has been recently re-erected to celebrate the opening of The Hepworth Museum, Wakefield, where is will remain until May 2012.

蓝屋
The Blue House

艺术家：珍妮·凡·希斯维克（荷兰）
地点：荷兰阿姆斯特丹艾瑟尔堡，房屋居住区 35 号别墅
推荐人：汉娜·皮尔斯
Artist：Jeanne van Heeswijk（Netherland）
Location：Villa in Housing Block 35, IJburg, Amsterdam,
The Netherlands
Researcher：Hannah Pierce

作品描述

"蓝屋"位于阿姆斯特丹艾瑟尔堡的一个住宅区中，这是一座人造岛，建造完毕后将增添 18 000 幢新房屋，容纳 4.5 万名居民，将于 2012 年完工。建造设计计划包括为这些居民使用的相关设施，但在过渡期间岛上居民还没有完备的设施。"蓝屋"坐落在居住区 35 街的中心位置，这里混合着社会部门的公租房和私有住宅。"蓝屋"因外墙为钴蓝色而得名，这个房屋被改建为文化活动、社区研究和艺术创作的公共空间，也是一个研究平台。这是一项持续性的项目，重点在于对人们使用、占领和改造公共空间的方式的探索和研究。

"蓝屋"房屋协会是一支来自当地和国际的实践者团队，包括国际策展人、艺术家、媒体活动家和社区成员。受邀者参加为期六个月的驻场项目，包括创作艺术品、拍摄电影、出版读物、举办展览及其他活动，相互之间积极对话并地与当地居民沟通。

"蓝屋"活动计划还包括为当地居民增添相关设施。当地一位图书管理员认为岛上缺乏儿童图书馆，他便与"蓝屋"合作发起了面向儿童的图书馆项目，每周开放一天。在这个项目的启发下，其他居民也开始与蓝屋协会合作，开办了一家公众图书馆，主要功能是在居民中间进行图书交流。其他合作项目还包括一个露天电影院和游船服务。通过这些活动逐步建立起新的基础设施，介入当地的日常生活。

解读

人类既是自然的产物，又是文化的产物，人类的发展离不开文化积淀与指引。作品"蓝屋"旨在为人工岛新建住宅区提供一个集文化、研究、发展为一体的社区试验中心，在严格规划外搭建一个可以自由交流的对话平台。邀请来自世界不同地区的文化先驱及学者到蓝屋居住、工作和交流，建立一项长期的文化交流机制，促使住宅区内的公众感受到世界先进思想、文化的勃勃生机。"蓝屋"作为媒介，令社区居民学习重塑公共空间，进而更好的发展社区自身的文化历史。"蓝屋"在构建社区文化平台的同时，给予大众各项福利措施，与居民合作，希望针对他们的需要和困难作出迅速直接的反应，令本来缓慢的过程加速，缓解因社区"硬件"发展过快与文化"软件"发展不足的不平衡关系。（马熙逵）

Artwork Description

The Blue House is located within a new housing community that is situated on a man-made island in Ijburg, Amsterdam. Upon completion in 2012 the community will include 18,000 new houses, accommodating forty-five thousand residents. The construction plans call for public facilities for the residents, but during the transition period there were no improved facilities ready for use by island residents. Blue House is located within a residential area comprised of 35 streets at the center of the community, and characterized by a mix of public and private housing. The building got its name because its exterior wall is cobalt blue. It has been converted for use as a public space for cultural activities, research and artistic creation, and also as a base for a research platform. This is an ongoing project, focusing on the use, occupation and the transformation of public space from an exploratory and experimental perspective.

Blue House housing association is comprised of a team of local and international practitioners including curators, artists, media activists and community members. Invitees participate in a six-month art residency, which may include painting and drawing, creating film, preparing a book for publishing, preparing exhibitions or other art activities, as well as interacting and communicating with local residents.

Blue House also plans to add facilities for local residents. After noticing there was no children's library on the island, a local librarian launched a children's library project in cooperation with Blue House, which is run every Wednesday. Inspired by this project, a group of other residents set up a book exchange program among residents in cooperation with the Blue House association. Other cooperation projects include an outdoor theater and a boat tour service. Through these activities an initial infrastructure has been created to positively affect local residents' daily life.

Artwork Excellence

Jeanne Van Heeswijk's long-term commitment to community organizing and social engagement, of which The Blue House is an exemplary model, was acknowledged in 2011 when she received The Leonore Annenberg Prize for Art and Social Change. The award recognises excellence in artistic practice that engages communities around critical public issues. The networked nature of Van Heeswijk's work allows it to transcend its immediate vicinity. In consequence, The Blue House not only stimulated debate about the quality and use of public space within its specific situated context, but on an international level among artists, architects and urban planning professionals, generating positive long-term impact both locally and further afield.

The Blue House demonstrates a unique and innovative model of an artist-initiated and led public project. The work was featured as a best practice model of durational public art work in the 2011 publication Locating the Producers, which stated that The Blue House "re-set the terms for experimental public art in this decade combining theoretical sophistication, practical ingenuity and a highly considered critical production methodology." The place-making ability of The Blue House is evident in the quality of exchange and collaboration between the practitioners and the IJburg community that was underlying in the successful of delivery of the project.

妥协屋
The ComproHouse

艺术家：曼斯 • 兰吉 / OMBUD 工作室（瑞典）
地点：瑞典斯德哥尔摩
推荐人：汉娜 • 皮尔斯
Artist: Måns Wrange/OMBUD（Sweden）
Location: Stockholm, Sweden
Researcher: Hannah Pierce

作品描述

在瑞典的摩尔特纳斯，"妥协屋"是一个正在进行的实验性房屋建造项目，吸收了瑞典历史及折中的建筑特点，探求愿望和现实的矛盾。此项目把"妥协"作为一个积极的、有益的概念进行探究，并据此而运用到项目的各个方面：从设计过程到它的审美、空间和社会各方面的综合结果。

"妥协屋"坐落于斯德哥尔摩郊区。和其他西方国家一样，在瑞典，小家庭是比较理想而普遍的。因此，在瑞典郊区的大部分住宅区的房屋为分离式。通过房屋特点创意性的阐述，"妥协屋"项目团队成功建造出一个复合的、折中于集体住宅和独身住宅的房屋。此屋是为社会某一类人而不是为传统小家庭设计的。它同时作为几个个体的私人房屋和一件公共艺术作品，并含有临时展览的小型画廊空间。它也可以是建筑意义上的雕塑作品；由于处于小山顶端，很远可以看见。

倾斜的屋顶和木质的覆层，房屋的形状是依照典型的瑞典独身屋所制。房屋有多而大的窗户，绿色的屋身，和山顶翠绿色的环境相融合。房屋的内部分割为数个私人空间，通过墙、地板和天花上多个迥异的缺口和相对公共的空间链接体现其空间分隔的创意性。

以下为"妥协屋"中体现折中建筑学与社会看法的一些举例：

"中间一半楼层"或"约翰•马尔科维奇楼层"：

"中间一半为楼层"为一个 2 米直径宽、1.6 米高的空间，此空间位于二楼的一张大餐桌上方；三楼的大工作桌下方。空间的相对下方部分由有开关控制的透明液晶玻璃做成，这样从下方餐桌处往上看感觉天花较高。当玻璃调控为磨砂的状态时，空间像一个大的光亮体对餐桌起到照亮作用。

旺德墙：

15 米长的墙沿着房屋的长生长，墙面被分割成一些有较小空间功能而组成的一个开阔空间，小空间隐藏有多个不同的房门、舱门和抽屉，就像一个书橱或一个大的日程表。

社交餐桌：

客人可以面对面地坐在这个圆桌的外面和里面的长椅上。桌子分为两个部分，里部缓慢旋转，这样客人慢慢地会被移动到可面对在桌子另外一边的新客人处。

旋转图书室：

环形的图书室整天地轻慢旋转。早上较早时面对着厨房开启，中午时面对客厅完全开放，傍晚半开着面对窗外风景，晚上面对墙壁时关闭。

解读

"瑞典模式"的一个突出特点是政治妥协和阶级合作，往往不采取激烈的方式。作品"妥协屋"吸收历史上瑞典的"妥协"特色，在设计领域积极探索，寻求一种融合审美、空间和社会需求为一体的设计解决方案。这种"妥协"的内容依然是为满足现实生活需要，采取一种折中的方式，在继承本身传统的形式基础上融合新的功能和视觉元素，在矛盾中寻求平衡和发展。实践性设计本身是在矛盾中寻求统一的困难任务，综合的过程需要时间去验证最后的成果，"妥协屋"追求相对公共空间分割的创意性和复杂性，外形与现代设计相比显得有些"畸形"，但随着时间和潮流的变化，不外乎这种建筑形式会与几百年前同样受嘲讽的巴洛克风格暗合，备受后世敬仰。（马熙逵）

Artwork Description

The ComproHouse in Mörtnäs, Sweden, is an ongoing experimental house project that draws on the Swedish history of architecture and compromise, and exploring the rift between aspiration and reality. The project investigates the concept of the compromise as a positive, productive principle, and accordingly applies it to every aspect of the house: from the design process to its aesthetic, spatial and social solutions.

The ComproHouse is situated in a suburb of Stockholm. In Sweden, as in most other western countries, the nuclear family is the ideal and the norm. Consequently, the building code of most suburban residential areas in Sweden only permits detached houses for one family. Via creative interpretations of the building code, The ComproHouse project group has managed to build a house that is a complex compromise between a collective house and a single-family house. The house is designed for social constellations other than the traditional nuclear family. It is simultaneously a private home for several individuals and a public artwork, containing a small public gallery for temporary exhibitions. It also functions as an architectural sculpture; widely visible due to its location on the top of a small hill.

With a pitched roof and wooden cladding, the form of the house refers to the archetypical Swedish single-family home. The house has multiple large windows and it is green in colour, allowing it merge with the verdant hilltop environment. Internally, the house is divided into a number of private spaces, which are connected to more public spaces through different openings in walls, floors and ceilings, as well as inventive room dividing systems.

Following are some examples of the compromises between architectural and social ideals made at The ComproHouse:

The Half-Size-In-Between-Floor, or "John Malkovich Floor"

The Half-Size-In-Between-Floor is a 2m diameter wide and 1.6m high room that makes use of the unused space above a large dinner table on the second floor and the space below a large working table on the third floor. When the lower part of the space, which is made of liquid crystal switchable glass, is transparent, there is an impression that the ceiling above the dinner table is higher. When the glass is frosted, the room functions as a large light source for the dinner table.

WunderWall

This 15m long wall runs along the length of the house. It divides the plan into one open space with a range of smaller room functions concealed by various kinds of doors, hatches and drawers like a curiosa cabinet or large advent calendar.

The Social Table

At this round dining table, guests can sit facing each other on benches on the outside and inside of the table. The table is divided into two parts, and the inner part of the table rotates slowly so that the dinner guests gradually will be facing new guests on the opposite side of the table as it turns.

The Rotating Library

The circular library room rotates slowly throughout the day. It opens early in the morning facing the kitchen, is fully open at midday facing the living space, is half open in the early evening facing the view, and closes at night facing the wall.

Artwork Excellence

The excellence of The ComproHouse has been recognised by the multiple private funders who have chosen to support the delivery of the project, in addition to the funding of the ecological building partners. The process of compromise is intentionally used as a creative method throughout the project to stimulate experimental and functional design. These compromises have resulted in new and unexpected solutions, which have formed the basis for the design of The ComproHouse.

The ComproHouse is situated in a suburban area where it would be unlikely for a publicly commissioned artwork to be placed otherwise. The project introduces innovative architectural, political and social ideas to a local audience, and contributes to the cultural life of Mörtnäs with a public gallery programme. The house itself is a striking architectural intervention within the landscape; the design referencing the history of Swedish architecture. The inventive design of the interior encourages a dialogue about how we use the space we inhabit on a personal level, and questions the building code that dictates the shaping of residential areas in Sweden, giving the project national, and indeed international, relevance.

厨师、农民、他的妻子和他们的邻居
The Cook, the Farmer, His Wife and Their Neighbor

艺术家：马杰提卡·波特尔克和怀尔德·维特森（斯洛文尼亚）
地点：荷兰阿姆斯特丹新西部罗德维克·凡·德赛尔斯拉特 61 号
推荐人：吉乌希·切克拉
Artist：Marjetica Potrč and Wilde Westen(Slovenia)
Location：Lodewijk van Deysselstraat 61,
New West Amsterdam, Amsterdam, The Netherlands
Researcher：Giusy Checola

作品描述

这个项目融合了社会结构和视觉艺术的因素，重新定义了绿色村庄的概念。该地区在二战之后得到兴建，今天是欧洲重新发展的最大型居住区之一。但是在 20 世纪 80 年代，这里曾一度成为无人之地。2004 年，城市把所有权转让给了房屋公司，但该公司在这块敞开的空间上看不到价值。最终导致一群年轻的艺术家、设计师、建筑师和社区居民合作，回归粮食生产。

在这个地区，农业和烹饪被人们视为分享知识和传统的方式，作为一种文化复兴和居住区重生的手段。在项目发展的一年中，居民们自己成为了参与项目的最主要生力军。2009 年 9 月，他们接管了管理工作，组成了一个

8 人委员会，负责两个空间，负责开放花园以供生产，敞开厨房烹制食物。这些空间也用作工作坊的场所，给当地居民们开展文化项目。在社区周边建立了一个中心，社区居民可以参与"建立一个场所"的整个过程。社区厨房给街道带来了安全感，这是项目给社区居民带来的另一个价值所在。

这个项目证明，居民不仅渴望参与设计他们的城市，而且这一渴望完全是可以实现的，他们有能力推进项目的进展，使公共空间永久地从企业控制下半自主的空间转换成一个充满活力的公共社区。这个过程不仅赋予社区居民以力量，还激发了他们的权利意识，而且也对其他社区产生影响。

解读

"厨师、农民、他的妻子和他们的邻居"是一个协同合作项目，艺术家们希望将设计的主体任务托付给当地社区居民，艺术成为公众改善生活环境和追逐幸福生活的媒介，借此恢复当地的人文气息。这个社区的居民来自22个民族，在没有任何雇佣关系的前提下，自发地参与到艺术项目中，他们通过"集体厨房"、"社区花园"的形式分享劳动果实，合理地利用了有限的空间，开放性地将社区居民凝聚起来。"厨师、农民、他的妻子和他们的邻居"为公众提供了一个共同参与、真情分享的场所，加强了公众交流和文化融合，启示人们通过参与公共生活增进彼此的联系。在快节奏的都市生活中，互信的邻里关系是一种奢侈，"厨师、农民、他的妻子和他们的邻居"是增强公众互信、启示人们善待彼此的一剂良药。（马熙逵）

Artwork Description

This project combines elements of visual art and social architecture to redefine the village green. New West Amsterdam was built after World War II and today constitutes one of the largest residential 'urban renewal' projects in Europe. During the 1980s, however, the district became a no man's land, and in 2004 the city transferred ownership to housing corporations. There was no perception that the district's open space had any value, however, until a group of young artists, designers, architects and community residents began a back-to-the-earth farming effort.

The participants view farming and cooking as a way for people to share knowledge and traditions, and as a path for cultural renewal and neighborhood rebirth. During the year in which the project developed the residents themselves became the most important people involved, and in September 2009 took over its management. They created a committee of eight people responsible for the kitchen and garden, and committed to actively continuing food production and cooking. In addition, the space serves as a location for workshops and neighborhood cultural programs. Altogether it functions as a core around which the community can engage in the process of "place-building" . A key value of this project is the sense of security that the community kitchen has brought to the residents.

The project demonstrates that the self-determination of residents in guiding the character of their city is not only desirable but feasible, even without their having any formal control of space. By reclaiming public space as part of a vibrant community, the participants not only endowed themselves with power and aroused their awareness of their rights, but empowered other communities to act in kind.

Artwork Excellence

The Cook, the Farmer, His Wife and Their Neighbor was selected for the 4th International Architecture Biennale Rotterdam (IABR) Open city: Designing coexistence – Squat city at the NAI from the 25th Sep 2009 to the 10th Jan 2010. It won third prize in the Squat City Competition. More importantly, because of this project, public space has been permanently transformed from a corporate-controlled, semi-private space into a vibrant community commons. In doing so, it empowered residents of the neighborhood, who have carried the project forward, and other communities who hope to replicate the results in their neighborhoods.

福克斯通美人鱼
The Folkestone Mermaid

艺术家：科妮莉亚·派克（英国）
地点：英国福克斯通
推荐人：汉娜·皮尔斯
Artist: Cornelia Parker（UK）
Location: Folkestone, UK
Researcher: Hannah Pierce

作品描述

受到了哥本哈根小美人鱼的影响，科妮莉亚·派克的"福克斯通美人鱼"是一尊真人大小的铜制雕像，坐落于一块岩石的顶部，能够俯瞰福克斯通阳光明媚的沙滩。在"2011年福克斯通三年展"上，这件作品被第二次安装在了临时的底座上。但如今已筹集到资金，足够使作品永久安放在福克斯通了。

裸体雕像有着独特的美人鱼姿势，座落在一块大岩石上望向海平面。她有一双腿而非鱼尾，那纤纤的海藻轮廓嵌入她的长发一直垂到脚面。

这件作品坐落于阳光沙滩的尽头，之所以选择这个地点是为了展示哥本哈根海港的小美人鱼姿态。这个海滩非常靠近福克斯通的海港，是这个小镇唯一较长的沙滩延伸处，在夏季吸引了大量游客，同时也引来很多人在雕塑前驻足欣赏。"福克斯通美人鱼"的计划构思于2009年，那期间哥本哈根的气候会议正在召开。艺术家借用了她以前作品中相同的气候问题为主题，比如2007年的作品"乔姆斯基抽象"，让美人鱼望向大海，凝视着低垂的水平线。从这一层面上来说，"福克斯通美人鱼"是一位守护者，她静静地观察着潜在的涨潮，表达了对潜在的气候问题的忧虑。

在画家简介中，派克建议使用当地妇女的形象用作小美人鱼的形象。2010年艺术家邀请福克斯通的全体妇女参与小美人鱼形象的公开竞选。于是在全镇散发了广告，呼吁潜在的小美人鱼们把自己身穿泳装的照片提供给艺术家挑选。唯一的参选规则是模特必须18岁以上，其形象不要和安徒生童话里的浪漫形象一样，她应该是更为现实的女性形象的代表，能够得到广泛的认同。照片提交过程开始之后，派克说道："这不是选美比赛。我不是在寻找理想化的哥本哈根小美人鱼，而是在寻找一个真实的人，一种自由的精神，所以不管怎样的体型与尺寸，都欢迎参加！"在50名候选人中，两个孩子的母亲，出生并仍旧生活在福克斯通的乔治娜•贝克被艺术家选为福克斯通美人鱼的原型。

福克斯通美人鱼项目是当地人的。虽然这尊雕塑与哥本哈根美人鱼有着类似的姿态，但福克斯通美人鱼更多地参考了H•G•威尔斯写的中篇小说中的当地人形象《海上夫人》，当时作者生活于福克斯通的沙门地区。进一步对比小美人鱼，本件作品是真人大小的当地母亲雕塑。真人雕塑铸造工艺使作品达到那种逼真效果是雕刻所达不到的。这件雕塑在一家当地的小型铸造厂里铸造，熔解于拉姆斯盖特，那里此前没有过制作如此规模作品的经验，但是厂家与艺术家和三年展的组办人员密切配合，成功提交了这件作品。

解读

传说美人鱼是以腰部为界，上半身是美丽的女人，下半身是披着鳞片的漂亮的鱼尾，整个躯体，既富有诱惑力，又便于迅速逃遁。而"福克斯通美人鱼"没有尾巴，造型接近现实中的女性，这种造型没有盲目追求理想化的哥本哈根小美人鱼形象，而是在寻找一个真实的人，一位坚实的守护者。雕塑作品的原型来自于当地的成年女性，为了挑选合适的模特，艺术家向全镇女性发出了邀请，这增强了作品的当地居民认同。艺术家在构思"福克斯通美人鱼"时，正值哥本哈根世界气候大会召开，会议希望人类的工业化发展应控制温室气体的减排，否则随着全球温度上升，冰川湖水冲破堤岸，将对人类造成灾难性的后果。"福克斯通美人鱼"代表着一种母亲的期盼，她静静地观察着海面潜在的涨潮，表达一种对未来气候问题的担忧。
（马熙逵）

Artwork Description

Drawing influence from Copenhagen's The Little Mermaid, Cornelia Parker's The Folkestone Mermaid is a life-size bronze body cast perched on top of a rock overlooking Folkestone's Sunny Sands Beach. The work was installed on a temporary basis for the second instalment of Folkestone Triennial in 2011, however funds are currently being raised to acquire the work for Folkestone permanently.

The nude statue is perched in the distinctive mermaid pose, on a large rock looking out towards the horizon. She is cast with legs rather a tail and subtle outlines of seaweed are laid into her hair and draped across her feet. The work is located at the end of Sunny Sands beach, which was chosen to reflect The Little Mermaid's position in Copenhagen harbour. The beach is in very close proximity to Folkestone's harbour and is the town's only long stretch of sand, attracting large numbers of visitors during the summer months and high footfall for the sculpture. The proposal for The Folkestone Mermaid was developed in 2009, during which time the Copenhagen Climate conference was taking place. Drawing upon the artist's ecological concerns that can be indentified as a motif in her previous work, as in Chomskian Abstract (2007), the mermaid looks out to sea, gaze falling upon the line of the horizon. The Folkestone Mermaid is a guardian in this sense, quietly observing the potentially rising tides and the potential climatic concerns that this presents.

In response to the artist brief, Parker proposed to use a local woman to model as the mermaid. In 2010 the artist invited all of the women of Folkestone to apply to be cast as the mermaid through an open submission. Advertisements were distributed throughout the town appealing to potential mermaids to send images of themselves in swimwear to the artist for consideration. The only specification was that the model was over the age of 18, and not of the romanticised figure of Hans Christian Andersen's fairytale, but a more realistic representation of a woman who could be more widely identified with. When the submission process was launched in Parker stated, "This is not a beauty contest. I am not looking for a look-alike of the idealised Copenhagen Mermaid, but for a real person, a free spirit, so any shape or size welcome!" From the 50 applicants, Georgina Baker, mother of two, born and living still in Folkestone was selected by the artist to be cast as The Folkestone Mermaid.

The Folkestone Mermaid is a celebration of the local. While the sculpture shares the pose of its Copenhagen counterpart, The Folkestone Mermaid makes reference to a more local literary figure, The Sea Lady, a H.G. Wells novella written by the author while he was living in the Sandgate Area of Folkestone. In further contrast to The Little Mermaid this is a life-size sculpture of a local mother. The life-casting process lends the work a level of verisimilitude impossible to achieve through carving. The sculpture was cast at a small local foundry, Meltdowns in Ramsgate, who had no prior experience of producing works of this scale but worked closely with the artist and the Triennial to deliver the project.

Artwork Excellence

Since its installation in Spring 2011 the work has quickly established itself as a landmark sculpture and an emblem of Folkestone. As a contemporary artwork, the sculpture attracts a number of national and international visitors, while the high footfall of the location makes the work available to numerous beach goers who are often keen to be photographed alongside the work. The Folkestone Mermaid was a widely used image in press coverage of the last Triennial, and is one of many new works that are contributing to a new legacy of cultural excellence in Folkestone. In response to the positive responses of local residents who identified with the local significance of the project, as well as the wider audience and press, the Folkestone Triennial and the artist are working to ensure that the work is installed on a permanent basis on Sunny Sands Beach.

森林人之家
The Forester's House

艺术家：西蒙·帕特森（法国）
地点：法国奥尔市
推荐人：薇拉·托尔曼
Artist: Simon Patterson（France）
Location: Route départementale of Ors，France
Researcher: Vera Tollmann

作品描述

位于法国北部城市奥尔的"森林人之家"是第一次世界大战的最后一次行动中，士兵和诗人威尔弗雷德·欧文试图穿越桑布尔—瓦兹运河（1918 年11 月 4 日）时牺牲前渡过最后一晚的地方。欧文现长眠于奥尔社区公墓中的英联邦区域。虽然鲜有法国人知道欧文，但在英国他仍被认为是那场战争十分重要的见证者。他的文字和信件强调了战争的"野蛮荒谬性"。

奥尔市长捷克·杜米尼注意到大量的英国游客前来寻找欧文的墓碑并要求参观森林人之家的地窖，因而对此产生了兴趣。英国有一个威尔弗雷德·欧文之友协会，于是法国也成立了该协会的一个分支。

杜米尼连同 Haute Sambre and Bois l'Evèque 的区议会主席马克·杜夫海纳，翻译欧文作品的泽维尔·韩诺特，法国威尔弗雷德·欧文协会主席贝诺·米松以及奥尔市居民希望在当地建立一个兼具艺术性和文学性的项目来纪念欧文，目的在于建立由地方当局和广大民众支持的政治性项目，同时能够促进当地经济和旅游业发展。为实现这个目标，法国威尔弗雷德·欧文协会与艺术链接机构合作，该机构致力于当代艺术的创作，并充当法国基金会发起的新赞助人计划的中间人。

该组织决定做成一个当代作品，而非一个博物馆或者纪念馆。2004年，英国艺术家西蒙·帕特森受到邀请接受该项目的委派。在奥尔市市长以及民众的支持下，他提出要对这栋森林中的房子加以关注，赋予这栋废弃的屋子以新生，不仅要把它建成向欧文致敬的地点，还要让它成为诗歌之地。

法国建筑师让·克里斯托夫·丹尼斯将艺术家的设想变为了现实。房子的外观保持不变，但墙壁和屋顶被改成了白色。艺术家要使这栋房子在黑森林的映衬下凸显得像一堆"褪色白骨"。

房子集雕塑、视觉艺术和声音作品于一身：屋顶被设计成一本打开书的形状，而由玻璃制成的"页面"可以最大限度的将日光投射到室内，而内部空间的墙壁上则是威尔弗雷德·欧文作品的动画投影。一个圆形坡道将参观者引入地窖，那里是欧文向母亲写下最后一封信的地方。地窖是作品中唯一属于原建筑的部分，艺术家完全保留了它的原样。

该项目于2011年11月4日正式对公众开放，这个日子是欧文去世93周年的纪念日。同时西蒙·帕特森也制作了欧文的"名字画"以及限量版的战争诗歌印刷集

解读

作品"森林人之家"是为纪念英国诗人雷德·欧文而设立的纪念馆，馆址所在地便是雷德·欧文牺牲的地方。雷德·欧文是一位富有同情心的诗人及军人，他的文字描述了士兵在战争中的经历，回忆战争的残酷以及它带给人们的噩梦。"森林人之家"本是一栋因年代久远而遭废弃的屋子，地方当局希望通过艺术改造使这里成为纪念诗人欧文的地点，同时能够促进当地经济和旅游业的发展。纪念馆中没有关于战争的文物、武器，只有一个房间和一屋子的诗歌。纪念馆希望通过诗人的文字来强调战争的野蛮、荒诞，人们在聆听诗歌的同时，沉思战争的徒劳和生命的可贵。纪念馆的墙壁和屋顶被改成了白色，不仅是美化，更是一种告诫，它时刻提醒着人们战争带来的只有混乱和悲剧。（马熙逵）

Artwork Description

The Forester's House (La Maison Forestière) at Ors in Northern France is where the soldier and poet Wilfred Owen spent his last night, before he was killed attempting to cross the Sambre-Oise canal on November 4, 1918, in one of the last actions of World War I. He now lies in the British Commonwealth section of the cemetery of the community of Ors. Still fairly unknown in France, Owen is considered in Britain a fundamental "witness" of the war. His texts and correspondence underline its "barbaric absurdity."

Having noticed that a great number of British visitors came looking for Owen's tomb and asking to visit the cellar of the forester's house,

Jacky Duminy, the mayor of Ors, became interested in Wilfred Owen. An association of Friends of Wilfred Owen existed in England, so a branch of the association was created in France.

Duminy—together with Marc Dufrenne, Chairman of the District Council of Haute Sambre and Bois l'Evèque; Xavier Hanotte, translator of Owen; Benoît Misson, Chairman of the French Friends of Wilfred Owen association; and the inhabitants of Ors—wanted to put an artistic and literary project in place that would commemorate Owen. The aim was to come up with a political project supported by the local authorities and the general population that also helped local economic development and tourism. To bring this about, together with the Association Wilfred Owen France they turned to artconnexion, an organization devoted to the production of contemporary art and mediator for the New Patrons initiative of the Fondation de France.

The organizations decided on a contemporary work, rather than a museum or memorial. In 2004, the British artist Simon Patterson was invited to respond to the commission. He proposed, with support of the mayor and the village of Ors, to focus on the forester's house, giving the disused building new life both as a hommage to Wilfred Owen but also as a venue for poetry.

The artist's design for The Forester's House was realized by the French architect Jean-Christophe Denise. The exterior of the house remains the same, but the walls and the roof were turned white. The artist's idea was for the house to stand out like a "bleached bone" against the dark forest.

The house is simultaneously a sculpture, a visual work and a sound piece: its roof is remade to represent an open book, the "pages" constructed out of glass to admit maximum daylight into the interior, while the internal space is filled with animated projections of texts by Wilfred Owen on the walls. A circular ramp leads down to the cellar where Wilfred Owen wrote his last letter home to his mother. It is the only original part of the building and the artist has preserved it exactly as it was.

The project opened on November 4, 2011, the 93rd anniversary of Owen's death. Simon Patterson has also produced a "name painting" for Wilfred Owen and a limited edition portfolio of prints, featuring the war poets.

Artwork Excellence

The Forester's House challenges the concept of a memorial. Patterson found a unique form to commemorate Wilfred Owen's work, in particular with the form of an open "book." The project was initiated by the mayor and the inhabitants in a small village in the North of France and realized in the framework of the New Patrons program.

觉醒的曼西亚幽灵
The Ghost of Manshia Awakes

艺术家：罗能•埃德尔曼（以色列）
地点：以色列边境
推荐人：里奥•谭
Artist: Ronen Eidelman (Israel)
Location: Border of Tel Aviv and Jaffa, Israel
Researcher: Leon Tan

作品描述

罗能•埃德尔曼带来的"觉醒的曼西亚幽灵"是一个为 Jaffa 和 Tel Aviv 地区特别设计的艺术项目。其目的是干预该地区所相对应的文化记忆。罗能•埃德尔曼在接受了 Ayam 协会组织关于当地市政项目——名为"城市自传"的访谈后，着手创立此艺术项目。该项目于 2007 年 9 月犹太住棚节期间开展，持续了超过 4 天。罗能•埃德尔曼和他的团队利用足球场标记设备和油漆滚筒，标示出曼西亚旧区的所在地，在这个区域一直到 1947 年为止，曾存在过一个阿拉伯居民区。就这样，一直好几周，曼西亚的旧街道便再一次呈现在 Charles Clore 公园的长廊与草坪上。

根据罗能•埃德尔曼所言，虽然许多 Tel Aviv 的居民相信他们所生活的城市是几个世纪前从沙土形成的现代希伯来人城，而实际上，这里却有着很长的阿拉伯文化历史。所以这个项目的目的是以通过不破坏人们现有生活活动的方式，展现出此地的真实历史。正因为如此，罗能•埃德尔曼回避建造纪念碑和传统的考古方法，受 Sharon Rotbard's 的著作《黑城，白城》启发，选择了一个极其短暂的理性的方式，而非庞大宏伟宣言式的方式去纪念这段历史。

罗能·埃德尔曼在项目进程中遇到很多居民曾与老曼西亚及其周边地区频繁活动。他们中也包括在 1947 年因为政治环境和战争离开该地区巴勒斯坦人和犹太人谁，以及 Etzel（伊尔贡）退伍军人，曾负责征服曼西亚并驱逐当地阿拉伯人（此人为前军事组织成员。该组织由英国授权巴勒斯坦进行活动）。虽然 Etzel 老兵有所反对，但整个项目还是得到了很大的支持和鼓励，从而引发了该地区的回忆叙述。值得注意的是，该项目设立是为了方便随机到来的游客了解该地区，而不是一个为招揽观众计划周详的艺术展览（有那些各种各样的展览等节目的艺术展）。以这样简朴的方式，罗能·埃德尔曼以意想不到的方式悄无声息地介入到当地人和游客的日常生活中去，虽然他们的记忆正在突然被当今看似应接不暇的忙碌生活快速埋葬。

这个短暂的艺术项目值得一提，并不仅仅是因为其精致入微的策划，更因为这个项目突出了这样一个事实：边缘和界限会持续不断地被建立和重建。这正如世界从历史角度看，所有经济、民族、各种联合体这样的社会实体以及他们所在的领地都只不过是稍纵即逝的。罗能·埃德尔曼让我们关注到这些边界从长远来看是怎样自然而然变化转换的。通过这样的方式，他将所观察到的那些僵硬的国界，种族和宗教的界限等视作有待解决的问题呈现出来，并建议不同的社区更多地相互渗透共存。不要再让民族主义和宗教原教旨主义及其在历史上的"划分国界"事件升温。在这里，他致力向人们演示如何用艺术化的形式去创造地点。这样做，一方面能将敏感的现实生活展现在大家眼前，另一方面也可以向人们建议另一种理念——一种分享和创造"地点"的观念。

解读

作品"觉醒的曼西亚幽灵"采取一种缓和的方式，悄无声息地介入到当地人和游客的日常生活中，表达一种对社会实体此消彼长的思考。艺术家借用足球场标记线，勾勒出曼西亚原有的房屋、街道网络，让人们铭记这片土地上被"遗忘"的房屋和文化。此项计划希望通过不破坏人们现有生活活动方式，展现此地的真实历史——在此曾经存在过一个阿拉伯居民区，由于政治环境的改变，原有的阿拉伯人已被驱逐。这种区域的标记，是世界经济、民族、各种联合体以及他们领地此消彼长的一个缩影。艺术家承认区域现状符合历史发展的必然性，但同时暴露出种族和宗教的界限等视作有待解决的问题，他呼吁不同文化、宗教、种族的人群应当花香渗透共存，不应再因民族主义和宗教原旨主义引发的争端升温。（马熙逵）

Artwork Description

Ronen Eidelman's The Ghost of Manshia Awakes is a site-specific intervention into the cultural memory of the region of Jaffa and Tel Aviv. Eidelman created it after the Ayam Association for Recognition and Dialogue invited him to take part in the The Autobiography of a City project. Taking place over four days in September 2007, during the Jewish holiday of Sukkot, Eidelman and his team used soccer field marking equipment and paint rollers to demarcate the older Manshia Quarter, an Arab neighborhood that existed until 1947. Thus, for a number of weeks, the former streets of Manshia were made visible again on the Promenade and lawns of Charles Clore Park.

According to Eidelman, while many residents of Tel Aviv believe that the city emerged out of the sands a century ago as a "Modern Hebrew

City," it is in actual fact a place with a much longer Arab history. The objective was to portray this history without at the same time destroying the lively present day activities of the site. For this reason, Eidelman eschewed both monumental and archaeological approaches, opting for an ephemeral and relational work inspired by Sharon Rotbard's White City, Black City that would engage in cultural memory making without grand statements of historical truth.

During the course of the project, Eidelman encountered many residents with active relationships with the old Manshia quarter and its surrounds: Palestinians and Jews who left the area in 1947 because of political circumstances and fighting, as well as Etzel (Irgun) veterans (former members of a military organization active during the British Mandate of Palestine) responsible for the conquest of Manshia and the expulsion of Arabs. With the exception of the Etzel veterans, the project on the whole received a great deal of support and encouragement, eliciting the narration of memories of the area. Significantly, the project was set up to facilitate random encounters with visitors to the area, and not as a special art event to which audiences would come (expecting an exhibition or performance of sorts). In this way, Eidelman's intervention functioned as an unexpected event in the everyday lives of visitors and residents, a sudden quickening of memory buried beneath the preoccupations of the present.

The ephemeral nature of Eidelman's project is worth mentioning, not only for its strategic subtlety, but also because it highlights the fact that borders and boundaries are continually being made and remade. Considered across the longue durée (long duration) history of world-economies, social entities such as nations and communities—together with their territorial markers—are but momentary fluxes. Eidelman draws our attention to the transitional nature of borders by taking the perspective of a longer duration. In so doing, he problematizes the perceived rigidity and permanence of state, ethnic and religious borders, suggesting the alternative of more permeable divisions between communities, concomitant with a rejection of the fevers of nationalism and religious fundamentalism characteristic of histories of state-making. Here he demonstrates a form of artistic place-making that is on the one hand sensitive to present day realities, and on the other, suggestive of different conceptions of the sharing and making of places.

Artwork Excellence

Ronen Eidelman's project stands out as a form of artistic place-making that is on the one hand sensitive to present day realities, and on the other suggestive of different conceptions of the sharing and making of places. It draws our attention to the violence of state-making (in this case Israel) and the accompanying repression of cultural memory related to the Arab quarter of Manshia. As a work that intervenes directly in repressed cultural memory, within a politically fraught site, it is commendable for its subtletly and tact, and excellent insofar as it garnered a great deal of public support and engagement with place.

热土
The Promising Land

艺术家：“空间实验室”（德国）
地点：英国利物浦布特尔
推荐人：汉娜·皮尔斯
Artist: Raumlabor (Germany)
Location: Bootle, Liverpool , UK
Researcher: Hannah Pierce

作品描述

"热土"由利物浦双年展委派，属于"都市 09"项目的一部分，"都市 09"是为期五天沿利兹 - 利物浦运河举行的节日，这期间分享了艺术家对这一地区进行的为时一年的进驻、干预和研究的成果。

"热土"项目在运河的两岸延伸开来：分别位于废弃的圣·怀恩弗里德和圣理查德小学和邻近的卡罗莱纳码头。经过对两个地点的仔细研究，空间实验室决定对这两处被忽略的公共空间重新想象和再创造。实验室非常重视当地居民的共同参与，空荡荡的学校被改造成了一座都市世外桃源，成为了"都市 09"节日的中心点。这个项目涉及以下这个问题："当目前这个被普遍接受的历史对现状没有帮助时，将一个历史悠久的地点重新改建，是一个好主意吗？""热土"这个标题就是指这个地区，它认可了对当地居民做出的各种有关开发当地潜力的承诺，而之前这些是从未付诸实现的。

空间实验室为运河两岸的地点制造了一个新的传奇，居民可以在校内的各个地点、场地和水道对面的码头穿行。有这么一个传说，说是大约在 1813 年时，利物浦勋爵打算委派建筑师威廉·纳什设计一座公园来纪念乔治四世的妻子——布朗斯维克 - 沃尔芬比特尔的卡洛琳公主和国王已故母亲的印度遗产继承者。因婚姻的解除导致了政治上的失败，建造卡洛琳公园的计划从未得到实施。后来这个传说通过《切割》杂志的报道在整个南赛弗顿地区传播开来。《切割》是利物浦双年展出版的分发给当地 40000 个家庭以推进都市化活动的一份刊物。借助"热土"的宣传，空间实验室通过把过去与现实、真实与都市传奇地融汇交织，建成了当初美梦未能成真的卡洛琳公园，把在残破的运河上流传这一未竟的历史幻想了广大观众。

"热土"的入口处是一个根据很多元素建造而成的大型迷宫，将观众带入改建后的圣·怀恩弗里德。这所学校如同展示这个传奇的博物馆，展示了地图、历史绘画和传说中的人物画。观众被带领着走过桥梁，走到观景窗和放映屏幕前，再到室内重建的运河景点处，那里有英格兰和印度的微雕模型，由水道隔开。离开学校穿越了运河后，参观者到达了严塔尔·曼塔尔，这个故事的尽头。这个真人大小的木结构作品是重建的印度赤道日晷，它是该地区新历史的巅峰式里程碑，只有坐船才能到达那里。

严塔尔·曼塔尔是"热土"的轴心，从迷宫的入口处到背面墙壁处的观景窗，它标志着故事的结束。要到达严塔尔·曼塔尔只有亲自穿越水道，这个过程中需要观众亲身与运河进行互动，以此实现该项目和更广大范围的都市化项目的目标。空间实验室还把学校大厅的重新设计纳入了这个项目。大厅成为了一个功能化的会议中心，利物浦双年展邀请参与贡献的艺术家和演讲者来此地召开座谈会，主要探讨如何在再建过程中充分利用当地资产。

解读

随着时代更迭，地区经济、文化中心也随之变迁，很多历史上繁华的地域也因自身落后的工业步伐变得相对贫困，利物浦的利兹市运河便是其中典型之一，这里在大约 40 年的疏忽下，已变成市中心平民区的荒地。"热土"项目旨在重新规划运河两岸的公共空间，将原有的废弃土地改造成都市中的"世外桃源"。艺术家们重视当地居民的共同参与，认为居民的主动参与才能实现项目的最大潜能。他们充分考虑了河岸功能性的丰富，以重塑历史为目标为公众搭建各种具有故事情节的建筑。这不仅是对区域地理旧貌的改造，更是一场鼓舞人心的思维革新运动。"热土"的实现，唤醒了当地居民以往的自豪感，重塑了他们作为主人翁的自信。（马熙迏）

Artwork Description

 was transformed into an urban arcadia that became the hub for the Urbanism festival. This project was approached with the question: "Is reinventing the history of a place a good idea, when the current accepted history isn't helping the present situation?" The title The Promising Land, refers directly to the site, recognising the multiple promises made to the residents about its potential that had never materialised.

Raumlabor established a new legend for the canal-side site that viewers engaged with by venturing through multiple stations within the school, its grounds and the wharf on the opposite side of the waterway. The story was told that around 1813 Lord Liverpool was to have commissioned the architect William Nash to design a park in honour of Princes Caroline of Brunswick-Wolfenbüttel, the wife of King George IV, and the Indian heritage of his late mother. Due to the political fall after the dissolution of their marriage, Carolina Park was never built. The legend was disseminated across South Sefton through the "The Cut," a publication published by Liverpool Biennial and distributed to 40,000 homes in the area to promote Urbanism activities. Through The Promising Land, Raumlabor recreated the unrealised Carolina Park through interweaving fragments of past and present, truth and urban legend, to deliver to the audience an unfulfilled historical fantasy across the ordinarily neglected and dilapidated canal.

The entrance to The Promising Land was via a large-scale labyrinth built from hoarding elements that led into the transformed St. Winifrede's. The school functioned as a museum to the legend, displaying maps, historical drawings and images of the characters to illustrate the tale. The audience were led over bridges, to viewing windows and film screenings, and alongside an indoor reconstruction of the canal site, with miniature models of England and India separated by the waterway. After leaving the school and crossing the canal, visitors reached the end of the story at the Jantar Mantar. This wooden structure, a life-size reconstruction of the Indian equatorial sundial, was the pinnacle landmark in the area's new history and accessible only by boat. The Jantar Mantar articulated the whole axis of the The Promising Land, from the entrance in the labyrinth to the viewing window of its back wall, which signified the end of the tale. As it was only possible to reach the Jantar Mantar by physically crossing the water, the process necessitated the viewer's physical interaction with the canal, fulfilling the aim of the project and the wider Urbanism programme.

Raumlabor also incorporated the redesign of the school hall into the project. The hall became a functioning conference centre from where a series of talks by contributing artists and a conference of speakers invited by Liverpool Biennial addressed issues of how to use local assets in regeneration.

Artwork Excellence

The success of Raumlabor's ability to implement artistic strategy in redefining urban space is endorsed by Liverpool Biennial's invitation to participate a second time in Urbanism 09 following their initial commission a year earlier. The Promising Land was well received by both local residents and visitors who saw the disused school and neglected canal transformed beyond recognition by intelligent, imaginative storytelling delivered though subtle and innovative design. Through physically and cognitively engaging the audience with a possible history of the site, viewers were able to rediscover pleasures alongside the canal, a site to which access is ordinarily restricted. British Waterways reported that anti-social behavior along the canal was reduced by 100% during the Liverpool Biennial's periods of activity there, to which Raumlabor made a substantial contribution. Following its installation as part of The Promising Land, the Jantar Mantar structure was commissioned several months later by The Forestry Commission for a one-day installation in nearby Southport to launch a new community woodland site.

贝鲁特的天空：城市徒步之旅
The Sky over Beirut: Walking Tours of the City

艺术家：托尼·恰卡尔（黎巴嫩）
地点：黎巴嫩贝鲁特
推荐人：里奥·谭
Artist: Tony Chakar (Lebanon)
Location: Beirut, Lebanon
Researcher: Leon Tan

作品描述

　　"贝鲁特的天空：城市徒步之旅"由黎巴嫩建筑师托尼·恰卡尔在 2010
年创作，属于家庭作业 5 个项目的组成部分，策展人是克里斯丁·托姆，
来自阿斯卡尔·阿尔万——黎巴嫩造型艺术协会。项目的灵感发源于有一
次恰卡尔在街边发现了一本笔记本，上面画了一条线条，笔记本上的内容
明显是一位建筑师写的："杰克逊·波洛克的画是为假想城市做的真实计
划。"在"贝鲁特的天空：城市徒步之旅"里，恰卡尔重新评估了城市，
对比着作者的文章和图纸，画了一条删除线，源于法国的杂志《巴黎匹配》，
曾经这样描述着贝鲁特——"地中海沿岸最丑陋的城市"恰卡尔让自己的
位置更明确，他说道："建筑师质疑开放的审议和问题，《巴黎匹配》的
记者被迫使结束这些议题，或者说是从他的观点中闭锁了。"

恰卡尔设计了他的旅途路线，不是一种传统的导览，而是作为特质边缘心理的冒险，结合个人轶事和历史信息的记忆。如此一来，他避免封闭的新闻报道，以及史学的类似形式，与他们的"行李"一起（提升了欧洲裔美国人的教规，达到美和成熟的巅峰）。这为游客提供了机会，超越了对于美和丑的评判，美和丑，为了吸取恰卡尔穿越贝鲁特之旅的意向和主观方面，去理解贝鲁特这座城市是如何成型的，如今又是怎样的风貌。恰卡尔自我的位置如同是统一理论和统一城市的敌人。通过他的旅途，他把主体看做是不固定、不稳定的物体，也包括人类，强调主体而不是作为一个零碎的组合。

主体"不再局限于可辨识的人类，而是包罗所有的生物存在，从胚胎到怪胎；它的力量不再存在于统一的模型，而在于碎片、肢解，打破的暗示"。当然主体从来不只是人类所有，但存在于有组织的社会整体，有着跨越时空的规模，从微生物到人类，到城市，甚至到整个世界经济。一个主体的部分，它的器官逐渐地衰竭死去，同时也在再生和重组。恰卡尔的旅途并不是为了呈现对于贝鲁特连贯的陈述，而是提供城市的片段，针对游客们，也愿意这样。城市片段，像游客的片段，如同一个组合，他们的联系和成果难以预料。

托尼·恰卡尔的"贝鲁特的天空：城市徒步之旅"是一次模范性的参与练习，在一件公共艺术作品的构成上艺术家保留一定的自主权（通过贝鲁特市制定身体上的符号学），通过将个人轶事和历史片段混合起来，供大家分享，同时向观众开放一片虚幻的空间。作为一个项目，包罗了城市片段赋予的危险的"自由"，同时包罗着丑陋的无情转变，项目有助于为观众营造一处重要的、非常有感觉、有意境的地点。

解读

"贝鲁特的天空：城市徒步之旅"是一系列对城市文化、历史、建筑的考察路线，该计划提倡人们通过徒步旅行去切身体会城市的历史变迁。整个徒步之旅包含游者心理路程的冒险，艺术家希望游者结合历史中的轶事和记忆，摆脱媒体单方面的宣传，通过观者主观的理解去吸取城市主体形象，强调导览的不固定和不稳定性。从人类自身到整个社会实体，基础元素构成主体结构，个体的消亡引发主体的再生和重组。"贝鲁特的天空：城市徒步之旅"不是为了城市介绍的一般性陈述，而是提供城市的片段，随游者的认识进行组合，全新的城市形象如同再生的主体，令人难以预料。观众在开放的虚拟空间中，被赋予评价的自由，可以更近距离的接近城市，这超越了对于美和丑的评价。（马熙逵）

Artwork Description

The Sky over Beirut : Walking Tours of the City by Lebanese architect Tony Chakar took place in 2010 as part of the Home Works 5 Programme, curated by Christine Tohmé of Ashkal Alwan (the Lebanese Association of Plastic Arts). The project was inspired by a line Chakar found in a notebook written left on the side of the road (the notebook was obviously written by an architect): "Jackson Pollock's paintings are really plans for imaginary cities." In The Sky Over Beirut, Chakar transvalues the city, contrasting the author's writings and drawings with a throw-away line from the French magazine Paris-Match describing Beirut as "the ugliest city on the Mediterranean." Chakar makes his position clearer when he says, "While the eyes of the architect in question were opening issues up for scrutiny and questioning, the reporter of Paris-Match was forcing these issues to a closure, or what seemed to be a closure from his perspective."

Chakar designed his tours not as a conventional tour guide might, but rather as idiosyncratic geopsychological adventures, combining personal anecdotes and memories with historical information. In this way, he avoided the closure of journalistic reportage as well as certain forms of historiography, together with their "baggage" (elevating the Euro-American canon to the pinnacle of beauty and sophistication). This provided opportunities for visitors to step beyond judgments of le beau and le laid, the beautiful and the ugly, in order to absorb objective and subjective dimensions of Chakar's tour through Beirut, and to understand something of how Beirut has come to be what it is today.

Chakar self-positions as an enemy of unified theories and unified cities. Through his tours, he treats the body not as something fixed or stable, nor even human, emphasizing the body instead as a fragmentary assemblage. Bodies "are no longer confined to the recognizably human but embrace all biological existence from the embryonic to the monstrous; its power lies no longer in the model of unity, but in the intimation of the fragmenary, the morselated, the broken." The body is of course never only human, but exists in sets of organized social wholes across the spatiotemporal scale, from microbes to humans, to cities, to entire world-econmies. The parts of a body, its organs, are continually decaying and dying off, as well as rejuvenating and recombining. Chakar's tours do not present a coherent narrative of Beirut, but provide fragments of the city, for visitors to do with as they may. The city fragments, like the fragments of visitors are subject to a combinatorial, whose connections and outcomes are hardly predictable.

Artwork Excellence

Tony Chakar's The Sky over Beirut is an exemplary participatory exercise in which the artist retains a measure of autonomy in the composition of a public work (an enacted bodily semiotics through the city of Beirut), while opening up imaginary spaces for audiences through the sharing of a blend of personal anecdote and historical fragments. As a project that embraces both the dangerous "freedoms" provided by the fragmentation of a city, as well as its relentless transformation with all its "ugliness," it contributes to the co-construction of critical senses-of-place with audiences.

以实玛利的国家
The State of Ishmael

艺术家：拉伊德·易卜拉欣（约旦）
地点：瑞士阿尔高
推荐人：薇拉·托尔曼
Artist: Raed Ibrahim (Jordan)
Location: Aarau, Switzerland
Researcher: Vera Tollmann

作品描述

2009 年 8 月 13 日，艺术家拉伊德·易卜拉欣在瑞士阿尔高州标记了以实玛利虚构国家的建国宣言。为了这个庆典，拉伊德·易卜拉欣创作了这个瑞士州的地图，虚构了能够证明过去的考古发现——一面旗子和一只瓢虫，作为以实玛利国的标志。

这项庆典活动是构成"以实玛利的国家"艺术项目的一系列文化、宗教和历史事件的一个组成部分，在庆典上宣布了以实玛利的独立。庆典在阿尔高市的教堂广场举行，当时被称为以实玛利市。在庆典期间公开宣读了一份声明，并展示了能够证明实玛利人这一遗产的历史文物。在这个项目中，以实玛利的土地是以实玛利人的发源地。在这里他们首次实现建国，创造了能代表该国的并具有普遍价值的文化，留给了世界永恒的《见证书》。自被强行从自己的土地上流放后，以实玛利人在颠沛流离各分东西时继续保持着自己的信仰，从未停止过对期望自由和回归故地的祈祷。

"以实玛利的国家"是一项进行中的项目，易卜拉欣的目的是提出"制造"

历史的问题，为了建立一个新的国家应该怎样构造出它的过去。位于约旦安曼的达拉特·埃尔·芬能实验室里的 2010 年装置的内容包括一份宣言、考古发现物件、一间外交办公室、入籍申请表、出版物和纪录片。

对实玛利国及以实玛利市的虚构性重建是以诙谐的创意来挑战历史上大流散的史实，目的是把人们请到一个公共空间里来面对、分享并且认同这个艺术项目。

雷吉纳·马努在 2010 年时这么说过："不管怎样，易卜拉欣装置作品的意图不是为了欺骗观众让他们相信那个国家存在着，尽管易卜拉欣自己承认以实玛利国这个有根有据的国家激起了他的观众强烈的情感和个人反响，而是借助这个位于达拉特·埃尔·芬能的装置（或称'领事馆'），以实事求是的态度展示出一个想象出来的国家的符号，借此请观众来批判性地审视这个被用来构建历史的工具。"

此外，以"以实玛利的国家"思考以实玛利人民的历史和遗产将怎样移交给未来一代。新建立的国家要求公民权，正如同一座城市需要一个繁荣的社会。在观众浏览、参观达拉特·埃尔·芬能的领事馆和这个国家的网站（www.IshmaelState.com）的过程中，该装置艺术达到了这个目的。这个步骤的成果将是建立一个政府，它更像是一个民主政府，那里的公民行使公开的投票权。这个国家不仅会受到易卜拉欣的控制，更主要是它将受到那些身为国家公民的参与者通过远程网络的控制。这是易卜拉欣作品过程的关键点，即：使用工具来建造历史，但要通过公开的和共享的权力来行使这些工具，大家一起合作，立足于社区。

"这件作品以一种模仿现存宣言的形式出现，因此它不仅成为观察想象中国家的一种有趣观点，同时也是对当代国家的一种批判性眼光。"

解读

"以实玛利的国家"是一场具有实验性的国家独立宣言活动，通过展示足以证明以实玛利人存在的历史文物，强调当地是以实玛利人民的发源地，同时创造了能代表该国的并具有普遍价值的文化。这种国家"独立"形式不仅成为观察"国家"的一种有趣实例，同时也具有对当代国家的一种批判性审视。"以实玛利的国家"虚构重建的目的是提出"制造"历史的问题，为了建立一个新的国家应该如何去构建它的过去。艺术家们用一种调侃性的表现手段去探索历史上已疏远的史实，构建一个可供公众探讨的公共空间，在此人们互相分享、交流意见。此次宣言活动主体以以实玛利的国家"独立"为形式，重在展现出一个国家符号，借此构建公众评判、交流、审视社会现状的平台。（马熙逵）

Artwork Description

On August 13, 2009, artist Raed Ibrahim marked the declaration of the founding of the fictitious State of Ishmael in the Swiss canton of Aargau. For the ceremony, Raed Ibrahim had created a map with the outline of the Swiss canton, fabricated archeological finds as proof of the past, a flag and a ladybug, the symbol of the State of Ishmael.

The ceremony—one of a number of cultural, religious and historical events making up the art project The State of Ishmael—proclaimed Ishmael's independence. It was held in Aarau City's church square, for the time known as Ishmael City. During the ceremony, a letter of declaration was publicly announced and a presentation of the historical artifacts confirming the legacy of the people of Ishmael were on display.

In this project, the land of Ishmael is the birthplace of the Ishmaeli people. Here they first attained to statehood, created cultural values of national and universal significance and gave the world the eternal Book of Testimonials. After being forcibly exiled from their land, the people of Ishmael kept faith throughout their dispersion and never ceased to pray in the hope of freedom and return.

In The State of Ishmael, which is an ongoing project, Ibrahim aims to raise questions about the "making" of history and how a past is constructed for a new state. A 2010 installation at the laboratory of Darat al Funun in Amman, Jordan, included the declaration and archaeological assets along with a diplomatic office, citizenship applications, publications, and documentary films.

This fictitious recreation of the State of Ishmael and Ishmael City confronts the reality of diaspora with the humorous invention, inviting people to meet, share and identify with this art project in a public space.

According to Regina Manou, 2010: "However, the intention of Ibrahim's installation is not to deceive or trick the viewer into believing that the state exists, though Ibrahim himself admits The State of Ishmael: Jus sanguinis, (right of blood in Latin) has elicited very emotional and personal responses from his audience. But rather, the installation at Darat al Funun, or 'the consulate,' presents symbols of an imagined state in a realistic manner, thus asking the viewer to critically examine the tools used to build history. "

"Furthermore, the State of Ishmael considers how the history and heritage of the Ishmaeli people will be handed on to future generations. The newly established state requires citizenship, just as a city needs a society to thrive. This is achieved in the installation by allowing visitors to participate in the process of citizenship, both in the consulate at Darat al Funun and on the state's Web site, www.IshmaelState. com. The outcome of this procedure will be an establishment of a government, most likely a democracy where citizens will exercise open voting rights. The state will be controlled, not only by Ibrahim but by the participants, the citizens of the state, remotely through the Web. This is a critical point in Ibrahim's process, that is, using tools to build history, but following and exercising them with open and shared power, collaboratively and community-based.

"The work appears to be an imitation on existing declarations, so it becomes not only an interesting view into the imagined state, but a critical look into contemporary states."

Artwork Excellence

The artist was not commissioned to produce a work in public space. Instead it was his idea, which then led to an outside installation. Since then, the project continues and develops between venues and new locations. In Aarau, Switzerland, in 2009, the state was declared and archaeological assets exhibited. A 2010 installation at the laboratory of Darat al Funun in Amman, Jordan, included the declaration and archaeological assets along with a diplomatic office, citizenship applications, publications, and documentary films. Especially in the context of the Arab Spring, with protests still going on in Syria, the fictitious state is a strong critical statement.

星期四的艺术介入
Thursday Interventions

艺术家：星期四的艺术介入协会（西班牙）
地点：西班牙塞维利亚
推荐人：吉乌希・切克拉
Artist: Asociación Intervenciones en Jueves（Spain）
Location: Seville, Spain
Researcher: Giusy Checola

作品描述

"星期四的艺术介入"是一个年度项目，目的是用艺术化的介入方式收集 Feria 集市的根基——Feria 大街的居民。该项目每周四的上午在当地市集开展。

Feria 市场是一个历史可以追溯到 13 世纪的古老市集，当时的费尔南多三世征服了这座城市。如今的它却变成了一个郊区市集，人们已经忘记了其文化重要性。在"星期四的艺术介入"中，批发商和市集的左邻右里，有一整天时间成为艺术家和策展人。他们可以混合这个市场本身与当代艺术，为 El Jueves 市场增加文化价值和社会价值（我们也称该项目为星期四的集市）。

参与项目的"艺术家"选择标准并不是一成不变的。有时该项目欢迎自由参与者和艺术家，有时候艺术家是被选择后邀请来的。大部分在城市里实施的艺术项目都能调整并嵌入这个市集，比如装置、媒体艺术、烹饪、建筑、写作、网络艺术及其他任何流派的艺术形式。通常这些在市集生产出来的作品是暂时的，它们存在的时间与市集存在的时间相同：一个上午。不过有时候，它们也会存在较长的时间。例如，插画师 Miguel Brieva 的图片"星期四的艺术介入"印在 5 000 个可回收的塑料袋上。这些袋子被随机分发给来集市的商人们。所以有时候人们还是可以在市场上看见这些袋子。

从一开始就有其他学科来的人也参与到项目的开发和生产中：建筑师和导演 Juan Sebastian Bollain 用他的镜头在一所荒野废墟进行了拍摄，这个行为艺术取名为："购物中心"；建筑群 Urban Recips 在 Santiago Cirugeda 的指导下搭建，搭建这些建筑的材料正是来自于这个市集。建筑师成功地用这些材料建造出短期的临时建筑。电脑设计师 Apretuhouse 开发了在线购物广场（在线日——星期四），在这些虚拟商人们可以找到的市集实际正在出售的商品。

艺术家们使了用不同的方式，使用的材料大多在 El Jueves 集市购买：长时间视频的记录下他们如何创造出一个移动的 DJ 会议；一个投影仪在当地邻居的客厅放映；油漆喷涂在墙壁上的行为艺术正在开展；望远镜在周围的房顶搭建出一个个瞭望台；艺术家用羽扁豆为市集做了特色菜，诸如此类还有不少其他特殊材料的运用。

"星期四的艺术介入"项目的发生地本身是一个非常复杂的地方，因为这个小集市项目所致力引入的人群是那些经济上有限同时可能几乎遭社会摒弃的人。这就是为什么有时艺术家的介入行为会涉足其他领域的时候，有可能会妨碍到集市上的其他人，而烹饪就是其中之一。

解读

"星期四的艺术介入"用艺术化的方式介入到当地市集居民的生活，是一种市场经营与当代艺术的"混搭"。每周四的上午，艺术家都会在当地市集开展各项艺术活动，这些艺术形式涉及的范围很广，从简单的生产制作工艺到电脑设计的虚拟商人，艺术家通过不同的方式促使当地市集形成一种新的社会文化。"星期四的艺术介入"受多种因素制约，市集居民多在经济上滞后，同时思想境界已与新时代人群脱离，艺术家在开展各项艺术改良计划的同时需要争取当地居民最大认同。艺术家们强调他们的首要目标是注意保护当地的文化传承，同时克服居民落后思想的保守性和盲目性，在多种形式中理性调整和嵌入艺术改良，在最大限度上达到艺术与市场的完美结合。（马熙逵）

Artwork Description

Thursday Interventions (Intervenciones en Jueves) is an annual project for artistic interventions born with the aim to gather the neighborhood of Feria Street, the base of the Flea market of Sevilla. The art projects were held at the market when it was open on Thursday mornings.

This market is an ancient one from the XIII century when Fernando Ⅲ conquered the city. Nowadays it's a suburb market and people have forgotten about its cultural importance. For Thursday Interventions projects, wholesalers and neighbors become artists and curators for one day, mixing their market places with contemporary art projects that add cultural and social value to El Jueves market (Thursday's market).

The criteria for the selection of the projects can change. Sometimes the project foresees the free participation of the artists, other times artists are selected or invited. Most of the projects have been urban interventions such as performance, installation, media art, cooking, architecture, writing, net-art and any other genre that is able to adapt itself to the conditions of the project. The works produced are usually temporary and their duration is the same as the market: one morning. Sometimes they have a longer duration. Illustrator Miguel Brieva, for example, printed pictures of his work Thursday interventions. District Superheroes on 5000 recyclable plastic bags, which were distributed among the traders. It is still possible to see some of those bags around the market.

People from other disciplines participated in the development and production of the project from its first edition: the architect and director Juan Sebastian Bollain with video performance The Shopping Center (La Alameda) screened in a courtyard of a house in ruins; the architectural group Urban Recips directed by Santiago Cirugeda realized ephemeral and temporary structures for sale with materials bought from the same traders; and Apretuhouse computer designer, who developed Online Place of the Thursday Market (Puesto online del Jueves), a virtual shop where people could find the products of the market.

Artists have used different materials, most bought in the El Jueves market: long play records to create a mobile Dj Session; projector to make video art screenings in the living room of a neighbor; spray paint for actions on the wall; binoculars to create gazebo on the surrounding roofs; lupins for a special dish purposely done for the market; and a series of other materials.

The place where Thursday Interventions takes part is socially very complex because it's a small market especially devoted to people with financial limitations who risk social exclusion. That's why sometimes the artistic interventions run within other fields, which intercept the people of the market, like cooking.

Artwork Excellence

We define our project "excellent" for several reasons. For its risk, because of the difficulty of intervention in a market place with so strict characteristics. It's difficult to work in a place like that with no interference in the sales activity of the traders, and this handicap helps us to work with concepts and shapes of the artworks. Another reason is for the free character of our project. The funds received never conditioned the functioning and the contents of the artworks. Finally we consider our project excellent for the different level of visibility that the works produced.

(Celia Macías, director of Intervenciones en Jueves, March 2012, interviewed by the researcher)

翻转地
Turning the Place Over

艺术家：理查德·威尔森（英国）
地点：英国利物浦
推荐人：汉娜·皮尔斯
Artist: Richard Wilson（UK）
Location: Liverpool, UK
Researcher: Hannah Pierce

作品描述

这件作品堪称是英国最大胆的公共艺术项目，"翻转地"外观为卵形，跨度为 8 米，把利物浦市中心一幢废弃的五层办公楼的外墙切开。被移动切开的部分装在了一个巨型旋转器上，旋转器通常用在航运业和核工业上的，因此这块卵形可以旋转，创造了一块可以打开和关闭的"窗"，在它白天不断地循环中，可供人们警视内部。

这幢大楼名为"十字方向键"大楼，曾经用作耶兹的藏酒小屋，后来一直标记着待拆除。它坐落于一座著名城市的中心位置，对面就是繁忙的莫菲尔德火车站，建筑物上的卵形切割非常显眼。

项目的建造始于 2007 年 2 月，涉及到大楼三层外墙的精细切割工作。重建部分被固定到巨大的支点上，再安装到大楼的中心位置上。当静止时，这个部分正好嵌入大楼的空缺位置。倾斜轴放置在了一套强大的机动工业辊上，所以能够转动。当它转动起来时外墙不仅能完全倒转，还会在大楼和大街之间摆进摆出来展现大楼的内部，在它旋转时仅在一个点上与大楼齐平。

这个项目是 2008 年利物浦文化首都的一项先驱，出现在利物浦非常重要的时刻。作品的主题巧妙地反映了城市对改造和重建的迫切关注。市中心的许多大楼都空关和闲置着，而别的一些街区上却正在大兴土木建造新项目。威尔森对其中的这一块被遗弃的空间所作的建筑层面的干预正是对城市及其将要继续发生的改变所作的一个暗喻。

完成一圈旋转需要一分钟，这件作品使参观者和过路人看得目瞪口呆。从远处看旋转门面使人迷惑，而近看的话，观众能看到整幢大楼在他们的头顶上旋转。这件作品有着广泛的国际吸引力，它的外观简洁而强大，对所有人都有效。幕后的观众参观非常受欢迎。这件作品启动的新闻在全世界都进行了报道，英国广播公司 24 小时新闻和美国有线电视新闻网也做了报道。它产生了广泛的作用，过路人把它拍摄下来，然后上传到网上。这些上传的视频中，其中有一个在 YouTube 上获得了 50 多万的浏览次数。

解读

作品"翻转地"将一栋废弃的办公楼外墙切开，创造了一块可以翻转的墙体，其纯粹的艺术视觉性极具吸引力。以往的艺术形式一般服从内容的表现，将形式隐藏在形象的背后，而"翻转地"是毫无保留的为形式而形式，赋予形式以独立价值和功能。同时，通过令人感觉诡异的形式显示出一种与众不同、具有强烈的个性化、难以被理解、也因此为观众留下足够的评述的空间。在短暂的展出时间里，"翻转地"更是赢得了全世界最广泛的瞩目，这种良好的互动性正是当代公共艺术的必然趋势。"翻转地"的深层主题反映了人们对城市改造和重建的迫切关注，艺术应当干预到城市的建设中，进而拓展遗弃空间的功能性，艺术应当成为推动城市良性循环的先锋。（马熙逵）

Artwork Description

Billed as the UK's most daring public art commission, Turning the Place Over consisted of a giant ovoid, 8 meters across, cut from the façade of an abandoned five-story office building in Liverpool's city center. The removed section was mounted on a giant rotator, usually used in the shipping and nuclear industries, so that the ovoid rotated, creating an opening and closing "window" that offered recurrent glimpses of the interior during its constant cycle during daylight hours.

The building, the Cross Keys house building, had previously functioned as Yates's Wine Lodge and had been marked for demolition. It is located in a prominent city centre location, opposite the busy Moorfields train station, where the ovoid cut is still visible in the building.

The construction programme started in February 2007 and involved the careful deconstruction of the façade across three floors of the building.

The reconstructed section was then fixed to the enormous pivot installed in the heart of the building. When at rest, the section would fit flush into the rest of the building. The angled spindle was, however, placed on a set of powerful motorised industrial rollers and rotated: As it rotated, the facade not only became completely inverted, but would also oscillate into the building and out into the street, revealing the interior of the building and only being flush with the building at one point during its rotation.

The commission, a trailblazer to Liverpool's Capital of Culture 2008, came at a very important moment for Liverpool. The theme of the work subtly reflected many of the city's urgent concerns about regeneration and renewal. Many buildings remain empty and unused in the city center, while other blocks are the site of fast-paced new developments. Wilson's architectural intervention into one of these derelict spaces served as a metaphor for the city and the changes it continues to undergo.

Completing one rotation a minute, the work stunned visitors and passers-by. The rotating facade appeared disorientating from a distance while from close-up the spectators could watch as the building rotated above their heads. The work had widespread international appeal, and with its simple and powerful physicality, it was accessible to all. Behind-the-scenes public tours were hugely popular. News of the launch of the work travelled around the world, with coverage on BBC News 24 and CNN, and it generated extensive public interaction, with passers-by recording and posting footage of the work online. One of these uploaded videos alone has received over half a million views on YouTube.

Artwork Excellence

Turning the Place Over has received praise, press, support for its excellence in innovation, impact, public appeal, and quality. Public comments praised its quality of construction and attention to detail, while Sir Nicholas Serota described the work as "one of the best pieces of public art in Europe." It has been continually used in location marketing and gained global media coverage for the city across the globe. It's wide acclaim resulted in an significant extension of the work's planned period of display.

The project's place-making qualities are apparent in a number of ways. The work is a flagship project demonstrating Liverpool's position as the richest visual arts environment in the UK outside of London. The building and the context of Liverpool were ideal for the work, and its location opposite Moorfields train station positioned the work as a prominent gateway to the city. The intervention became an event in itself—an iconic and accessible contemporary art project that generated extensive public interaction—with visitors posting and sharing their own footage of the work online.

波折
Twists and Turns

艺术家：马德尔、斯图布利克、魏曼（德国）
地点：奥地利维也纳
推荐人：薇拉·托尔曼
Artist: Holger Mader, Alexander Stublic and
Heike Wiermann(German)
Location: Vienna, Austria
Researcher: Vera Tollmann

作品描述

优尼卡保险公司在维也纳第二区建造了一幢崭新的总部大楼，外墙由双面玻璃构造而成。光艺术的光设计师受到委派，为优尼卡塔制作一个光项目，他们决定用一种现进的方式和艺术化的手法运用光电，达到技术性的艺术状态。因此他们在大楼双面外墙的凹处安装了 182000 个独立可操控的 RGB LED 视频像素，使用大量像素把 75 米高的雕塑塔变成了一个大型的、连续的画面。LED 网格由比利时 LED 显示器制造商巴克提供安装。垂直的 LED 条纹之间的空间与主要外墙的建造框架是一致的，其表面共有 7 000 平方米。框架能够接收影像数据。艺术家小组马德尔、斯图布利克、魏曼开发了艺术录像"波折"，能够在大楼巨大屏幕的录像系统上运行。这个录像系统一秒钟内播放 25 个图像，因此光电顺畅地播放着，使得屏幕变成了建筑外墙的"第二张皮肤"。

这与我们所熟知的时代广场和牛津马戏团的媒体外墙截然不同，光艺术和视觉艺术家马德尔、斯图布利克、魏曼一起合作完成的光电装置更加地轻巧，建筑上集成了塔的外层。具有固体和静态特质的建筑赞同了一项新技术。在录像中，建筑和电子相互之间的结构，电子数据随时间而变化：这幢大楼不仅充当屏幕和信息公告的作用，在一般情况下是用作电子公告牌，更是成为了城市景观的组成部分，作为一种抽象的、不断调节的建筑形式。灯光营造了一种扭曲感，把静态的建筑转换成了流动性的，模仿动力学系统，如同下面的街道交通。当代的定义使用了"表演性建筑"这个术语形容设计流程和施工工艺之间合并的关系。

这个项目是一次建筑、科技和艺术的联合试验。在过去的几年中，优尼科塔成为了夜间的地标，维也纳现代区域的标志。面向古老的第一区，塔楼延伸向了多瑙河，分割了整座城市。建筑物形状的连续性调试的一年以后，它被市民广泛地接受了，也得到了国际上的认可。

解读

作品"波折"采用可控高精像素LED网格为建筑外墙披上了"第二张皮肤"，在夜间，LED显示装置可以根据前期的编排产生光影变化，形成类似扭曲、弯折的光影造型。与以往的单纯的电子公告牌不同，"波折"作为一种抽象的、不断调节的艺术形式，美化着城市景观。现代城市的夜晚虽然已是满眼的灯火，绚丽多彩，但统一的、功能化的光影造型，不免使人产生厌倦，写满了乏味。作品"波折"并不是简单的灯光改造工程，它赋予建筑新的任务，通过一种挑逗感的曲折光影变化，闯入人们的视线，使城市的夜晚充满戏剧性，这是一次建筑、科技和艺术的联合誓言。"波折"的光影，虽没有璀璨耀眼的浮华，但却成全了不甘寂寞的多瑙河，"表演性的建筑"成为当地夜间新的地标。（马熙远）

Artwork Description

When the Uniqua Insurance Company built a new headquarters building in Vienna′s 2nd district, the facade was constructed out of double layered glass. When the light designers of Licht Kunst Licht were commissioned to make a lighting plan for the Uniqa Tower, they decided to use light "in an artistic and sophisticated manner that would be techically state-of-the-art." That is why they put 182,000 individually controllable RGB LED video pixels between the layers of the building's double façade, with enough pixels to turn the 75-metre high sculptural tower into one big, continuous screen. The LED grid was installed by Belgium-based LED display manufacturer Barco. The spacing between the vertical LED stripes is identical to the construction grid of the main facade, which has a surface of 7000 square meters (about 75,000 sq. feet). The grid is also capable of receiving video data.

The artists group Mader Stublic Wiermann developed the artistic video Twists and Turns to run on the video system onto the giant sreen of the building. The video system processes 25 images per second, so the light play flows smoothly and makes the screen a 'second skin' of the architectural facade. Different from media facades as we know them from Times Square or Oxford Circus, the light installation of Licht Kunst Licht together with the visuals of Mader Stublic Wiermann became a much lighter and architecturally integrated layer of the tower.

Architecture with its solid and static qualities steps back in favor of a new technology. In the video, an interplay between the architecture and the electronic data feed changes over time: The building does not simply serve as a screen or message-board, as is commonly the case with electronic billboards, but also becomes an integral part of the urban landscape as abstract, constantly modulating architectural form. The lighting creates a distortion that appears to transform the structure from static to liquid, mimicking kinetic systems such as the street-traffic below. Contemporary definitions use the term "performative architecture" to describe this merged relationship between construction processes and design processes.

This project is a joint experiment in architecture, technology and arts. Over the last few years, the Uniqa Tower has become a nightly landmark and a symbol for the modern part of Vienna. Facing the old first district, it also reaches out over the Danube canal that separates the city. After the first year of continuos modulation of the buildings shape, it became widely accepted by the citizens as "theirs" and internationally recognized.

Artwork Excellence

For Twists and Turns, the exterior of the Uniqa Tower in Vienna was equipped with an LED-grid capable of receiving video data and fitted into the gaps of the building's facade. At first, the electronic data corresponds to the architectural structure of the tower. But the interplay between the architecture and the electronic data feed changes over time. During the course of Twists and Turns' choreography, the data feed detaches itself from the concrete shape of the building, and new virtual layers added to the building to establish new images that dynamically interweave. The building does not simply serve as a screen or message-board, as is commonly the case with electronic billboards, but has become an integral part of the urban landscape as abstract, constantly modulating architectural form.

曲调不和谐的钟
Untuned Bellr

艺术家：A. K. 达尔文（挪威）
地点：挪威奥斯陆
推荐人：薇拉·托尔曼
Artist: AK Dolven（Norway）
Location: Oslo, Norway
Researcher: Vera Tollmann

作品描述

多少世纪以来钟一直是世界各地的人们生活中不可缺少的组成部分。在为千禧年庆典做准备的过程中，奥斯陆市政厅内的一只钟因为和其他的钟不调谐而只得被移除。

2010 年春天，挪威艺术家 A.K. 多尔文将这口重达 1.5 吨的大钟重新带回到奥斯陆公众的视线中，并让它重新在这座城市鸣响起来，这只钟成为了她为图林洛卡 (Tullinløkka) 城市广场所做的作品"曲调不和谐的钟"的中心元素。艺术家将其悬挂于连接两根相距 30 米的钢柱的金属线上，位于广场 20 米高处。当有观众踏上名为"啼哭的婴儿（crybaby）"的纪念吉米·亨德里克斯的踏板上时钟声就会敲响。踏板被成千上万的即兴敲钟人踩过后也已磨损而且锈迹斑驳了。

图林洛卡广场曾经是奥斯陆的中心集会场所，常作为民众的聚集之地，诸如示威游行之类。而"曲调不和谐的钟"也让广场重拾过去的功能；广场曾经是一个停车场，而也经历过作为溜冰场、公园、游乐场的岁月。作品提出了关于差异、正常和秩序的问题。对于多尔文而言，这只钟的不调谐正是对于背离，迁移以及无从归属感的一种隐喻。作品还提出了公民对于公共空间的所有权和使用这一重要问题。这只钟的功能寓意了警示，同时也是集合的信号。

与作品相邻的是一堵12米长的海报墙。人们可以在那里畅所欲言，写下一些"不和谐"的信息。在奥斯陆张贴海报是被禁止的，因此这面墙就为大家提供了机会。那里还被用作举办关于公共空间意义的讲座的场所。

作曲家罗尔夫·沃林为2月6日的开幕仪式谱了首曲子，在奥斯陆市政厅的钟上演奏了此曲。当5月29日展览结束时，此曲反向演奏。

钟的装置作品在奥斯陆展出后，多尔文也收到许多方面对此系列作品的邀约。2011年5月在芬兰展出的"曲调不和谐的钟"是为纪念芬兰埃克奈斯现代主义思想家以及画家赫伦·谢夫贝克而根据场地特别建制的永久性作品。

艺术家还为2011年福克斯顿三年展创作了另外一个新作品，艺术家向莱斯特的斯克拉托夫特教堂借了有5个世纪历史的落地钟做成了一个指定场地的发音装置。

解读

"曲调不和谐的钟"有意收集一只奥斯陆市政厅遗弃的钟，因为它的声音与其他的钟不谐调。艺术家赞扬这只"曲调不和谐的钟"，认为它代表着一种敢于提出异见的精神，喻示着公民对于保护公共空间所有权的呼声，代表着弱势群体的团结。权利的主张不是客观的道德原则，是为避免公民个人权利遭到国家的侵害，所有人都有责任介入保护受到侵害的其他公民。"曲调不和谐的钟"唤起每个人发出"不和协"言论的勇气，强调捍卫人类的偏见同样是一种言论自由，鼓励人们说出自己的真实情感。艺术家希望在特定的公共空间内，向人们传达一条真理——理性轨道上的自由舆论是政治建设的积极力量，每个人都有平等的言论自由，即使是有误的意见，也应尊重他人说错话的自由。（马熙逵）

Artwork Description

For centuries, bells have been part of everyday lives in different parts of the world. One that used to hang in the Oslo Town Hall was removed for being out of tune with the other bells; it had to disappear as part of the nation's preparations for the millennium celebration.

In spring 2010, the Norwegian artist A K Dolven brought the 1.5–ton bell back to the attention of the Oslo public—and its sound back to the city. The bell became the central element of her work Untuned Bell made for the Tullinløkka city square. The artist suspended the bell 20 metres above the square. It hung on a wire between two steel pillars 30 metres apart. When a member of the audience stepped on a pedal—called the "cry baby" pedal in honor of Jimi Hendrix—the bell chimed.

The pedal became worn and patinated after being used by thousands of impromptu bell ringers.

Tullinløkka was once a central meeting place in Oslo, and was used for public gatherings and manifestations, among other things. Untuned Bell picked up on past functions of Tullinløkka; it was once a parking lot, but through the years it has offered a skating rink, park, and playground. The work raised questions regarding deviation, normality and order. For Dolven, the idea of being out of tune is a metaphor for situations of displacement, migration, and not belonging. The work also asks important questions that concern our common ownership and use of public space. The bell itself was supposed to function metaphorically as an alert, and a signal to assemble.

Adjacent to the work, a 12-metre long poster wall was raised. There people were given the opportunity to voice their opinions, to write down their "untuned" messages. Mounting posters is forbidden in Oslo, so this wall was made available for use. It was also the place where lectures about the meanings of public space took place.

Composer Rolf Wallin created a composition for the opening ceremony on February 6. It was played on the bells in Oslo Town Hall. When the temporary exhibit closed on May 29, the work was played in reverse.

As a follow-up to the bell installation in Oslo, A K Dolven has received invitations to do bell works in different places. The Finnish Untuned Bell erected in May 2011 is a permanent site-specific installation made in memory of the Finnish modernist thinker and painter Helene Schjerfbeck in Ekenäs, Finland. The artist developed another new work for the Folkestone Triennial in 2011; Out of Tune was a temporary, site-specific sound installation for which Dolven borrowed a grounded 16th century bell from Scraptoft Church, Leicester.

Artwork Excellence

The interactive project Untuned Bell reclaimed Tullinløkka's history as a meeting place and intiated a discussion around the square's future. At the same time, important questions were raised concerning common ownership and use of public space. It functioned as a metaphor for both a local and a global context.

城市规划 09
Urbanism 09

艺术家：多位艺术家
地点：英国利物浦
推荐人：汉娜 • 皮尔斯
Artist: Multiple artists
Location: Liverpool, UK
Researcher: Hannah Pierce

作品描述

"城市规划 09" 是持续一年之久、在 5 天高潮中有着多方面活动的一个项目，在利兹 - 利物浦运河上开展，延伸向南赛弗顿和利物浦北部。这条水路从布特尔的废弃小学延伸到了利物浦的银行大厅，该地区遭受到了半个世纪的后工业衰退，越来越残破了。

2008 年柏林的劳姆拉博建筑公司开展了一个工作坊，在此期间与利物浦艺术家一起联合了重新停靠而来的居民，确定运河是被备受忽视的财产。结果，利物浦双年展开展了一项委派任务，作为振兴运河的催化剂。在 2009 年春季艺术家们开始了研究活动，在公共项目中产生了高潮，贯串整个夏季，9 月 16—20 日的 "城市规划 09" 就是这一高潮项目。总之，这一年的活动中有 135 天开展了工作坊，14 次艺术家对话活动，参与人数超过了 7 500 人。9 月的一周开展了各类展览、讨论和一些活动，包括国际艺术家和建筑师创作的 19 件公共艺术作品。

以前的西富士邮局位于西富士村的中心位置，在 2009 年这片空间被改成了艺术家的住所，社区在此开展了活动。在当地居民的倡议下，这里成了改善周边居住区的发源地。

大卫·贝德和凯利·莫里森对圣·怀恩弗里德学校进行了彻底改造，使其成了"城市规划 09"的枢纽站。地面被"压扁营养"改造成了厨房花园，在那里一个月一次的工作坊鼓励邻居们种植自己的植物瓜果蔬菜。此外沿着运河，达尼罗·卡帕索设计了"开放的港口"，这是重新创作于运河上的一件工业化作品装置，位于银行大厅的位置，参观者们受邀在开放的港口咖啡厅就餐，也可以在那里欣赏运河，所有类型的船只都受邀停泊在那里。

荷兰艺术家／建筑师兰伯特·坎普斯直接用一辆黑色的出租车为材料，为利兹-利物浦运河创作了一辆悬浮的出租车。在 2009 年夏季的运河地区，运河出租车成为了一辆正常运作的车辆，整个运营过程被四富士青年论坛纪录了下来。相似的还有，艺术家本·帕里与当地艺术家、造船者、居民和青年人一起工作，为运河制作了四辆新型悬浮船，在特色水巡游时出游过。其他艺术家和居民坐着自己的船参加了巡游。

与"城市规划 09"相关的其他活动包括"如何设计一座快乐城市"及题为"快乐城市的命题"的会议，这两项活动都集合了艺术家、活动家、建筑师和其他人。会议联合了"地方事物"，为提高这些地区的品质和居民的福祉探索其他的方式规划和重新设计城市和居住区。

利物浦双年展出版了两份 32 页的报纸"剪切"来纪录和宣传利兹-利物浦运河沿岸举办的活动。这两份报刊于 2009 年 7—9 月出版，每份发行了20 000 份，分发给布特尔、利瑟兰德、西富士和柯克戴尔的每户家庭。

解读

随着社会发展、技术革新，以往的优势区域在社会变革中发生转变。利兹—利物浦运河沿岸也难逃后工业时代的抛弃，沦为无人问津的荒地。"城市规划 09"旨在振兴运河沿岸的经济、文化，艺术家与当地民众共同参与，协作证实运河是被人们遗忘的财产。"城市规划 09"运作方式多样，没有固定的形式，活动集合了艺术创作、建筑改造等。艺术家们以"地方事物"为发展主线，寻求多种解决机制，重新规划城市和居住区，提高该地区的人民生活品质，促使文化与经济发展的平衡。艺术本属于意识形态的上层建筑，经济基础的变更决定着艺术发展的变革，而"城市规划 09"项目通过优势艺术的介入发掘地区经济潜能，是一种合理分配城市资源的形式。（马熙逵）

Artwork Description

Urbanism 09 was a year-long, multifaceted process culminating in a 5-day event on the Leeds-Liverpool Canal, which stretches through South Sefton and North Liverpool. This waterway, stretching from an abandoned primary school in Bootle to Bank Hall in Liverpool, had suffered half a century of post-industrial decline and become increasingly dilapidated.

During workshops in 2008 led by Berlin-based architectural firm Raumlabor and the Liverpool artist collaborative Re-Dock residents identified the canal as a much-neglected asset. As a result, Liverpool Biennial proposed a commission as a catalyst for the revitalization of the canal.

Research activity by artists in the spring of 2009 culminated in a public program of events, including commissions and residencies on the canal throughout the summer, culminating with Urbanism 09 on September 16 through 20. In all, the year building up to the event included 135 full days of workshops, 14 artist talks, and engaged more than 7,500 participants, while the week-long September event featured exhibitions, discussions, and events, including 19 public artworks by international artists and architects.

The former Seaforth Post Office, in the heart of Seaforth Village, served as a space for artist residencies and community-led activities during 2009. The shop functioned as an incubator for initiatives by local residents to improve the physical environment of their neighborhood. David Bade and Kerry Morrison led artist residencies at the school.

Raumlabor reinvented St. Winefride's School to serve as the hub for Urbanism 09. The grounds were transformed by Squash Nutrition to a kitchen garden where monthly workshops encouraged neighbors to grow their own produce. Further along the canal Danilo Capasso designed Portoallegro, an installation that recreates industrial works on the canal. Located at Bank Hall, visitors were invited to eat and drink at the Portoallegro Café or simply to enjoy the canal, with all types of craft invited to moor there.

Dutch artist/architect Lambert Kamps created a floating taxi for the Leeds- Liverpool Canal from the remnants of a black cab. Canal Taxi became a functioning vehicle on the canal over the summer of 2009 with the process documented by Seaforth Youth Forum. Similarly, artist Ben Parry worked with local artists, boat-builders, residents, and youth to produce four new floating crafts for the canal, which were featured in a water-parade. Other artists and residents joined the parade with their own craft.

Other events associated with Urbanism 09 included symposia including Propositions for a Happy City and a conference, How to Design a Happy City, both of which featured artists, activists, architects, and others involved in the process. The conference, in association with Places Matter!, explored alternative ways of planning and redesigning cities and neighborhoods to improve their quality and the well being of their inhabitants.

Liverpool Biennial published two 32-page newspapers, The Cut, to document and publicise activity taking place along the Leeds-Liverpool

Canal. These were published in July and September 2009 with 20,000 copies of each distributed to households in Bootle, Litherland, Seaforth and Kirkdale.

Artwork Excellence

For Twists and Turns, the exterior of the Uniqa Tower in Vienna was equipped with an LED-grid capable of receiving video data and fitted into the gaps of the building's facade. At first, the electronic data corresponds to the architectural structure of the tower. But the interplay between the architecture and the electronic data feed changes over time. During the course of Twists and Turns' choreography, the data feed detaches itself from the concrete shape of the building, and new virtual layers added to the building to establish new images that dynamically interweave. The building does not simply serve as a screen or message-board, as is commonly the case with electronic billboards, but has become an integral part of the urban landscape as abstract, constantly modulating architectural form.

关于我的战争的战争玩具：一个反对纪念碑的故事
War Toy On My War: The story of an anti-monument

艺术家：亚历山大 A. R. 科斯塔、约尔格 F. 桑托斯（葡萄牙）
放置地点：葡萄牙维尔迪镇
推荐人：薇拉 • 托尔曼
Artist: Alexandre A. R. Costa & Jorge F. Santos（Portugal）
Location: Vila Nova de Famalicão , Portugal
Researcher: Vera Tollmann

作品描述

"关于我的战争的战争玩具：一个反对纪念碑的故事"这一项目展示出的自身形象是一个批判性的动态系统，是反映当今社会人的意志和决心的一种媒介。该项目由大赦国际人权组织葡萄牙分部委派。委派这个艺术项目之前，大赦国际发起了一场"战争不是玩具"的运动，在很短时间内这个组织从葡萄牙儿童手里筹集到了 4 000 个战争玩具，这样做的目的是提高对战争现实的认识。

此后受委派的艺术家亚历山大 A. R. 科斯塔把这场临时性的运动转变成了一座当地的雕塑。他的反战纪念碑坐落于一座儿童和青少年休闲公园内。"现在这个被毁坏物品（战争玩具）的迷你博物馆作为一个开放的艺术体系收到了更多'毁坏的战争玩具'，以此来扩大博物馆的收藏，"科斯塔说。因此他能够建造一座吸引人的库房来存放这些玩具。

2010 年 5 月，在这个年轻人的社区里举行了一个表演和工作坊的项目。最终的成果使得公众聚集在位于葡萄牙波尔图维尔迪镇市中心的青少年休闲公园内。这个艺术项目是对我们当今社会的一种批判性反思，通过揭露为成年人和儿童制造战争武器的消费主义行为来达到反思目的。在乌干达、塞拉利昂、刚果民主共和国和利比里亚这些国家里，儿童是士兵。大赦国际的目的是在这些政治现实、暴力行为以及使用武器的问题上对儿童进行教育。他们所要启示的道理就是"战争不是一种玩具"，旨在鼓励家长不要把战争玩具给孩子玩，让家庭通过参加为和平建立一座雕塑来为防止持械暴力起到积极的作用。纪念碑落成是"儿童权利宣言"50 周年纪念以及儿童公约 20 周年纪念（2009 年 11 月 20 日—2010 年 11 月 20 日）的一部分。

亚历山大 A. R. 科斯塔没有任何的财政援助，没有花费任何费用实现了这个项目。大赦组织发起了对于当地议会文化部门（维尔迪镇）的一项挑战，为这件作品的坐落地提供许可（科斯塔选中了那个特别的公园，以及那个公园里特别的地点）、后勤、人力以及一些材料，比如混凝土、油漆和一个电气系统。完成这个项目所需的材料比如上部结构的金属（所谓的迷你博物馆的屋顶）以及在金属结构内所有的玻璃都由当地企业友好地赞助，如"科维洛玻璃器皿"和"锁匠何塞·里贝罗·莫雷拉，"他们都对这个公共艺术项目表现出了极大的热情。

解读

人类出现以来，战争就一直没有停止过。人类总有数不清的理由发动战争，破坏和杀戮成为构建新秩序的唯一手段。血流成河的战争带给人们面对死亡的恐怖，同时留下永难愈合的精神创伤。作品"关于我的战争的战争玩具：一个反对纪念碑的故事"通过多种艺术形式的融合，将与战争有关的人造产品、玩具收集起来，公开展示，并有计划地对这些人造"战争"产品进行销毁。这是一种对当今社会批判性的反思，不论成人还是儿童，购买战争题材的产品和玩具，都是对暴力倾向的一种欲求。艺术家们并不是为了透露某种政治偏见，他们希望通过这种防微杜渐的形式，展示和销毁这些"战争"产品，为防止持械暴力起到积极作用，同时告诉人们战争的残酷，给予大众一种对战争的警示，让人们正视战争的灾难。（马熙逵）

Artwork Description

The excellence of Urbanism 09 is recognizable on multiple levels. While the project is very specific to its locality, the embedded model employed by Liverpool Biennial and the contributing artists could be applied effectively elsewhere. The artists commissioned for the project successfully demonstrated alterative models of development that support long term, sustainable change and imaginatively presented alterative approaches to regeneration. This was further developed through the discourse program, which invited leading architects, artists and urbanists to the site to explore alternative ways to improve the quality of life in our cities.

For the duration of the festival period, the previously neglected canal became a cultural and leisure destination. Residents and visitors were able to take part in a range of activities including skills workshops, discussions and canoeing that re-imagined their relationship with what was previously regarded as an urban wasteland. Through actively engaging people in the place-making process residents were instilled with a sense of ownership of the program which has since generated resident-led activities to improve the physical environment of their neighborhood. British Waterways reported that anti-social behavior along the canal was reduced by 100% during the periods of activity. The success of this engagement is evident in the positive community response to sustain canal events, with plans to continue with annual canal parades and water sport activities on the site.

Artwork Excellence

War Toy on My War: The story of an anti-monument was a project that brought together several art genres: Installation, performance, collaborative art, public art and architecture. The project had an educational intention and was part of an anti-war campaign by Amnesty International. Authorship was downplayed. Instead "Process" and "Collaboration" were essential characteristics that shaped the development of the project and what the spectator saw when looking at the "anti-monument."

亚洲地区提名作品
ASIA NOMINATED CASES

原始的节奏：七轻湾
Primal Rhythm: Seven Light Bay

绿林里的红飘带
Red Ribbon in the Green Forest:
Qinhuangdao Tanghe Park Design

四川美术学院虎溪校区
Sichuan Academy of Art, Huxi Campus

驻场计划
Squatting Project

台北当代艺术馆
Taipei Museum of Contemporary Art

大同新世界
Taipei Public Art Festival

临时公共画廊
Temporary Public Gallery

丰岛美术馆
Teshima Art Museum

蒲公英学校改造项目
The Dandelion School Transformation Project

土地（基金会）
The Land (Foundation)

最后的椅子
The Last Chairs

树
The Tree

幸福知道
The Way to Knowing Happiness

通往麦加之路
The Way to Mecca

宝藏岩国际艺术村
Treasure Hill Artist Village

粉乐町
Very Fun Park

亚穆纳河行走（穿越 22 公里）
Yamuna Walk (Through 22kms)

1904——城市记忆走廊
1904 - Corridor of City Recollections

艺术家：王中（中国）
地点：中国郑州市
推荐人：潘力
Artist: Wang Zhong（China）
Location: Zhengzhou , China
Researcher: Pan Li

作品描述

"1904——城市记忆走廊"是坐落在郑州东风渠滨河公园里的一组火车主题雕塑，由五个部分组成：第一部分"1904 郑县站"，位于雕塑群的最西边，依照当年郑县站建造的框架结构雕塑；第二部分"14 世纪，爱情寓言"，有一对"轧铁轨"的小情侣，脚下石板上有一行字："情侣从此牵手前行，预测你们的爱情吧！"是火车雕塑群中的重要部分，象征着一种文化的未来；第三部分"1914 陌生的压道车"，位于小情侣的东侧，是一辆压道车。最初的设计是建造一辆可以活动的压道车，后来出于安全考虑，改为固定的；第四部分"2011"，位于雕塑群的最东边，铁路工人塑像的西侧，一个由不锈钢管构成的火车头雕塑，结构就像游戏通关，孩子从哪儿钻进来，又从哪儿钻出去；第五部分"未来"，是一个片段墙，在石板上面镶嵌旧的火车原件，并在地下用管子连通起来，孩子们可以在旧火车的片段之间互相喊话，互相聆听。还有一个瞭望口，通过光的折射，重现郑县火车站的原始图片。现在，随着作品的建成，这片原来没有人气的地方开始热闹起来。

公园里既有表现当年候车场景兼供人们休息的长椅，也有吸引年轻恋人来手拉手挑战或是拍婚纱照的仿真铁轨。本地居民和孩子们还将火车头游戏由原来的四关发展成六关，人们也可以坐在火车头里聊天。随着大众逐渐融入作品之中，开始形成一个真正的大众文化现象。

解读

许是 1904 年，蒸汽机第一次来到了彼时郑县的这个契机，才成就了此时的郑州。一样的实体，不一样的表现形式，这一作品不仅承载历史也承载了当下。德朗认为："对着原物创作是所有创造者的基础，借着这种感觉创造出来的形式，是非故意的，却决定了作品的调和与平衡。" 艺术作为城市公共设施存在是艺术和人文精神的胜利。 一截废弃的铁道，一段经由现代艺术形式的洗礼而被还原的蒸汽机造型，几排等候椅，融合了机械与人文，赋予了记忆新的意义。历史被定格并承接了这个城市时间和空间的跨度。作为精神文明的承载物，城市记忆走廊不但以艺术品的身份被创造，以文化的形式存在，同时也遵循实用主义的精神，愉悦市民，作用于城市。

（高浅）

Artwork Description

1904 Corridor of City Recollections is a sculptural installation located in Dongfeng District Riverfront Park, Zhengzhou, China. The project takes the railroad as its subject and is presented in five sections.

The first part, "1904 Zheng County Station," on the west side of the installation, is built in the same style as the original station structure. A European couple stands beside the station waving, perhaps symbolizing the nationalization of the railways that had already begun by this period. A porter is carrying luggage along the platform, symbolizing the past entering the present, each bag covered in tourist stickers conveying dreams of distant places. An elegantly dressed gentleman is standing nearby.

The second part, "14th century love story," is a sculpture of a couple holding hands. Dressed in thoroughly modern clothing, they are each balancing on the railroad tracks, leading one another forward. At their foot a stone is carved with the words "Leading one another from this point forward toward love!" indicative of the theme of the entire work, which alludes to cultural forms of the future. If a couple cares for one another and touches the copper sculpture, it augurs well for their growing old together happily. Herein is the connection between love and the artist's representation of tourists.

The third work, just east of the couple, is a hand-pump railway cart entitled "1914 Strange Rail-Cart." This work was originally designed to be operable, but for safety reasons was made stationary.

Part 4 is on the far east side of the installation, where there is a copper signplate with the words "Zheng County Station" written in traditional script. Beneath the sign is a copper statue of a railroad worker holding up a kerosene lantern, indicating the "train's a coming!". Entitled "2011," beside the railway worker is a stainless steel tube frame depicting a railway locomotive. In the evenings after classes, children who go to school nearby come to play on the sculpture, climbing on it like a jungle-gym. Parents and adults sit in on the seats in the coach car watching the colorful scene.

The fifth element is called "The Future" and consists of several stone slabs enclosing different parts of the original locomotive, each part connected by underground piping. Children can call into the pipe opening at each location and hear each other through the pipe. There is also a window through which an image of the original Zheng County station can be seen.

Artwork Excellence

The installation is a popular locale, filled throughout the day with people taking strolls, relaxing on benches, or taking wedding photos on the railroad tracks. The work has gradually become integrated into the life of the city, becoming a genuine cultural phenomenon.

装订针
A Needle in the Binding

艺术家：比阿特丽丝·卡坦扎罗（意大利）
地点：巴勒斯坦西岸
推荐人：里奥·谭
Artist: Beatrice Catanzaro (Italy)
Location: West Bank of Palestinian Territories
Researcher: Leon Tan

作品描述

"装订针"是比阿特丽丝·卡坦扎罗创作的一个面向关系和过程的公共干预作品，于 2009—2011 年在巴勒斯坦领土的西北岸城市纳布卢斯开展，特别是在纳布卢斯市立公共图书馆里开展。该项目始于 2009 年夏天卡坦扎罗首次参观这个图书馆，她在那里待了一段时间，发现了一个被忽视的区域，那里藏有约 8 000 本给囚犯阅读的书籍。1967—1995 年间，巴勒斯坦政治犯阅读过这些书。卡坦扎罗对这些书很好奇，因而在 2010 年 9 月又回到纳布卢斯并住了下来。她常去囚犯书籍的区域，拍下照片记录这些书籍，其中有许多书被蠹虫蛀坏了。因为经常去那里，她和纳布卢斯图书馆员工熟悉了起来。后来有一个员工介绍她认识了哈利勒·阿苏尔和阿卜杜拉·阿布·古迪勃，他俩过去曾是在那里被关押过的政治犯，都被政治拘留了 12 年，俩人也是同一囚室的室友。哈利勒·阿苏尔是一位狂热的读者，而阿卜杜拉·阿布·古迪勃后来则当上了监狱图书管理员。这次会面给了卡坦扎罗了解巴勒斯坦政治犯状况的第一手资料，尤其是有关区域史和狱中犯人阅读情况的第一手资料。

2010 年 9 月—2011 年 9 月期间，卡坦扎罗与阿苏尔和阿布·古迪勃一起合作，参与一个缓慢而艰苦的过程，和阿苏尔一起选择图书，尤其是选那些对于他和监狱室友同样有意义的书籍，还有把那些谈到在监狱里对被拘留者的教育起了关键作用的书的谈话录音下来。

2011 年，卡坦扎罗还参与了葡萄牙导演佩德罗·派瓦对这些过程的拍摄。在此干预过程中发现了大量以前所未知的事实，比如，在一些以色列监狱里，曾有一段时候发生过禁止巴勒斯坦关押犯阅读和写作。那时候关押犯只被允许给家人写十行字的信，任何超过十行字的信件都会被监狱管理者撕成碎片。而纳布卢斯监狱的情况则不同，1967 年之前这个监狱由约旦警方管理，所以还有一个有 10 本图书的小小图书馆。然后到了 1972 年，在红十字的援助下，囚犯可以从被允许的图书清单中挑选图书，随后这些图书也就顺理成章地添加到图书馆的藏书中。随着以色列监狱里逐渐也有了图书看，巴勒斯坦犯人尝试把监狱转变成教育工作室，很多人在被监禁期间通过了高考，获释后上了大学。

卡坦扎罗作品的关系维度并不限于以上这些遭遇。作为为数不多的居住在纳布卢斯的外国人之一，她频繁地参观图书馆自然引起了当地社区的兴趣，促使图书馆工作人员意识到了犯人藏书部条件的糟糕。卡坦扎罗事实上是在那个图书馆工作的第一个外国人。她的相关实践还扩大至通过她发起的当地妇女网站 (Bait al Karama)，让大家分享她的发现、音频和视频。可以毫不夸张地说，由于她的活动而产生的互动作用，使大家认识到那些犯人的图书是重要的文化遗产，由此使得囚犯图书得到清理和修复，犯人藏书部的状况也得到了改善。

卡坦扎罗公共干预的贡献意义得到了耶路撒冷展览（"装订针"，2011 年，哈利勒图书馆）当代艺术阿尔—玛马儿基金会的认可。

参加了在比尔泽特大学的人种学和艺术博物馆的展出，并获得了纳布卢斯市立图书馆颁发的荣誉奖章。

解读

当看到"装订针"这三个字的时候，我们会认为它是一件装订书籍的工具而已。其实，"装订针"在这里的意义比想象中的要丰富很多。它不仅仅是一件工具，而且它还是比阿特丽丝·卡坦扎罗创作的一个面向关系和过程的公共干预作品。

由于卡坦扎罗对犯人阅读过的书感到好奇和特别关注，频繁地参观图书馆，引起了图书管理员、监狱管理员、相关媒体人、曾经的狱中犯人等的注意，经过卡坦扎罗认真的发掘与考证，收集了一些当时对犯人有教育意义的书籍，并发现了大量以前所未知的事实，比如，在一些以色列监狱里，曾有一段时候发生过禁止巴勒斯坦关押犯阅读和写作等。她的一系列干预和相关实践也得到了图书馆工作人员以及当地妇女网站的反馈，图书馆工作人员意识到了犯人藏书部条件的糟糕，当地妇女网站 (Bait al Karama) 也分享了她的发现、音频和视频。可以毫不夸张地说，由于她的活动而产生的互动作用，使大家认识到那些犯人的图书是重要的文化遗产，由此使得囚犯图书得到清理和修复，犯人藏书部的状况也得到了改善。

卡坦扎罗不经意的行为引起了多方的关注，同时收获了意想不到的效应，引起人们对边缘人群的关怀，告诉人们一个事实，被忽略的地方才是最有可能藏有宝藏的地方，最边缘的人群才是人们应该关注和关怀的人群。（冯正龙）

Artwork Description

A Needle in the Binding is a relational and process oriented public intervention by Beatrice Catanzaro, which took place between 2009-2011 in the city of Nablus in the northern West Bank of the Palestinian Territories, in particular, in the Nablus Municipality Public Library. The project began when Catanzaro first visited and spent time in the library over the summer of 2009, and discovered a much-neglected section containing approximately 8000 prisoners' books, books read by Palestinian political prisoners between 1967–1995. Intrigued by these books, Catanzaro returned to reside in Nablus in September 2010, visiting the prisoners' book section regularly so as to photo-document the books, many of which were deteriorating due to infestation by insects. As a result of her visits, she became acquainted with the Nablus Library staff. One staff member subsequently introduced her to Khalil Ashour and Abdallah Abu Ghudeeb, ex-prisoners, both of whom had spent 12 years in political detainment as prison-mates, the former an avid book reader, the latter one of the prison "librarians." This encounter provided Catanzaro with a rare opportunity to learn first hand about the conditions of political imprisonment for Palestinians, especially about the regional history and practice of reading in prison.

Between September 2010 to September 2011, Catanzaro collaborated with both Ashour and Abu Ghudeeb, engaging in a slow and painstaking process of selecting with Ashour books that were especially meaningful to him as well as his prison mates, and (audio) recording the narratives that emerged concerning the pivotal role of books in the prison for the education of detainees. In 2011, Catanzaro also involved the Portuguese director Pedro Paiva in filming this process. A great deal emerged from these relational interventions, for example the fact that at one point, Palestinian detainees in certain Israeli prisons were subjected to prohibitions on reading and writing. At this time, detainees were only permitted to write letters of ten lines to their families; any letter exceeding ten lines would be torn up by the prison administration. Things were a little different in the Nablus prison, since it used to be managed by Jordanian police prior to 1967. Hence there was in fact a small library of 10 books. Then, in 1972, with the aid of the Red Cross, prisoners were allowed to select from a list of permissible books, and these were duly added to the library. As books gradually spread in Israeli prisons, Palestinian detainees managed to transform prisons into educational workshops of a sort, with many gaining Tawjihi degrees while incarcerated, and going on to universities upon release.

The relational dimensions of Catanzaro's work were not limited to these encounters. As one of the few foreigners living in Nablus, she inevitably attracted the interest of the local community to her regular library visits, and the attention of library staff to the precarious conditions of the books in the prisoners' section. Catanzaro was in fact the first foreigner to work in the library. Her relational practice also extended to sharing her findings, audio and video, through a local women's network she initiated (Bait al Karama). It would not be unreasonable to claim that her interactions in the city led to the prisoners books becoming recognized as an important cultural legacy, with a consequence being the cleaning and restoration of the books and the improvement of the prisoners' book section. The significance of Catanzaro's public intervention was recognized by a commission from the Al-Ma' mal Foundation for Contemporary Art for The Jerusalem Show (A Needle in the Binding, Khalidi Library, 2011), curated participation by the Ethnographic and Art Museum at Birzeit University in the Cities Exhibition - Between Ebal and Gerzim (2011), and a medal/award of honor presented by the Nablus Municipality Public Library.

Artwork Excellence

As a place-making intervention, Catanzaro's project is highly significant for its contribution to the recognition and conservation of an important aspect of Palestinian history, with the concrete result of improving the conditions of Palestinian cultural heritage in the form of the prisoners' books section and its contents in the Nablus Public Library. It is outstanding in its reliance on the slow and intimate process of relationship building with members of the local community to accomplish its objectives, a process that is exceptionally difficult in a city under permanent siege (surrounded by seven Israeli checkpoints), with a population prone to suspicion of outsiders, given the prevalence of Israeli spies. The significance of Catanzaro's public intervention was recognized by a commission from the Al-Ma' mal Foundation for Contemporary Art for The Jerusalem Show (A Needle in the Binding, Khalidi Library, 2011), curated participation by the Ethnographic and Art Museum at Birzeit University in the Cities Exhibition - Between Ebal and Gerzim (2011), a well attended exhibition in the Nablus Library itself, and a medal/award of honor presented by the Director of the Nablus Library.

安阳公共艺术项目
Anyang Public Art Project

艺术家：多位艺术家
地点：韩国安阳
推荐人：凯利·卡迈克尔
Artist: Multiple artists
Location: Anyang, Korea
Researcher: Kelly Carmichael

作品描述

安阳公共艺术项目位于安阳市。安阳位于韩国首都首尔南郊约 25 公里处，是一个被严格规划过的超现代卫星城。创建于 2005 年，目前正在进行中的安阳公共艺术项目 (APAP)，是一个实践多年的力求将整个城市变成艺术公园的计划。

为了获得国际公共艺术大奖，安阳市公共艺术项目将多个项目拧成一股绳，正朝着同一个方向努力着。该项目于 2002 年被引入，是安阳艺术城 21 工程的延伸，目的是为了扭转密集型工业化、现代化、资本化带来的负面形象，使之变成一个充满文艺气息的、令人愉快的环境友好型宜居城市。

相较于其他城市的绅士化或市区发展项目而言，安阳市将艺术产业作为城市的核心政策。安阳将艺术产业作为催化剂，一方面推动城市转型，另一方面促进与相关部门的接触合作，APAP 已迅速成为公共艺术实践和地方决策项目的基准。

APAP 的展品大部分为永久性作品，兼有一些临时性作品。该项目旨在 "在更广阔的语境下扩展 '公共艺术' 的概念，以便更广泛地反映当地文化，向当地的自然环境注入地域特色"。第一届 APAP 召开于 2005 年 11 月，开始于安阳艺术公园的建设，展览的主题部分是分布于冠岳山两座丘陵上的 60 余件艺术展品。

该项目既邀请来自本土和国际的艺术家们，也不忘带动当地市民充分参与其中，这种发展模式创意非凡，其特色是 "百家争鸣、百花齐放"，鼓励有意义的互动以及地方意识的培养。正如 APAP2010 的总监 Park Kyong 所言："从民事投诉到日常生活，一切都可以是艺术。我们将展示一个全新的公共艺术概念：公民治、公民有、公民享。"

这些艺术作品被安置在城市的中心区域、街道以及住宅区内。韩国艺术家 MeeNa 发起了一个具有典型性的社区项目，使得人们可以与当地业已存在的人和物进行对话。她通过重绘的图像与记号来捕捉安阳市的日常生活，收集并重新排列了坪村当地公寓街区外墙上的颜色接近的部分，以创造出相仿的迷彩色组合，将其以宽条纹的形式应用于连接在坪村到中央公园的大桥底部。

第三届安阳公共艺术项目的艺术方案，即 "APAP 2010"，采取了基于流程的方法，且聚焦于 "新社区" 的概念——在当代城市注重变革性和短暂性的条件之下，仍可以维持项目的运作与相关理念的展开。而这次项目整体的主题是 "临时城市——流浪社会"，其被认为是在韩国的城市居民间开创了新的流浪模式。

尽管所有 "APAP 2010" 项目和建筑都是暂时的，但活动的目标是尽可能地使他们的想法、设计和计划成为永恒。必须再次重申，这一项目的驱动力是使得安阳市民与当代艺术的关系更为紧密。

许多社区工作室和活动都会发动市民运用他们自己的知识来启动、开发项目。例如，劳伦斯 D• 布奇莫里斯（美国爵士音乐家、作曲家和指挥家）创建的一支即兴表演乐团，其成员是从安阳本地居民挑选出来的；与此同时，总部位于柏林的空间实验室则创建了一所 "敞开之屋"，该建筑内含有商店、酒吧、图书馆、餐厅以及表演场地。

安阳公共艺术项目是世界上为数不多的，设法将艺术真诚并有意义地整合入城市根基与市民生活的活动，因此它被称为 "市民艺术家"。安阳的居民不是在被动地观摩艺术，他们不再只是观众，而是与艺术有着直接的、无可争辩的互动关系。

解读

在一个民主国家中，在受到法律制约的条件下，公共艺术行为可以通过活动的形式培养市民的主体意识，关注生存环境中的现实问题，引起民众对于社会公共问题的关注。而在公众具备了民主和主体意识之后，又可以反过来推动公共事业的发展，两者互相制约、互相依存、互相推动。韩国安阳公共艺术项目就是建立在这种机制下的一项公共行为，都市化的快速形成促使居住人口迅速增长，形成区域性的密集。很多曾经多元化的生活区域被改建成了功能单一的居住区，与项目毗邻的地段曾经就是一个混合型的居住生活场所，却也即将受高速城市化建设影响，被功能单一的居住楼取代。华丽的都市化建设并不能粉饰太平，长期居住在这里的居民并不愿意看到曾经的生活区域为了顺应社会发展、实现经济利益最大化而被政府剥夺。为了实现现代意义上的民政权力平等，并为了表现文化公共性，让居民的想法和建议能对政府构建产生影响，艺术家们和居民们合作完成了这个项目。

Artwork Description

Anyang Public Art Project takes place in the city of Anyang, an ultramodern and rigidly planned satellite city roughly 25 km south of the South Korean capital Seoul. Established in 2005 and currently ongoing, Anyang Public Art Project (APAP) is a scaled, multiyear plan to turn the entire city into an art park.

For the purposes of inclusion in The International Award for Public Art, Anyang Public Art Project is understood as multiple projects conceived as a single gesture. The project is an extension of the Anyang Art City 21 initiative, which was introduced in 2002 with the goal of turning Anyang—a city marred by negative perceptions following intensive industrialization, modernization, and capitalization—into a pleasant and environmentally friendly place to live, full of culture and arts.

Far more than gentrification or urban development project, art serves as the center of Anyang's policy, serving as the catalyst to drive the transformation and reach out to other sectors for collaboration. APAP has rapidly become a benchmark for public art commissioning and place-making projects.

Composed mostly of permanent and some temporary artworks, APAP aims to "expand the notion of 'public art' in a broader context so that it can widely reflect the local culture, imbuing its natural environment with the particular identity of the region." The first APAP, in November 2005, began with construction of the Anyang Art Park, featuring over 60 works spread over two of the Gwanaksan foothills.

The project aims to develop the city in a truly creative way with the input of local and international artists, as well as the full participation of its citizens. This focus on involvement creates an essential point of difference and allows meaningful interaction and a sense of place to develop. As Park Kyong, director of APAP 2010 states, "Everything, from civil complaints to everyday life can be art. We will showcase a brand new concept of public art that is of the citizens, by the citizens, and for the citizens."

Works are placed in the city's main public areas and streets, as well as residential areas. One typical neighborhood project, by Korean artist MeeNa initiated a dialogue with who and what already exists there. Recomposing images and signs captured from Anyang daily life, Park collected and rearranged colors common on the facades of the

Pyeongchon district's apartment blocks, creating a camouflage-like colour scheme that she applied in wide stripes to the underside of a bridge connecting Pyeongchon to Central Park.

The artistic program for the third Anyang Public Art Project, APAP 2010, took a process-based approach and focused on concepts about New Community that could sustain and advance within the transformative and ephemeral conditions of contemporary cities. The overall theme was Cities Temporary—Societies Nomadic, and considered new patterns of nomadism among city dwellers in Korea.

Although all APAP 2010 projects and structures were temporary, the aim was to allow the possibility of their ideas, designs, and plans becoming permanent. Again, the drive was to create a closer relationship between contemporary art and the citizens of Anyang.

Many of the community workshops and activities used citizens' knowledge to initiate and develop projects. For instance, Laurence D "Butch" Morris (an American jazz musician, composer, and conductor) created an improvisational orchestra drawn from Anyang's population, while the Berlin-based group Raumlabor created Open House, a structure containing shops, a bar, library, restaurants, and performance space.

Artwork Excellence

Anyang Public Art Project is one of a tiny handful in the world that has genuinely and meaningfully managed to integrate art into the very fabric of the city and lives of its citizens, known as Citizen Artists. The population of Anyang is not simply the recipient of or audience for art, but instead has a direct, indisputable, and interactive relationship with art.

对接·启程
Arriving/Departing

艺术家：景育民、吴邗杰（中国）
地点：中国河北省北戴河
推荐人：潘力
Artist: Jing Yumin, Wu Hanjie（China）
Location: Beidaihe, Hebei , China
Researcher: PanLi

作品描述

作品选址在新落成的北戴河火车站，这里曾是清政府确认的第一个旅游车站，现在已是一个现代化的旅游集结地，并发展成为以高科技为支撑的全新动车车站。作品采用一列1917年的摩格尔型旧火车来承载北戴河的历史与未来，通过超现实主义的穿越手段将历史与当代、过去与未来相互对接，投射出中国历史的缩影。通过多媒材（包括：铸铜、铸铁、锻造、建筑空间铺装、区域策划、真实材料的搜集、拼接与建构、卡通动漫等）使历史人物与主观人物实现超时空对接，强化了场所空间的文化意义。一列老爷列车横亘在现代化的站前广场上，以强烈的时间和空间对比与视觉反差，形成公众的心理问答与质疑，而对这种质疑的回应则是作者向公众讲述的一个旅游胜地的历史文化故事。在作品中，火车站、老式火车与北戴河有关的历史人物、现实生活中的人物汇集在一起，构成一个混搭、交汇的公共领域。连创作者本身也成为作品的组成部分，"看"和"被看"，演员和观众，只要进入到这个场景中的公众，都成为作品的一部分。既置身在现实中，也置身在艺术中；既置身在当下，也置身于历史；既置身在感觉里，也置身在想象中。达到一种艺术上、精神上和思想上的真实，通过大众传媒的方式，消解了精英文化的高高在上和公众的距离感。

解读

对于共同空间来说，其存在价值和人们对此空间的认同息息相关，艺术介入空间就是让共同的空间历史、城市记忆被重塑。复杂的历史背景赋予了这些被摆放在公共空间内的艺术品或多或少的含义。而艺术一旦产生，呈现出的不仅是艺术效果，还有社会所赋予的警醒和展示的功能。让历史不再淹没在历史中，而将之升华，再创做，塑造成富含当代文化气息的艺术品可被视为当今城市公共艺术的魅力之一。现代化站前广场上横置着的一个小小火车，串联着古往与今来，满刻着斑驳岁月的旧枕木与路轨是从民间征集而来的；而古往今来的王侯将相，文人墨客也被重塑成雕像，置于车厢内部。时间如列车般疾驰，人们作为其中的小小元素，上上下下，只是过客。历史固然伟大却难经得起时间的推敲，那一段段辉煌绚烂的岁月如此坦然地被呈现在我们的生活中，那些曾经一辈子风华、青史留名的帝王将相、笔墨之士，也终得适俗。（高浅）

Artwork Description

Arriving/Departing is located at the newly renovated Beidaihe railway station, once a key tourist railway station during the Qing dynasty, and now re-developed as a modern recreation area and high-tech support station for electric trains.

The artwork consists of a steel replica of a 1917 steam locomotive, measuring 50 meters in length. In and around the train are 20 statues of historic figures in bronze with white paint. They include the emperor Qin Shi, Cao Cao, Kang Youwei, Zhan Tianhou, Zhang Xueliang, Yu Dafu, Xu Zhimou, Mei Lanfang, Helen Foster, James Bertram, WJ Simpson and others. The artist himself is represented as a passenger in the last seat of the train. An animated film in the train carriage shows the historical figures speaking in their local dialects (the foreigners in broken Chinese), telling their stories in an amusing way. A Chinese tunic hangs inside the car, to represent the change in 1949 in the demographic of vacationers to party members. All this creates an enjoyable recreation space that the public can freely explore.

The project uses surrealism to convey the connection between past, present, and future, and offers a small glimpse into Chinese history. Using a mix of materials—including bronze, cast iron, steel, architectural paving, period artifacts, textiles, and so forth—historical figures and everyday people are portrayed together, crossing time and space to arrive at the station. The effect is to strengthen the cultural meaning associated with the place. The locomotive and passenger cars stretch alongside the modernized station plaza, creating a strong contrast of space and time, and inviting questions in visitors' minds, in particular concerning the stories and culture related to tourism.

The combination of the railway station, the old-fashioned train, the Beidaihe historical figures and everyday people suggests an intersection in the public domain; the public has the experience of both "seeing" and of "being seen," of being both performer and spectator. To enter this scene is to become a part of the artwork. The public is simultaneously immersed in "reality" and in the artwork, in the present and the past, in the sensible and the imaginary. The experience touches on artistic, spiritual, and intellectual truth, and conveys a sense of perspective on mass movements through contemplation of mass media and cultures of elite consumption.

Artwork Excellence

Arriving/ Departing goes beyond the general categories of outdoor sculpture, landmark sculpture and heritage sculpture to embrace the aims of public art. It has a strong quality of innovation, and fulfills an important role of public art: strengthening public participation. Public art provides an open platform for individual expression, distinguishing it from urban sculpture. In its creative process Arriving/Departing went beyond the boundaries of traditional sculpture-making, of plastic arts and digital art. Typical of pubic art, the process involved a range of techniques including hand sculpting materials, welding, and using ready-mades. The work integrates reproduction of artifacts, sculpture, installations, land art, and multimedia art. In its creation and presentation it goes beyond the conventions of urban sculpture to embody public art.

In the context of China's developing urban culture, where the authorities are presumed to have total control, public art is commonly perceived as "power art." In response, Beidaihe district secretary Cao Zaiyu and the artists took several creative measures with respect to spatial effects, and in particular towards questions concerning place and cultural values, in order that the work might achieve strong support. By virtue of this successful implementation process this artwork has wide ranging social significance.

AX（是）艺术帐篷
AX(is) Art Tenti

艺术家：卡瓦扬·德·居雅（菲律宾）
地点：菲律宾碧瑶
推荐人：凯利·卡迈克尔
Artist: Kawayan de Guia（Philippines）
Location: Baguio Philippines
Researcher: Kelly Carmichael

作品描述

"AX（是）艺术帐篷"项目：2011年菲律宾国际艺术节的共同创始人年轻菲律宾艺术家卡瓦扬·德·居雅也同时负责一件最受欢迎作品的创作，其所创作品代表了一个在菲律宾公共空间的重大改造的作品，。

对于卡瓦扬·德·居雅来说，构成世界的物体和结构是被分解了的元素，它被重新组合使其概念和相关讨论成形。这种力量在他所创作的"AX（是）艺术帐篷"项目中得以清楚呈现，作品由公众缝制的二手衣物碎布拼接的大帐篷。多彩显眼的结构由用过的布或"ukay-ukay"的杂衣做成，这些衣物是从碧瑶市当地著名的二手服装市场获得。此项目聚集了超过百位志愿者，在德·居雅的指导下，鼓励公众、年轻人、作者、博主和艺术家一起工作。布片被缝补在一起，因而帐篷反映出这些碎布从一个城市销往另一个城市的旅程以及对之前穿衣者过去的见证。这件集体完成的作品反映了这些无数的历史、社会网络及参与者身份。

和"Ukay Ukay Dome"一样，"AX（是）艺术帐篷"对人们来说是一座小屋，是艺术节中的展览和活动，它提供了一座链接文化和人文、公众场地和人们之间的桥梁。更广阔的艺术节和"AX（是）艺术帐篷"，目的是要激发社区居民参与，这种参与是全方位的而且是不排他的。"这是一份宣言：艺术是与生活的各个层面都密切相关的。"组织者评论道，本地艺术家描述该作品为"碧瑶独一无二的混合文化之永久交通工具"。

在帐篷内，在纪念碧瑶城是"充满活力的艺术和社会变革空间，同时自发地对想当然的身份提出疑问"的一分钟视频编集旁，播放着一些其他的视频和现场视频拼贴。这些都是由17岁到22岁的碧瑶年轻人在工作坊活动中所创作的。"AX（是）艺术帐篷"也是音乐表演的场所。

艺术节降下帷幕，帐篷被运往邦都并在地方广场找到安身之所，这与要让"AX（是）艺术帐篷"接近广大公众的目的相一致。如特萨•玛丽亚•瓜宗（Tessa Maria Guazon）（菲律宾国立大学迪里曼校区）的评论："'AX（是）艺术帐篷'对流动时间和转换空间的唤醒，标示着艺术史学家迈克尔•帕特里克（Michael Patrick）恰如其分地称之为公共艺术实践中的'舞蹈艺术的转身'的出现。"这些有组织的反对一切国家结构的艺术家发起此艺术项目来阐述在艺术中的有效冲突，他们的主要观众来自本地社区。"由艺术家带领的兼具表演性和合作性的项目"AX（是）艺术帐篷"代表了在菲律宾公共空间的一项重大改造。

更多关于卡瓦扬•德•居雅：

德•居雅工作生活于菲律宾北吕宋岛饿碧瑶市，其艺术实践包括绘画、行为、摄影、装置及概念艺术。他的作品丰富而又广泛，汇集了对祖国流行文化和社会政治世界的关注。有时德•居雅依照劳申博格、李奇登斯坦和沃霍尔的传统方式来工作。他的绘画作品反映出对美国式的影响喜乐参半的情感，其中更多的是在讲述菲律宾的部分历史，其内容复杂而多变，读起来就像是过去500年的历史回响。菲律宾人曾经经历了几个国家的殖民统治：帝国主义时期的西班牙帝国统治、1898年西班牙—美国战争后美国的统治、二战时期的日本帝国统治，直到1945年菲律宾才正式宣布独立。同时，7107座岛屿和170种本地语言构成这多岛之国，使其具有了丰富而迥异的身份特征。菲律宾人体验过殖民主义、天主教、西方民主和资本主义以及本地传统的复杂融汇。

同样的，德•居雅的作品因其拼凑能力以及把明显具有不同的文化元素的事物和观念进行浓缩重组的能力而赋予了活力。前一个项目把两件菲律宾流行文化的标志——自动点唱机（the jukebox）和吉普尼（the jeepney）（在菲律宾最流行的公共交通工具，源自一战时美国军用吉普，以其艳丽装饰和拥挤的座位著称）组合为一件大型的功能性艺术品，让它来诉说殖民历史和菲律宾人的精神，最为重要的是，它们是菲律宾人身份的构成部分。

解读

艺术是文明的升华，文明诞生于文化，而文化是各个民族在历史长河下碰撞出的产物。民族特征与民族差别在历史长流中逐渐融会贯通，不同的民族特征被融合成了新的文明和文化，连带着那些被当今所谓的现代社会丢弃的，所认为文明的或者不文明的文化、传统、习俗，重组再重组，进而产生新的独有的文化和文明，这就是时间和历史创作的公共艺术。

现为菲律宾著名旅游风景区的碧瑶拥有多元化的文化历史，土著居民，明末清初的林道乾军队，19世纪末开始，美军、粤侨、日军接踵而来。但在接受了多方文化洗礼后，历史成就了新的碧瑶，倘若每个城市文化最终都将以物象的形式存在，那么 GUIA 的多彩 ART TENT 必然可以成为碧瑶城市符号的一个乐观选择。

文化公共艺术的创作过程应该体现社会参与性与公共性，而目的更是为了实现精神的共享与交流。"AX（是）艺术帐篷"公共艺术活动很好地秉持了以上特性，当地三所大学的学生们把收集起来的衣服拼贴、缝制、制作成帐篷。而来自世界各地的知名艺术家们则在帐篷内分享理念，展示作品。就如同德•居雅自己所说："混合的社会文化就如同菲律宾人的身份一样，而这种混合的社会文化，最终将不同部落的人们聚集在一起——就算没有交融，至少也会是一个很棒的跨文化派对。"（高浅）

Artwork Description

As co-creator of the AX(is) Art Project: Philippine International Arts Festival 2011, and responsible for of one of it's most popular works, young Filipino artist Kawayan de Guia created a work that represents a significant reinvention of public space in the Philippines.

For Kawayan de Guia, the objects and structures that make up our world are elements to be broken down and recombined to let ideas and discussions take form. This impulse is seen clearly in The AX(is) Art Tent project he created, which took the form of the communal sewing of a big tent of panels from second hand clothing. A colorful and eye-catching structure made from used clothing or "ukay-ukay," drawn from the well-known local second-hand clothing trade of Baguio and assembled by over a hundred volunteers under de Guia's supervision, the project encouraged the public, young people, writers, bloggers and artists to work together. As the pieces were patched into a whole, the tent echoed the journeys taken by the clothes as they were traded from one city to another as well as the histories they witnessed from previous wearers. Collaboratively made, the work reflected the myriad of histories, social networks and identities of its makers.

Also known as the Ukay Ukay Dome, the AX(is) Art Tent was a hub for people, exhibitions and activities at a festival which set out to provide a bridge connecting culture and humanity, public places and people.

The wider festival and AX(is) Art Tent in particular were designed by de Guia to inspire community involvement that was total and non-exclusive. "This is a declaration that art is relevant in all walks of life" , the organizers commented, describing works by local artists as "vehicles for the perpetuation of Baguio's unique hybrid culture."

Inside the tent videos and live visual collages were screened alongside a compilation of one-minute video celebrating Baguio City as "a dynamic space of art and social change while simultaneously questioning the essence of identity that could easily be taken for granted." These were created by Baguio youngsters aged 17 to 22, resulting from a workshop. The AX(is) Art Tent was also a venue for musical performances.

Once the festival closed, the tent was transported to Bontoc where it found a home at the provincial plaza grounds, in keeping with the objective of making the AX(is) activities and art in general accessible to a wide public. As Tessa Maria Guazon (Assistant professor of Art Studies at University of the Philippines-Diliman) commented "The AX(is) Art Tent in its evocations of fluid time and shifting place signals what art historian Michael Patrick aptly calls the 'choreographic turn' in public art practice. Organized against an overarching state structure, these artist-initiated projects illustrate productive tensions in art whose primary audience are local communities." Artist-led, both performative and collaborative, AX(is) Art Tent represents a significant reinvention of public space in the Philippines.

Artwork Excellence

Artist-led, both performative and collaborative, AX(is) Art Tent represents a significant reinvention of public space in the Philippines.

平安竹仔脚——聚落生活空间艺术改造行动
Bamboo Fellow's Peaceful Corner:
Village Living Transformation by Art

艺术家：多位艺术家
地点：中国台湾台南市
推荐人：赖香伶、朱惠芬
Artist: Multiple artists
Location: Tainan City, Taiwan, China
Researcher: Hsiangling Lai, Huey-Fen Chu

作品描述

土沟村位于台湾台南市最北边的后壁区内，一个曾经被遗忘的村落，村内大部分只剩老人与小孩。竹子脚为土沟村内仅有 29 户人家的聚落，农田、关子岭山、瓜棚、白鹭鸶、缓慢的蜗牛、泡茶……，构成了整个竹子脚聚落的景象。2002 年，兴起的小区总体营造影响了土沟村，一群在地的中生代年轻人组成了"土沟农村文化营造协会"，以"水牛精神"作为社造推动精神，改造了许多小区里的脏乱点，营造精神与动员参与的号召，将土沟村居民凝聚起来。2003 年，台南艺术大学建筑艺术研究所小区营造组进驻土沟，开启了小区与艺术的连接，这些研究生一直进驻至今。

本案之规划设计理念为：

1. 平安：聚落里的各个角落隐藏了平安信息的符号，如：门联、喜幛、平安符、灯笼……本案藉由简单的艺术手法，将这些在地的平安符号转化，成为家家户户幸福生活的一部分。

2. 在地、手感：利用在地元素及材料，手感的创造，创造出具农村感的作品，如：铁车轮、废弃红砖、红被单等，都变成生活中的艺术品。

3. 点线面的整体聚落改造：以小点的生活对象艺术创作，到家户的景观艺术改造，连接家户与家户间的生活路径，慢慢串联起整体艺术聚落。

4. 艺术行动：除了艺术介入空间、将空间艺术化外，还结合了表演艺术、歌曲创作的艺术，亦包括了村民的艺术教育活动。

解读

美国著名考古学家 Gordon Randolph Willey 认为："聚落是人类对自己所居住的地面上所做的各种处理方式，包括了房屋的形式，房屋的安置模式，以及与社区生活有关之建筑物的性质及安排方式。"也就是说聚落并不是简单可以用地理面积去衡量的场所，它不同于传统意义上的居民点，而是住宅区、生产和生活设施的集合体。再换句话说，聚落的公共空间是精神的载体，公共空间的利用和改善与居民的生活质量以及生活状态息息相关。公共艺术项目的进驻，可以极大地丰富居民的精神生活，而公共艺术作为公共空间的改造和建设手段，为居民参与其中提供了很大的可能性。台湾台南的平安竹仔脚——聚落生活空间艺术改造行动就是一项团结居民改造生活空间的集体艺术活动，也是一种文化的交汇。艺术改造项目带来的不光是艺术，还有年轻人的活力。居民们积极地参与到这项活动中，就地取材，将生活元素融入艺术。项目结束后，焕然一新的不光是生活环境，还有居民们的精神生活。（高浅）

Artwork Description

TuGou Village, with its farmland, Guanziling Mountain, protected white egrets, and tea, is located inside the northernmost wall of Tainan City, Taiwan. It had once been a relatively forgotten village, populated almost entirely by children and elderly. But in 2002 a local group of young people organized the "TuGou Agricultural Town Cultural Development Association" to invigorate the local spirit, clean up parts of the district, build civic participation and galvanize everyone in TuGou Village.

In 2003 the Tianan National University Institute of Art & Architecture Research Institute stationed a study group in TuGou Village, experimenting with connections between the community and art. It's called the Bamboo Fellow settlement. The graduate students have continued their involvement with the community ever since.

Bamboo Fellow's Peaceful Corner project has four parts:

1. Peace: Messages and symbols of peace were placed in unobtrusive corners around the community, including door curtains, talismans, peace symbols, lanterns and so on. Introducing these simple art pieces was a way of bringing lightness and happiness to each household.

2. Touching native materials: Creating artworks using local materials and implements such as iron wheels, red brick from abandoned buildings, blankets and such.

3. Overall transformation of the settlement: Starting from single projects within the households to more collective participatory programs, developing connections between people, their household spaces and their life paths.

4. Art events: Besides bringing art into spaces and fostering spatial aesthetics, the project integrated performance art, music and songwriting into the community through events and other arts education initiatives.

Artwork Excellence

1. Providing an opportunity for young people to live in the countryside and participate in the transformation of rural communities

2. Using local resources in new ways to improve the living environment

3. Long term programming that interfaces with the economy

4. Mutually beneficial exchanges between artists and residents

"艺术让生活更美好"
——上海曹杨新村公共艺术活动
"Better Art, Better Life"
- Caoyang New Village Public Art Project

艺术家：上海大学美术学院（中国）
地点：中国上海
推荐人：潘力
Artist: Shanghai University, Academy of Fine Arts
Location: Shanghai, China
Researcher: Pan Li

作品描述

"'艺术让生活更美好'——上海曹杨新村公共艺术活动"中"曹杨新村公共导示艺术"是革命样板戏《红色娘子军》中的芭蕾造型的公共导向设施，放置在社区入口及公共区域；"被单文化"是艺术家邀请居民把美好的愿望剪成文字缝制在自家的被单上，并挂在自家阳台外和社区内的晒衣架上进行展示，约有 1 900 位居民参与到这件作品的创作。"美好时光"是卡通造型的灯光装置，安装在社区主干道两侧的 16 栋建筑入口的屋檐下。

解读

贫困时候的幸福其实很简单，一口米饭就是幸福。寒冷时候的幸福也很简单，一床暖被便是幸福。当今社会，当今城市，鲜少缺衣少食，也不再食不果腹，但社会的浮华，却让幸福感距离我们越来越远。在世事蜕变、年华老去的时候，唯有灵魂的羽翼永远丰满，记忆的拾起是精神上能够得到的最大满足。曹杨新村的这一群老人并不需要简单的经济援助，也不需要生活物品的慰藉，对于曾经满载光荣搬进这个新村的他们来说，时间的流逝带不走那些引以为傲的荣耀，他们满怀着对于生活的期冀参与到艺术活动之中，重拾起曾经的记忆，晾晒起对于生活的热忱，他们的快乐和热情因为艺术的介入而更加的闪亮。在他们眼中，生活一直如此美好和明媚，生命也如是。（高浅）

Artwork Description

The Better Art, Better Life project was planned by the Academy of Fine Arts at Shanghai University for the 2010 Shanghai Work Expo, whose theme was "Better City, Better Life." Its aim was to introduce public art into Caoyang Village, an aging community built in 1950. Ten well-known artists and curators were invited to participate, along with about 1,000 art and engineering students from Shanghai University. What follows is a summary of some of the most representative works in the project.

• In Xie Jian's Caoyang New Village Space Regeneration, the artist made design changes to local residential buildings through consultation and interaction with the residents.

• For Laundry Culture, artists Marjolijn Dijkman and Zhao Lei invited residents to write words expressing hopes and joy on their linens then display them outdoors on drying racks. Approximately 1900 residents participated in this project.

• In Zhang Lili's Caoyang New Village Public Pathfinding Art signposts in the shape of "Red Pioneer" girls were placed at the entrance to the community and in public areas.

• Jin Yinuo and her team created a work called Color, Play, Life using bright colored paints to bring life to the neighborhood's public open-air clothes hanging racks.

•Han Feng's Good Old Days used lighted signs made in a cartoon-style affixed above the building entranceways of 16 apartment buildings lining the main road through the community.

In addition to these projects, the artists organized performance art projects that elicited resident's memories, such as an opening ceremony in which 5 foreign artists offered flowers to former "model workers" of the community, stirring recollections of past glory. Wearing green army caps, red bandanas, navy shirts and white running shoes, the students marched into Caoyang Village during the opening ceremony, inviting residents to play old childhood games, which everyone enthusiastically participated in.

Artwork Excellence

The project is notable for its organization and implementation. The Shanghai Bureau of Education provided funds for the project, while Putuo District Government, Shanghai University, and Caoyang New Village Neighborhood Committee formed a tripartite organizing coalition. To establish the working parameters and criteria for projects the curators, artists and organizers themselves first completed field trips and discussions with the residents of the community. After this preliminary step artists freely designed their different projects, and subsequently the organizing committee and a panel of scholars selected the 20 top proposals. The project period was one year, and all works were created specifically for the local conditions and created on-site with a wide range of different kinds of resident involvement and interaction. Also worth mentioning is the fact that the district government has taken responsibility for management and maintenance of the building structures and spaces created during the project, thereby improving the community environment over the long term.

重庆通远门城墙遗址公园
Chongqing Tongyuan Gate Ancient Wall Park

艺术家：陈雨茁（中国）
地点：中国重庆
推荐人：潘力
Artist: Kawayan de Guia（Philippines）
Location: Baguio Philippines
Researcher: Kelly Carmichael

作品描述

重庆市七星岗"重庆通远门城墙遗址公园"系 2004 年重庆市渝中区重点建设项目，由渝中区政府投资 6 800 万元，于 2004 年 4 月 1 日开工，2005 年 2 月建成。项目源于政府希望保护重庆古城——通远门城墙遗址和延续城市文脉的愿望。七星岗通远门是重庆古城九开八闭十七门中唯一还残留着的一段古城墙，因以前这里是古战场，死伤无数，所以也称为"乱葬岗"。重庆直辖以来，经济日益繁荣。为打造一段巴渝文化风情景观带，2004 年 4 月，渝中区人民政府计划从一号桥开始，新建一座山城步道，经华一坡，上七星岗，穿过通远门，到金汤街、领事巷，与石板坡的山城步道相连，形成一段文物古迹和历史建筑相结合的传统街区。2005 年，政府拆迁了七星岗附近的一片旧房，建成现在的"城墙遗址"。当初讨论时，有人认为通远门是重庆当时通向成都的重要关口，具有商业价值，而发生在这里有影响的战斗并不多，因此雕塑应表现商贸主题。采用攻城雕塑。经多方调查和采访群众意见，攻城雕塑更符合通远门周围的环境，也满足老百姓潜在的心愿。所以，当通远门的一座山变成了古战场，一块大理石的《通远门赋》，解释了大型历史雕塑的由来。七星岗的旅游音阶一下子提高了八度。许多公交车站就设在通远门城墙下。通远门及古城墙遗址位于渝中区七星岗中山一路段。原来的重庆城区的边缘随着历史的变迁和城区的扩展，现已位于市区主干道上，四条公路在这里交汇，每天车水马龙，人流不断，不同历史时期的民居建筑附着在城墙上修建。城上城下市井生活悠然自得，原来守护城区的关隘如今成为人民安居乐业的场所。

解读

公共艺术项目的功能之一是为了传播城市文化并得到市民的认同。作为巴文化中心，对于自古以来就是军事要塞的重庆来说，城市的历史就是战争的历史，虽然当年金戈铁马、气吞山河的气势已云飞泥沉，猿鹤沙虫也已作古。但为了保留这段以兵去兵的城市史，让游客和市民对于重庆历史的了解更加直观，重庆市政府依附古城墙，投资建设了"重庆通远门城墙遗址公园"，这组名为"攻城"的铜质雕塑，就坐落于古城墙前。但凡是经过此地的人们，都会被极具画面感的雕塑强化记忆。这是现代和历史的结合，也是艺术和生活井然有序的共存。（高浅）

Artwork Excellence

The Chonqing Tongyuan Gate Ancient Wall Park is located in the Qi Xinggang neighborhood of Chongqing's Yuzhong District. It is a key construction project of Yuzhong District in which the district government invested 68 million RMB. The project includes the old city wall, sculptures, plazas and other elements.

Tongyuan Gate is one of Chongqing's two remaining city gates, and its surrounding walls are the only remaining intact ancient walls. Formerly the east gate to the town of Shudu, the wall was originally constructed by the city protector Li Yan in the 4th year of the Jian Xing reign during the Three Kingdoms period (CE 226), and was fortified by Peng Daya in the 2nd year of the Jia Xi reign of the Southern Song Dynasty (CE 1238) to fend off the invading Mongolian cavalry. At the beginning of the Hongwu period of the Ming Dynasty the base of the old wall was reconstructed with brick and a ding (bronze cauldron) erected there. There are 12 trees in the Ancient Wall Park that serve as structural retaining elements for the wall.

A group of sculptures showing the old fortifications, the city defenders, the historical story of a siege, an army general reigning in his rearing horse, foot-soldiers erecting a siege ladder, and other related people, gives the public an opportunity to return to the scene of an attack on the ancient city.

In addition two folk sculptures, Fried Rice in Sugar Water and Chair Litter, convey the peacetime life of the old city of Chongqing, visions from the old marketplace and its vanished customs. A set of cannons is situated in a plaza above the park wall, alongside a seating area amongst shady trees and landscaping where people can relax.

At the intersection of Zhongshan Road and Jinyang Street is a small square where wall comes to an end, and a pathway ascends to a small plaza on top of the wall. On the west side of the wall is a replica of an ancient building called Tongyuan House, which offers panoramic views of the city and Ancient Wall Park, and is a focal point of views from the surrounding area. At the other end of the wall is a reconstruction of residential buildings that are used as a teahouse, and a calligraphy and painting studio. At this point the wall merges with the modern-day built environment.

Artwork Excellence

Designed with input from the community, this extensive project combines historic preservation, interpretive history, and recreational/leisure spaces using a dramatic set of sculptures depicting historic events at the Chongqing Tongyuan Gate Ancient Wall Park.

门
Doors

艺术家：崔贞华（韩国）
地点：韩国首尔
推荐人：凯利·卡迈克尔
Artist: Choi Jeong-Hwa (Korea)
Location: Seoul, South Korea
Researcher: Kelly Carmichael

作品描述

"门"（2009 年）是一件庞大的公共艺术装置，动用了 1 000 扇门，由韩国艺术家和设计师崔贞华创作。该作品使用了色彩鲜艳的废弃的家庭房门，把它们加装在建筑脚手架上。艺术家将首尔一幢中层大楼平淡的外部进行了转变，做了大胆的声明，起到了一种幻动的效果。在这件 10 层楼高的艺术品中，崔贞华尝试用特殊的方式重新使用日常生活的物品，"门"项目是艺术家最雄心勃勃的作品之一。

艺术家以他那吸引眼球的视觉、放任般的作品而闻名，他的作品看上去情绪高昂、充满乐趣、具有装饰性而且色彩缤纷，它们既吸引人又具有娱乐性，毫不掩饰地要与公共空间的使用者发生联系。崔贞华的作品如从近处观看更能够激发人内心深处的共鸣。与许多同一年龄段的当代韩国艺术家一样，崔的艺术作品常常以幽默的手法强调生命的短暂本质。在这个不断变化的世界里，艺术家被塑料那具有讽刺意味的不变和永恒的本质所吸引。

"我用了很长时间去思考一朵真正的花和一朵塑料花之间的区别，"他解释道，"真正的花朵在很短暂的时间内就凋零、衰败了，而塑料花却是不朽的。同时，有趣的事是有时假花看上去很真实，而真花看上去很假。我反复思考塑料、真实和虚假这些概念，塑料系列作品就这样开始了。"

受到当地室外市场混乱环境的启发，崔贞华在他作品的原材料中使用了令人印象深刻的光谱，包括视频监视仪、模压动物、真假食物、所有日常生活中的物品、灯光和电线。尤其是大量的塑料，那些廉价、色彩鲜艳、大量生产、可塑性塑料。在任何的街道市场或五金商店里都能见到这些塑料。

崔贞华使用日常物品创作自己的装置作品，使他的艺术吸引广大的观众，促使他们以不同的眼光来看待包围着他们的种种物体和杂乱。他的作品——诸如"高高兴兴"（2009 年）——请求观众不仅将普通的街道、市场里的材料和塑料家居用品作为一种新事物来看待，同时对他们视觉景观中的脏乱现象提出质疑。他在高尚艺术 / 低俗文化之间玩弄着模糊概念，已知的作品是建造了一座巨型的迷宫装置，用来"迷惑"人们，让他们在这块创造的空间里四处移动，要求观众以新的方式来看待这些日常用品和这块营造的空间。

在对建筑、艺术、自然和环境营造之间的边界扩散进行探索的同时，崔贞华还质疑了废弃物的观念。我们通常认为在当代生活中收集塑料和废弃物品是很负面的，崔贞华则有意对这种传统观念提出挑战。他为 2008 年"首尔设计奥林匹克"创作了一件作品：他把一千万名参观者留下的废弃垃圾收集起来，悬挂在蚕室体育馆内。

"这个创意关注点是认识到垃圾与艺术之间的区别并且提出这一问题：什么引发我们的情绪和情感？谁有权决定什么有价值什么没有价值？整座体育馆内覆盖满了垃圾，但是当灯光投射上去时，作品变得很美丽，闪闪发光，让人难忘。这基本上是一场运动，强调使用不值钱的材料进行创作。我正在尝试发现艺术和垃圾之间的区别，去发现是什么感动着他们（公众）"。创作这件大规模的装置使用了各种各样的塑料制品，比如牛奶盒、水瓶和包装箱。

解读

门是空间的代言人，门的启合界定了空间的尺度。门与门形态乍看无甚不同，但却难以知晓跨入门之后的隐喻。我们能够掌控空间的实体，却无法掌握作为空间附属品之进入所带来的命运，童话故事里面的爱丽丝追随着大白兔子进入了奇幻世界，经历美好的让人神往；而在现实生活中劳碌生活的人们，何曾不渴望着童话并不仅仅只是童话？由 1 000 多扇七彩斑斓的门组成的作品象征的也是七彩斑斓的人生和梦想。人们无法掌控自己的人生，却大可以幻想每扇门背后的故事，下一步跨入的空间中面对着你的会是谁？等待着你的又是什么？一步的迈进会不会构筑人生的转折？作为一个在社会中活动着的自由人，我们有限的生命中要走进的门数不胜数。那是否已经做好十全的准备去迎接每一扇门的敞开？（高浅）

Artwork Description

Doors (2009) is an enormous public art installation of 1,000 doors by South Korean artist and designer Choi Jeong-Hwa. Made from brightly colored, reused domestic doors attached to a construction scaffold, the work transforms the bland exterior of a mid-rise Seoul building, making a bold statement and giving a pixilated effect.

The ambitious, 10-story artwork captures the essence of Choi Jeong-Hwa's practice. The Doors project is one of the artist's most ambitious attempts to reuse everyday items in extraordinary ways. Known for works of eye-popping visual indulgence that turn an abundance of objects into joyful cascades in public places, his works have immediate appeal. His work is high-spirited, playful, decorative and colourful, it engages and entertains, overtly setting out to connect with users of public space.

Jeong-Hwa's works resonate more deeply on closer inspection. In common with many contemporary Korean artists of his generation, Choi Jeong-Hwa's practices focuses, often humorously, on the ephemeral nature of life. In a constantly changing and decaying world he is attracted to the ironic permanence and eternal qualities of plastic.

"I contemplate the differences between a real flower and a plastic flower a lot," he explains. "The real flower can be ripped and will deteriorate in a short time, but the plastic flower is immortal. And the interesting thing is that there is such thing as a fake flower that looks real and a fake flower that looks fake. I played around with these concepts of plastic, real, and fake; and that's how the plastic series started."

Inspired by the chaotic milieu of local outdoor markets, Jeong-Hwa uses a impressive range spectrum of raw materials in his work including video monitors, moulded animals, real and fake food, all manner of everyday objects, lights, and wires. And especially plastic—tons of cheap, colourful, mass-produced, malleable plastic, the sort available at any street market or hardware shop.

Jeong-Hwa employs everyday objects in his installations to make art that engages with a wide audience and to impel viewers to approach the objects and clutter that surround them in a different way. His works—such as Happy, Happy (2009)—ask their audience to not only see common street market materials and plastic household necessities as something new, but also to question the detritus of their visual landscape. Playing with the ambiguities of high art/ low culture, he has been known to build giant maze-like installations to trap people, move them around the created space, and ask them to see not only the everyday objects but also the space in a new way.

While exploring the diffusion of boundaries between architecture, art, the natural and the built environment, Jeong-Hwa also questions notions of waste. While we typically think of the accumulation of plastic and disposable items in our contemporary lives as negative, Jeong-Hwa seeks to challenge that assumption. For a 2008 work at the Seoul Design Olympiad he gathered the trash thrown away by the 10 million people in attendance and hung it in Jamsil Stadium.

"The point was to see the difference between trash and art and ask the questions: What makes us feel emotion and affection? Who is to decide what's worthy and what's not? The whole stadium was covered with trash, but it became beautiful and sparkly and memorable when light was projected on it. It's basically a campaign that emphasises working with worthless materials. I'm trying to find the difference between art and garbage, to see what moves them 'the public' " The large scale installation was created using various types of plastic products such as milk containers, water bottles and crates.

Artwork Excellence

Choi Jeong-Hwa's Doors reinvents found-object artwork on a colossal scale. By skinning a bland, mid-rise building in Seoul with a multi-story curtain of reused domestic doors, Jeong-Hwa creates a marriage of beauty and practicality to a wry and approachable commentary on consumer/throw-away culture.

浮力
Flotage

艺术家：薇薇安·孙达拉姆（印度）
地点：印度新德里
推荐人：里奥·谭
Artist: Vivan Sundaram（India）
Location: New Delhi, India
Researcher: Leon Tan

作品描述

该作品由 48°公共艺术生态（德里的第一个公共艺术节）策展人普加·苏德委派，"浮力"是薇薇安·孙达拉姆创作的一件雕塑装置作品，为德里北部的古代莫卧儿遗址罗莎纳拉·巴格（花园）所作。该作品设计、建造，展示于 2008 年 12 月，"浮力"的创作工作由艺术家与城市拾荒者、环境研究和 Chintan 行动小组共同完成，他们在罗莎纳拉陵墓笔直视线的凹陷区域内收集组装一万个塑料瓶子，在曾经装水的地面凹陷处安装了一个格框，带着明亮粉红色瓶盖的塑料瓶子铺开了一个鲜艳色彩的区域，将废弃的塑料有效地转换成了一种视觉奇观，原本这些废弃的瓶子很快就从瓶装水消费者的视线里消失，通常会被拾荒者踩扁，送到回收站去换取现金。"浮力"由两个蛋形的氦气球组成，漂浮在粉红色塑料瓶盖区域附近，起到了信号灯的作用，让很远处的观众一眼就能看到凹陷的波尔卡虚线彩色区域。气球信号灯和用来存贮已被日益商品化的公共资源"蓝金"（水）的一次性容器加在一起，构成了作品的布局。"浮力"可以被理解为对大规模的

223

社会经济变化过程,土地共有权的私有化,消费主义的商品一次性丢弃文化、以及由此产生的废弃物堆积问题(其中有许多废弃物的分解过程非常缓慢)的静心反思。

事实上,瓶装水会导致每年产生大约 150 万吨的废塑料,而这些塑料则每年需要高达 47 万加仑的石油才能生产出来。瓶装水不仅使生态系统耗费巨大,对瓶装水依赖的不断增加会对有利于公益的市政供水系统安全的发展与维护造成损害。作为一件持续性的装置,"浮力"完成了 48°公共艺术节的关键目标:质问全世界污染最严重地区之一的新德里脆弱的生态系统,吸引大量观众并约他们一起在花园内参与休闲活动以期振兴长期被忽视和遗弃的市内罗莎纳拉地区。因此,作为该项目的一个延伸,孙达拉姆把堆放到 4 英尺高的瓶子由一队拾荒者骑自行车穿越城市街区,运送到被忽视、污染的亚穆纳河。这次穿越街道的游行引发了公众的广泛关注,促发人们对拾荒者提出了大量的问题,因而引起大家对这种到处可见的塑料水瓶应该如何循环利用问题的关注。在河上,这些瓶子被组装成一个 45 英尺长的筏子,沿着亚穆纳河漂流而下,最终彰显了这个项目标题的内在含义。

孙达拉姆由"浮力"的这一主题制作了一个影像作品,该作品后来在葡萄牙贝拉多博物馆(2009 年)、芝加哥威尔士画廊(2009 年)、亚穆纳河—易北河公共艺术项目(2011—2012 年)展出过。贝拉多博物馆获得了"浮力"作品的一部分,与此同时,剩余的瓶子由拾荒者转移至粉碎厂,在那里塑料被转化为羊毛纤维,作为冬季外套里的御寒填充物对外出售。

"浮力"最终,按艺术家的说法,成为"一个循环游戏",它不仅涉及印度的塑料回收,纪录片像带的回收利用也扩大到了世界各地。它是一场好玩又严肃的游戏,宣告了一场充分开发利用水资的战争即将开始。

解读

我们得益于自然的恩赐,拥有了惊人的成就。现今却受困于自然,难以继续发展。当艺术品可以作为语言的表征,代替声音阐述思想的同时,设计师们也在不断探讨当代艺术家、艺术作品的社会和理论价值。每个艺术家在设计艺术作品的时候都应该希望自己的作品能够触动群众心灵,影响群众行为或者生活方式。美国设计理论教育家 Papanek 认为:"设计的最大作用并不是创造商业价值,也不是包装和风格方面的竞争,而是一种适当的社会变革过程中的元素。"他带来了绿色设计和生态设计概念的思潮,旨在宣扬低碳节能的 3R 思想——Reduce, Reuse ,Recycle,即减少能源消耗、重复使用、回收。公共艺术设计作为肩负着政府行政话语权的艺术表达形态,必然被这种设计方式影响,借由作品传播 3R 概念,意在影响观者行为。

桑德拉姆是 Papanel 理论的忠实追随者,她舍弃了传统的创作材料,例如青铜、泥土、木头、石料等,转而使用了废弃的塑料瓶这一普通又饱受争议的化学产物,这是 Reduce 和 Recycle;再将这些早已经失去实用价值的废弃塑料瓶打造成艺术品,浮在水面,承载观客,传递艺术理念,再次发挥价值,这是 Reuse;最后销毁,回收,这是 Recycle。这一过程形铸就了良性循环。即便她丢弃传统的自然创作原材料是由于这些常规自然原料基于自身物理属性的限制,无法具备这个艺术作品所需要的特征,例如良好的密闭性,隔水性和浮力,但选用废弃的而非全新的塑料瓶进行创作就是证明她是宣扬绿色环保理念的生态设计师的最好证明。(高浅)

Artwork Description

Commissioned by Pooja Sood, Curator of 48°C Public Art Ecology (Delhi's first public art festival), Flotage is a sculptural installation created by Vivan Sundaram for the ancient Mughal site Roshanara Bagh (Garden) in North Delhi. Designed, constructed and exhibited in December 2008, Flotage involved the artist working with the city's waste-pickers and the environmental research and action group Chintan, in order to collect and assemble 10,000 plastic water bottles in a recessed area in direct line of sight of Roshanara's Mausoleum. Organized as a grid within the recess in the ground that once contained water, the plastic bottles with their bright pink caps laid out a vibrant color field, effectively transforming the waste plastic into a visual spectacle, where otherwise they would usually disappear rapidly from the view of consumers of bottled water, hunted down and delivered to recycling sites in exchange for cash by waste pickers.

Flotage also consisted of two large egg-shaped helium balloons floating in the vicinity of the pink field of plastic caps, functioning as beacons signaling the presence of the recessed polka-dotted color field to visitors from afar. Taken together as a composition, the signaling beacons and the disposable containers for a public resource (water) that has increasingly become commodified as "blue gold," Flotage may be understood as a meditation on large scale socio-economic processes, privatization of the commons, consumerist throw-away culture, and the resultant problem of accumulating waste, much of which has a very slow rate of decay. In fact, bottled water is responsible for approximately 1.5 million tons of plastic waste per year, requiring up to 47 million gallons of oil per year to produce. Bottled water is not only costly for the ecosystem, increasing reliance on it detracts from the development and maintenance of safe municipal water systems as a public good.

As a durational installation, Flotage accomplished key objectives of the 48 °C public art festival, namely, to interrogate the fragile Delhi ecosystem, one of the most polluted in the world, and to revitalize the Roshanara area, long abandoned and neglected within the city by drawing in numerous visitors and engaging them in leisure activities within the garden. As an extension of the project, Sundaram had the bottles (stacked 4-feet high) transported by a team of waste pickers on bicycles through the city streets to the neglected and polluted Yamuna River. This procession through the streets stimulated a great deal of public attention, prompting many questions to the waste-pickers, thus drawing attention to the life cycle of the ubiquitous commodity of the plastic water bottle. At the river, the bottles were assembled into a 45-foot-long raft, which was then floated along the Yamuna, finally becoming worthy of the project title.

A video-work was produced by Sundaram based on this aspect of Flotage, which was subsequently exhibited in the Berardo Museum, Portugal (2009), the Walsh Gallery, Chicago (2009) and in the Yamuna-Elbe public art project (2011-2012). Part of the Flotage work was acquired by the Berardo Museum, while the rest of the bottles

were transported by waste-pickers to a shredding factory, where the plastic was converted into fibre-wool and on-sold as stuffing for puffer winter jackets for guarding against the elements. Flotage ultimately became "a recycling game" to quote the artist, involving not only the recycling of plastic in India, but also the recycling of documentary video across international sites, a game at once playful and serious, raising the specter of a coming war for the exploitation of blue gold.

Artwork Excellence

The international excellence of Flotage is attested to by the acquisition of a part of the sculptural installation by Portugal's Berardo Museum, as well as by the exhibition of the video-work deriving from Flotage at local and international sites including the Yamuna-Elbe project, the Walsh Gallery, Chicago, and the Berardo Museum, Portugal. As a place-making intervention, Flotage must also be recognized as a success for the manner in which it revitalized the neglected Roshanara area, and the heritage site Roshanara Bagh, by attracting numerous visitors to take part in leisure activities within the garden. Additionally, in stimulating public questions and thinking on large-scale processes such as the commodification of water and the life cycle of waste products born of consumerist habits, Flotage stands out as a public art intervention that contributes to the raising of ecological consciousness, in a mega-city that is plagued by persistent problems of pollution.

足球场
Football Field

艺术家：梅德·洛佩兹（西班牙）
地点：阿联酋沙迦美术馆广场
推荐人：吉乌希·切克拉
Artist：Maider López（Spain）
Location：Sharjah Museum Square, Sharjah, Arab Emirates
Researcher：Giusy Checola

作品描述

艺术家在沙迦美术馆广场创建一个足球场，提出广场的新用途和一个城市空间的新结构，广场上的街道设施也被带进此足球游戏中。她漆上红色的球场结构线，并在两边加上球门。由于之前就存在如长椅和街灯这些无法改变的广场元素，广场成为一个综合多用途的公共空间。她的作品都是以特定地点的创作为基础的，从对空间颠覆性的干预到完全推翻空间的通常运作方式的介入行为，暂时改变该地点的意义，并作为改变其与公众关系的结果。

通过把足球场的娱乐用途覆盖在广场多功能结构上，创造了共同存在的新互动方式。因为这是通过人们实践而意在重新改写城市观念，居住在空间附近居民和其经历活跃了并推动了该项目的气氛，从而表现人类改变城市空间及建造更适于其空间使用的能力。"足球场"的网格线将公共广场转化为艺术街区。

这是一个简易轻松、基于人类关系的多种活动融合的模式。该项目的成功之处在于，从跨学科的角度考虑艺术的社会、政治和文化意义，作为临时的规划已经被社区作为永久城市项目而接受，为大众和自然、环境的关系创造出更好的理解途径。

解读

让居民以最简单快捷的方式融于社会，让居民行为能够直接作用于生活是城市公共艺术设计之道。在很多时候，我们宣扬的公共空间的营造并不需要巨大的雕塑或者华美的装置，更不需要另辟天地，只要有其真正的存在价值，便能坦然存在于人群之中，独立分割，孑然存在。城市文化可以增强居民的幸福感，因此，多数以升级居民生活水平为目的的公共艺术项目都会被支持。沙迦美术馆前广场上的足球场没有隔离网，没有进入门，简简单单的几条线和两座球门为足球爱好者们开辟了意识空间。而公众设施——灯柱，则毫无顾忌地成为了"足球场"艺术项目的一个部件，设计师没有因为广场上的灯柱而缩小足球场的面积，她只是提出了一个新的城市构建理念，设计并不需要拘泥于环境，设计师们的构想也许并不需要受到场地的限制，只要是能为城市居民带来欢乐的公共设计项目都是好的设计项目。（高浅）

Artwork Description

In the center of Sharjah Museum Square the artist painted the red lines of a football field and placed goalposts at either end. Existing benches and streetlamps remained in place, incorporated into the playing field. In effect, the square became a hybrid multi-purpose public space, taking on new uses and an altered significance. Site-specificity underpins all of Maider Lopez's work, with interventions ranging from subversive tampering with spaces to completely overturning their everyday functionality. These interventions temporarily change the meaning of the place, and as a consequence change the public's relationship to it.

The project grafted an entertainment use over the formal function of the square, creating co-existing modes of interactivity. Nearby residents were able to experience the space anew, and through new practices to reaffirm the city as a notion that is always being revised. It reflects humankind's capacity to alter the spaces of our cities in order that they may be more suitable for living. The football field's stripes transformed the public square into an area of art.

This is a model of a simple, low-impact intervention which integrates different human activities. The project succeeds from a social sciences point of view in addressing the role of art in society, politics and culture; and in terms of direct experience, the project has been accepted by the community as a permanent feature of the city, enhancing understanding of the relationship between nature, environment and the public.

Artwork Excellence

The project suggests a model for easy and relaxed integration of different activities, based on human relations. It's informed and activated by the inhabitants of the space and their experience, since it attempts to rework the idea of the city through the practice of the people. It aims to show people's capacity for transforming urban space and building up the city with their own use of the space. One measure of its success is that what was planned as a temporary project has been adopted by the community as a permanent urban intervention.

猴猴先生
Hohosan

艺术家：李明道（中国）
地点：中国台湾新北市
推荐人：赖香伶、朱惠芬
Artist: Li Mingdao（China）
Location: Xinbei City, Taiwan, China
Researcher: Hsiangling Lai, Huey-Fen Chu

作品描述

位于台湾新北市瑞芳区之侯硐，地名的由来是因附近山洞曾有猿猴群居栖息之故。这里曾是采金热潮的发源地之一，也曾是台湾煤矿最大的产地。现今煤金产业虽荣景不再，却因风景灵秀成为热门观光景点。

本案作品"猴猴先生"的概念来自于平埔族的地名"侯硐"，艺术家李明道在考察过学校环境以及建筑特色之后，便有了"猴猴先生"的创作构想。多年来艺术家活跃于设计界及艺术界，作品具有非常鲜明的个人风格。早期为歌手所设计的唱片封面具有一种糅杂了科技感和古早位的台式普普趣味，兼有前卫和怀旧的质感；而近期为人所熟知的数字艺术创作、动画或立体雕塑，更是源自艺术家个人的收藏喜好（旧式公仔、机器人以及各式车与枪等玩具）以及对于大众文化的敏锐嗅觉。艺术家作品中较常出现的动物和机器人，都带有拟人化的表情个性，造型既讨喜又兼有创意质地，是台湾难得成功跨越商业艺术和纯粹造型艺术的创作者。

艺术家以童心、爱心以及绝佳创意，为当地之侯硐小学打造了兼具观赏（雕塑）、游憩（树屋、滑梯）以及具备实用功能的公共艺术作品。侯硐小学的公共艺术规划，源于为学校营造一个"家"的氛围，设计一位陪伴孩子成长的"家人"之构想。由于侯硐地名与猴群栖息地的渊源，艺术家因而选择猴子作为学校和孩子的家人及忠实伙伴，并为其取了与自己日语小名Akibo（明仔）有着相同昵趣的名字"猴猴先生"。本件作品造型极富动漫趣味，透过艺术家的巧思构想，远观是雕塑（或大型公仔），近看是游具（溜滑梯＋树屋），头部设计为树屋与瞭望台，身身规划为溜滑梯，后背则化为阶梯，尾巴为扶手，以方便孩子上下攀附。本作品于创作期间和落成后，已成为学校的鲜明标志，并累积发展出师生和家长对于校方的认同感。

解读

公共领域与公共艺术之间有着密不可分的关系。在哲学与社会学中，公共领域是介于国家和社会之间的公共空间，而从字面意义上看，公共领域直接与私人领域相对。哈贝马斯认为公共领域必须具备如下几个特征：首先，如字面意义，公共领域是公共的，而非私人的；其次，公共领域是一个可以产生公共意见的场所；最后，公共领域的利益（无论是实际利益还是精神利益）都属于公众。而公共艺术可以被认为是依附在公共领域特征之下的、诞生于此空间内的产物。其产生的目的就是在实现物质性艺术的前提下，体现大多数人的文化态度、生活态度和精神需求，在公共领域中激起民众的共鸣，实现其公共领域的价值。公共艺术作品应当具备文化性，历史性、符号性、场域性、空间性等，这些附加在公共艺术中，存在于公共领域内的性征，引发的深思和争议应该远远超过艺术本身。

李明道先生的"猴猴先生"就是这样一个具有实用功能和公共领域特征的艺术作品，新北市瑞芳区之侯硐曾经是猿猴群栖的地方，而现在当地的隔代教养风俗又是一个特殊的家庭形态。为了给在特殊家庭环境中生活的小朋友们提供一个玩耍交流的场所，创造一个快乐和谐的氛围，李先生在当地小学内设计了这个富有动漫趣味的"猴猴先生"。而此设计方案正和哈贝马斯提出的观念不谋而合。（高浅）

Artwork Description

Houtong Elementary School is located in Houtong Town, which is situated in Xinbei City's Juifang District in Taiwan. Its name is derived from a story about a society of apes that once lived in nearby hill caves. It is known for a gold mining boom that occurred here, and for being the birthplace of Taiwan's coal mining industry. While the mining industries have since moved on the area's scenic landscape has brought a new wave of tourism.

During planning meetings the school emphasized the need for a "home-like" atmosphere for the kids, and for a partner who would be like "family and friend" to accompany the children in their growth. After visiting the local Houtong Elementary School grounds, the area's former ape habitat, and reflecting on the Pingpu ethnic minority's name for the place, Houtong, artist Li Mingdao chose a monkey to serve as the school children's loyal partner, naming him Hohosan ("Ho" is the Chinese word for "monkey").

On behalf of the school the artist created a work that combines the ornamental qualities of sculpture with the functional qualities of a playground or tree house, embodying playful innocence, love and good values. The artist came up with an ingenious design, which from a distance looks like a sculpture (or large mascot for civic responsibility) and up close becomes a playground (with slide and tree house). The monkey's head can be used as a tree house lookout, with slides down the front and a ladder with handrails on the tail for kids to cling to. It succeeds in expanding the limited space on the school campus, providing both a useful meeting place for teachers and students and an excellent recreational space.

Li Mingdao has developed a very distinctive style over many years working in the art and design worlds. Early on he designed album covers for pop musicians, mixing a technological sensibility with an old-fashioned visual aesthetic, both avant-garde and retro. More recently he is known for his digital art creations, animations and 3-D sculpture (drawing on his personal collection of old dolls, robots, and all kinds of cars, guns and other toys). He has a keen feel for movements of mass culture. His works most commonly take the form of animals and robots with anthropomorphic expressions of personality, and have achieved rare success in both the commercial and arts domains.

Artwork Excellence

Through its creation and since its completion, Hohosan has become a distinctive symbol of the Houtong Elementary School, and has galvanized teachers, students and parents. It successfully conveys elements of the local culture and history, while incorporating animation shapes adored by elementary school children, and fits perfectly with the surrounding campus. While accommodating priorities of safety and practicality, the work reflects the artist's concerns for "art in life" and "art and public participation," and serves as a model for schools to create shared interactive spaces through public art.

拉尔集市
Lal Chowk

艺术家：尼基·乔普拉（印度）
放置地点：印度斯利那加
推荐人：里奥·谭
Artist: Nikhil Chopra (India)
Location: Srinagar, India
Researcher: Leon Tan

作品描述

"拉尔集市"是尼基·乔普拉的一件公共行为作品，于 2007 年 11 月在克什米尔的斯利那加市中心拉尔集市（红色广场）公开表演。

KHOJ 国际艺术家协会 2007 年工作室项目的一部分，包含了许多艺术家的作品，乔普拉的作品毫无疑问是其中最大胆的。

在印度和巴基斯坦之间有一个出名的地区，那里充斥着政治冲突、恐怖活动和抗议，那里的武装部队无处不在，行为艺术从乔普拉的夸张表演开始，通过戏服转换为一种艺术上的"字符"，即：约格·拉吉·契特拉卡尔，艺术的过程是艺术家徒步走在拉尔集市中心，这一路上他画了周围的建筑，广场中心的钟楼和周围的楼房，他用碳笔直接画在马路表面。

穿着花花公子的式样，他穿着马裤、花呢夹克，戴着领带，反映了在印度独立前印度精英吸取了英帝国主人的时尚风，契特拉卡尔打定主意要吸引当地观众的注意，在他画画时很多人在旁边围观。

这个过程立即召来了军队的注意，1990 年颁布的"军队（查谟和克什米尔）特别权力法"禁止超过五个以上的群体在公共场所聚集。

军事人员让围观者在一个点上排队，开始对他们进行人身搜查，同时要求艺术家离开。

乔普拉依旧保持沉默，坚持作画。值得注意的是，市民开始干预，敦促他继续："画吧，"他们说，"继续画，我们和你在一起，我们不会让任何人阻止你的。"

市民持续的干预行为在三小时后乔普拉离开广场后结束。

有意义的是，军方并没有强行提前结束这一公开事件，他们认为或许艺术家和当地人正在一起参与非暴力抵抗活动，设法收回一个公共空间，尽管有戒严限制，但是民众和艺术家至少有一段时间能获得他们自己的空间。

不像乔普拉在美术馆、博物馆空间内更加安全的行为艺术表演，这次的干预非常难忘，因为在当地人共同加入的情况下测试了公共空间这个概念的最大限度，并且把人民的主权和公共地区的所有权这类问题放到了一个充满政治因素的区域。

这个项目印证了它的长久意义，纪录片"约格·拉吉·契特拉卡尔：拉尔集市"在众多展览中展出，包括：2010 年纽约的新博物馆，2011 年澳大利亚墨尔本国际艺术节，费喜利当代乌托邦呈现，亲密公众作品。这个项目于 2010 年由艺术评论家吉利什·夏哈恩和南希·阿达加尼亚撰写过相关的文章。

解读

不知从何时开始 "Irreducible"、"force" 和 "art" 这三个词开始被频频同时提及。不可约的、武力、艺术，这看似无关联的词似乎成为了年轻艺术家们新的心头之好。并且，艺术不再与政治毫无关联，反而政府行为似乎配合着艺术发生某些改变，或干预或置放。

这个地点本就混乱，当地政府 1990 年还颁布了 "禁止超过五个以上的群体在公共场所聚集" 的法律。因此当乔普拉身着完全不符合当地人衣着习惯的行头出现在街头，这是一场预知的混乱，人们聚集起来，观看他的行为，这聚众本身就是在抗拒着法律和社会。随着政府军队的驱散行为，市民们开始了意识的觉醒，坚定地站在了画家的身边，成为了这项公共行为艺术中的公共艺术行为的参与者。而更值得一提的是官方军队随即默许了这种聚集行为，没有继续驱逐，他们也许认为艺术并不会对社会造成不安定影响，人群聚集也就是单纯的人群聚集，而并非会发生冲突导致地区的不和谐。

艺术家也许是想要通过其公共艺术行为试探政府干预市民生活的程度，试探政府对于公共艺术行为的态度。而政府的态度似乎也非常明确，并且并不如所预计那般消极。而这项行为所导致的直接结果就是，证明了公共艺术行为在拉尔集市中的存在形式是可以被群众和政府接受，并起到宣传引导作用。（高浅）

Artwork Description

Yog Raj Chitrakar: Lal Chowk is a public performance by Nikhil Chopra that took place in November 2007 in Lal Chowk (Red Square) in the center of the city of Srinagar, Kashmir. Part of the KHOJ International Artists' Association's 2007 workshop involving a number of other artists, Chopra's work was without doubt one of the most daring. In a region well-known for political conflict between India and Pakistan, terrorist activity and protest, as well as the ubiquitous presence of armed forces, the performance began with Chopra's metamorphosis through costume into an artistic "character," namely, Yog Raj Chitrakar, and proceeded with the artist making his way on foot to the center of Lal Chowk, where he created drawings of the surrounding architecture, the clock tower in the center of the square and surrounding buildings, in charcoal directly on the road surface.

Dressed in the manner of a dandy, his attire of breeches, tie and tweed jacket echoing the adoption of the fashions of British imperial "masters" by Indian elites in pre-Independence India, Chitrakar could not fail to attract an audience of locals, many of whom congregated around his drawing activity. This of course immediately drew the attention of armed military forces, as the "Armed Forces (Jammu and Kashmir) Special Powers Act, 1990" prohibits the assembly of groups of more than five in public spaces. Military personnel at one point lined the audience members up and subjected them to physical searches, and also ordered the artist to leave. Chopra-Chitrakar however, kept silent and persisted in his drawing activity. Remarkably, members of the public intervened, urging him on, "Draw, they said, keep drawing, we're with you, we won't let anyone stop you."

The durational intervention came to an end after 3 hours, when Chitrakar finally left the square. Significantly, the military did not forcibly bring the public event to a premature end, suggesting perhaps that the artist and the locals engaging in non-violent resistance managed to reclaim a public space, to make a place for themselves, despite conditions of martial law, at least for a period of time. Unlike Chopra's much "safer" performances in gallery and museum spaces, this intervention is highly memorable precisely because it engaged locals in testing the very limits of the notion of a public sphere, inserting into a politically fraught zone the question or problem of the sovereignty of a people and its ownership of a commons. As a testament to the project's long term significance, documentation of Yog Raj Chitraker: Lal Chowk has featured in a number of exhibitions including at the New Museum (2010, New York) and Utopia presents Intimate Publics at

Fehily Contemporary (2011, Melbourne International Festival, Australia). The project has also been written on by art critics Girish Shahane (2010) and Nancy Adajania (2010).

Artwork Excellence

While claims are frequently made concerning the political nature of many public artworks, not as many of such works end up in direct confrontations with armed military forces. Even fewer take place in sites as politically contentious as Srinagar, where martial law has prevailed since 1990. Nikhil Chopra's durational intervention in Lal Chowk stands out as one in which the possibility for public placemaking was tested to its political limits. In spontaneously galvanizing members of the public this artistic intervention became a form of non-violent resistance, enabling Kashmiris to reclaim for a spell a shred of sovereignty over a commons, despite the opposition of armed military forces. It demonstrates a rare case where art and politics meet in unexpected circumstances (no one could have anticipated the manner in which events unfolded), so as to empower audiences to lay claim to public space over-run by martial law.

现场张江
Living Zhangjiang

艺术家：现场张江学术委员会（中国）
地点：中国上海
推荐人：潘力
Artist: Living Zhangjiang oversight committee (China)
Location: Shanghai, China
Researcher: Pan Li

作品描述

丁乙"翼桥"：

这是一件极简主义的抽象雕塑，+符号的交叉结构在为雕塑提供了美学轮廓的同时，也成为桥梁两侧的护翼，为完成桥梁本身的结构提供了视觉和力学的完整性。这座桥的创作目的是让艺术渗入最日常的生活，使雕塑作品可以被行人使用。

郭伟"投降的熊"：

作品让人们在嬉戏的氛围中去体会人类应该与自然和谐相处的道理。如果违背了和谐原则，今天投降的是熊，明天投降的就可能是人类自己。雕塑做成蓝色，体现出作者希望眼前的一切都会成为美好又永久的游戏。

刘炜"胡萝卜"：

胡萝卜与超市购物车，都是日常生活中司空见惯的纯粹符号，艺术家试图赋予它们新的生命力和内在思考，让普普通通的物体弥漫出对于生命力的幻想和回忆。植物与土地、生态与消费，构成了一组荒诞、幽默而又生动的现实图景。

曲丰国"回"：

鲸，原本是生活在陆地上的动物，它们什么时候能够回来？在平静的草地上，一只尾鳍仿佛正随巨大的身体而移动。鳍的造型，象征着美丽、力量和方向，而张江的土地仿佛已化身为创意与梦想的海洋。

王玉平"鸟·人"：

艺术家试图通过表现人与鸟之间友好互动的瞬间，体现出对生命的尊重和感动，进而表达对人类与自然和谐相处的美好愿望。作品的构思来源于艺术家童年时期的养鸟经历。

薛松"数字圆桌"：

以"数字"为作品符号，是因为张江园区有着强烈的高科技特征，在数字化时代，科技、信息与数字是紧密相连的。艺术家希望能够模糊观赏性和实用性的界限，让人更加亲近雕塑，所以把数字符号做成了一组街头家具。

叶永青"红鸟笼"：

艺术家通过改变普通鸟笼的尺度、材料和色彩，使陌生与熟悉的视觉经验相互冲突。另外，鸟笼特有的关于自由与囚禁的寓意也是创作动因之一。作品试图显现一种不确定的空间关系：矛盾、戏谐、轻快却暗含深意。

张恩利"肥皂剧"：

艺术家把老式的普通椅子罗列在公园中，但原有的木质已被转化为会不断锈蚀的钢材，材质的改变阐述出作品的内在概念——把时间凝固在那个时代。这件作品和今日张江的现代化建筑群形成了鲜明的对照关系。

赵能智"表情1号"：

多年来，艺术家试图通过个人化的语言和视角去关注人在纯粹自我状态下的独特表情。借助雕塑的空间感和体量感，作品以局部放大的方式，揭示出当代都市人复杂而多元的内心世界。

周春芽"草地上的绿狗"：

狗是人类最早驯化的哺乳动物，也人类最好的朋友。《绿狗》是艺术家作品中延续多年的经典创作题材，成就了他的许多代表作品。把狗和绿色结合在一起，是为了体现人与动物在大自然中的和谐相处。

周伟"上海张江当代艺术馆"：

上海张江当代艺术馆位于一个30 000余平米的城市公园中，公园像我们常见的那样是以模拟自然的手法营造，有景可望，有径可寻，有境可感。设计的意图即从一个简单的事实抑或理由开始：如何在人工化的城市自然空间和社会生活中小心介入一个适当的"插件"？以资实现一个休憩"可驻"之地，一个精神"可寄"之所。"亭者，停也，人所停集也。"于是，"亭"成为这个"插件"介入的可能途径，见水构亭，与艺术馆基地不远处的一条河道建立一种现实的交叉关系构成了"十字亭"的意象基础及环境关联，将此基础平移至基地垂直交叠并依园中交汇路径为向旋转，既得亭之形态，又获艺术馆之特殊功能诉求。在这个平移、交叠、旋转的过程中，生成了室外——非内非外——室内——内在外中——亦内亦外等复杂多样的关联，而"亭"、"馆"之有限空间的无限性也因此被感知和体悟。

施慧"停"：

在古汉语中，"亭"和"停"是通假字，亭子是为停留而设计。艺术家充分

238

发挥出对金属材料刚柔并济的使用能力，将两张钢丝网进行一系列不同的折叠组合，构造出丰富的形态。不同的亭子对应着不同的椅子，共同构成了繁忙都市中的一个暂停空间。

刘韡"积木"：

艺术家通过对儿童积木进行直接而单纯的改造，产生出双重含义：一方面作者将一个玩具放大到非人性化的尺度，对艺术家权利的界限进行了夸张的表述；另一方面用"玩具"来偷换"公共艺术"的概念，来质疑艺术在公共空间中的完整性和有效性。

马岩松"菌"：

菌，寓意着艺术家个体创造力的勃勃生机，这些良性"疯狂菌种"的存在和生长，保证着整个社会文化肌体的合理运转和健康发展。这些蘑菇形状的座椅有着不规则的基本形态，在空间中可以无限组合，扩张成更大的造型。

莫尼卡·邦维奇尼（意大利）"厕所"：

"厕所"是意大利艺术家莫妮卡的著名作品，她通过单面玻璃的形式把私密空间暴露在公共空间中，引起对于私密行为的焦虑，从而引起人们对空间、性别和权利这三者之间关系和问题的考虑。

孙良"∧"：

艺术家借用波普艺术的方式放大了一个带着吸管的易拉罐，使之成为一个滑梯。游戏的儿童从罐底钻进又从吸管滑出，仿佛重新经历了从母体中诞生的过程，同时，这件作品也隐喻着当今儿童受快餐文化强烈影响的生存状况。

杨旭"站"

艺术家用混凝土复制了三百年前的张江古地图，既作为地基支撑着车站的钢构架，又作为侧板供人倚靠；巧妙地将一个地方的历史移植到都市运动的节点（车站）上，从而形成一个"停"与"流"的混合空间。

钟山"基因的记忆"：

艺术家以数字铺就的道路表达着一种疑问：当下数字化社会中，个体生命的基因是什么？世界的根基究竟何在？这条"数字小道"不仅体现了张江园区的特色，同时也向人们传达出哲学化的思考：时间在流逝，我们在行走。

陈底里"千年虫"：

艺术家把金属装置放大成户外建筑作品，除整体造型表达出"千年虫"的寓意外，一系列如飞碟般造型的圆盘连接组合，又成为极具未来感的回廊和亭子，强悍有力的材料与轻盈梦幻的形态共同构成了作品的梦幻气质。

苍鑫"萨满图腾"：

艺术家多年来沉迷于对萨满信仰的研究中，通过对祖先崇拜和自然崇拜精神的发掘，他找到了更加宏大的主题——人类与自然的关系。这个双面的多种生命混合体既表达了艺术家对萨满教的理解，也是人类敬畏自然、与自然和谐相处的形象化体现。

丰江舟"八音"：

"八音"的造型如同一滴水珠的异型金属盒子，这个盒子上有三个座位和一根天线。观众坐在上面启动按钮，就可以听到从盒子内部发出的声音。这些声音是艺术家自己创作并预先设置在电子芯片中的。不同时刻的启动，会发出不同的声音；不同数量的观众坐上去，也会产生不同的混响效果。

"现场张江"大型公共艺术活动介绍了参加活动的中外艺术家和他们创作的作品，探讨在高科技园区开展艺术活动的意义和影响。作为新开发的文化产业基地的张江，是个充满着想象和创造力的土地，吸引着更多专业的战略投资者和国内外的文化创意企业入驻，一个崭新的文化科技创意产业研发和生产基地已然形成，张江正在孕育着巨大的潜质。现场张江利用公共艺术的强大魅力，塑造其文化形象，以国内外优秀艺术家的作品来吸引公众关注，让公共艺术贴近公众、干预公众的日常生活，不仅符合张江的文化定位，而且无形中使其成为浦东的一张具有内涵的文化名片，一个重要得文化窗口。"现场张江"从整体上体现了公共艺术介入公众日常生活和公共空间的作用，从而让公众感受到了张江的活力与魅力。 （冯正龙）

Artwork Description

The Living Zhangjiang project was started in 2006, one of the Zhangjiang High Technology Park's programs to support public art. It was conceived with the aim of bringing about closer connections between society and art and developing a stronger exchange between everyday people and art, while at the same time building better corporate citizens and improving the local environment.

The theme in 2006 was "The Shape of Urban Progress," with a concern for expressing transformations and related questions pertaining to urban development in China in the context of globalization. With the funds provided, several distinctive artworks were created within the Zhangjiang community.

In the following year, the project theme was "Poetry of Stillness and Motion" —an investigation of the connections between the spaces of modern cities and people. The project sought to elicit instants of poetic transformation in the everyday urban scene, offering the public an opportunity to pause briefly and acquire a more personal awareness of their surrounding environment.

In 2008, was the theme was "Levitating Quickness," which focused on the opening of the first magnetic levitating high speed rail-line in the Zhangjiang region, as inspiration for finer attention to human fellowship.

The artworks in Living Zhangjiang are permanent sculptural installations located along main thoroughfares, within parks, public plazas, or inside buildings. To date, 40 outstanding works of public art have been created, including: a sublime pavilion, Staying, by Shi Hui; a comical soda-can playground slide by Sun Liang; remarkable "mushroom chairs" of Ma Yansong; and the mysterious "genetic pathway" of Zhong Shan.

In December of 2006 the Zhangjiang Museum of Modern Art was established as part of the Living Zhangjiang project. Totalling approximately 760 square meters, the museum is two stories, with 380 square meters of exhibition space on the second floor. The lower floor is an "art" coffee shop. The museum is a civic cultural institution created by the Zhangjiang High Technology Park authority as a service to the Zhangjiang region and in adherence to the principles of lifelong learning and public interest. Its function is to hold exhibitions, collect works of art, cultivate aesthetic education, and advance cultural exchange. Connected to the museum is a 20,000 square meter sculpture park where several well known artists from Shanghai, Beijing, Sichuan Province, and other places around China have built 10 scuptures, commissioned by the authorities and their designated representatives, the Shanghai 863 Safety Information Company, Ltd, and the Zhangjiang Art and Culture Company, Ltd.

Artwork Excellence

The Living Zhangjiang project was conceived with the aim of facilitating closer connections between society and art and developing a stronger exchange between everyday people and art, while at the same time building better corporate citizens and improving the local environment. Created in July, 1992, the Zhangjiang High Technology Park is situated on 25 square kilometers in the center of Shanghai's Pudong New District. In August, 1999, the Shanghai city government published "Rising Zhangjiang," a policy planning document, which noted that 5,359 enterprises had located within the tech park, with an average annual revenue of 39.7 billion RMB. The plan initially placed an emphasis on the park's functionality, without a full consideration for meeting people's basic needs for commercial space, nor for providing a full complement of living services. Forty-eight percent of people working in Zhangjiang leave the area after work. Under these conditions, Zhangjiang became a city without a soul. To improve the cultural life of this young community the Pudong district government and the Zhangjiang authority initiated the "Living Zhangjiang 2006" public art program, which has been continued every year thru 2008. Living Zhangjiang is the second largest publicly subsidized art program in Shanghai after the Shanghai Biennale. Many notable notable curators have been invited to participate, and the Zhangjiang authority has committed tremendous capital and resources to the project. As a result Living Zhangjiang has become a sustainable public art program which has gradually become institutionalized.

玛哈卡德：市场中史诗般的艺术
Mahakad: Epic Arts in the Market

艺术家：纳文·罗旺柴库尔（泰国）
地点：泰国清迈
推荐人：凯利·卡迈克尔
Artist: Navin Rawanchaikul（Thailand）
Location: Chiang Mai , Thailand
Researcher: Kelly Carmichael

作品描述

在"地方营造（placemaking）"还是一个词，或开始创造城镇社会团结之前，简陋的、随处可见的街市就已经存在。全世界的大多街市都是公共空间的经典形式，特别是在非西方国家。市场是本地社区的连接点，它提供重要的社会联系，是独一无二的本地身份创建要素。别具一格的、吵闹的、挤满了当地人和购物者（无论一个星期中的哪一天的哪个时间）的街市全然描述出小生意和个体商店的情状。

北吕宋最繁忙的市场为建立于 1910 年的清迈瓦洛洛市场。市场有着很长远的历史，它位于一个文化大熔炉的社区中，其中有中国神庙、印度锡克教徒庙宇和佛教庙宇、清真寺和天主教教堂，在清迈市民的生活中有着很重要的地位。白天，这里是巨大的三层楼市场，购物者几乎可以在那里买到任何他们所想要的东西，包括食物、衣物、布料、鞋子、化妆品和农具。黄昏过后，这里便成为开放的夜市，马路两旁兜售各种食物、裙装、首饰及本地制的手工艺品。

泰国艺术家纳文·罗旺柴库尔在此市场中呈现作品"玛哈卡德：市场中史诗般的艺术"（2010—2011），这是一件史诗般的、充满活力的、协作性并具实验性的艺术节日和沟通平台，让大家认识到社区参与的潜力。"玛哈卡德：市场中史诗般的艺术"通过当代艺术和传统手艺人融入到市场迥异的文化集体中，意在提升本地的身份和其独一无二的文化意识。

纳文制作室所制的节日新作中有一幅黑白单色印刷的全景壁画，描述瓦洛洛市场超过 200 名成员集中于建筑之前，有些人去世已久，他们的样子由老照片重现。在本地艺术家和市场自身之间的介入与协作是显而易见的，以及一些邻接的项目：妥善保存的纪实影像和市场历史的记录以及清迈的过去在附近展示。泰国摄影师道·瓦希其利（Dow Wasikiri）和市场本身一起共同创作连接过去与现在的视觉对话，作品包含卡纸修剪和粗糙的塑料制成的背景，创作市场和市场人的当代摄影插画。同时推出项目有关的市场观光、电影放映、行为表演和公共艺术讨论会。还有清迈大学建筑系和堪希彦小学学生们的工作坊活动，同时，《指南针》杂志为"玛哈卡德：市场中史诗般的艺术"制作并发行特别双语刊，公众可免费取阅。

作为对此区域探究的激励，一本"护照"模样的小册子让参observer收集他们所到之处而得的印章。集齐一整套印章意味着参观者将获得一本"玛哈卡德：市场中史诗般的艺术"历史漫画书，它是模仿泰国过去卖 10 萨当（泰国货币 100 萨当等于 1 泰铢）一本的低俗插画小说。

恰逢市场和社区百年，纳文·罗旺柴库尔的想法是举办庆典并使瓦洛洛市场和相关清迈人民生活的角色再次充满活力。

当地社区的墙壁或者街道错综复杂与便于行人使用，这些因素更是决定了街市的规模和人流，而不是去迎合或接受工业化设计的产品布局或商场格局规划的控制。公共空间，或许尤其是街市，把每件事物联系在一起。不需要大规模的"城市复兴"或"经济发展"项目规划，传统街市已经完成了这些任务，它把不同种族和不同经济条件的人吸引到一起，提供吸引人且安全的公共环境，支持小型经济体并提供新鲜的、高质量的产品。作为对平民百姓日常生活保持一直兴趣的艺术家，纳文·罗旺柴库尔看在眼里并着手开始创作也许是公共艺术项目里最体现公共性的项目——在他童年记忆里的市场大厅中间举行节日。"这个市场是我成长的地方。我的父亲在此仍有一间商店。"纳文·罗旺柴库尔说道。

纳文·罗旺柴库尔代表泰国参加 2011 年威尼斯双年展后，他的作品因其共享性、特定场域性和经常融合社区和通常把社区和个体经验融合进不同寻常的虚构故事而被世界所知。他的艺术实践是对当地情况和全球化趋势之间的对话的一个探究过程。居住在一个传统价值被西方价值、城市化及工业化快速取代的世界中，罗旺柴库尔的实践对泰国农村普通人的生活和身份变化发表了他的看法。他的作品时常追求缓和并解决文化变化问题并且消除城市中的年轻人和接受过教育的人与居住在泰国农村的老一辈人之间的隔阂。

在原先名为"纳文曼谷画廊"（Navin Gallery Bangkok）（1995—1998）的作品中，罗旺柴库尔把一辆出租车改变为一个移动画廊，并邀请泰国和全球的艺术家前来展出作品，他利用出租车的公共空间为当代艺术世界与对当代艺术知之甚少的本地东南亚观众提供了一座相连接的桥梁。受此激励，"玛哈卡德：市场中史诗般的艺术"把在国际性舞台展示的艺术家带回到自己的故乡。

解读

"Public Art"直译是公众艺术，即为有民众参与的都可谓 PUBLIC ART，而公众与人有关公共与场所和人皆有关。出于人们天生的对于场所、场域等和地理、块块的敏感，我们如今常用公共艺术取代公众艺术，即为了更好地宣扬公共艺术的特性，其真实存在，被置放于社会某一角，作用于市民。

具有百年历史的居民市场是承载公共艺术项目的最好平台，此聚落生活空间虽然鱼龙混杂，但却经得起细细揣摩，生活区的内在人文精神对于社会和艺术的辐射远比其外观的单纯表现要丰富得多，千百年沿袭下来的出行方式映衬着这个地区的居民文化。艺术源于生活，数以百万计的人们生活的痕迹是功效性最强的空间设计，应点而出的商铺，各种便于生活的公共设施构筑了这个不依靠现代化工业设计却实用性极强的生活空间。艺术家唯一要做的就是在此基础上将现代设施与此街区完美融合，消除固有的观念上的隔阂。让此市场中居民的生活在充满生活痕迹的同时还充满活力。(高浅)

Artwork Description

Long before "placemaking" was a word, or initiatives to create social cohesion within towns and cities commonplace, the humble and ubiquitous street market was already doing just that. Street markets are the classic form of public space throughout much of the world, especially the non-Western world. Markets are connection points for local communities, provide vital social contact and are essential in the creation of unique local identities. Eclectic, noisy, thronged with locals and shoppers—no matter what the time or day of the week—street markets are the very definition of small business and independent shop keeping.

Northern Thailand's busiest marketplace, the Warorot (or Kad Luang) market in Chiang Mai, was established in 1910. The market—in a neighborhood that's a cultural melting pot complete with a Chinese shrine, Sikh and Buddhist temples, mosque and a Catholic church—has a long history and important role in the lives of Chiang Mai's citizens. During the day, this is a large three-storey market where shoppers can buy almost anything they want including food, clothes, fabrics, shoes, cosmetics and farming tools. After dusk, it becomes an open-air night market with roadside stalls selling a wide variety of food, garments and accessories and local handicrafts.

Within that market, Thai artist Navin Rawanchaikul, launched Mahakad: Epic Arts in the Market (2010-2011), an epic, dynamic, collaborative and experimental festival and communication platform recognizing the potential of community engagement. By integrating contemporary art and traditional artisans into the market's diverse cultural community, Mahakad aimed to promote awareness of local identity and its unique cultures.

New artworks were produced for this festival by the Navin Production Studio, such as a panoramic print of a monochrome mural a year in the making depicting more than 200 members of the Kad Luang market community gathered in front of its building—some long dead, their likeness recreated from old photos. Interventions and collaborations between local artists and the marketplace itself were evident and several adjacent projects, among them a documentary video with faithfully preserved and recorded histories of the market community and Chiang Mai's past, were held nearby. Thai photographer Dow Wasikiri collaborated with the market itself to create visual dialogues linking the past to the present, a project involving cardboard cut outs and kitsch plastic backdrops from history creating inserts for contemporary photographs of the market and its people. The programme also featured fresh-market tours, films, performances and a symposium on communal art. Workshops with Chiang Mai University's Faculty of Architecture and Kamthieng elementary school students were held, while Compass magazine produced a special bilingual issue published for Mahakad and made freely available to the public.

As encouragement to explore the area, a leaflet that acted as a 'passport' allowed visitors to collect stamps wherever they went. Obtaining a complete set of stamps meant the visitor was given a comic book of Mahakad history, emulating the illustration of the 10-satang pulp fiction novels from Thailand's past.

Timed to coincide with the centennial of the market and its community, Navin Rawanchaikul's brainchild recognized, celebrated and reactivated the Kad Luang market and its role in connecting the lives of Chiang Mai's people. With halls or streets that are intertwined and pedestrian friendly, local communities define the scope and flow of street markets, rather than fit into or become manipulated by industry designated product placement or store layout patterns. Public spaces, and perhaps street markets particularly, connect everything together. Without needing major "urban revitalization" or "economic development" project plans, traditional street markets accomplish this already, bringing people of different ethnic groups and incomes together, providing inviting and safe public spaces, supporting small-scale economic activity and providing fresh, high-quality produce to urban populations. As an artist with an on-going interest in the daily lives of ordinary people, Navin Rawanchaikul could see all this and set out to create perhaps the most public of all public art projects, a festival within the very market halls of his childhood. "This marketplace is where I grew up. My father still has a shop here," commented Rawanchaikul.

Since representing Thailand at the 2011 Venice Biennale, Navin Rawanchaikul is internationally known for works that are participatory, site-specific and often integrate community and individual experience into eccentric fictional tales. His practice is a process of exploring the negotiation between local circumstances and trends of globalization, often engaging in direct social commentary and referencing local or personal identity. Inhabiting a world in which traditional values are being rapidly supplanted by Western values, urbanisation and industrialisation, Rawanchaikul's practice has commented on the changes to village life and identity for the ordinary people of rural Thailand. His work often seeks to mitigate and broker cultural change and bridge the gap between the younger, educated people in cities and the older Thai living rurally.

For an early project titled Navin Gallery Bangkok (1995-1998), Rawanchaikul converted a taxi into a mobile art gallery and invited artists from Thailand and around the world to exhibit, using the public space of a taxi cab to provide a bridge between the contemporary art world and a local South East Asian audience who had little knowledge of contemporary art practice. Instigating the Mahakad: Epic Arts in the Market project brought this internationally exhibiting artist back home.

Artwork Excellence

Timed to coincide with the centennial of the market and its community, Navin Rawanchaikul's brainchild recognized, celebrated and reactivated the Kad Luang market and its role in connecting the lives of Chiang Mai's people. Public spaces, and perhaps street markets particularly, connect everything together. Without needing major "urban revitalization" or "economic development" project plans, traditional street markets accomplish this already, bringing people of different ethnic groups and incomes together, providing inviting and safe public spaces, supporting small-scale economic activity and providing fresh, high-quality produce to urban populations. As an artist with an on-going interest in the daily lives of ordinary people, Navin Rawanchaikul could see all this and set out to create perhaps the most public of all public art projects, a festival within the very market halls of his childhood.

室生艺术森林
Murou Art Forest

艺术家：丹尼·卡拉万（以色列）
地点：日本奈良
推荐人：凯利·卡迈克尔
Artist: Dani Karavan(Israel)
Location: Murou Village, Nara Prefecture Japan
Researcher: Kelly Carmichel

作品描述

　　"室生艺术森林"是一个环境项目，由以色列艺术家丹尼·卡拉万在日本奈良室生村创作而成。该项目坐落于室生寺附近，这座寺庙建造于延历时代 (780—805)。寺庙里有一座五层高的塔，它是日本最小的户外塔，这座寺庙矗立于山谷中，被公认为奈良最宝贵的寺庙之一。它和金色大厅，大雄宝殿一起已被认定为国宝，室生完好地保存着早期平安时代的艺术品，还有一尊久坐木制佛像，堪称寺庙里众多佛像中最优秀的一尊。

　　公园的几次山体滑坡严重破坏了水流并危及到这个地区，因此卡拉万的项目应运而生。为了保护该地区，室生村发展了一个"艺术世外桃源计划"，旨在建立一个 21 世纪的世外桃源。"世外桃源"计划的核心是建立一座公园，由卡拉万负责设计。

室生艺术森林是一座大型公园，坐落于室生寺后面的高山上。丹尼·卡拉万担任设计，他是环境艺术的先锋。森林中有很多他的雕塑作品。除了重建水稻梯田提供一种怀旧的日本风光的体验，参观者还可以沉浸在自然与艺术相融合的感官享受之中。

"室生艺术森林"是一个完整的项目，是一件景观雕塑作品。"室生艺术森林"是对古迹保护的一种回应，同时也是丹尼·卡拉万的一件重要作品。在持续建造的 8 年过程中，这个项目延绵一公里长，其 "目的在于揭示人与自然之间的关系，力求与它的精神一致，" 艺术家这样说道。"理解这件作品需要在古迹周围游览，游客要积极地参与，刺激所有的五个感官：视觉，听觉，味觉，触觉和嗅觉。"

关于丹尼·卡拉万的更多介绍：

丹尼·卡拉万是最多产的公共雕塑家之一，以创作融合于特定地点的纪念雕塑和纪念碑而闻名，他也在艺术和建筑方面做出过贡献。20 世纪 60 年代，卡拉万为玛莎葛兰姆舞蹈团、巴特舍瓦舞蹈团，以色列室内乐团设计舞台布景。他代表以色列参加了 1976 年的威尼斯双年展，参展作品是雕塑 "耶路撒冷和平之城"。和平的主题、人与人之间和谐共处以及自然与文明的和谐，弥漫于他的作品中。他的装置则融合了雕塑、建筑、景观和城市规划。在选择与周围环境产生共鸣的形状和材料之前，卡拉万先要对建造地点进行深入细致的研究，同时考虑到当地的历史、自然和建筑造型。卡拉万早期作品 "内盖夫沙纪念碑"（1963—1968）位于以色列，是一座有着建筑质感的大型实验雕塑作品，而 "和平的环境" 则确立了他的国际声誉，作品包含了水、橄榄树和风琴。艺术品的坐落地点是艺术家实践的关键，不仅在于景观方面，也有历史的原因。他的创作过程和身体密切联系，而观众也需到现场 "去感受，去听，去闻，去摸，去走"，艺术家这样解释道。

解读

"清晨入古寺，初日照高林。曲径通幽处，禅房花木深。山光悦鸟性，潭影空人心。万籁此俱寂，但余钟磬音。" 这首诗在当今社会也就仅仅存留纸上。现代社会，难得觅到这样一块远离尘嚣的世外桃源，更无缘于古寺、山林、曲径、禅房、山鸟、清潭这等朴素庄严的旷凉之物。我们期盼艺术能够改变生活，还心底一块净土，那么是否该先探讨艺术之路究竟该如何处行？这曲折的问号不光是对于艺术行为、创作和理念的询问，更是对于冗繁世事的询问，艺术是什么？自然元素如何与人文艺术相融合？城市公共艺术介入到自然中是一个新鲜的主题，创作于森林中的公共景观艺术作品带来了现代与原始交织的美。简单的线条与周围的环境完美搭配，更显静谧。人类本身就是自然的元素，拥有人文气息的大自然更加美好，回归到原始，不损生态，和谐并存，这许是艺术家的最终目的。（高浅）

Artwork Description

Murou Art Forest is an environmental project by the Israeli artist Dani Karavan in Murou village, Nara Prefecture, Japan. The project is located by the Murouji temple constructed in the Enryaku era (780-805). The temple, whose five-storied pagoda is the smallest outdoor pagoda in Japan, stands in a mountain valley and is considered one of Nara's most treasured temples. It has been designated a National Treasure, along with the Golden Hall and the Main Hall. Murouji possesses a fine collection of early Heian Period art works, and a wooden sedentary statue of Buddha considered to be the most outstanding among the number of Buddhist images kept at the temple.

Karavan's project came about after landslides in the park damaged water flow and seriously endangered the site. In order to preserve this area, Murou Village developed an "Art Arcadia Plan" aimed at establishing a 21st century Arcadia. Central to the Arcadia was the establishment of a park to be designed by Karavan. Murou Art Forest is a vast park is located on a tall hill rising above the Murouji Temple. Designed by Dani Karavan, a pioneer of environmental art, the forest features a major work. It consists of a reservoir with 3 islands and a circular pond in a restored paddy field where an astronomical tower stands 8 metres tall. Three small ampitheatres, a bamboo spiral and tunnel, waves of glass, a dome planted with a tree and a line of water form a course in a natural woodland mountain site.

Alongside rice terraces recreated to offer a nostalgic experience of Japanese scenery, visitors are offered the opportunity to immerse their senses in a fusion of nature and art.

Conceived as a complete project and environmental sculpture, The Murou Art Forest was an answer to the preservation of the site but equally a major art work for Dani Karavan. Created over the course of eight years, this project stretches over a kilometre long and had "the aim to reveal the links between man and nature" and "to serve his spirit," according to the artist. An understanding of the work requires movement around the site and "active participation from the visitor by stimulating all five senses: sight, hearing, taste, touch and smell."

More about Dani Karavan:

Dani Karavan is one of the most prolific public sculptors best known for site-specific memorials and monuments which merge into the environment, though he has made other significant contributions to art and architecture. In the 1960s Karavan designed stage sets for the Martha Graham Dance Company, the Batsheva Dance Company and the Israel Chamber Orchestra. He also represented Israel at the 1976 Venice Biennale with the sculpture Jerusalem City of Peace. Pervading his works is the theme of peace, the harmony of people with each other, as well as the harmony of civilization with nature. For his installations Karavan conceives a fusion of sculpture, architecture, landscape, and city planning. Prior to selecting shapes and materials that resonate with their surroundings, Karavan conducts a patient, in-depth study of the site, taking account of its history and its natural and built forms. Karavan's early Negev Monument (1963-1968) in Israel is a large-scale essay in the architectural qualities of sculpture and his international reputation was confirmed by Environment for Peace, which included water, olive trees and wind organs. Location is key to the artist's practice, not just in terms of landscape, but of history too. The physicality of his creative process is matched to the site "to feel, to listen, to smell, to touch, to walk through" as the artist explains. In the work Passages, Homage to Walter Benjamin (1990-94) Karavan evoked relationship between place – the cultural theorist Benjamin committed suicide at the site – text (some of Benjamin's own words) and life, in the form of a single carefully-sited olive tree.

Artwork Excellence

Conceived as a complete project and environmental sculpture, the Murou Art Forest was an answer to the preservation of the landslide-damaged site but equally a major art work for Dani Karavan. Created over the course of eight years, this project stretches over a kilometre long and had "the aim to reveal the links between man and nature" and "to serve his spirit," according to the artist. An understanding of the work requires movement around the site and "active participation from the visitor by stimulating all five senses: sight, hearing, taste, touch and smell."

纳尔帕和皮拉·谷迪斯
Nalpar and Pilla Gudis

艺术家：纳弗约特·阿尔塔夫（印度）
放置地点：印度恰蒂斯加尔巴斯塔地区
推荐人：里奥·谭
Artist：Navjot Altaf（India）
Location：Bastar region of Chattisgarh, India
Researcher：Leon Tan

作品描述

这是两个极具意义的社区公共艺术干预实践项目，在印度中部地区持续进行了十年，所关注的是被国家忽视的公民问题。阿迪瓦西是印度的原住部落居民，近几十年来，他们的各种生活权益不断受到印度政府和私有企业的侵犯。这两个项目强调了在完全不同的社会经济背景下（都市和郊区）的各群体之间的相互对话。最有意义的是向社区居民提供卫生的生活用水，不仅对改善当地的生活条件起到了积极作用，而且对妇女和儿童而言意义深远。

七名阿迪瓦西女艺术家参与了"纳尔帕"项目，即当地手动水泵地点的艺术化改造。这些公用取水点原先被废弃而年久失修，周围经常是郁积的死水潭，成为各种疾病的温床。第一个纳尔帕完工于 2001 年，到 2007 年又陆续建成七个，每个"纳尔帕"都由大型混凝土平台组成，由于取水是当地妇女的职责，因此在很大程度上改善了妇女的劳动条件和生活状态。

"皮拉·谷迪斯"的字面意思是"献给儿童的寺庙","纳尔帕"的混凝土墙上有描述阿迪瓦西生活的装饰图案,不仅为儿童提供了全新的有益体验,还成为他们放学后聚集玩耍的空间,可以通过直接的感官体验获取审美知识。这块地点也被用于非正式艺术工作坊和开展融入阿迪瓦西传统文化活动的场所,以及当地人的信仰和实践以及对现实问题的对话交流空间。

解读

这是发生在印度的公共艺术案例,通过公共艺术的介入为社会矛盾和公民问题的解决架起了一座桥梁。

印度恰蒂斯加尔巴斯塔地区,"纳尔帕和皮拉·谷迪斯"是艺术家实施的两个具有深远意义的社区公共艺术干预项目。阿迪瓦西是印度的原住部落居民,近几十年来,他们的各种生活权益不断受到印度政府和私有企业的侵犯。这两个项目强调了在完全不同的社会经济背景下(都市和郊区)的各群体之间的相互对话。卫生的生活用水的提供对于当地的妇女和儿童而言意义深远,同时对改善当地的生活条件起到了积极作用。七名阿迪瓦西女艺术家参与了"纳尔帕和皮拉·谷迪斯"项目,即当地手动水泵地点的艺术化改造。"皮拉·谷迪斯"的字面意思是"献给儿童的寺庙","纳尔帕"的混凝土墙上有描述阿迪瓦西生活的装饰图案,不仅为儿童提供了全新的视觉体验,而且成为玩耍的公共空间。同时,这块地点也被用于非正式艺术工作坊和开展阿迪瓦西传统文化活动的场所,以及当地人的信仰和实践以及居民之间对话互动的空间。(冯正龙)

Artwork Description

These are two major public art-in-community intervention projects implemented in central India during the past ten years. The central concern of these projects is the marginalized social circumstances of the Adivasi, a group of indigenous tribal peoples of India whose livelihoods, land and rights have been steadily eroded in recent decades by the activities of the Indian government and private firms. Both these projects are exemplary for their emphasis on continual dialogue between collaborating artists from very different socio-economic contexts (urban and rural). Most notably, they have resulted in improved access to clean water for the community residents, of profound significance for women and children in particular.

Seven Adivasi women artists participated in the "Nalpar" project, involving the functional and aesthetic transformation of local hand pump sites (or "Nalpar"). These communal sites for drawing water were previously neglected, in disrepair, and often surrounded by stagnant pools of muddy water, breeding grounds for malaria-carrying mosquitoes and and other disease vectors. Each nalpar consists of a large level concrete platform. The first one was completed in 2001, and by 2007 seven others had been constructed. Water-carrying is exclusively the work of the local women, and consequently the project has succeeded especially in improving women's working and living conditions.

At Pilla Gudis, literally "Shrine to the Children," the concrete walls of the structure form traditional Adivasi decorative motifs. The structure has become an after-school playground for children, passing on aesthetic knowledge through direct sensory experience. The site is used for informal art workshops and traditional Adivasi cultural activities, as well as for open exchange and dialogue among the community.

major "urban revitalization" or "economic development" project plans, traditional street markets accomplish this already, bringing people of different ethnic groups and incomes together, providing inviting and safe public spaces, supporting small-scale economic activity and providing fresh, high-quality produce to urban populations. As an artist with an on-going interest in the daily lives of ordinary people, Navin Rawanchaikul could see all this and set out to create perhaps the most public of all public art projects, a festival within the very market halls of his childhood. "This marketplace is where I grew up. My father still has a shop here," commented Rawanchaikul.

Since representing Thailand at the 2011 Venice Biennale, Navin

Artwork Excellence

The excellence of the project is attested to in many ways. Concerning the Nalpar strand, this may be considered a significant public intervention insofar as it directly addresses the problem of State neglect of civic issues, here concerning the hygienic provision of basic necessities of life, water, to under-privileged groups, namely the Adivasi (indigenous) communities of Bastar. It has resulted in a significant improvement of living conditions, in particular, for the women of the region, at least in two ways. Firstly, it has improved the health of women through the ergonomic interventions in the structure of the Nalpar, lessening strain on the spinal cord. Secondly, it has provided a space of respite from the male scopic regime, allowing women to experience privacy. Concerning the Pilla Gudis strand, this is significant insofar as it provides for new and beneficial experiences of place for children, where such spaces did not exist before. The Pilla Gudis is in fact a resurrection and transformation of the Adivasi Ghotul institution, repressed by the Indian government because Hindus considered it immoral as it allowed for unsupervised interactions between boys and girls, sometimes including sex. (See Grant Kester, 2011: 84 for details). In facilitating aesthetic knowledge acquisition through direct sensory experiences, Pilla Gudis is outstanding as a communitarian public artwork, all the more so for its non-tokenistic engagement of an often neglected social group, children. Details in the additional information section provide further evidence of its international excellence.

台北捷运南港站
Nangang Rapid Transit Station

艺术家：几米、陶亚伦（中国）
地点：中国台湾台北市
推荐人：赖香伶、朱惠芬
Artist: Ji Mi, Tao Yalun（China）
Location: Taipei, China
Researcher: Hsiangling Lai, Huey-Fen Chu

作品描述

台湾台北南港站原位于兴华路与兴中路交会处，此为南港交通发展之重要区位，然因却受限于平面铁路之阻隔。在铁路地下化后，南港站与台铁转运站的联合开发，将使当地的交通产生换乘及连贯的便利性。而捷运车站的设立，增加旅客行经该地的机会和意愿，无形中带动了小区的繁荣发展。本区公共艺术计划从捷运车站的设立，由人文思考的角度出发，虽然捷运站区位周围腹地狭小，但因邻近地区多项人文历史古迹，例如南港茶园、胡适纪念馆等，具有观光资源潜力。因此，公共艺术的规划善用区位人文特色，以启动周边各项开发计划，期使该区朝多元化现代都市脚步迈进。

公共艺术设置理念本站的设计主题定为"怀旧与科技"，因为该站附近原出产茶叶、煤矿及桂花，且与诸多南港区的人文历史古迹相邻，而未来与对面的台铁南港站连接，将成为交通枢纽，具有多元化的开发潜力。因此车站设计除以科技表现时代感之外，也呼应当地旧有产业的辉煌历史。根据此理念，车站装修主题色选用湖水绿色，以配合昔日的地区意象，并大量应用在墙面、地坪及天花板上，其中地坪缀以古朴的陶砖。在付费与非付费区之玻璃栅栏上以喷砂方式勾勒出昔日的影像，配置可变化色彩的光纤。本站也首创将自动扶梯侧板从不锈钢板改为珐琅板及喷砂玻璃，使站体之必要硬件设施也成为艺术展现的舞台。

几米作品说明：

本站与知名插画家几米合作，将其瑰丽奇幻的画作真实地呈现于捷运中。其中月台两侧各有长达 62.5 米的艺术墙板，一侧表现车厢外观、一侧是车厢内的景象，展现真车与画中车、真人与画中人呼应的趣味。另在穿堂层及出入口的侧墙也都有几米为本站特别挑选或绘制的画作。几米这六幅大型绘本作品《地下铁》，其创作主题不仅吻合基地特质，作品的设置更将整座捷运站体包纳于艺术的梦想空间中，并形塑出迥异的乘车经历。（数据源：台北市政府捷运工程局）

陶亚伦作品说明：

以怀旧与科技为主题，唤起南港旧产业之辉煌记忆，重起小区营造计划的新生命，以舟、茶、桂、煤为核心印象，将南港站"捷运站台"视为"当代的转运码头"，呈现早期"南港仔"码头的历史印象，表现昔日南港码头为丰盛农产、矿产出口港的历史氛围。（数据源：文建会公共艺术官方网站）

解读

几米的作品总是能引起反思与共鸣的，犹记得《地下铁》在大陆售卖的那天，还在上高中的我第一时间跑到了上海地铁一号线陕西南站出口的季风书院抢到了一本。未到家就迫不及待的翻开绘本，幻想着地铁中擦肩而过的命运，幻想在穿过地下通道的时候，会不会有大象兔子和绿色的森林在我眨眼的一瞬间闪现。这个地下世界是那么的昏暗，但在几米的世界里却那么的清晰和充满了爱。

时间一晃，十几年之后，我们来去匆匆，在地下迅速的穿寻，从一个地铁站到另外一个地铁站台，忙碌着走路，忙碌着学习，忙碌着工作，忙碌间忘记了用耳朵去聆听，忘记了用眼睛去注视，忙碌到忘记了周围的风景。

这是一个让大家忙得失去了幻想能力的世界，所以才需要几米的故事填补我们空洞的灵魂。匆匆碌碌间，擦肩而过的人在过着怎样的人生？来来往往的人，有多少曾经满怀着梦想和希冀，就如同《地下铁》中那个盲眼的小女孩一样，期望着幸福哀伤掉落在自己身上？在我们小时候，每个孩子都拥有自己的童话故事和美好世界，却终于在社会的打磨中慢慢地忘记了最初的纯净，而成为了迷失在庸碌社会中，奔波促碌群体中的一员。几米曾经说过："世界如此艰难，我们一起唱歌跳舞吧"。（高浅）

Artwork Description

Taipei's Nangang Station is located at the intersection of Xinghua Road and Xingzhong Road, which was previously an at-grade railway crossing. The new underground Nangang station combines a municipal rapid transit hub with a Taiwan Railway transfer station, greatly improving regional transportation convenience. The construction of the station created an influx of passengers, leading to the development and growing prosperity of the local community.

The station's public art project began with a view toward extending the new Nangang Station development toward the periphery, making use of the district's cultural strengths to lay strong foundations for a diversified, modern urban environment. While the surrounding area is small, it is close to a number of cultural and historic monuments, including the Nangang Tea Garden, Hu Shih Memorial Hall, and other potential tourist destinations.

The design theme for the project was "Nostalgia and Technology" in recognition of the nearby cultural monuments and Nangang District's history as a center of tea, coal and scented osmanthus production, all of which was juxtaposed against the high-tech Nangang station. The design focused not only on evoking a sense of modern technology but also on echoing the colorful history of the district's former industries. Accordingly, the color chosen for the station was deep sea green suggestive of the harbor, and a large number of rustic ceramic tiles were used for floor, wall and ceiling coverings. Images of these traditional industries are sandblasted onto the glass dividing fence for ticket-holders and non-ticket holders, changing color according to the light. Escalator side panels of stainless steel plate, enamel and sandblasted glass also suggest a mix of technology and nostalgia, incorporating the design theme into the facility's basic infrastructure.

The well-known illustrator Ji Mi collaborated on the project, creating magnificent fantasy paintings for the station walls. 62.5-meter panels line both platforms of the rapid transit tunnel, one side depicting the outside of a passing train, the other side depicting its interior. Looking alternately at real trains and painted trains, real people and the people in the artwork has a delightfully amusing effect. The entrance corridors to the platform are also lined with artworks specially selected or created by Ji Mi. His giant murals are called "Sound of Colors." They are not only expressive of the basic functionality of the station, but give the rapid transit tunnel the quality of a dream-space, completely altering the travel experience! (Source: Taipei Municipal Mass Rapid Transit Authority)

Nangang Station is conceived as a "contemporary trans-shipment terminal," successor to the early Nangang Pier, where the movement of boats carrying tea, spices, and coal created a thriving community. Tao Yalun's installation along the rapid transit platform creates an historical impression, evoking the former South Terminal filled with agricultural produce and mineral exports.

Artwork Excellence

There are two public art installations in this case: the "Sound of Colors," consisting of six large murals on enamel tile designed by the well-known picture book illustrator Ji Mi, who was invited directly by the rapid transit authority; and "Rapid Transit - Pier" by the artist Tao Yalun, who was selected in an open competitive procurement process in accordance with state funding for public art. The excellence of this case rests firstly on the successful integration of public art into the core project planning for the station, achieving both an environment permeated by art, and art that corresponds to the environment. Additionally, the project invites reflection on the region's industrial history, blending everyday life scenes with popular picturebook imagery, and encouraging in passers-by feelings of kinship and compassion, promoting an aesthetics of connectivity with people's lives.

日本新潟大地艺术三年展
Nigata Land Art Triennial

艺术家：北川富朗（日本）
地点：日本新潟县
推荐人：潘力
Artist: Kitakawa Franku（Japan）
Location: Niigata, Japan
Researcher: Pan Li

作品描述

　　"新潟大地艺术三年展"自 2000 年 7 月开始在日本新潟县越后妻有地区六个市、町、村方圆七百六十余平方公里的地域举办。大地艺术节以三年展的形式邀请日本艺术家及国际知名当代艺术家参加，通过艺术媒介探索和重新认识人类与自然地域的关系。六个市、町、村扬长避短，凸显当地特色，共同体现区域丰富的资源和魅力。作品形态主要有户外雕塑、装置、环境改造，在荒地、废弃屋、山坳、梯田、森林、河流及田边地头展开。大部分作品不随展览结束被拆除，而是作为地区的永久设施形成独特的公共艺术景观。"空家·废校工程"是三年展的一个主要内容。组委会首先寻找可利用的空置民宅和校舍，征得所有者同意之后，由指定艺术家提出设计方案，并邀请建筑师进行专业技术指导，不仅将其开发为作品展示场所，还分别作为工作室、展览馆、餐厅和旅店等，募集管理员进行长期经营。东京多所美术类高校均有师生在这里建立工作室，常年参加活动。同时，越来越多的普通市民也参与到以区域振兴为目的的艺术活动中，对当代艺术有了逐步明晰的理解。当代艺术借此走出封闭的象牙塔，成为开拓大众思维方式的媒介。至 2009 年，大地艺术三年展已举办四届，成为当地经济文化的新动力。

解读

在 20 世纪六七十年代以来，由于日本当地农业政策的失调，大批青年人流向都市，使得新潟地区人口密度日趋下降，并趋于高龄化，经济上在全日本也处于滞后状态。为推动地区经济和社会的复兴，经长期运筹，一个前所未有 "大地艺术" 为表现方式的 "新潟大地艺术三年展" 公共艺术活动，在新潟的六个市镇的 760 余平方公里的区域得以实施。来自 32 个国家的约 140 余位艺术家与日本本土艺术家一道，在这个距今 4 500 年前的绳文时代便产生了灿烂的古典文明的地域，进行了当代文化与传统的乡土文化之间、个性化的艺术创作理念与具有历史普遍价值意义的文化遗产和自然遗产的对话。同样在现代语境下使得地方乡土文化资源为背景的公共艺术，起到了不可替代的文化和社会的振兴作用。日本西北部 20 世纪末的新潟展开的以乡土文化资源为基础，并予以现代理念相整合的 "大地艺术" 活动，不仅保护和传承了当地的传统文化遗产和人文文化，而且使用公共艺术介入的方式探索性地对当地特色的民族、环境、生态、宗教等一些问题进行了对话与回应，维护和尊重当地民众的个人利益，积极邀请他们参与公共艺术活动，感受公共艺术活动带来的快乐，由于前来参观的旅游者越来越多，也推动了当地经济的快速发展。（冯正龙）

Artwork Description

The Nigata Land Art Triennial was first held in July 2000 in Nigata Prefecture, Echigo-Tsumari District, Japan, which encompasses 6 cities plus towns and villages in an area of 760 square kilometers. Japanese and internationally renowned contemporary artists were invited to the Triennial explore the relationship between humans and the natural environment through art. The Triennial was held four times, the last in 2009. Each Triennial drew around 200,000 people, constituting a unique kind of cultural tourism, and helping to revitalize the local economy.

Created among the forests, rivers, terraced fields, old farmhouses, and wastelands of the region, artists build earthworks, outdoor sculptures and installations highlighting the rich local resources. The works are mostly permanent, which has resulted in a unique landscape of public art. Today, approximately 160 artworks by artists from all over the world are dotted across the Echigo-Tsumari region.

"Snow-Land Agrarian Culture Center," "Forest School," and other large projects, have become key local cultural attractions. The largest project, located in Tokamachi City, is named "Snow Corridors." It's a meeting hall with a floor area of 5,138 square meters, on a 19,147 square meter site, housing cultural, artistic and commercial activities, which has become a tremendous cultural asset to the community.

Due to a declining birth rate and out-migration, Echigo-Tsumari district has experienced an aging of its population, and greater incidence of vacant homes and schools. Over 300 vacant homes were recorded in one statistical report. As a result, the "Empty House & Abandoned School Project" is one the central projects of the Triennial. The festival committee looks for vacant homes and school buildings that may be re-designed by designated artists. With the consent of the owner and guidance from architectural consultants the buildings are given new life not just as exhibition space but also as work studios, galleries, restaurants and inns, catalyzing businesses that are viable in the long-term.

Through 2009, hundreds of vacant homes and 10 abandoned school buildings have been used or re-developed. For example, the Yugoslav artist Marina Abramovic's "Dream House" is a re-use of a centuries-old building. The four rooms on the second floor were lighted in red, yellow, blue and purple respectively, and a wooden box shaped like a coffin was placed in each. Visitors were invited to wear spacesuit-like pajamas designed by the artist and spend a night in the wooden boxes. The next day they were required to record their dreams in a "dream book" on their bedside table. In this artwork a traditional Japanese house coexists with the fantastical, and through the rituals of daily living the visitor experiences a surreal world.

Teachers and students from several Tokyo fine arts colleges and universities have over the years established studios in the region because of the "Empty House & Abandoned School Project." At the same time, local citizens started becoming increasingly involved in art practices as a way of revitalizing the region, and have a clearer understanding and appreciation for contemporary art.

tunnel, waves of glass, a dome planted with a tree and a line of water form a course in a natural woodland mountain site.

Alongside rice terraces recreated to offer a nostalgic experience of Japanese scenery, visitors are offered the opportunity to immerse their senses in a fusion of nature and art.

Conceived as a complete project and environmental sculpture, The Murou Art Forest was an answer to the preservation of the site but equally a major art work for Dani Karavan. Created over the course of eight years, this project stretches over a kilometre long and had "the aim to reveal the links between man and nature" and "to serve his spirit," according to the artist. An understanding of the work requires movement around the site and "active participation from the visitor by stimulating all five senses: sight, hearing, taste, touch and smell."

More about Dani Karavan:

Dani Karavan is one of the most prolific public sculptors best known for site-specific memorials and monuments which merge into the environment, though he has made other significant contributions to art and architecture. In the 1960s Karavan designed stage sets for the Martha Graham Dance Company, the Batsheva Dance Company and the Israel Chamber Orchestra. He also represented Israel at the 1976 Venice Biennale with the sculpture Jerusalem City of Peace. Pervading his works is the theme of peace, the harmony of people with each other, as well as the harmony of civilization with nature. For his installations

Artwork Excellence

Created as part of a "village revival plan" in the Nigata Prefecture, which has been experiencing a declining population, the Nigata Land Art Triennial was a successful placemaking project. It brought hundreds of artists to the region, many of whom established studios there. It brought 200,000 visitors to each event, helping to boost the local culture and economy. At the same time, local citizens started becoming increasingly involved in art practices as a way of revitalizing the region, and have a clearer understanding and appreciation for contemporary art.

乌鸦巢
Oush Grab (Crow's Nest)

艺术家：亚历桑德罗·佩蒂、珊迪·希拉尔（巴勒斯坦）和埃亚尔·威兹曼（以色列）
地点：巴勒斯坦鸦巢山
推荐人：吉乌希·切克拉
Artist：Alessandro Petti, Sandi Hilal（Palestine）and Eyal Weizman（Israel）
Location：Oush Grab, Palestine
Researcher：Giusy Checola

作品描述

作品位于伯利恒之间的阿拉伯地区和死海沙漠南部边缘的巴勒斯坦最高山的顶部，几个混凝土建筑形成了军事基地的核心。尽管军队已经撤离，但高地区域仍然由军方指定作为巴勒斯坦的限制区域，要塞遗迹成为定居者、以色列军队、巴勒斯坦组织和国际活动家之间对抗的主要站点。消弭冲突的关键在于改变这些建筑空间的潜在破坏力，促使个人和社会组织在文化和政治领域进行多元化合作。

这个山顶另一个独特的原因是它还是五百多万只候鸟的庇护所，每年欧洲东北部和非洲东部之间的鸟都会进行季节性迁移，它们利用约旦谷地的耶路撒冷山作为迁徙路线。因此，每年秋季和春季里的一段时间，山顶上和周围地区的树林里会降落成千上万的鸟，于是在它们周围还形成了丰富的小型食肉动物和其他野生动物微生态循环系统。

该项目的意图并不是翻新和改造军事设施，而是为了加快退化和解体的进程，使它重返大自然。第一阶段，不同的建筑物内外墙上被打出等距离的孔洞，形成一种统一的视觉形式。环保和动物学家预计这些孔洞将吸引一些体型较小的候鸟逗留，也可能会有一些本地物种长期居住。这不是一个单纯意义的建筑项目，它是一门以艺术视野为立足点的"未来考古学"。以另一种方式来看待和使用军事要塞，是对政治问题的艺术干预。

解读

乌鸦总是远人烟、喜空旷。曾经强大的军事要塞随着战争的远去而逐渐荒废，如今只是一片荒凉，艺术家或许只想要烘托一种寂寥的废墟美。百年过去，这里也许还是寸草不生，人们甚至会完全忘记这片土地上曾经悲壮的历史，忘记枪林弹雨曾经在这片土地上横飞。但候鸟记得，乌鸦记得。年复一年，百万只候鸟在这里栖息，它们相比我们似乎更愿意记住这块土地。

艺术家用布满孔洞的建筑，强调着这里悲怆的过去，如果无法在短时间内用美丽的建筑包装这块布满枪眼的区域，那么就这样直接地放大战争的残酷吧，让这里成为乌鸦巢，让动物的记忆代替人类的记忆，这是最慈悲的行为。（高浅）

Artwork Description

The artwork is located on the highest hill in Palestine between Bethlehem and the Dead Sea. The concrete bunkers of a former Israeli military installation ring the summit, which despite the army's withdrawal is still designated off-limits to Palestinians. The site has become a hotspot for confrontations between settlers, the Israeli military, Palestinian organizations and international activists. The key to eliminating the conflict rests in changing the destructive associations of these architectural spaces, by including individuals and civil society organizations in various forms of cooperative cultural and political activities there.

The hilltop is remarkable as well for its importance as a stopover for some five hundred million birds on their twice-annual migration between northeastern Europe and east Africa. For a few days each autumn and spring, tens of thousands of these birds land on the hilltop and the surrounding area. Around them a rich micro-ecology of small predators and other wildlife also gathers.

The intent of this project was never to renovate and convert the military base, but rather to accelerate the processes of deterioration, disintegration and return to nature. In the first phase, the artists punctured the external walls of the different buildings on the hilltop with a series of equally spaced holes. Environmentalists and zoologists expect these holes will be inhabited by some of the smaller-sized migratory birds as well as some local species throughout the year. The project is less an architectural adaptation than an instance of "future archaeology" , as the artists put it. It demonstrates an alternative vision of a military installation, and a possible way for art to intervene in political life.

Artwork Excellence

The innovational character of Oush Grab project, as like as the entire work of DAAR in Palestine, deals with the psychological status of the people, who are strongly resigned and affected by daily humiliations, a life in which it does not seem possible to plan anything. The idea is to challenge the way to see things, to 'decolonize the mind' before the physical places, to generate images of the future before the architectural planning and the construction of buildings.

In this case the mind is the only place where one can act, and the architeture and the urban planning are the main tools, since here— as in other parts of the world—they're used in formal and functional ways to decide the time, the way and the space where the people can go and can be controlled.

Future development of the buildings are in the hand of Palestinian people, but DAAR's Crow's Nest project fed the debate about "what to picture out," when two colliding sides, dealing on one hand by ownership and on the other by the sense of common good, are stuck in a struggling present.

The idea of "future archeology" and "design by destruction" has attracted many researchers, professionals and curators worldwide. The work has been presented in seminars, presentations, debates and expositions inside important cultural and art events (i.e. at REDCAT Los Angeles).

原始的节奏：七轻湾
Primal Rhythm: Seven Light Bay

术家：森万里子（日本）
地点：日本冲绳县宫古岛
推荐人：凯利·卡迈克尔
Artist: Mariko Mori（Japan）
Location: Miyako Island, Okinawa Japan
Researcher: Kelly Carmichael

作品描述

作品坐落于宁静的宫古岛旁的七轻湾，离冲绳海岸约 180 英里，上面矗立着一座纪念碑式的装置，作品由日本著名艺术家森万里子创作。这件作品名为"原始的节奏：七轻湾"（2011），由一座高耸的太阳能柱（或称之为太阳支柱）构成，在岛屿的海湾处有单独的岩石丘，上面有一块发光的月亮石。受到了日本冲绳洞穴的启发，月亮石是一块空壳，当它漂浮在海湾时，光线透过空壳而射入，会转换颜色，潮水浅时是红色，涨潮时变成了蓝色，颜色转换之间有许多层次。非常优雅，具有强烈、简洁的线条，这件作品犹如布朗库西的极简主义风格。

试图去"培养创造力，探索艺术和自然之间的对话，""原始的节奏：七轻湾"如同森万里子的大多数作品，有着将艺术融合于科技的欲望。将东方神话与西方文化、佛教并列起来，天地的精神意识的观念在森万里子的作品中是一个非常强大的、反复出现的主题。灵感往往来源于数个世纪以来的古老仪式和符号，森万里子为我们这个后现代世界创作了众多无比美丽的视觉作品。作为一种特定地点的作品，为这块遥远、未受污染的宫古岛而创作，"原始的节奏：七轻湾"唤起古老文化的生育仪式，突出现在的人性和自然之间微妙的平衡。让人联想起凯尔特异教徒和其他古代文化，他们的生存完全受到不同季节的控制，他们进行公共仪式来标记着这些变化，森万里子建立了一块纪念碑，面向重大的天文事件。

森万里子为她在宫古岛的作品选择了精确的坐标，因此在冬至时刻，太阳支柱的影子被拉长了，将穿透安装在海湾上的月亮石，将天体与陆地结合起来，犹如男性与女性。

这是一个生育的寓言，最近的灾害事件导致国家满目疮痍，"原始的节奏：七轻湾"是重生的一种重要象征。2011年3月11日周五，日本太平洋沿岸发生9级强烈地震，这是至今为止已知的对日本破坏力最强的一次地震，也是自1900年以来现代记录中，全世界最强的五次地震中的一次。地震导致了海啸，引起一系列核反应；福岛核电站中的三个反应堆发生了七级崩塌。"311大地震"使得日本几乎在每种方式上都动摇了。数百上千的日本人失去了自己的家庭和房子，据统计至少15 800人死于这场灾难。

这个国家是科技最发达先进的国家之一，遭遇了地震、海啸和核事故的三重打击而面临着燃料、电力和食物的短缺。2011年冬至对于日本人的心理来说是一个重要的日子。世界各地许多文化中，都会纪念那一天，冬至标志着复与新生，同时也标志着人类集体的团结。冬至，尽管它有着最长的夜晚，同时是一年中的转折点，在这一夜过后，太阳在天空中变得更强烈了，白日逐渐变得更长。因此，冬至也是重生的一种纪念。古老文化利用了这个强有力的符号，如今已经没有这种庆祝了。

森万里子的"原始的节奏：七轻湾"庆祝着宇宙原始的节奏，城市和现代生活很大部分从我们的意识中被擦除了。这件重要的作品目是将人类与自然世界重新连接起来，这样做会引起人们的讨论，这提供了一个愈合创伤的机会。森万里子非常善于通过当代公共艺术作品和古老的智慧，提高公共认知意识，"原始的节奏：七轻湾"可以称得上是她最强有力的作品。

解读

艺术的胜利等同于观念的胜利，当冗繁复杂的古旧纹样形状铺天盖地，掩盖了素净原始的本质，随着科技进步而产生的极简形态反而似乎成为了未来主义追随者们喜闻乐见的精神表达方式。但是这种思潮本身是矛盾的，人们无法切断故去和未来之间的联系，无法完全摈弃古旧的，也就意味着否定了一直试图坚信的未来。号称是未来主义追随者的日本艺术家森万里子似乎也是这样，在她的"原始的节奏：七轻湾"公共艺术项目中，时间好像永远停滞在了那句著名的"时间和空间已于昨天死亡"上。

或许是我对作品的误读，在我的眼里，设计师试图用科技元素去讨论生命的形态。《太始经》云："昔二仪未分之时，号曰洪源。溟涬濛鸿，如鸡子状，名曰混沌。"徐整《三五历纪》云："天地混沌如鸡子，盘古生其中。"海天交际的地方，一直充满着神秘感，在科技力量下发光并不断变换色彩的月亮石正如鸡子，代表一切生命的孕育。而另一边是最古老的计时工具日晷，这是人类文明开始的标志之一。一个虚幻的传言，一个实在的古器，两者在几千年后科技发达的今天，在饱受争议的混沌海天交界处，和科技融合，重新被制作和展出，就是为了引起人类对于原始生命形态的思考。（高浅）

Artwork Description

Situated at Seven Light Bay on tranquil Miyako Island, some 180 miles off the coast of Okinawa, stands a monumental installation by celebrated Japanese artist Mariko Mori. The piece is titled Primal Rhythm: Seven Light Bay (2011) and consists of a towering solar column or "sun pillar" and a glowing Moon Stone on separate rock mounds in the island's bay.

Inspired by the caves of Okinawa in Japan, the Moon Stone is a hollow shell through which light enters as it floats in the bay, shifting colour from red at low tide to blue at high tide, with many gradations in between.

Starkly elegant with strong, simple lines, the work has a Brancusi-like minimalism about it. Intended to "nurture creativity and explore a dialogue between art and nature," Primal Rhythm, like much of Mariko Mori's work, draws upon a desire to fuse art and technology.

The juxtaposition of Eastern mythology with Western culture, Buddhism, and the idea of universal spiritual consciousness is a strong and recurring theme in Mori's practice. Often drawing from centuries old rituals and symbols Mori has created many strikingly beautiful visions for our postmodern world.

As a site-specific work for the remote and unspoiled landscapes of Miyako Island, Primal Rhythm evokes the fertility rites of ancient cultures and highlights an existing and delicate balance between humanity and nature. Reminiscent of pagan Celtic and other ancient cultures whose lives were controlled absolutely by the seasons and who marked their change with communal rituals, Mori has built a monument that is oriented towards major celestial events. Mori has chosen exact coordinates for her work on Miyako Island so that at the moment of winter solstice, the lengthening shadow of the "sun pillar" will penetrate the moonstone installed in the bay, uniting the celestial with the terrestrial, the masculine with the feminine. An allegory of fertility, Primal Rhythm: Seven Light Bay acts as an important signifier of rebirth and life in a region and country shattered by recent events.

The earthquake off the Pacific coast of Japan, a magnitude 9.0 on Friday 11 March 2011, was the most powerful known earthquake ever to have hit Japan and one of the five most powerful earthquakes in the world since modern record keeping began in 1900. The earthquake triggered powerful tsunami waves and caused a number of nuclear reactions; primarily the ongoing level-seven meltdowns at three reactors in the Fukushima Nuclear Power Plant complex.

The events of 11 March left Japan shaken in almost every way. Hundreds of thousands Japanese lost their families and homes and over 15,800 deaths have been recorded. The country known as the one of most technologically advanced faced shortages of fuel, electricity, and food due to the three-fold tragedy of earthquake, tsunami, and nuclear accident.

The Winter Solstice of 2011 was an important event in the Japanese psyche. Celebrated in many cultures throughout the world, solstice honours love and new birth as well as the collective unity of humankind. Winter Solstice, although it is the longest night, is also the turning point of the year, as following this night the sun grows stronger in the sky, and the days become gradually longer once more. Thus, Winter Solstice is also a celebration of rebirth. A powerful symbol harnessed by ancient cultures, this celebration no less valid today.

Mariko Mori's Primal Rhythm: Seven Light Bay celebrates the primary rhythms of the universe, which urban life and modern living largely erases from our consciousness. This important piece intentionally reconnects people with the natural world and, in doing so one might argue that it offers an opportunity to heal. Mori is no stranger to engaging public awareness through ancient wisdoms and contemporary public art works, and Primal Rhythm: Seven Light Bay may just be her most powerful work yet.

Artwork Excellence

Mariko Mori's Primal Rhythm: Seven Light Bay uses a sophisticated understanding of celestial cycles to celebrate the winter solstice on Japan's Miyako Island, some 180 miles off the coast of Okinawa. In the wake of the devastating hurricane, tsunami, and nuclear disaster of March 2011, Primal Rhythm provides a sense of healing through the unchanging cycles of the earth and sun.

绿林里的红飘带：秦皇岛汤河公园设计
Red Ribbon in the Green Forest:
Qinhuangdao Tanghe Park Design

艺术家：俞孔坚、凌世红、宁维晶（中国）
地点：中国秦皇岛市
推荐人：潘力
Artist: Yu Kongjian, Ling Shihong, Ning Weijing（China）
Location: Qinhuangdao, China
Researcher: Pan Li

作品描述

秦皇岛市汤河公园位于中国著名滨海旅游城市秦皇岛市区西部，坐落于汤河东岸，长约 1 公里，总面积约 20 公顷。场地自然环境良好，水位稳定，水质清澈，两岸植被茂密，但部分河岸坍塌严重。具有城郊接合部的典型特征，"脏乱差"的人为环境已经开始威胁水源卫生；大部分遗留构筑物破损、陈旧或废弃。用途比较复杂，存在安全隐患，可达性差。基于场地情况，结合城市居民的功能需求，设计首先保护和完善一个蓝色和绿色基底。严格保护原有水域和湿地，严格保护现有植被设计，避免河道的硬化，保持原河道的自然形态，对局部塌方河岸，采用生物护堤措施，在此基础上丰富乡土物种，包括增加水生和湿生植物，形成一个乡土植被的绿色基地。其次，建立连续的自行车和步行系统。沿河两岸都有自行车道和步行道，并与城市道路系统相联系，增强场地的安全性与可达性。第三，设计一条融合座椅、照明、植物展示和解说系统为一体的红飘带——绵延于东岸林中的线性景观元素。五个以乡土野草为主题的节点和两个专类植物园使红飘带成为寓教于乐的科普展示廊。本设计强调对原有的自然河道和植物的尊重，并在此基础上叠加新的设计，在城市与自然之间，在人与生物之间，在历史与现代之间，建立一种界面，这种界面便体现为一种设计的景观。

解读

对于中国的园林艺术家们来说，造曲是造园中的重要步骤，百转千回，层叠曲婉，将景观错落分隔开，意图用园林设计塑造文人诗词中的景致，使园林景观更加的动态，美观和诗意。

如若就此照搬也并无新意，有趣的是俞孔坚先生本人在借用了"曲"的概念的同时，还推崇生态极简主义这个现代观点。500米的红色飘带在妖娆引路的同时，还兼并了座椅，光源，展示场所等功能，坐落在栈道边，供游客们欣赏休息。将人文元素带入自然，在自然中保护自然，这才是人类与自然相处的最佳法则。（高浅）

Artwork Description

Red Ribbon in the Green Forest in China's Tanghe Park exemplifies the creation of urban green space through use of an existing site with a minimum of design and intervention. The design integrates ecosystem restoration with urban green space, combining a pathway for a leisurely stroll with an environmental interpretation system using native specimen planting, lighting and other functions.

Tanghe Park is located in the western district of the famous coastal city of Qinhuangdao. It is situated along the east bank of the Tang River, with about 1km of shoreline and a total area of about 20 hectares. The area has rich natural resources, stable water levels, and clear water quality with dense vegetation, but there is severe erosion along the riverbank. As in many outlying districts of large cities, intense human activity has begun to threaten the integrity of the water supply; most of the original waterworks structures are now damaged, obsolete or abandoned. As a result of safety risks, accessibility to the area is poor.

Based on the environmental conditions and in the context of the functional needs of the urban community, the first design priority at Tanghe Park was to protect the basic conditions of the water and vegetation. This meant strict protection of existing wetlands and strict protection of existing vegetation. Further measures called for the creation of biological berms in order to maintain the shape of the river and prevent its gradual erosion. The berms themselves were made from native species, both aquatic and wetland plants. Second, a continuous bicycle and pedestrian system was established. Bike paths and walking trails were created along the riverbank, connected with a road system making the area more accessible and secure. Third, an integrated seating, lighting and botanical viewing installation was created; called the Red Ribbon in the Green Forest, it weaves through the landscape along the forest line of the east riverbank.

Red Ribbon is more than 500 meters long, composed of a fiberglass reinforced hollow cube of varying width. The shape and dimensions of the ribbon change to reflect the terrain as it winds through the trees. It has five nodes corresponding to five ecological zones and includes explanatory signs and plant identifiers. Each node has one "cloud" where there is an enclosed seating area and cover, providing protection from both sun and rain. One of the zones features a planting activity, where native plant species including Pennisetym, Andropogon, Lard Mans, Phragmites Australis, Imperata and others can be distributed. Because only native plant species and indigenous resources are used there is minimal required grounds maintenance, consistent with low-carbon design and the park's aesthetic of balance.

At night the ribbon provides an enchanting spot for stargazing, the night sky reflected in its surface. Overall, the effect is of a designed landscape that conveys respect for the river and plant life, while suggesting the interface between city and nature, the man-made and the biological, history and modernity.

Artwork Excellence

Red Ribbon exemplifies the creation of urban green space through use of an existing site with a minimum of design and intervention. The design integrates ecosystem restoration with urban green space, combining a pathway for a leisurely stroll with an environmental interpretation system using native specimen planting, lighting and other functions. Overall, the effect is of a designed landscape that conveys respect for the river and plant life, while suggesting the interface between city and nature, the man-made and the biological, history and modernity.

四川美术学院虎溪校区
Sichuan Academy of Art, Huxi Campus

艺术家：郝大鹏（中国）
地点：中国重庆市大学城
推荐人：潘力
Artist: Hao Dapeng (China)
Location: Chongqing University Town, Chongqing
Researcher: Pan Li

作品描述

"四川美术学院虎溪校区"是重庆大学城规划中的一部分，原址是虎溪镇伍家沟村七社，占地面积 1 000 亩。校区规划与建设坚持传承地域文化，体现出人文校园建设的理念和特色，基本保持自然地形地貌，注重依山就势、顺势而为的园林布局，不铲一个山头，不填一个池塘，完整保留了 11 个山丘，建筑群散落其间，建筑形态和表现贯彻粗材细作的原则，以丰富的形态和朴实的材质体现合乎原地形地貌的生长状态。许多建筑都有一种创意工厂的氛围和感觉。不同的建筑使用不同的材质，尽量保持本质和原生状态，刻画历经风霜的表面，外墙色调体现重庆这座工业重镇的历史记忆。

整个校区充满静谧、清幽的自然氛围，将地域文化与校园个性联系在一起。校区还保留了部分原有的农舍、水渠和农田，具有地方特色的乡村长廊构成连接校区的网络，农家生活在安静的校园中悉如从前，各种农具散布在池边回廊之中成为一种记忆符号，旨在保留大学城的城市原点。建筑布局充分考虑到重庆的自然条件，贯彻"十面埋伏"的设计理念，整体上以聚落的方式在山坡周围自由散布，而群落正是典型的重庆山城建筑传统。四川美术学院虎溪校区既为建设具有自身文化特色的校园探索了一条艺术与技术相结合的路径，也对当代城市建设具有导向性意义。

解读

"人应诗意地栖息在大地上。"海德格尔如是说。大地是人类的起源和归宿，但当我们在地理面积上无法拥有更多，唯一能做的就是不去减少原本的面积和损坏原本的地貌。雕琢自然空间就是雕琢艺术，大自然的鬼斧神刀已经为设计师们构建了设计轮廓，顺势而建，无损形貌，将自然空间有效地和人类工业结合在一起，这对于艺术家们来说是值得挑战的设计项目。

大地艺术家罗伯特·史密斯对于艺术、自然、生态和工业化建设的关系给了如下评述：艺术可以成为自然法则的策略，它使生态学家与工业家达成和解，生态与工业不再是两条单行道，而是可以交叉的，艺术为它们提供必要的论证。在人们对于城市建设过度发展带来的负面效应反思时，公共艺术项目和公共艺术行为向自然领域渗透，更能凸显当今社会有识之士对于自然环境的敬畏。（高浅）

Artwork Description

The Sichuan Academy of Fine Arts, Huxi Campus, is located on a 1,000 mu (165 acres) planned section of Chongqing University Town, which was originally known as "Huxi Wujiagou Village Community 7." Planning and construction of the campus adheres to the locality's geographical and cultural heritage. Reflecting the humanistic philosophy of the school, the campus plan leaves the natural topography undisturbed, using the slope of the hillsides as a basic context for the site design. The school buildings were constructed throughout the 11 hills in the area, without any excavations or landfills. The basic principal of the architecture was to render fine forms with rough materials, in the spirit of the rugged landscape. Many of the buildings have the feeling of a spacious creative studio. Different materials were used for the buildings, reflecting the variation in the landscape, with weathered surfaces and exterior walls painted in tones reminiscent of the industrial legacy of Chongqing.

The entire campus exudes a quiet, peaceful, natural aura in harmony with the regional culture and campus activities. The campus also retains a portion of the original farmhouse, wetlands and fields, with these agrarian features forming a corridor connecting the campus network. Farm life continues as before, interwoven into the life of the quiet campus, and a collection of farm implements adorn the cloister as a remembrance of University Town's origins. The layout of the buildings reflects the natural conditions of Chongqing, adhering to the traditional "ambush on all sides" architectural design approach, in which a whole settlement scatters around the hillside freely. In its exploration of how art may be combined with technology the Sichuan Fine Art College Huxi campus has a distinctive character, and is a noteworthy example to guide contemporary city construction.

Artwork Excellence

Sichuan Academy of Art, Huxi Campus supports the continuity of traditional culture and natural geography, exemplifying a humanistic approach to campus planning and construction. The beautiful ecological planning not only provides an outstanding environment for teaching and learning, but it has made the Sichuan Academy of Fine Arts a popular public space, attracting visitors from near and far who come to enjoy the grounds. The campus succeeds in affirming the academy's own cultural character through a combination of art and technology, while also indicating possible new directions for urban construction for China as a whole in its period of rapid modernization.

驻场计划
Squatting Project

艺术家：绿洲（金康、金渊还，韩国）
地点：中国香港
推荐人：凯丽·卡米切尔
Artist: Oasis (Kim Kang and Kim Youn Hoan，South Korea)
Location: Hong Kong, China
Recommender: Kelly Carmichael

作品描述

自从 20 世纪 90 年代末期，韩国开始出现大量不同风格的、更加吸引人的公共艺术形式。这些作品唤起了激进的理想和社会参与，不仅反映了韩国社会的变迁，也促成了韩国公共艺术实践特点的形成。金康和金渊还对于韩国公共艺术实践的重新定义起到了关键的作用。2004 年，他们成立了绿洲团队，目的是为不同形式的小组创意活动构建一个占据的平台。在公共艺术和社会发展的风口浪尖上，集体的实践包括艺术、研究及向资本主义提出质疑和进入都市空间的各种活动。

英国《卫报》这样评论道："对于很多有创意的和雄心勃勃的个人来说，做出驻场的决定——为了工作或生活或兼顾两者——更多是出于理想主义或 DIY（自己动手做）的精神，同时也出于经济上的权宜考虑。很少有年轻艺术家能够同时付得起房租和工作室的租金。更少有年轻艺术家能够负担起自己能随意使用的大型展览场所的巨大开销。"

绿洲行动包括了占领韩国艺术家中心来抗议艺术家缺乏创作空间。2011 年

7 月开始，团队在香港举办一个项目，香港是世界上最昂贵、人口密度最高的城市之一。由香港艺术团队"低音十"主持，绿洲探索了驻场城市空余空间的潜力。虽然香港有违章占地建筑的丰富的历史，但出于艺术或政治目的的驻场行为尚属新生事物。

"低音十"发现韩国十年以来建筑物的空置情况和香港西九龙文娱区有着很多相似处，西九龙文娱区那里经受了多年的停滞和当局规划的修订。

在居住期间，绿洲举办了驻场及其历史的工作坊，把许多参与者召集起来，聚集在地图周围，用一台笔记本电脑串联，编写可能驻场的地点。其中一个潜在的地点是在观塘棚户区的屋顶。其他的地点包括一座空置的豪宅，由于被围栏锁住了而无法接近。还有马氏铺的一个乡下村子，当地社区组织者希望艺术家驻场者占领这片土地，以阻止房地产开发。

在北角以前的油街艺术家社区，一些工作坊的参与者翻过大门进去探索。该社区仅运行了一年之后就在 1999 年关闭了，艺术家重新搬迁到马头角道的当前地点。油街站点从此以后就空置了。

为了完成这个作品，绿洲运用了驻场地理信息系统（SGIS），这样能够更好地帮助他们调查大财团或政府团体拥有的房地产，找出哪些空间没有被使用或被忽视了，以便给这些空间做出地图，用于国际性的驻场计划。

解读

在通常情况下，我们认为公共艺术传播了掌权者的意识，或许还如同中世纪的那些私人艺术作品一样，谁出资就是谁的话语工具。但这并不意味着平民们就没有自主表达权力。在民主社会，普通市民同样可以以公共艺术的形式向政府和社会提出控诉。 驻场计划就是这样的一个公共艺术项目。

生活空间的创建是基于自然空间的基础，而自然空间为何要用金钱作为交换？当一方房间空置，而另一方竟无遮风避雨之处时，社会上提倡的经济法则就显得尤为可笑。倘若无法以人为本，一切商品物质利益，经济形态都是空谈。在驻场计划项目中，艺术家未经房主的允许直接占用长久未有人使用的房间抑或被废弃的建筑物，在里面打造生活区域就是为了引发社会对于个人土地霸权的思考。（高浅）

Artwork Description

Since the end of the 1990s public art works of a vastly different and more engaged style began to appear in South Korea. These evoked activist ideals and social engagement, and not only reflected changes in Korean society, but also contributed to the characteristics of what is now known as Korean public art practice. Kim Kang and Kim Youn Hoan were essential to this redefining of public art practice in South Korea. In 2004, they set up Oasis, a collective intended to make squatting the platform for different form of group creative actions. On the cusp of public art and community development, the collective's practice includes art, research and daily activities that question capitalism and access to urban space.

As the British newspaper The Guardian comments, "for many creative and ambitious individuals the decision to squat - for working or living or both-comes as much from an idealistic, DIY ethos as it does from financial expediency. Few young artists can afford to pay rent on housing and studio space at the same time. Fewer still can afford the luxury of a vast exhibition space they can do what they want with."

Among Oasis' actions include the occupation of the Korean Artist Centre to protest against the lack of physical space for artists to work. Starting in July 2011, the collective recently undertook a project in Hong Kong, one of the most expensive and densely populated areas of the world.

Hosted by Hong Kong community arts group Woofer Ten, Oasis explored the potential for squatting in the city's vacant spaces. Although Hong Kong has a rich history of squatter settlements, the use of squatting for artistic or political aims remains relatively new. Woofer Ten saw many similarities between the Korean situation of buildings empty for decades and Hong Kong's West Kowloon Cultural District, which has endured years of stagnation and bureaucratic planning revisions.

While in residence, Oasis held workshops on squatting and its history, bringing together many participants to gather around maps and a laptop to collaborate and compile possible squatting locations. One of the potential sites was a rooftop shantytown in Kwun Tong. Other sites included a vacant mansion inaccessible behind a locked fence, and a rural village in Ma Shi Po where local community organizers hoped artist squatters might occupy land to prevent real estate development. At the former Oil Street Artist Community in North Point, some of the workshop participants climbed over the locked gates to explore. The Community closed in 1999 after just one year of operation and the artists relocated to the current site on Ma Tau Kok Road. The Oil Street site has been vacant ever since.

To do this work, Oasis runs the Squat Geography Information System (SGIS), which helps them investigate real estate owned by big conglomerates or government organisations to discover which spaces are not used or neglected in order to map space for squatting projects internationally.

Artwork Excellence

Oasis, a community-minded squatting initiative, offers more than simply tea and anarchy—it carefully explores and forefronts contemporary relationships between power, access, control, social submission and domination and urban space. Via their squatting projects, Oasis is changing geographic involvement and awareness on a local level. They do this through a system they designed that's similar to the well-known "geographic information systems" (GIS). The Squat Geography Information System (SGIS) integrates, stores, analyses, shares and displays geographic information to inform decision-making. By taking the academic practices of GIS and mapping to local areas to promote knowledge production, Oasis provides an empowering and inclusive tool for the average citizen who has little voice in the public arena and even less involvement in policy making decision for public space.

台北当代艺术馆
Taipei Museum of Contemporary Art

艺术家：多位艺术家
地点：中国台湾台北市
推荐人：赖香伶、朱惠芬
Artist: Multiple artists
Location: Taipei, China
Researcher: Hsiangling Lai, Huey-Fen Chu

作品描述

"台北当代艺术馆"（简称："当代馆"）乃是台湾台北市重要的艺术殿堂之一，又因为其坐落于历史文化聚落及捷运交通便捷之区块，使得当代馆邻近之捷运中山站内一公里长的地下街成为年轻一代及亲子活动的大本营，其中设有地下书街、商街以及街舞练习特区等，造就了本区充满了新旧世代与文化不断交响的特色。近四年来，"当代馆"并以"新艺术活化旧小区"项目来开拓馆外业务，而此策略迅速地从点、线、面的建构升级到"艺术一条街"的打造，藉此将当代艺术的影响力扩展并根植于都会环境与小区生活中。在此项目之推动中，主要包括四种实践案例：

1. 永久性公共艺术设置，如：李亿勋及李明道之公共艺术作品；

2. 暂时性公共艺术展示，如："气象万千 AirSupply"策划项目；

作品描述

"台北当代艺术馆"（简称："当代馆"）乃是台湾台北市重要的艺术殿堂之一，又因为其坐落于历史文化聚落及捷运交通便捷之区块，使得当代馆邻近之捷运中山站内一公里长的地下街成为年轻一代及亲子活动的大本营，其中设有地下书街、商街以及街舞练习特区等，造就了本区充满了新旧世代与文化不断交响的特色。近四年来，"当代馆"并以"新艺术活化旧小区"项目来开拓馆外业务，而此策略迅速地从点、线、面的建构升级到"艺术一条街"的打造，藉此将当代艺术的影响力扩展并根植于都会环境与小区生活中。在此项目之推动中，主要包括四种实践案例：

1. 永久性公共艺术设置，如：李亿勋及李明道之公共艺术作品；

2. 暂时性公共艺术展示，如："气象万千 AirSupply"策划项目；

3. 节庆式的公共艺术，如："2010 台北灯节公共艺术北区项目"

4. 艺术介入既成空间的公共艺术，如："捷运忠孝复兴站墙面美化艺术项目"

李亿勋作品"台北传奇"：

本作品系列透过六种主题的结合，幽默的视觉表现，传达出小区与场域精神。金库破墙而出的超现实画面象征了本地的商机与潜力；拟人化的可爱小动物群，呼应了本小区中的各种商业活动与民众的行为，原本冷灰单调的捷运机房墙面与出口，因为这些元素的加入而活化，变身为具有美感及话题性的艺术作品，既增进了公共空间的亲和性，也激发民众与环境美学的对话。

李明道作品"音响机器人"：

"音响机器人 BIGPOW"是个乐团，包括主唱 BIGPOW 和合音 TWINPOW-L & TWINPOW-R 共三个成员。他们超喜欢音乐，喜欢自己听，更喜欢把音乐分享给大家。

台北当代艺术馆作为一个专业的美术馆，其不仅自许成为一个创意、艺术及生活的媒合平台，同时希冀能开展企业、小区和艺术共享荣和的三赢局面。在每一年不同之项目推动累进下，当代馆已然串联并扮演成"公共艺术推广"、"文化创意分享"、"生活美学创造"及"小区环境改造"的中介角色，并更进一步将当代馆区块成为"小区生活、时尚商圈、创意聚落"的新景观亮点。

解读

"台北当代艺术馆"悉心致力将艺术、古迹、科技三者相结合，呈现于"台北当代艺术馆"这个文化空间中，同时也策划了多档精彩绝伦的当代艺术展览，开拓了参观者的视野，也促进了不同文化、甚至不同国籍间的多层次对话。台北当代艺术馆是为台北市艺文发展的一个重要据点，将提供更具互动性且整合度强的展览平台，以鼓励多元风貌的艺术创作与呈现，并藉此开拓出具前瞻性的新文化思维，并希望提供当代城市源源不绝的丰沛创意活力。

近四年来，"台北当代艺术馆"跟随当代艺术发展的步伐，策划和参与了相关公共艺术项目，如永久性公共艺术作品的设置，暂时性公共艺术作品的展览、节庆式公共艺术以及艺术介入即成公共空间的项目等，使当代馆成为了年轻人及儿童活动的大本营以及民众喜爱的艺术殿堂。（冯正龙）

Artwork Description

The Taipei Museum of Contemporary Art (hereafter "the Museum") is one of the key art exhibition spaces in Taipei. It is located in an historical cultural district with convenient subway access and proximity to the Zhongshan Road regional transit hub, a complex of underground pathways filled with youth and family-oriented retail activity, a book row, a hip-hop dance space and other attractions for young and old.

To "infuse the old neighborhood with new art," the Museum has over the past four years extended its activities beyond the museum walls, acting strategically from single points of intervention, to corridors, to programs that involve the whole neighborhood in the creation of an art-infused district. This extension of contemporary art's influence is rooted in the urban environment and community life.

There are four broad programs that are highlighted in this case study:

1. Permanent public art installations (such as Li Yixun and Li Mingdao's public art works);

2. Temporary public art exhibitions (such as AirSupply meteorological curatorial project);

3. Festivals (such as 2010 Taipei Lantern Festival);

4. Site-specific installations (such as Zhong Xiao's Fuxing Metro station wall murals).

Li Yixun's Taipei Legends:

Li Yixun's work is a series of five different sculptures in mosaic tile. Using humor he conveys the spirit of the locality and community. Amusing anthropomorphic creatures echo a variety of commercial activities in the community and convey people's behavior. The cold, gray wall of the metro station entrance is animated by these elements, and transformed into a compelling and culturally relevant work of art. These works not only enhance the identity of public spaces but also stimulate a dialogue about environmental aesthetics.

Li Mingdao's Audio Robot:

The artist has for many years practiced animation design in his art. To engage with the many young and creative people living and working in the vicinity of the museum, he created a group of interactive robots that sing. Audio Robot BIGPOW is the name of a music band made up of the lead singer BIGPOW, and TWINPOW-L and TWINPOW-R as the chorus. The installation is informed by contemporary forms of media exchange and social networking. People are not only able to listen to the robots' pre-programmed music, but using their interface can upload their own music to share with everyone. The artist uses the roles of the different robots to reflect the free experimentation of young people with different cultural forms, while giving the public a highly interesting art experience.

bringing together many participants to gather around maps and a laptop to collaborate and compile possible squatting locations. One of the potential sites was a rooftop shantytown in Kwun Tong. Other sites included a vacant mansion inaccessible behind a locked fence, and a rural village in Ma Shi Po where local community organizers hoped artist squatters might occupy land to prevent real estate development. At the former Oil Street Artist Community in North Point, some of the workshop participants climbed over the locked gates to explore. The Community closed in 1999 after just one year of operation and the artists relocated to the current site on Ma Tau Kok Road. The Oil Street site has been vacant ever since.

To do this work, Oasis runs the Squat Geography Information System (SGIS), which helps them investigate real estate owned by big conglomerates or government organisations to discover which spaces are not used or neglected in order to map space for squatting projects internationally.

Artwork Excellence

The Taipei Museum of Contemporary Art public art project is an exemplary case of "art intervening in spaces," using the nearby MRT station facilities, parks and open space, and inviting artists to direct their attention to different places and implement community planning activities. Besides offering people an aesthetic experience, the project highlights the connection between art and citizens, and art and life. It accentuates the interactivity and approachability of artworks—encouraging people to interact with the works—and draws people's curiosity.

The Taipei Museum of Contemporary Art is a specialized art museum that not only has become a platform for creativity and the arts, but also aspires to develop the dynamic interrelationship between business, community and art. In its yearly, arts interventions in spaces the museum threads together its aims of "promoting public ar,t" "sharing creative culture," "building life aesthetics," and "transforming the community environment." The museum exerts itself to help transform the nearby area into a "living community with a rich mix of fashion, business, and creativity," and a new bright spot on the cultural scene.

大同新世界
Taipei Public Art Festival

艺术家：多位艺术家
地点：中国台北市
推荐人：赖香伶、朱惠芬
Artist: Multiple artists
Location: Taipei, China
Researcher: Hsiangling Lai, Huey-Fen Chu

作品描述

本案例之公共艺术设置缘起乃是台湾公共艺术百分之一经费的要求，然而因为兴办单位"台北市政府工务局卫生下水道工程处"，与委办单位"台北市政府文化局"因合作而发展出一种比较周延且可累积的新版政策，让"依法必要执行"的行政业务进阶成为一种"创意"行政的指标。

本案例藉由地方节庆的设计和民众参与活动之增强，将一个所费不赀的公共艺术设置案，从污水联想／闲人莫入的工厂区域，挪移至一种可以亲水近土，真正属于公众的空间中，并将之置换为更有趣味话题和鼓励更多群众参与的一种社会节庆活动；其中的用意和预期效益，正如当时"文化局"长廖咸浩所言："期许公共艺术得以在节庆的推动下，激发群众的注意与讨论，并从中培养市民的艺术思维，展现不同区域的文化特质。"

本案例明确设定了三个工作重点，对后续上场的节庆设计、艺术策展及方案执行，提供了相当有用的参考和依据：

1. "先公共、后艺术"的基本主张：即"以艺术吸引民众参与"的政策思考；

2. 让"当代艺术与传统文化结合"的具体要求：着眼于迪化污水处理厂周边丰富的历史文化资源，而寄于本案的相关创作者及活动设计者的一种信息与期待；

3. 经由公共艺术活动"开发地方资产，恢复历史记忆，重整空间关系，再造小区认同"的设置理念，对本案后续的短期性展演及各项小区参与式活动，乃至于大型节庆活动的规划执行，隐约也发挥了路线导引和期约创作的双重作用。

本案例虽是基于"台湾政府"采购法的规定，委由公开征选得标的团队负责策划和执行，但兴办单位在征选策展团队之前置作业阶段，实际已着眼于将本案从公务机关制式化的"财物购置"流程作业，转化提升为"公共艺术节"的概念演绎和一个鼓励多方参与形塑的社会性活动；这个"创意行政"的动因，对策展的内容构设和执行方式，虽预设了一些规范与框架，但它所期许产生的，也正是能大幅超越"依法行政"的有形与无形之果。

解读

"大同新世界"是巴哈依教教义。虽说人之初性本善，但社会上少有人不在争夺自己的利益，无关正确或是错误；也有不少人为了各自利益而引发各种冲突，国家的、家庭的、个人的。但巴哈伊教却拥有最淡泊的名利观——种族是共存的，文化也是值得宽容的，这个世界上所有的一切都应该是共享的，包括环境、资源、艺术等。我们共同生活的环境需要一起维护，共同的文化需要齐心协力的传承。即人类一家，世界大同。

在中华文化中，指纹不光代表相同的意志与承诺，还反应了参与者们对于这项公共艺术所要表达的理念的认同。在这个由公共艺术创造的新世界里，所有的个体组成联合体，为同一个目标而努力，如真若如此，便无愧于此活动的期盼 --- 大同新世界。（高浅）

Artwork Description

The Taipei Public Art Festival was originally conceived as part of Taiwan's program for allocating one per cent of public funds for public art. The agency in charge of implementing the project, Taipei City Works & Sanitation Department, and the agency with administrative authority, the "Taipei City Government Cultural Affairs Bureau", cooperated successfully in implementing this new policy, and advancing the goal of developing creative thinking within the cultural administrative sector itself.

This major public arts initiative was undertaken to re-position a Taipei city sewage treatment plant zone from a restricted access "no-man's land" to a true public space that reflects the value of cherishing water. The project had a further objective of stimulating greater civic participation in local activities and festivals. The purpose and intent of the project as stated by Liao Xianhao, the head of "Cultural Affairs

Bureau" at the time, was: "to use public art in the context of a festival, in order to stimulate awareness and discussion among the public, to foster artistic thinking, and to celebrate the cultural qualities of different regions."

The project clearly illustrated three important principals, providing a useful guide for subsequent festival and exhibit design:

1. Public comes first, art second. That is, focus on policies that use art as a means of drawing people to participate in public affairs. By first setting up community workshops and inviting cultural workers and artists to station themselves in the community, conduct research and interact with the public, the case exemplifies this principle.

2. Consider how to wed contemporary art with traditional arts and culture. The program at the Dihua Sewage Treatment Plant focused on the rich surrounding environment of cultural and historical resources, enlisting all the participating artists and activity planners in this effort.

3. In terms of the short-term follow-up efforts of this project, as well as planning for new programs, focus on developing local resources, recovering historical memory, re-imagining places, and rebuilding community identity.

The Taiwan Public Art Festival exemplifies these principles by incorporating long-term works and temporary works, and a variety of forms of cultural expression (e.g. poetry, music, dance and theater). Through seminars, workshops and documentary film shooting, the project fostered the reexamination of local cultural assets, and collective learning about ways to unite community awareness. Through these programs the project invigorated public participation and served as a catalyst for civic spirit. (The above extracted from records of the Taiwan Public Art Award 2nd Award Ceremony).

Artwork Excellence

The care devoted to forming the selection processes demonstrates the substance of the idea of a "creative administration," whose tangible and intangible benefits exceed anything that could be achieved by acting on the basis of the procurement law alone. The case raises questions about the relationship between public art and community building, the social function of art, the artist's social role, artistic creation and place-making, and provides a possible model of contemporary art as a social practice. More important, as a large-scale year-long public art festival, the Taipei Public Art Festival created the possibility for artworks coming out of the community to linger in the community, over time contributing to greater community cohesiveness and identity. By integrating contemporary artistic creation with the local historic and cultural resources the case laid the groundwork for long-term development of community spirit.

临时公共画廊
Temporary Public Gallery

艺术家：螺旋桨艺术小组（越南和美国）
地点：越南胡志明市
推荐人：凯利·卡迈克尔

Artist: The Propeller Group artist collective (Vietnam and USA)
Location: Ho Chi Minh City , Vietnam
Researcher: Kelly Carmichel

作品描述

2010 年螺旋桨小组在胡志明市租用了一块公共告示牌，放置在一个公共汽车候车亭里，为期三个月举办了这个公共艺术项目。

在一个国家里精明、有见地的举动，所有的视觉元素都在风景中呈现了——从公共艺术到广告——通过不同的审查机构而控制着，"临时公共画廊"意图是探索胡志明市这些空间监管下，政治/审查和公共空间、公共艺术、私有化商业空间之间的关系。

这个项目始于小组的兴趣：城市景观的视觉元素是怎样反映当地社会政治的变迁，同时也可以影响视觉元素本身。

表达出愿望要见识一下视觉元素怎样能够在快速变迁的越南景观中有影响作用，这个集体通过出租广告空间，在公共地区展出艺术品，试图找到一个系统中的漏洞，对于越南的公共空间、广告和公共艺术的概念提出了挑战。

这个艺术团队以反对对抗边界而知名，螺旋桨小组的倡议强调对了越南公共艺术的期望，并标明了框架。以欧洲为中心，对于公共艺术的理解并不存在于东南亚的大多数地区。基于集体活动，社会活动和履行，任何公共干预总是非常短暂的。

正如团队的解释："越南的公共艺术在最近几十年以来是非常有限的，只有公园的大理石雕塑和一些老的宣传标牌，贴在各大城市的各种墙上。但是景观从宣传画开始转移了，如今变成了越来越多的广告，此时公共艺术依然是无害的，人迹罕至的，受到审查机构的限制。"

螺旋桨小组的实践有一种折中主义和自发性，越南有序的社会主义国家的对立面。他们的角色是挑战，激发这种现状，团队和临时公共画廊项目作为催化剂而存在，正在匍匐前进，但强大的社会变革——不仅仅是视觉化景观，也存在于人们如何使用和回应他们的城市。

由纽约的艺术事务组织拨款资助，这个组织协助艺术家在美学和社会之间破土动工，这个项目要打破审查障碍，从来没有实行过它的完整形式。

螺旋桨小组成立于 2006 年，是一个艺术团队和媒体制作公司，由 Phu Nam Thuc Ha，马特·卢塞洛和团·安德鲁·阮——这些来自西贡和洛杉矶的视觉艺术家所组成。螺旋桨小组帮助人们实现合作声明，重新定义了社会和政治对于当代亚文化和通俗文化的理解。

利用自身能力吸引到电视、电影、录像和网站，使信息更具吸引力，并期望规模扩大，螺旋桨小组是媒体语言的操控者，希望能够有越来越多的观众，参与世界画廊博物馆空间的艺术表达活动。

对这个存在于政治意识形态和经济实现之间模糊的空间感兴趣，该小组利用大众文化、影院、电视、广告、互联网、画廊、博物馆、写作、采访和对话来发布和传播这个艺术项目。他们的作品不仅仅展示于主流电视台和国际电影节上，同时也在国外主要的博物馆和画廊里展示。

解读

无论是临时的还是永久性的画廊，其目的都是为了向特定的群众传播艺术的某种信息。这是一个大胆的尝试，让画廊存在于市井中，说明人们有意愿和决心让其成为生活环境的一部分，让其融入当地人的生活。从一定意义上来说，这种行为其实已经起到了教化作用。胡志明市的这个"临时公共画廊"暂时借用了车站，很难知道它最终到底是否会作为艺术设施而存在，但无论如何，其所要展示的元素都面向公众，以艺术的形式存在，成为城市公共空间的一部分。（高浅）

Artwork Description

In 2010 The Propeller Group rented a public billboard at a bus shelter in Ho Chi Minh City for 3 months to stage a public art project. A shrewd and insightful move in a country where all visual elements in the landscape— from public art to advertising—are controlled through different censorship bodies, Temporary Public Gallery was intended to explore matters concerning public space, public art, privatized commercial space and the politics/ censorship behind the regulation of these spaces in Ho Chi Minh City. This project began with the group's interest in how the visual elements of a landscape not only reflect the socio-political changes of that locale, but can have an affect on it as well. Expressing a desire to see how they could contribute to this affect in the rapidly changing landscape of Viet Nam, the collective attempted to locate a loophole in the system by renting out advertising space to curate artworks in public, challenging notions of public space, advertising, and public art in Viet Nam.

An art collective known for pushing against boundaries, The Propeller Group' s initiative highlighted the framework for and expectations of public art in Vietnam. The Euro-centric understanding of public art does not exist in much of South East Asia. Any public interventions tend to be often very temporary, based on collective activity, social activation and performance. As the collective explain: "Public art in Viet Nam has been limited in the last few decades to marble sculptures in the park and some old propaganda signage attached to various walls throughout different cities. But the landscape has been shifting from having more propaganda to now having more advertising, while public art remains innocuous, inaccessible, and limited by censorship bodies."

There is an eclecticism and spontaneity to the practice of The Propeller Group, the antithesis of Vietnam's ordered socialist state. In their role of challenging and provoking the status quo, the collective and the Temporary Public Gallery project exist as a catalyst for a creeping but powerful social change—not only in the visual landscape, but also in how the people use and respond to their city. Funded by a grant from Art Matters, the New York based organisation created to assist artists breaking ground aesthetically and socially, the project hit censorship hurdles and was never realised in its full form.

Established in 2006, The Propeller Group is an art collective and media production company comprised of Phu Nam Thuc Ha, Matt Lucero and Tuan Andrew Nguyen, visual artists from Saigon and Los Angeles. The Propeller Group helps to realise collaborative statements that re-define the social and political understanding of contemporary sub-cultures and popular cultures. Drawn to television, film, video and the web for its ability to make information attractive and desired on a mass scale, The Propeller Group are manipulators of media language and keen to reach a larger audience that takes the presentation of art beyond the world of gallery spaces and museums. Interested in the fuzzy spaces that exist between political ideologies and their economic implementations, the group uses popular culture, cinema, television, advertising, the internet, gallery, museums, writing, interviews and conversations to distribute and disseminate art projects. Their work has not only been shown on mainstream television and international film festivals but also in major museums and galleries abroad.

Artwork Excellence

Temporary Public Gallery was intended to explore matters concerning public space, public art, privatized commercial space and the politics/ censorship behind the regulation of these spaces in Ho Chi Minh City. In their role of challenging and provoking the status quo, The Propeller Group and the Temporary Public Gallery project exist as a catalyst for a creeping but powerful social change—not only in the visual landscape, but also in how the people use and respond to their city.

丰岛美术馆
Teshima Art Museum

艺术家：雷·内藤、西泽立卫（日本）
地点：日本丰岛
推荐人：凯利·卡迈克尔
Artist: Rei Naito, Ryue Nishizawa（Japan）
Location: The island of Teshima, Japan
Researcher: Kelly Carmichael

作品描述

丰岛是 2010 年濑户艺术节开展活动的七座岛中的一座，目标是让濑户海附近的岛屿重新焕发生机。艺术节期间向公众开放的 "丰岛美术馆" 由日本建筑师西泽立卫和艺术家内藤礼合作设计的。这座空灵的建筑矗立在丰岛的小山上，俯瞰濑户内海。内藤礼以在精心准备的空间中布置精致的小物件而闻名，这件作品的标题意味着一种在孕育生命的母爱形象，从地上渗出的水滴构成了这一主题。

"丰岛美术馆" 是一个 25 厘米厚的白色混凝土壳，没有任何柱子或承重结构。覆盖了 40 米 ×60 米的空间。没有直接的清晰入口，整个结构也几乎是空的。流动、有机的内部似乎像活的有机体一样膨胀与收缩。建筑外形像水滴，从其开口处飘来阵风，泉水涌出地面。建筑本身与周围环境相互作用，勾勒出人工建筑物与自然之间的界线。天花板向天空敞开，在轻轻转动的形态中消失于地面。混凝土膜覆盖着地面，包裹着阴影的边缘，跨度为一个低矮通畅的圆顶，传达了固有材料和固定地点的自然美。

整个建筑的氛围随着太阳的运行和时间的改变而变化，通过建筑、人类与大自然的融合，捕捉与自然的和谐关系。参观者在整洁的空间里独自留下，细细体味其结构与自然环境。"丰岛美术馆"在艺术节闭幕之后还将继续运营，并举办包括艺术、建筑、食品和环境等在内的多种活动。

解读

在中国哲学观中，"天人合一"大的概念举重若轻。自古以来，人们就期望能与自然亲密无间。而受中国汉文化影响至深的日本多少也传承了这些哲学观念。近现代以来，随着工业的发展、科技的进步、城市构建的改变，自然和人类渐离渐远。受西方极简主义和未来主义的影响，艺术家们试图将现代工业和材料制作成简单但富有线条的雕塑或者景观类艺术作品，融入到自然中，再现人法地、地法天、天法道、道法自然。

中国的《易经》讲究阴阳结合。日本对于易学的推崇也是有目共睹的，相信设计师在设计的时候必然也遵从了这方面的考量。依山傍水是人尽皆知的风水宝地的象征，在这个艺术项目中，我们已知山属阳，而缺水，水滴形的形状正弥补了水属阴的特征，阴阳结合，背山面水，藏风聚气，这许是艺术家们对于人和自然相处模式最深刻的考虑。（高浅）

Artwork Description

Teshima is an island in Japan's Seto inland sea, and was one of the seven island venues of the 2010 Setouchi Art Festival. The festival was created to bring renewed vitality to the region. Designed by Japanese architect Ryue Nishizawa and artist Rei Naito, the Teshima Art Museum was opened to the public during the festival. The spacious structure is located on a hill overlooking the sea. Rei Naito is renowned for his use of carefully arranged space to situate small, delicate objects. In this case the objects are water droplets which enter the space naturally through openings in the roof.

The Teshima Art Museum is comprised of a white concrete shell 25 cm thick, enclosing a column-free space of 40 by 60 metres. The structure is almost completely empty and there is no obvious point of entry. Its interior is fluid, organic, and seems to expand and contract like a living, breathing being. Its overall form is that of a water drop, echoing the drops of water that form on the floor carried by the wind through the openings. The ceiling which opens to the sky, and the floor which curves with the ground's gently shifting morphology establishes a dynamic interaction with the surrounding environment, blurring the boundary between manmade structure and nature. The concrete membrane covers the ground and curves seamlessly into the low clear-span dome overhead. Form is stripped down to its essence, expressing the natural beauty inherent in the materials and location.

The whole building atmosphere changes as the sun moves across the sky. Individual visitors linger in the spacious room, savoring its structure and connection with the natural environment. After the closing of the arts festival, Teshima Art Museum has continued to operate, hosting various activities related to art, architecture, food and the environment.

Artwork Excellence

Teshima Art Museum helps transform an island known as a dumping ground for toxic waste into a destination for contemporary arts. The features of the sculpture demonstrate a unique sensitivity to the island's environment—in striking contrast to the waste scandal that lies in the island's history.

蒲公英学校改造项目
The Dandelion School Transformation Project

艺术家：叶蕾蕾（美国）
地点：中国北京市大兴区蒲公英中学
推荐人：潘力
Artist: Lily Yeh (USA)
Location: Dandelion Middle School, Beijing, China
Researcher: Pan Li

作品描述

蒲公英中学建立于 2005 年 5 月，是北京第一所官方注册的非营利性学校，特别为外来流动务工人员的孩子而建。当学校刚建造时，学校周围的区域非常脏，墙壁一塌糊涂，地面坑坑洼洼高低不平。学校的校长郑洪在一次会议上遇到了费城大学艺术系教授兼环境艺术家叶蕾蕾。2006—2009 年此后的四年里，艺术家与蒲公英中学的教职工和学生一起合作，共同创作了一个校园项目，名为"通过环境来转变心灵"。

叶蕾蕾指导该项目的设计与执行工作，灵感来源于学生在教室里自己画的画，强调天空、大海、森林和花朵的自然色彩。

校园由教学大楼和环绕着的沥青操场以及长条形的平房所组成，校园所有的墙壁都是转换计划潜在的画布，能让年轻的参与者们发挥无边无界的自由想象，也是为了通过整个校园全面的彩绘而产生一种凝聚力。陕西民间

293

艺人库淑兰的剪纸成为整个项目主题的灵感来源。壁画受到了这些剪纸中所描述的师生们熟悉的故事的启发，尤其是"生命之树。"

"生命之树"借鉴了大马士革清真寺的马赛克，壁画主体的蓝色和绿色是伊斯兰艺术的特色。"生命之树"也描绘在学校门口的支柱上，使用了马赛克瓷砖，它的树枝蔓延到了柱子的顶端，包围着代表了孩子们的希望的花朵与星星。两只美丽的凤凰的图画使用马赛克镶嵌制作而成，还有玻璃弹珠和黑色的水泥点缀在操场的观礼台上。台上的老虎和龙的图案基本来自汉代篆刻。通过中心位置的教学大楼的大门漆成白色，还画着辉煌的彩虹，如果遇到雨水，颜色便会折射。

通过艺术品的创作，师生双方均接触到了各式各样的传统工艺及文化形式，并藉由对自然科学、历史、地理、文学、艺术以及其他学科的整合，获得了不断学习和建立个人知识体系的经验。凭借对使用素材的独立搜索，每一个学生都得以认真思考社会互动的本质，每一类学说也都获得了更广阔的发展空间。总而言之，艺术创作的过程给学校带来了更为多样化的教育体验。

关于蒲公英中学：

因其创新性的教育方式和透明化的管理实践，蒲公英中学获得了来自政府、基金会、社会团体以及不计其数的个人志愿者的大力支持。如今，学校已有来自中国 25 个不同省、市、自治区的 59 名教职员以及 678 名学生。学校现有 37 个教室，其中包括物理、化学、生物的专用实验室，音乐教室，艺术工作室，宽敞的展厅以及新近翻修、藏书超过 15 000 册的图书馆。与此同时，学校还拥有足以容纳超过 500 名学生的宿舍区，为广大学生提供了在家所无法享受到的良好生活条件及积极的学习环境。而这一切都是通过个人及企业捐款得以实现的。

解读

艺术家为特殊的空间而做出的设计是这个空间特殊的表现符号。其代表此空间里生活的人们相同的人生观和文化观以及对于此空间存在形式的认同。生活未必一帆风顺，艺术却能让生活增添姿色。作为第一所为来京务工的低收入家庭子女创办的公益性中学，即便如今的蒲公英学校已经受到了社会的广泛关注，改造方法可以很多元化，但是叶蕾蕾女士的方案却能够让孩子们切实参与到实践中来，一砖一瓦，一笔一画，用艺术的手法设计装点自己的学校，这才是这个项目本身最大的意义。（高浅）

Artwork Description

Dandelion Middle School, founded in May 2005, is Beijing's first officially registered non-profit school created specially for the children of migrant workers. When it was built, the area around the school was dirty, walls were in poor condition and the ground was potholed and uneven. The school's principal, Zheng Hong, met the environmental artist and Philadelphia University College of Art professor Lily Yeh (Ye Lei Lei) at a conference. For four years thereafter, from 2006 to 2009, the artist collaborated with Dandelion's faculty and students on a campus project called "Spiritual Transformation through Environment." Yeh guided the design and realization of the project, taking inspiration at times from the students' own classroom artworks, and emphasizing natural colors of the sky, sea, forests and flowers.

The campus is comprised of an asphalt playground surrounded by school buildings, and a long strip of bungalows. All the walls on campus were potential surfaces for the transformation plan, in order that participants' youthful spirit might have free range, but also in order that a cohesiveness might emerge throughout the entire campus. The papercuts of Shaanxi folk artist Ku Shulan served as inspiration for the theme of the entire project. Murals were inspired by stories familiar to students and teachers portrayed in these paper cuts, in particular that of the "tree of life." The "tree of life" is a reference to a mosque mosaic in Damascus, and the blue and green colors that dominate the murals have an Islamic quality. The "tree of life" is also depicted in mosaic tile on the pillars at the entrance to the school, its branches twisting up to the top of the pillars that are encircled with flowers and stars representing the wishes of the students. An image with two beautiful phoenixes made with inlaid mosaic, glass marbles, and black cement adorns a stage in the playground. The tiger and dragon motif at the base of the stage comes from Han dynasty seal carvings. Through the main gate of the campus stands the central school building, painted white with a brilliant rainbow, colors refracting as if seen through rain.

Through making such artworks students and faculty came into contact with a variety of traditional arts and cultural forms, and experienced a kind of life learning process, building knowledge through the integration of nature, history, geography, literature, art and other subjects. By letting each student search independently for materials to use, they were given the opportunity to reflect on the substance of social interactions, and each class was given a wider space for the development of ideas. Overall the process of art creation brought to the school a more diverse educational experience.

About the Dandelion Middle School:

Because of its innovative educational approach and transparent administrative practices, Dandelion Middle School has received tremendous support from government, foundations, civil society organizations, and countless individual volunteers. Today, the school has 59 faculty members and 678 students from 25 different cities, provinces and autonomous zones all over China. The school now comprises 37 classrooms including dedicated physics, chemistry and biology lab rooms, a music room, a studio art room, a spacious exhibition space, and a newly renovated library housing more than 15,000 volumes. In addition the school has dormitory space for more than 500 students, offering them good living conditions and a positive learning environment, which may not be available in their homes. All this is made possible through the donations of individuals and businesses.

Red Ribbon is more than 500 meters long, composed of a fiberglass reinforced hollow cube of varying width. The shape and dimensions of the ribbon change to reflect the terrain as it winds through the trees. It has five nodes corresponding to five ecological zones and includes explanatory signs and plant identifiers. Each node has one "cloud" where there is an enclosed seating area and cover, providing protection from both sun and rain. One of the zones features a planting activity, where native plant species including Pennisetym, Andropogon, Lard Mans, Phragmites Australis, Imperata and others can be distributed. Because only native plant species and indigenous resources are used there is minimal required grounds maintenance, consistent with low-carbon design and the park's aesthetic of balance.

At night the ribbon provides an enchanting spot for stargazing, the night sky reflected in its surface. Overall, the effect is of a designed landscape that conveys respect for the river and plant life, while suggesting the interface between city and nature, the man-made and the biological, history and modernity.

Artwork Excellence

Thanks to the participation of all the students and teachers both in the classroom and during after-school time, the "Transforming the Spirit through Environment" project has changed the school environment over the course of its four years. Working with artist Lily Yeh, students and teachers participated in the design and creation of murals. Besides sourcing scrap tiles from the wholesale market for the multi-colored mosaic murals, the teachers and students actively rummaged through demolition sites for materials. These activities awakened the imagination of students, involving them actively in the happy experience of creation. Participating in this creative activity afforded the students an opportunity to develop their self-esteem, strengthen their imagination, and further develop their capabilities.

土地（基金会）
The Land (Foundation)

艺术家：里尔克利特·提拉瓦尼亚、卡明·勒特查普雷瑟（泰国）
地点：泰国清迈附近
推荐人：凯利·卡迈克尔
Artist：Rirkrit Tiravanija, Kamin Lertchaipraser（Thailand）
location：near Chiang Mai, Thailand
Researcher：Kelly Carmichael

作品描述

1998 年，两位艺术家在泰国清迈附近的村庄收购了一块土地，这里没有电和自来水。他们要打破画廊和美术馆的陈规陋矩，反抗制度系统对艺术家的权力操控，这就是"土地计划"的出发点。他们把艺术活动的地点搬到了一片田野上，要为当地农民、学生和国际艺术家创建一个充满活力的社会。

这个项目最具特色的艺术行为之一是为前来参观的人烹饪，他们还建造了一个镀铬的厨房控制台和一个用玻璃和钢铁打造的现代派家庭复制品，并邀请人们在里面做任何使他们高兴的事情：睡眠、阅读、乐队排练、听音乐、辩论政治问题等。他们还用泰国报纸做成 366 张制型纸物体，并在每个对象上绘制了从佛教和道教哲学中摘录的广为人知的谚语，引发人们对当代泰国文化所面临问题的沉思。

"土地计划"的环境是自我维系的，它的功能是一块种植水稻的农业空间，同时也是一块社会空间。"土地计划"无异于一个"乌托邦"，行动主体是艺术家和艺术学生，这是一个结合了当代艺术干预和传统农业观念的正在进行的合作项目，它脱离现有的框架，撇开"财产"或所有权的概念，开始创建一种别样的社会和生活方式。"把土地当成一个大家都可以接触到的开放空间"的思想是这个项目必不可缺的部分，其终极目的是"寻找在一起的新方式"。

解读

我们本是赤裸裸地来，在世上虚晃几十载，而后又赤条条地走，何故非要将自己的一生束缚于物质的枷锁？两位艺术家买下了空荡荡的一片土地，然后开始构建没有物质的世界。这块土地面向公共开放，任何人都可以在这里进行非物质的思索。只要忠于自身情感，任何话题都可以被大家接受，这是一群具有特殊思想的人在一个特殊的公共空间中进行的与现代社会格格不入的事情。人们沉思、冥想，维持着最基本的人生需要的同时却也在进行着最深刻的思想交流。（高浅）

Artwork Description

In 1998, two artists bought a parcel of village land near Chiang Mai, Thailand, with no electricity or running water. Their intent was to break out of the rigid conventions of the gallery and museum system which constrains artists' imaginative potential. This was how their "land" project began. Moving the locus of art programs to an open rice field, they set about building a lively community for local farmers, students and international artists.

One notable aspect of this project is the hospitality offered to anyone who may come to visit: cooked meals are prepared in a specially designed kitchen and visitors are welcome to sleep, read, play and listen to music, and discuss world affairs. An installation prepared by the artists provided a stimulus for discussion: 366 papier maché objects were created from Thai newspapers, with a Buddhist or Daoist proverb inscribed on each.

The Land is a self-sustaining environment operating as a farm where rice is cultivated, as well as a social space. The Land is essentially a utopian community whose chief participants are artists and art students implementing ongoing collaborative projects all of which combine contemporary art intervention and traditional agriculture. Putting aside the idea of "property" and the concept of ownership, they are creating a distinctive community and lifestyle. Above all the artists maintain the land "as an open space which everybody can enter into." The ultimate purpose is to find new ways of being together.

Artwork Excellence

The Land is a self sustaining environment that functions as an agricultural space growing rice. It is open to the neighbouring community and also acts as a social space where artistic practices are discussed and tested. Ultimately the project is about "finding new ways of being together" Tiravanija says, a humble space emphasizing interdisciplinary and collaborative practices, which functions as a hybrid of innovation and traditionalism, contrasting contemporary materials and technologies with ancient forms of agriculture.

最后的椅子
The Last Chairs

艺术家：里昂·林姆（马来西亚）
地点：马来西亚槟城州乔治城
推荐人：凯利·卡迈克尔
Artist: Leon Lim（Malaysia）
Location: George Town, Penang Malaysia
Researcher: Kelly Carmichael

作品描述

装置艺术"最后的椅子"（2010）坐落于槟城州的历史古都乔治城中，出自一位马来西亚出生、纽约成长的艺术家里昂·林姆之手。此作品位于乔治城内历史悠久的旧城区的路边，由林姆用 100 多个废弃的椅子混杂堆砌，形成大约 9 英尺 ×13 英尺的巨大几何形。其每到夜晚就点亮照明以使某些部分显现出来，而另一些部分则被隐去。而这堆无序组装的本土椅子让任何踱步经过的人感到既有趣又困惑。尽管组成这一装置的各部件或许略显普通，但那看似微不足道的标示意义却非常丰富。

除了具有美学上的优点之外，"最后的椅子"的不同凡响之处还在于它是这位艺术家的第一件公众装置作品，同时又是乔治城第一件当代装置艺术。虽然，目前在这块历史悠久地区也有其他的展示作品，但它们雕塑风更趋传统。此作品及其所涉及的委派和实施这件作品的种种努力促进改变了一个国家长期以来局限于传统陈旧风格的艺术框架，用当代艺术揭开了历史古都的新篇章。

高度城市化和工业化的槟城是马来西亚最先进和繁荣的城州之一，也是旅游胜地马六甲海峡的一站。其首府乔治城有着维多利亚风格的外墙、典雅的宗教地标和战前海峡兼收并蓄的建筑架构，可谓是具有栩栩如生的历史和文化之地。2008 年，因它异于其他东南亚的独特建筑和文化景观被授予联合国教科文组织世界遗产称号。

马来西亚的艺术和文化有着深厚的历史底蕴，但这现代国家也同时是一个伊斯兰民族，其艺术和其他表现风格都长久受到宗教传统禁忌对有生命物体进行描绘的影响。在如此丰富的传统背景下，"最后的椅子"一如林姆的其他风格，延续着他对历史和社会变迁进行的探索之路。此作品为现代公共艺术在乔治城的委派和展出开辟了新的路径，以快照式的手法描述了其周围地区并反映了城中深厚的社会历史。

林姆的作品通常以其透辟的观察力将物体抽离其原有的语境和日常功能，以此达到生动展现文化和社会的目的。林姆经常使用废弃的物件和破旧的街道，如他之前用的废灯泡、破玻璃杯、金属罐、玩具及电子和塑料媒体。他的尝试目的不在于这些被发现的物体，更多是在于转变的可能性以及一种文化物体或标志如何变成另一种。

林姆这样评论"最后的椅子"："它代表了艺术在乔治城中逐渐变化的历史和其未来的道路。"他进一步描述之为："一个生动的作品，它受到了人民参与的影响，并进一步受到了城市文化影响。"这座雕塑强调了平常物体对于乔治城的标志意义，也陈述了它们那种能够为城市带来社会和美学框架改变的影响力。在所谓"公众艺术还是一个新概念"的马来西亚，里昂·林姆认为他还没有简单的把旧椅子变成艺术，而是把陈旧而毫无起色的期待变成了新的现实。

解读

装置艺术是一个场地、材料与情感的综合展示艺术。位于马来西亚首府槟城中的古都乔治市是一个融合历史之城也是世界文化遗产名录。林姆选择这个地方很大程度上来说就是因为这个城市的历史性。这虽是马来西亚的城市，但历史上却不完全属于马来西亚。这些数量众多形态不一的椅子似乎就代表着这个城市历史的变迁。马来西亚不光是穆斯林国家，还有着华人的血脉，来自欧洲的殖民者们也曾在这里统治和生活，在这个多文化混杂的国家里，古往今来的掌权者们一代一代地沿袭着他们的传统和规则，条条框框束缚着这个国家，却又在某种程度上成就了这个国家现代的风华。椅子的英文是 Chair，而掌权者的英文也和椅子有直接的关系 "Chairman"，层层叠叠的椅子就如同那些一代又一代的掌权者们，尽管他们早已作古，却无法抹去他们曾经在这里生活过的印迹。（高浅）

Artwork Description

Located in the state of Penang's historical capital George Town, The Last Chairs (2010) was an installation by Malaysian born, New York–based artist Leon Lim. The work in two parts saw Lim create a towering jumble of more than 100 discarded chairs, stacked in a large geometric shape some 9 by 13 feet, by the side of the road in George Town's historic inner city. Lit by night so certain components are revealed and concealed, the chaotic assemblage of piled domestic chairs amused and baffled those who stumbled across the sculpture. Though the objects in the assemblage may have been ordinary, its gesture—however small—was radical.

Aside from any aesthetic merit, The Last Chairs broke new ground as it is both the artist's first public art installation and the first contemporary installation in George Town. Although other artworks are present in the historic area they are of more traditional sculptural form. Marking a new chapter for contemporary art in the capital, this work and the efforts of those involved to commission and deliver it helped to alter the artistic landscape in a country where art in the public realm has until recently been infrequent and often limited to traditional and banal representations.

Highly urbanised and industrialised, Penang is one of the most developed and economically important states in Malaysia, as well as a thriving tourist destination on the Straits of Malacca. Its capital George Town, complete with Victorian facades, quaint religious landmarks and pre-war Straits Eclectic architecture, is home to a vivid social and cultural history. In 2008 the town was given UNESCO World Heritage Site significance for its unique architectural and cultural townscape without parallel anywhere in East and Southeast Asia.

In Malaysia art and culture has strong roots, but this modern country is also an Islamic nation where art and other representational forms have historically been subject to traditional taboos regarding the depiction of living things. With such rich context to build from, The Last Chairs, like much of Leon Lim's practice, continues his exploration of configurations of historical and social narratives. Simultaneously breaking new ground for the commissioning and exhibition of contemporary pubic art in George Town, the work took a snapshot of its surroundings and the reflected on town's dense social history and past.

With a keen sense of observation Lim's work often takes objects out of their normal context and separates them from their everyday function so the objects instead become living representations of culture and society. Often working from found objects and the detritus of city streets, Lim has previously employed such discarded objects such as light bulbs, broken glasses, steel cans, toys, and electronic and plastic media in his work. His practice is not simply about the found object, but more strongly about the possibility of transformation and how one cultural object or gesture may turn into another.

The Last Chairs, Lim comments, "represents the graduation of art in George Town throughout history and its path in the future" . Described by the artist as "a living work that was inspired by people's participation in the influence of urban culture" the sculpture emphasised ordinary objects emblematic of George Town's past but also spoke of their transformative power to bring change to the social and aesthetic landscape of this city. Inn Malaysia where "public art is still a new concept" according to the artist, Leon Lim has not simply turned old chairs into art, but old and stagnant expectations into new realities.

Artwork Excellence

Artist Leon Lim created not only a surprising and disorienting installation-a boxed, assemblage of ordinary household chairs-he also took an extrodinary, historic step for Malaysia. The Last Chairs broke new ground as it is both the artist's first public art installation and the first contemporary installation in George Town. Although other artworks are present in the historic area they are of more traditional sculptural form. Marking a new chapter for contemporary art in the capital, this work and the efforts of those involved to commission and deliver it helped to alter the artistic landscape in a country where art in the public realm has until recently been infrequent and often limited to traditional and banal representations.

树
The Tree

艺术家：农场（艺术团队，新加坡）
地点：新加坡
推荐人：凯利·卡迈克尔
Artist: Farm (collective, Singapore)
Location: Singapore
Researcher: Kelly Carmichael

作品描述

"树"（2009 年）坐落于新加坡国家博物馆附近，这是一件雕塑作品，借鉴了几百年历史的榕树的神话故事，在 21 世纪重新诠释其形式。由新加坡的农场团队进行创作——这是一个以社区为中心的艺术组织，进行跨学科的设计实践——"树"使用了不同尺寸的钢铁长方形与方形的框架制作，从地面升起，交错在一起，创造了一片三维的空间。这件作品对一棵树作了一种抽象化和城市化概念的处理，这件雕塑被设计用作聚集的场所，这里可以让到博物馆的参观者坐下放松一下，欣赏博物馆的风景和作品本身。受到了附近著名的地标榕树的启发，雕塑悬挂着麦克风，代替了树的标志性气根。夜里，"树"有 LED 发光管子的"分枝"，灯光把金属框架缝合在一起，回应着周围的声音，轻轻地跳动着，每当游客的声音高低不同时，灯光便会调光或变亮。如果你站在树下，抬头望去，光线给你留下一种印象，你在观看着遥远的星座。新加坡是一个微小的共和国，面积只有 700 平方千米，是一个包含多个民族的社会大熔炉，是一个吵闹、具有活力、全球化的都市枢纽，每到晚上就焕发出勃勃生机。本作品回应着城市国家的夜间文化，但同时也创造一个深入的、更多的精神层面和隐喻，"树"把新加坡人与他们的社会和当地的历史相连接。虽然它的形式可能是城市化的、当代性的、生硬的，雕塑用丰富的含义由细微之处展开多个层面。

依据其设计说明，农场团队受启发于"博物馆门前草地上庄严竖立着的巨大古老的榕树，似乎在守护着这块有着丰富历史、故事和魔幻的地方"。对于新加坡当地人，"树"和榕树之间的联系一看便知。榕树的形象在亚洲和太平洋地区很突出，在南亚神圣，有很多关于它的神话，尤其是印度教和佛教。巨型树木的形状很显著，有着雄伟的树冠与气根，气根从树枝蔓延到地面。随着树龄的增长，这样的根就与主干连为一体，引申到树成为对于永恒生命和复兴的一种隐喻。榕树有着将人们聚集在一块的历史。在许多传统社会中，这种树经常被用作会面的地点或是村庄中心的标志，围绕在树干举行许多活动。树的造型往往启发了寓言化的象征主义，无数条气根被视作为宗教或文化统一的象征——一个实体有着许多遥远的根源。

唤起了神秘榕树的力量，"农场"不仅为公众创造了一个沉思的聚会场所，同时实际回应了使用者以及他们如何与城市及公众一起互动。拥有 100 平方米的空间，"树"是一种整体空间的体验，在一种全新的灯光下重新创造了亲密感。在一种有机自然形式的启发下创作了该雕塑，通过当代对于榕树的解读，雕塑暗示着丰富的文化和象征主义。这个催眠视觉景观作品通过其形式和互动反应，在新一代新加坡人和他们的文化记忆中制造了新鲜的、令人回味的链接。它不再是一棵被复制的树，它本身就是一种真实。

解读

草地上庄严竖立的古老榕树，似乎在依恋、守护着这块具有丰富历史、故事和魔幻的地方——新加坡。健壮的树木外形，雄伟的树冠气根，根与枝的紧紧相连，隐喻着对永恒和复兴的渴求与希冀。百年历史的榕树神话故事脉脉相传，时代潮流下的重新诠释与追忆，凝结成"树"这件多精神层面的当代作品。

"树"是一种公共空间的互动体验，在一种全新的形式下重新创造了亲近感，拉近了人与公共艺术作品的互动与交流。根深蒂固的文化记忆，使"树"不再是一棵被复制的树，而成为了一颗承载文化内涵的真实存在的榕树。(冯正龙)

Artwork Description

Situated near the National Museum of Singapore, The Tree (2009) is a sculpture that draws upon the centuries-old mythology of the Banyan tree and reinterprets its form for the twenty-first century.

Created by the Singapore based collective Farm—a community-centered arts organization and cross disciplinary design practice—The Tree is made from varying sizes of steel rectangles and square frames that rise from the ground and interlock, creating a three-dimensional space. The work is an abstracted, urban idea of a tree and a sculpture designed as a gathering place, a point for museum patrons to relax, sit, and enjoy the sights of the museum and work itself.

Inspired by a well-known local landmark Banyan just nearby, the sculpture has dangling microphones substituting for the tree's characteristic aerial roots. At night, The Tree glows with LED light tube "branches." The lights are sutured together between the metal frames and pulsate gently in response to ambient sounds, dimming and brightening as visitors voices vary. If you stand under it, looking up, the lines of light give you the impression you're looking at constellations of stars faraway.

A tiny republic of just over 700 square kilometers, Singapore is a rich ethnic and social stew and buzzing, dynamic, globalised urban hub that comes alive at night. Responding to the nocturnal culture of the city-state but also creating a deeper, more spiritual dimension and metaphor, The Tree lets Singaporeans connect to their social and local history. Though its form may be urban, contemporary, and angular, the sculpture unfolds many layers rich with meaning and nuance.

According to its design statement, FARM was inspired by "the huge old Banyan tree that sits majestically on the museum's front lawn, seemingly holding fort to a place full of histories, stories and magic." For Singapore locals, the connections between The Tree and a Banyan would instantly be recognizable. Banyan trees figure prominently in Asian and Pacific religions and myths and are sacred in South Asia, particularly to Hindus and Buddhists.

The shape of the giant tree is unmistakable, it has a majestic canopy with aerial roots running from the branches to the ground. With age, such roots can become indistinguishable from the main trunk, leading to the tree becoming a metaphor for renewal, or eternal life. The Banyan has a history of bringing people together. In many traditional societies the tree is used as a meeting spot or a marker for the center of the village and many activities take place around its trunk. The tree's form also inspires allegorical symbolism; its numerous arterial roots are seen as a symbol of religious or cultural unity—one entity with many far-flung roots.

Evoking the powers of the mythical Banyan tree, FARM has not only created a contemplative gathering place for the public, but also one that responds physically to its users and how they interact with each other and their city. With a footprint of 100 square meters, The Tree is a holistic spatial experience that re-invents the familiar in a whole new light. Inspired by and created in the image of an organic natural form, the sculpture alludes to cultural richness and symbolism through a contemporary reading of the Banyan tree. The hypnotic visual landscape the work creates through its form and interactive response creates fresh and evocative connections between a new generation of Singaporeans and their cultural memory. It ceases to be a copy of a tree, but a truth in its own right.

Artwork Excellence

For the form of this twenty-first century, sophisticated sculpture in Singapore, designers from FARM chose an ancient, culturally specific model—the Banyan tree, a traditional meeting ground. The result is a forward-looking structure using sophisticated modern techniques that has been adopted by the local community as a natural extension of traditional cultural habits.

幸福知道
The Way to Knowing Happiness

艺术家：江洋辉（中国）
地点：中国台北市
荐人：赖香伶、朱惠芬
Artist: Chiang Yang Huei (China)
Location: Taipei, China
Researcher: Hsiangling Lai, Huey-Fen Chu

作品描述

台北捷运小碧潭站的公共艺术设置乃是采取"邀请比件"方式；然而，不若一般之台湾公共艺术设置案，以设计兼施作之统包方式执行，本案例乃是征选艺术家之设计构想方案，由原站体工程施工单位执行施作部分。

一走出捷运小碧潭站，首先映入眼帘的是：采用钢构造的双弧线造形捷运站体外观，屋顶覆盖着银灰色的复合金属板和采光天窗，而钢架和柱子则是漆上柔和的芥末黄，并且柱子上有冉冉上升的云朵（作品"云在跳舞"及"时光冻"），强烈的对比色差增加了整个站体的光亮，再加上宽敞的空间、开阔的视野，让每一位进站的人立即感受到轻松愉悦的气氛。

而在西侧广场以"候车"为本区之主体意象，设有作品"幸福的预言"，包括了座椅"幸福一号"及不锈钢站牌"下一站就到了"；同时设有咖啡座，民众可在广场上远眺碧潭风景区及新店河滨公园全景。另外，以南方松木板所铺设而成的休闲广场，更是成为一个全新的公共休闲场域。

而东侧广场则有作品"甜蜜的模样"：包括有座椅"幸福一号"，主要以造形街灯和具特色之 LED 光雕玻璃护栏（作品"浮"），交织成有趣的艺术空间。而南侧广场虽是较狭小空间，然而衬托有简洁、具科技感之立体作品"我们都是全家福"，倒是成就另一番现代感风貌。整体而言，占地约有 1 500 坪大之空中户外广场的捷运小碧潭站，是台北大都会捷运系统中，难得具备有户外开展、宽敞的空间，可让乘客及附近居民稍做停留、舒展身心。

解读

人作为城市的主体，所有的艺术都应该与人相关，而公共艺术与城市生活又密切相关。工业革命以后，生活虽然便捷，但却被工业化建设过度占据，人类一辈子不停地穿梭于钢铁塑成的大小不一又形状各异的空间中。地铁就是一个盒子，载着人们往返于出发点和目的地，我们似乎一直在移动，却并不是自身在移动，静默着从起点到终点，从未曾深思过生活除了奔波忙碌还需要什么元素才能构成幸福。"幸福知道"的理念简单完美。艺术家就是想在市民日常生活区域中开辟出一个崭新又美观的交流场所。由于公共艺术作品的时效性，带有生命的艺术元素比如花鸟虫鱼都由于受到了生命的限制而很难被直接运用到公共艺术中，而江洋辉先生却借鉴自然的概念，将具有自然颜色的元素用现代建筑材料添入我们的生活区域，让生活环境变得不那么的单一、冰冷和无趣。（高浅）

Artwork Description

The Way to Know Happiness is part of the entire structure of the XiaoBiTan Station in Taipei, Taiwan. It includes metal sculptures, interactive works, public seating with lighting and landscape design.

The roof of the double-arch steel framework of the station structure is covered in silver metal plating and daylight screens. The steel frame and pillars themselves are painted a warm yellow hue, floral clouds capping each pillar. The striking contrast of color accentuates the brightness of the whole station and creates a sense of spaciousness. Visitors entering the station immediately experience the effect of the airy, spacious and cheerful atmosphere of the place.

In the west waiting room area, corresponding to the function of that space, is the work "Forecast of Happiness", including a seating area "Happiness No. 1" and station signpost "Now Arriving at The Next Stop". There is also a coffee stand that affords a panoramic view of the

Bitan landscape, looking out over the Xindian River Shore Park. The southern pine flooring further compliments the relaxed atmosphere of this pleasant public space.

In the station's east plaza stands the work "The Sweeties", which also includes seating areas and a lighting display with LED illuminated glass shades woven together in an interesting art composition. The south plaza, while relatively small, contains an eye-catching work of 3-D art, "We Are All One Family" , expressing a different kind of modern style, evocative of scientific and technological achievement. In all, the 1500 square meters of open-air plazas that comprise the Xiao Bitan Rapid Transit station is outstanding in the entire Taipei metropolitan rapid transit network for its use of outdoor space, and its success in providing passengers and nearby residents a place to temporarily relax and unwind.

Artwork Excellence

Generally, an aspect of Taiwanese public art that has great room for improvement is the successful interplay between artworks and their environment. Most completed works of public art are limited by the conditions of the facilities in which they are situated, and suffer from being additive elements. The excellence of this case is fundamentally in the planning of a space whose rigid requirements of engineering is wed from the very beginning to a vision of public art, that is, to take the idea of public art and fashion a kind of culture of art-in-engineering. Today, what people see is a perfectly integrated Rapid Transit station design, in which art and infrastructure are seamlessly interwoven, resulting in a delightful and cheery space for the pubic relax in their leisure.

通往麦加之路
The Way to Mecca

艺术家：布西·施瓦兹（以色列）
放置地点：以色列埃拉特
推荐人：里奥·谭
Artist: Bucy Schwartz（Israel）
Location: Eilat, Israel
Researcher: Leon Tan

作品描述

作品"通往麦加之路"是 1999 年由艺术家布西·施瓦兹和景观建筑设计师 Zvi Dekel 开始设想的公共艺术作品。最近这个设想开始建造并且接近完成。根据1999年的提案，该项目选址在以色列的埃拉Wadi Shahamon (峡谷)。这个项目被形容为"雕塑 / 考古景观工程"。因为作品记录了公元 7 世纪以来穆斯林朝圣者去麦加朝圣的景象。这些朝圣者的朝圣旅途总是经过这条长度为 400 米的路，树立了超过一百座真人大小的朝圣者雕塑。这些雕塑不断地提醒人们：在某些地区犹太人和阿拉伯人民共存了几百年。

埃拉特是一个有争议的地区，千百年来以色列人、罗马和穆斯林竞相争夺。埃拉特实际上是在希伯来文圣经"出埃及记"中提到的、是一个古埃及和琳的贸易伙伴。埃拉特阿拉伯语的名字 Umm Al-Rashrash，历史上来看是组成 Darb EL 朝觐或被称之为麦加之路的一部分，这条线路由非洲经过埃及通往麦加。作为一个公共艺术项目，"通往麦加之路"对长时间持续地途经埃拉特的穆斯林朝圣活动具有重要意义。尤其是二战后以色列国家版图形成和此后持续不断的冲突中。根据 1947 年联合国分治计划，埃拉特地区成为以色列的一部分。然而，1906 年奥斯曼帝国的政令指定该区域为埃及的一部分。该地 2008 年确实发现一个埋葬穿着军服的穆斯林（看起来像埃及人）万人坑。

解读

穆斯林教是全球第二大宗教，每年来自世界各地 70 多个国家的信奉者们为了自己的宗教信仰，马不停蹄地赶往朝圣之地麦加，完成他们的朝圣之旅。这是自公元 627 年就开始的传统宗教活动，一直延续到了现在，教徒们年复一年，乐此不疲。这些身披穆斯林长袍的教徒们虔诚地活着。在他们的内心深处，灵魂便属于他们的神、他们的《古兰经》。无论麦加之途几多风险，都永不停歇和放弃，只为抵达目的地。这是信奉宗教信仰的狂热和那一份与生俱来的对于宗教的忠诚。而肉体、欲望等，就如同空气一样虚幻缥缈，因为精神永恒才是真的永恒。

布西·施瓦兹在千古不变的通往麦加的必经之路上，放置了 100 座穆斯林教徒们朝圣的雕塑，这项艺术作品在集合了所有穆斯林教徒的意志后，向世界其他人们展示着穆斯林教徒的思想状态和精神寄托。它跨越了种族，跨越了世俗的生活，是人类宗教、信仰、精神寄托的集合体。（高浅）

Artwork Description

The Way to Mecca was originally conceived as a public artwork by artist Buky Schwartz, and landscape architect Zvi Dekel, in 1999. It has recently been under construction and is close to completion. According to the 1999 proposal, the project was to be sited at Wadi Shahamon (ravine) in Eilat, Israel. Described as a "sculptural / archaeological landscape project," the work pays respect to Muslim pilgrims journeying to Mecca since at least the 7th century. These pilgrims frequently passed through Wadi Shahamon on their religious travels. 400 meters in length, The Way to Mecca is composed of over one hundred life-size sculptures representing the pilgrims to Mecca. It is a continual reminder that the Jewish and Arabic peoples have coexisted in certain regions for several centuries.

Eilat is a contentious region given competing claims over the centuries by Israelites, Romans and Muslims. Eilat is actually mentioned in the Book of Exodus in the Hebrew Bible. It was a trading partner of Ancient Egypt and Elim. The Arabic name for Eilat is Umm Al-Rashrash, and historically the city formed part of the Darb el Hajj or Pilgrims' Road from Africa, through Egypt, to Mecca. As a public project, The Way to Mecca is significant for its references to the long duration of Muslim pilgrimage through Eilat en route to Mecca, given the ongoing conflicts engendered by the "formation" of the State of Israel after the world wars. According to the 1947 UN Partition Plan, the Eilat region was to form part of Israel. However, a 1906 Ottoman decree designates the area as Egyptian, as does the discovery of a mass grave of Muslims in military uniform (likely Egyptian) in 2008.

"Researcher's note: Buky Schwartz passed away in 2009. While the nominator states that the project was completed in 2011, Buky Schwartz's wife Ziva reported that the project is in fact not complete, but will be in a matter of months. I have attempted to locate more information on this project, including contacting Zvi Dekel (the collaborating landscape architect) and Amon Yariv (Schwartz's gallerist). Neither have responded to requests for information."

Artwork Excellence

As a public project, The Way to Mecca is significant for its references to the long duration of Muslim pilgrimage through Eilat en route to Mecca, given the ongoing conflicts engendered by the 'formation' of the State of Israel after the world wars. According to the 1947 UN Partition Plan, the Eilat region was to form part of Israel. However, a 1906 Ottoman decree designates the area as Egyptian, as does the discovery of a mass grave of Muslims in military uniform (likely Egyptian) in 2008. Historically, Eilat was part of the Darb el Hajj or Pilgrims' Road from Africa, through Egypt, to Mecca. Its Arabic name is Umm Al-Rashrash.

宝藏岩国际艺术村
Treasure Hill Artist Village

艺术家：多位艺术家
地点：中国台湾台北市
推荐人：赖香伶、朱惠芬
Artist: Multiple artists
Location: Taipei, China
Researcher: Hsiangling Lai, Huey-Fen Chu

作品描述

台北市宝藏岩历史聚落，发展肇始于日据末期，因其坐落于小观音山与新店溪，为台北市自来水建设之水路要地，遂规划成保护区，并设置军队警戒。初期仅有六户闽南籍家庭因宝藏岩寺之宗教信仰簇群定居；台湾光复后，至 20 世纪 60 年代逐渐因水源功能移转、地方造桥以及军营撤离等措施而取消管制。再经 20 世纪 80 年代的急速蓬勃的现代化经济变迁，退伍荣民及其配偶、城乡移民等弱势族群陆续迁入。之后透过长期自力造屋，聚落规模大幅扩张，其间约有百来间的违建房舍被搭建，住户间亦形塑出成熟绵密的生活网络及互助合作的生活方式。今日，宝藏岩聚落于 2004 年由台北市"文化局"正式登录为台北市第一处聚落型态之"历史聚落"，并于 2010 正式由"台北国际艺术村"团队进驻营运。宝藏岩聚落的地景特殊性，在于其巧妙刻画出巷弄蜿蜒、阶梯缓坡起落、沿着山城构筑错落的风貌。在既有居民环山簇居之外，活化保存的政策同时引入"艺术驻地"与"青年会所"创意，希冀"宝藏家园"、"宝藏岩国际艺术村"及"国际青年会所"三大计划，结合"生产、生活、生态"，期能以"共生"精神，创造出聚落美好未来。（部分内容节录自网站资料：http://www.beautiful.taipei.gov.tw/blog/plan-8/view/769-3947 & http://www.artistvillage.org/thav）

解读

这如同哈里波特电影中的建筑物并不那么和谐倚在葱郁的山坡上，凌乱、毫无章法的层叠在一起。但就是这外表看上去破破旧旧的房子，却见证了台湾几代移民族的历史。现如今，在政府的扶持下，这片山形聚落形态的建筑群再次改变了聚居属性，已然成为了艺术家们的聚集地。自然本身就是承载艺术品的最佳场景，这些曾经在不经意中形成的建筑风格是得天独厚的艺术基底，现代思想、艺术与历史建筑的结合，让这曾经仅仅由于生存才存在的生活空间被重新打造，成为新一代居民的多是艺术家，生活和工作共处一室，创作和日常生活被聚集在了一起。就如他们的艺术风格，风格各异却又和谐统一。（高浅）

Artwork Description

Casting Light and Shadows on Treasure Hill History installation project:

This installation is a site-specific community-driven project created by Pan Yuyou for the Taipei Treasure Hill historic village.

The development of the village began after the end of Japanese rule. Located on Small Guanyin mountain beside Hsintien Creek, the area was originally part of Taipei's water district and was protected by the army. Originally, there was only a small cluster of six families from south Fujian Province living around the Treasure Hill Buddhist temple. After Taiwan's recovery of self-government, until 1960, the army gradually retreated from the area, and Taipei's water supply system was re-engineered away from the mountain. By the 1980s, with rapid economic changes, migrant groups from the countryside had begun to populate the hillsides. After years of improvised housing construction, the community grew to more than a hundred households, living in a mutually supportive network.

In 2004, the "Taipei Bureau of Cultural Affairs" registered Treasure Hill as an historic settlement—the first such designation in the city—and in 2010 the Taipei Artist Village was established. The Treasure Hill landscape has a distinctive rambling quality; its paths and stairways wind about the gentle slope of the mountain through the neighborhood's eclectic structures. In addition to conserving the neighborhoods of the existing residents, the government policy for Treasure Hill provided for the establishment of a local arts camp and youth club, the Treasure Hill Community Center, and the Treasure Hill International Arts Village and International Youth Club. These programs were conceived as three mutually reinforcing components of a settlement with a bright future.

This project focuses on the International Arts Village and its related public art projects, which have brought artists' visions to bear on the challenge of conserving the historic architecture of the place while enriching life of the settlement. Treasure Hill has historically been a patchwork of different generations, cultures, and ethnicities; and after the renovations it continues to fulfill the function of cultural interchange. Creative people and international travelers venture out of the downtown to mix together, while at the same time Treasure Hill has become more integrated into metropolitan Taipei, and has become a major cultural attraction.

The premise of Casting Light and Shadows on Treasure Hill History is that lighting can be used in the evening to accentuate the unique atmosphere of Treasure Hill while respecting the original appearance of the historical section. Using a combination of landscape art, lighting design, and public participation, the work expresses the simple nature of life in the settlement. From the foot of the slope of Treasure Hill extending across the wide grassy field of Hsintien Creek, soft lights shine through the clear darkness to illuminate the settlement, giving the scene a poetic quality. When people walk along a stage that has been erected on the grass their shadows are projected on the historic building facades, turning the spectators into performers.

City Scenery mural project:

The mural projects of Pan Yuyou have been a vital, cultural centerpoint for the Taipei Treasure Hill historic village.

City Scenery captures the urban character of Treasure Hill by portraying key components of its streetscape and residential spaces together. The work records the diverse elements of urban life through vivid hand-painted murals, which offer the public an opportunity to consider the city in novel, modular compositions. The artist first conducted a survey of the settlement and surrounding area, and through interviews and questionnaires established eight visual subjects: "Tingzhou Road development," "Water Market," "Little Guanyin Mountain Ecosystem," "Toad Mountain and Coast Trail," "Live House," "Library and Café," "Bicycle," and "Treasure Hill collective life."

Artwork Excellence

By using the play of light on the hillside, the project illuminates the historical narrative of the Taipie's Treasure Hill community. For the installation, the artist Pan Ruzhou collaborated with violinist Lin Zhongyi, who performed selections of his choice while Ruzhou controlled the lighting display. The weekend presentations, combined with the public's own interactive movements, became like a painting in continual progress, showing the unlimited creative potential of performance art and interactive dance. This public art installation succeeds through a process of subtraction, highlighting certain fundamental aspects of the built environment and human culture. The night lighting scheme, the dance stage, and handling of landscaping elements together create a public space for performance and general recreation.

"City Scenery"

For the artist, Treasure Hill was an improvised assemblage of different generations and different ethnic groups who arrived from unspecified homelands. After its renovation and integration into the city this cultural function is perpetuated through the mix of creative people and international visitors. Since its creation, the mural project has become a point of pride for the community, and a draw for cultural tourists throughout Taipei.

粉乐町
Very Fun Park

艺术家：多位艺术家
地点：中国台北市
推荐人：赖香伶、朱惠芬
Artist: Multiple artists
Location: Taipei, China
Researcher: Hsiangling Lai, Huey-Fen Chu

作品描述

想象一下：当周末悠闲的午后，带着轻松愉悦之心情，在与三五好友约喝下午茶的时尚空间；或是恣意闲情逛街购物之时，都能与艺术品不期而遇，那该是多么令人惊喜之事！财团法人富邦艺术基金会即是开发此艺术生活化的前锋者，而其主导之艺术活动"粉乐町"首发于 2001 年，并自 2007 年起宣示每年策展不同主题的决心。连续五年来，"粉乐町"不仅成为富邦艺术基金会之年度业务重点外，亦是台北东区商业空间之年度艺术盛会。

"粉乐町"乃是由台湾策展人张元茜，带着台湾年轻的艺术家与部份国外艺术家，挑战城市不同空间与艺术结合的可能性。因此，策展人提出巷弄美学的梦想蓝图，透过史无前例的场地空间洽谈，意外地在人文荟萃、商业繁荣的台北东区之数十处商业空间展出作品，让"粉乐町"成为艺术无所不在的美好想象落实项目。而这亦实现了富邦艺术基金会未曾改变的初衷："艺术就该在生活当中发生与遇见，美好的、愉悦的、观念、行为、多重材质、装置、设计，都可以让创作者自由自在的与空间对话，没有门槛没有压力。"

解读

公共艺术理当存在于公共空间之中，世俗的繁华无法掩盖人类精神上的空虚。很多时候，我们需要这样的一种生活方式：艺术存在于生活空间的任何角落,我们幻想着与艺术不期而遇。艺术作品在点滴中影响着我们的生活，指导着我们的思维和生活方式。

公共艺术分为很多种类，这些高档商业场所出现的艺术品本身就不是单纯的艺术，他们的种类和出现的空间已经决定了他们的使命即为了创造经济利益。在商业中出现的这些披上了公共艺术的外衣的艺术品，并不是亵渎了艺术，因为确实是为了需要而产生。他们影响了部分人的生活，改善了部分人的思考方式，因此，虽然特别却异常成功。商业化的公共艺术，在一定程度上来说反而是一个公共艺术领域中的一个新的突破。这样的购物空间，让购物并不是精神疲怠后的唯一选择。（高浅）

Artwork Description

Imagine a leisurely afternoon stroll with friends, stopping in teahouses and shops and meeting by chance works of art in a place where customarily there is only merchandise. It's possible in Taipei. Starting in 2001, the Fubon Art Foundation has sponsored the arts event Very Fun Park, a "museum without walls," in the commercial spaces of East Taipei. Since 2007 the foundation has announced different annual curatorial themes for the event.

Very Fun Park was conceived by the Taiwanese curator Zhang Yuan Qian, in association with several young Taiwanese and foreign artists who were all inspired by the possibilities and challenges of combining art with urban space. The curator proposed examining the aesthetic of alleyways, negotiating a previously unexplored cultural territory of accidental encounters in the dozens of shops lining east Taipei's thriving commercial district. Very Fun Park is meant to infuse the entire area with art to provoke the imagination. The original intent of the Fubon Art Foundation, which has not changed over the years, was "to generate opportunities in everyday life for encountering beauty, new ideas and behaviors, different materials, assemblies and designs, and provide the creator opportunities to freely engage in dialogue with the space without any pressure or conditions."

The theme for the 2010 Very Fun Park was "Looking Forward," focusing on "taking a broad perspective on cultural diversity," and fostering values of learning, respect and cooperation. Nearly 40 artists and groups from Taiwan, Europe, America and Asia were invited and asked to engage in a mutual exchange of ideas with the local shop-owners to develop installations appropriate for their environments. The results cut across a diversity of media including painting, sculpture, site-specific installations, video, sound and theater. Together they formed a varied and complementary public art experience. The program's emphasis on communication, respect and cooperation was reflected in the mutual assistance provided by artists and locals together to complete the project.

Artwork Excellence

Very Fun Park exemplifies the notion of a "museum without walls," bringing art outside the museum and into daily life. Through its intervention the project fosters familiarity with spaces that blend with art, and provides an experimental ground for developing aesthetics and civic sensibility. Very Fun Park has become an annual highlight in East Taipei's bustling business district, and has now a rich history of promoting public arts education, supporting young artists, and helping businesses promote their brands through creative use of space. In recent years the event has become one of the most important large-scale temporary public art events in Taiwan. Very Fun Park is also an exemplary case of private enterprise in Taiwan participating in the arts. Through the planning of the annual theme enterprises are given the opportunity to consider carefully their social context and cultural environment and develop as socially responsible corporate citizens.

亚穆纳河行走（穿越 22 公里）
Yamuna Walk (Through 22kms)

艺术家：阿图尔·布哈拉（印度）
地点：印度新德里
推荐人：里奥·谭
Artist: Atul Bhalla（India）
Location: New Delhi, India
Researcher: Leon Tan

作品描述

亚穆纳河的名字源于印度女神亚穆纳，她是太阳神苏里耶的女儿，死神夜魔神的姐妹。它流经新德里市，是印度神圣的恒河的最大支流。按民间传说，在亚穆纳河沐浴能够抵御死亡，或者将人们从痛苦的折磨中解脱出来。具有讽刺意味的是，亚穆纳河自身正在为生存而奋斗，它是当今世界上污染最严重的河流之一，每天有百万升未经处理的污水流入河中，这无数升污水来自工业和商业活动的有毒化学品。也许因为这条河的污秽，新德里数百万居民几乎意识不到亚穆纳河的存在，很多人从没有见过这条河。在新德里的莫卧儿和英国殖民时期，在亚穆纳河旁边建造了大量楼房。但是独立后，人们就不再将建筑物建造在河附近。有人可能会说，这座城市将主要航道空留了出来。

阿图尔·布哈拉行为作品"亚穆纳河行走（穿越 22 公里）"正是因为以这条河为背景，它的内在含义才得到了最大的彰显。

艺术家受邀参加 KHOJ 国际艺术家协会的 2007 生态 + 艺术住所，布哈拉的构想是把亚穆纳行走作为一种手段，让新德里居民关注这条被忽视的河流。他将河流的状况记录下来，加深了他自己与城市、河流之间的关系。

"穿越 22 公里"涉及到贯穿新德里的亚穆纳河的都市延伸区，尽管艺术家实际的行走距离总共长达 56 公里。

这次行程从帕拉村出发，这是亚穆纳河流入新德里的起始点，穿越了奥克哈拉坝，走出城市直到北方邦。那里不可能沿着河流继续行走了，于是布哈拉搭船前行。项目始于贾噶特普尔村的野餐，那里正位于河岸边。选择这个地方是因为比起河岸的其他地点，这儿的污染更少，可以进行野餐，野餐的原因是"从来没有人在新德里的河边野餐过"。

野餐者们开始期待一场传统的"表演"，相反的他们享用了茶点，并且坐船出游。整个亚穆纳河行走花了布哈拉 4 天时间完成，住在河边的当地社区民众广泛地参与进来，包括渔夫、农民，洗衣和种西瓜的村民。

起初，部分公众对于拿着相机的布哈拉表现出了些许的疑虑。当地人开始猜测起来，我们做了什么不对的事？然而，当艺术家继续行走，和他遇到的人们一起谈话交流，疑虑让位给了好奇心，很多人对他的活动产生了很大的兴趣。

除了拍摄亚穆纳河和周边的居民，布哈拉还写了日记，他用零星排列的文本把旅途记录下来，类似一种纪录诗篇。举个例，1 月 23 日（上午 11 点 30 分）的开篇，包含了片段诸如"帕拉—北部边缘—德里？—沙子—和沙子—空间—我的—空间—流动……"

"亚穆纳河行走（穿越 22 公里）"作品还包括了一件雕塑装置，在河岸边放置了一系列厕所 / 卫浴制品，排列成螺旋型。布哈拉延展的纪录档案在 KHOJ 工作室以照片叙事的形式展出过，同时在外面有一个投影机投射出一部录像片，内容是一个满溢的马桶，让观众看到马桶时脸上流露出厌恶感。

还有许多照片充分证明了亚穆纳河的严重污染，一些照片被形容为风景如画。

风景如画的图片传达了艺术家的感情，或许这一切并没有完全消失，亚穆纳河并没有死去，只要有足够强的意志力，新德里居民会对河流产生不同的思索，认为河流的重要性高于城市，河流将在人们的意识里重新获得一个立足点。

解读

一项有力度的公共艺术作品，总是会给观者带来极强的视觉震撼和精神冲击，从而引起人们的深思。此种艺术作品并不一定作为观赏品而存在，还有可能作为有声语言的代替方式，阐述着艺术家的思想，影响着人们的行为以及生活方式。河流是上帝给我们的馈赠，是生命之源，但为什么会浑浊不堪？为了让新德里居民关注到这条被污染的河流，艺术家在 22 公里步行的途中，和当地的居民聊天，获取第一手资料，成功地引起了居民们的好奇心，当地居民积极参与到这个过程中，使之不再是一项由艺术家独立完成的作品，而成为了一项全民参与的活动。除此之外，艺术家将不同的马桶摆放在一起排成螺旋形，还为了警醒人们周围河流被污染的事实以及激发起人们对于保护环境的意识与积极性。试问谁会饮用洗刷马桶的水？谁又会在看到马桶时不曾想到污秽？（高浅）

Artwork Description

The Yamuna River takes its name from the Hindu Goddess Yamuna, daughter of the Sun God Surya and sister of the God of Death Yama. It flows through the city of Delhi and is the largest tributary river of India's sacred Ganges. Legend has it that bathing in the Yamuna wards off death, or frees one from its torments. Ironically, it is the Yamuna itself that struggles for life today as it is one of the most polluted rivers in the world, receiving millions of liters of raw sewage per day along with countless liters of toxic chemical effluents from industrial and commercial activities. Perhaps because of its filthiness, the Yamuna barely exists in the awareness of Delhi's millions of inhabitants, many of whom have never ever seen the river. In Delhi's Mughal and British colonial periods, buildings used to be constructed with views of the Yamuna. Post-independence however, buildings adjacent to the river would inevitably be constructed facing away from it. One might say that a city had turned its back to its major waterway.

The significance of Atul Bhalla's performative intervention Yamuna Walk (Through 22 kms) is best appreciated against this backdrop. Invited to participate in KHOJ International Artists' Association's 2007 Eco+Art Residency, Bhalla conceived of the Yamuna Walk as a means of engaging residents of Delhi with the neglected river, documenting its conditions, and deepening his own relationship with both the city and river. Through 22kms refers to the urban stretch of the Yamuna through Delhi, though the artist in fact walked a total of 56kms. The journey proceeded from Palla village, the Yamuna's point of entry into Delhi, through to Okhla Barrage, and out from the city into Uttar Pradesh. Where it was impossible to walk alongside the river, Bhalla resorted to taking a boat. The project began with a picnic at Jagatpur village, along the banks of the river. The site was chosen as one less littered than other parts of the river, and the event of a picnic, for the reason that "no one ever picnics on the river in Delhi." Picnickers came expecting a conventional "performance," and instead were treated to refreshments and a boat ride.

The entire Yamuna walk took Bhalla 4 days to complete, and involved extensive engagement with local communities living along the river, including fishermen, farmers and villagers washing their clothes and planting watermelons. Initially, some of the public responses were a little apprehensive due to the presence of Bhalla's camera. Locals appeared to wonder, What have we done wrong? However, as the artist persisted in his walking, and spent time conversing with those he encountered, apprehension gave way to curiosity, with many expressing a great deal of interest in his activities. Apart from photographing the Yamuna and the communities alongside it, Bhalla also kept a written diary, recording the journey with a sparse arrangement of text, a sort of documentary-poetry. The entry for 23 January (11:30am) for example, contains fragments such as "palla-northern border-delhi?-sand-and sand-space-mine-space-flow..."

Yamuna Walk concluded with a sculptural installation of toilet/sanitary products arranged in spiral formation on the riverbank. Bhalla's extensive documentation was presented at the KHOJ studios in the form of a photo narrative, together with an overflowing toilet rigged with a video camera feeding expressions of disgust on the faces of visitors to the toilet through to a projector located just outside. While many of the photographs provide ample evidence of the pollution of the Yamuna, several have been described as "picturesque." The picturesque images relay the artist's sense that perhaps all is not entirely lost, that the Yamuna is not altogether dead, and that with sufficient will, Delhi residents might begin to think differently about the river such that it regains a foothold in the consciousness and priority of a city.

Artwork Excellence

As a work that confronts a core problem in the mega-city of Delhi—namely, water pollution—Bhalla's work is commendable for engaging residents with the river and facilitating new perceptions for the river's future. As the population of Delhi grows, exerting more and more pressure on fragile municipal infrastructures and an intensely polluted river, Yamuna Walk stands out as a placemaking intervention, in which new perceptions are important for the generation of new and viable solutions, new imaginations for the possibility of a sustainable Delhi ecosystem. The excellence of this project is evidenced by its review by one of India's foremost art critics, Geeta Kapur (The KHOJ Book, 2010, p. 65-66), by its invited exhibition at the 4th Fukuoka Asian Art Triennale (2009), and a forthcoming book Yamuna Walk: Atul Bhalla with an essay by the Norma Jean Calderwood Curatorial Fellow in the Department of Islamic and Later Indian Art, Harvard Art Museum/Arthur M. Sackler Museum, Maliha Noorani, (due for release by Sepia Eye New York and University of Washington Press at the 4th edition of the India Art Fair, New Delhi, 2012). The significance of the project has also led to a commissioned work by Bhalla in 48 ° C Public Art Ecology (2008 – Delhi's first public art festival) and the artist's participation in the Yamuna-Elbe project between Hamburg and Delhi (2011-2012).

北美洲地区提名作品
NORTH AMERICA NOMINATED CASES

2501 个移民
2501 Migrants

无名氏的神光
Anonymou's Auras

熊
Bear

在我死之前
Before I Die

布拉多克马赛克公园
BraddockMosaic Park

数字统计
Census

鱼
F. I. S. H

她的秘密是耐心
Her secret is patience

光电瀑布
Light Showers

纽约市空中步道公园
NYC High Line Park & Art

开渠之流
Open Channel Flow

通道
Passage

铁路站公园
RailyardPark and Plaza

反省缺失
Reflecting Absence

索拉尼桥和广场
Soleri Bridge and Plaza

卡戎兄弟广场
Square des Frères-Charon

献给堪萨斯城的歌舞女神
Terpsichore for Kansas City

鸟
The Birds

何塞·福斯特的陶瓷
The Ceramics of José Fuster

无声的进化
The Ceramics of José Fuster

水上空中花园
WaterSkyGarden

沃兹房屋计划
Watts House Project

波形
Wave Forms

维恩伍德艺术区协会
Wynwood Arts District Association

西佩·托堤克
Xipe Totec

2501 个移民
2501 Migrants

艺术家：亚历杭德罗·圣地亚哥（美国）
地点：最初作品坐落于墨西哥蒙特雷的文化论坛大学
推荐人：彼得·莫拉雷斯
Artist: Alejandro Santiago(USA)
Location: Initially the work was loacted at the University of Cultures
Forum, Moterrey, Mexico
Researcher: Peter Morales

作品描述

在墨西哥，移民是个经常被讨论的大话题。每年都有无数人背井离乡，不顾一切地到外面的世界去寻找更好的工作和生活。作者本人也经历过风雨飘摇的移民生活，他在 9 岁之际随父亲离开家乡。后来他成长为一位职业画家，依然往返于欧洲和美国之间。当他回到故乡时，发现那里只剩下越来越少的老人、妇女和儿童。家乡人口大量流失的现象使他深感震惊，他由此决定以塑造真人尺寸的陶俑的形式，以填补出走的乡亲并反思这一现象。他统计了家乡的移民人数，包括他自己在内总共是 2 500 人。他的孩子出生在作品制作期间，使总数又增加了一个，即 2 501 个。

陶俑由粉红色黏土烧制，他以非写实的夸张手法，赋予每尊塑像以不同的体貌特征，逐一制作纹理、着色，并使用各种标志性元素，如面具、头盖骨、木棍和骨头等，使之具有强大的生命力。陶俑虽然都是站立姿势，但不同的细节体现出各自不同的人生经历。冷漠的神情流露出前途的不确定性所造成的心理压力，不规则的躯体造型无言地诉说着他们在旅途中所遭受的身体和情感上的冲击，作者以最直观的语言表达出移民在他们所向往的生活途中所遭遇的所有苦难。这些塑像曾被放置在美国与墨西哥的边境线上，遥望南方，那是他们家乡的方向，透露出震撼人心的气息。

解读

在当代社会中，移民是一种特别普遍的现象，这种现象背后的原因是多元的，生活、学习、工作通常是人们选择移民的主要原因。"2501个移民"这件公共艺术作品所反映的并不是什么原因造成的移民这种现象，而是反映移民背后对一地区即精神归属地所产生的影响。

这个作品的艺术家也经历过风雨飘摇的移民生活，当某一天回到阔别已久的故乡时，看到稀少的孤寡老人、可怜的留守儿童和妇女时，他深感震惊与难过，反思之余，作为职业画家的他，创作了2 500个真人尺寸的陶俑，由于作品制作期间，孩子出生，因此增加了一个，即"2501个"。

他把这些形态各异的陶俑放在面对家乡方向的美国与墨西哥的边境线上，期盼着远方的民众能回到故乡，找到回家的路。艺术家用公共艺术的方式方法关注处于社会边缘的弱势群体以及社会问题，在边缘视线中体现着人文关怀，并传达着震撼人心的心声。（冯正龙）

Artwork Description

In Mexico, immigration is a frequently discussed topic. There are countless people who leave their hometown every year to go in search of a better life and work. The artist had himself experienced the stormy life of an immigrant, leaving his hometown with his father when he was 9 years old. He later became a professional artist, traveling sometimes between Europe and the United States. When he finally returned to his hometown he found there were very few people remaining, only the elderly, women and children. Shocked at this incidence of large population loss, he decided to create a sculptural work of life-size figures to populate the spaces of his hometown and to invite reflection on this phenomenon. He calculated that the number of migrants from his hometown totaled 2,500, and that his daughter, who was born and grew up during the project, represented number 2501.

The artist created the figurines using pink clay, exaggerating features to give each statue a different physical characteristic. Each sculpture is individually textured and colored, and adorned with various symbolic elements, such as masks, skulls, sticks and bones, which express strong vitality. Although all the figures are rendered in a standing posture, different details reflect their different life experiences. They share a faraway look expressing the uncertainty of the future. Their bodies, contorted in irregular shapes, convey the physical and emotional stress suffered along their journey. Through a subtle intuitive language the artist evokes the difficulties these migrants encountered in their search for a better life.

Artwork Excellence

This project is large in scope, and the number of pieces is considerable. It goes beyond the conceptual, causing one to feel as though she is in a crowd of absent beings. Migrant rights is a huge, often-discussed issue in Mexico—and should be discussed more in the United States and other parts of the world. Ultimately, the artist wanted to line the sculptures of the migrants along the US/Mexico border looking south, but funding wasn't available. Regardless of where the work is displayed, 2,501 Migrants creates a special haunting space. The work has long lasting impact.

无名氏的神光
Anonymou's Auras

艺术家：贝亚特里斯·冈萨雷斯（哥伦比亚）
地点：哥伦比亚波哥大
推荐人：卡利·英格布里特森
Artist: Beatriz González（Colombia）
Location: Bogota, Colombia
Researcher: Cally Ingebritson

作品描述

在波哥大中心市区有一座殖民广场，那里有许许多多的教堂、世界上最大的黄金博物馆和曲折的鹅卵石路。 在这一历史文化中心三英里外坐落着哥伦比亚最大的墓地——中央公墓。它于1837年建成，1984年被定为国家纪念公墓，随之成为无数亡魂的安息之所。尽管有很多颇具影响力的哥伦比亚人长眠于此，其中不乏总统、作家和科学家，但大部分被埋葬在这里的还是普通人，他们中的许多人都死于与毒品相关的暴力。事实上，每年需要被埋在这里的普通人人数已经超过了公墓能够提供的墓地数量，为了给新的安葬腾出墓地，在下葬七年之后，很多普通逝者的遗体被送回到他们的家中。中央公墓保持着这样的动态调整一直到2005年。那一年，所有普通人的遗体被移出，那些无名的壁龛空落落地敞开着。随后，城市的管理者们批准了一项计划，要将壁龛拆除，来修建足球场和溜冰场。哥伦比亚艺术家贝亚特里斯·冈萨雷斯得知此事后，她提议公墓应当转变成公众艺术项目，以此来纪念那些长眠于此的人们。

解读

人类生命是不可逆的河流，而记忆是人类试图回望最用力的尝试，当生命迹象消失，公共祭祀就成为记忆唯一的避难所。对个体生命消逝的悼亡，对残破建筑物的保留，贝亚特里斯·冈萨雷斯创作的公共艺术作品"无名氏的神光"，用艺术重建了公众记忆的路径。

在"无名氏的神光"开放式的公共空间中，"里面"与"外面"的界限变得模糊，历史与今天的对峙逐渐消减，旁观和见证不再是公共艺术的全部意图。面对艺术家大量手绘个体形象的复制模板，参观者不得不努力辨识这一公共艺术作品中所悼亡群体的个体特质，由此，惊醒旁观者置身事外的梦境。农民与手上死亡的猎物，士兵与脚边阵亡的敌人，受难与施难，或者，生命的终止才更真实？（吴昉）

Artwork Description

Central Bogotá is home to colonial plazas, countless churches, the world's largest gold museum and winding cobblestone streets. Just three miles outside the historic center of Bogotá lies one of Colombia's largest cemeteries, Cementerio Central. Founded in 1837, the cemetery was declared a national monument in 1984 and became the resting place of thousands. A significant number of influential Colombians—including presidents, authors and scientists—were buried here, yet the majority were common people, many killed in drug related violence. In fact, the number of common people needing to be buried outnumbered the number of available plots—so much that seven years after interment, the remains of the deceased were returned to their families to make room for new burials. Cementerio Central remained an active cemetery until 2005 when the bodies were removed and the unnamed niches were left empty and open.

When Colombian artist Beatriz González learned city legislators had approved a plan to tear down the columbariums to build a soccer field and skating rink, she proposed the cemetery be transformed into a public art project in memory of those buried there. Having spent time in the cemetery, González says she thought there were a lot of energies or "auras" of the deceased floating in the air since the niches had been left open. She titled the project Anonymous Auras (Auras Anónimas) because it's unknown exactly who was buried there. "It seemed like an insult to the memory (of those buried here)," Beatriz states. "The historical, high-class cemetery is well preserved, but this one, the common people's cemetery," was slated to be torn down.

Accompanied by a team of five assistants, González used white acrylic plaques to seal the nearly 9,000 niches. "I really think the act of using this space and sealing the niches removed the gloominess that was here before," she explains. Each plaque was printed with the silhouettes of figures carrying dead bodies. Six such designs were repeated throughout the cemetery, showing the various ways corpses have been carried in Colombia after being killed in crossfire: using hammocks, covered by plastic tarps or completely uncovered. Beatriz states it is tragically ironic the "elite of Colombia were once carried on people's backs across Colombia and now those have been replaced with corpses."

Fellow artist Humberto Junca says, "She's telling us that this war is ours and these deaths are ours." As a child Beatriz would ask herself why the violence in Colombia seemed endless. As an artist she says her work "isn't meant to be a protest, but simply document what's taken place in Colombia. I truly hope things don't continue to repeat themselves."

Artwork Excellence

This piece honors the memory of thousands of deceased working class Colombians, many of whom were killed due to the drug related violence in Colombia. Beatriz González insisted their graves located in Cemeterio Central remain intact once she heard Bogotá legislators planned to tear them down while leaving the elite graves well preserved. Through sealing and decorating the graves, González invites us to respect and reflect so as to prevent further deaths from happening.

熊
Bear

艺术家：布劳尔·海切尔（美国）
地点：加拿大基隆拿
推荐人：绍娜·迪伊
Artist: Brower Hatcher (USA)
Location: Kelowna, Canada
Researcher: Shauna Dee

作品描述

　　"熊"是一尊 8 英尺 x 14 英尺的透明雕塑，坐落于加拿大不列颠哥伦比亚的基洛纳，靠近奥卡纳干湖。美国艺术家布劳尔·海切尔把金属片和自己用模具做的小物件系在一起创作了这一作品。熊体内的小物件——如果实和花朵——是代表基洛纳的历史时期的一种象征。像船一样形状的基座旨在提示该湖在社区的历史演变和身份形成的过程中起到的作用。海切尔采用了合作的过程来创作作品。他说："'熊'的创作是一种适合于 21 世纪的创造性方法，在尊重相互差异的同时我们一起合作，达到了对各自及对社会有利的结果。"熊"的一系列创作实现步骤也适用于创业和商业模式。作为一种原创，它成为了真正创作经验自我产生的源泉，然后成为了社区的资源。独创性的经验是主观的、集体性的。我／我们之前没有这类经验，但它与我（主观上），我们（社区）产生了共鸣，作为社会意象的一部分它的意义被展示了出来。"

解读

加拿大洛肯基山脉脚下的奥肯纳根湖,是印第安神话传说中水怪出没的地方,从来吸引着好奇者追随的目光,而水怪却如羚羊挂角,无迹可寻,如今,像龙非龙、似蛇非蛇的水怪雕塑倒成了基隆拿市的一个特殊符号,伫立在奥肯纳根湖边。布劳尔·海切尔的公共艺术作品"熊",正与水怪遥相呼应,神话与生活,分守湖泊的两岸。

三维骨架式的线条,空气通透,无阻隔,无遮挡,不负重,灰熊巨大的体量感,四两拨千斤,猛虎嗅蔷薇。天空的颜色就是作品的颜色,游人的视线穿过玲珑的结构,看到的是自然。天地人物,空气般天然融合,海切尔的"熊",轻松躲过了公共艺术环保主题的各种沉重,自如的存在,恰是公众性最好的注脚。(吴昉)

Artwork Description

Bear is an 8 x 14 foot transparent sculpture next to Okanagan Lake in Kelowna, British Columbia, Canada. To create it, American artist Brower Hatcher fastened together sliced metal and small objects that he created from molds. The small objects within Bear's body—like fruit and flowers—are symbols that represent periods of Kelowna's history. The boat-like shape at the base is a reference to the role the lake has played in the evolution of the community and the shaping of its identity.

The sculpture is a tribute to Kelowna's settlement on the shores of Okanagan Lake. The theme of the work is a grizzly bear: Kelowna is an English translation of the Okanagan/Syilx First Nation word for Female Grizzly Bear. Bear was installed in September 2010, in conjunction with the grand opening of Stuart Park, which was named for former long-term Kelowna Mayor James Stuart. While it has a serious historic theme, the twinkling, sparkling Bear also seems to suggest the skating rink over which it is positioned is a place for anyone and everyone to play.

Hatcher used a collaborative process to create the work. As he says, "Bear is a model of a creative approach appropriate to the 21st century where, in valuing each other's differences, we collaborate together to arrive at a mutually and socially beneficial outcome. The series of stages in the process of realization of the bear could also be applied to entrepreneurial and business models. As an original, it becomes its own generative source of an authentic creative experience, which is then a resource for the community. The experience of originality is subjective and collective. I/We have not had this experience before, yet it resonates with me (subjective), us (community), and its meaning unfolds as part of the social imagery."

The configuration of the bear (the outer bear shape) evolved from a 3-dimensional digital scan of an image of a grizzly bear. The data from this was used in another program to arrive at a geodesic triangulation in order to arrive at a buildable matrix structure. Hatcher's approach to sculpture is to create geometric frameworks and layered matrixes containing narrative embedded objects. The computer model was altered numerous times to satisfy various criterion and aesthetic concerns. Once satisfied with the outer matrix form, Hatcher worked with others to create the interior geometry and narrative objects that would be embedded into the work.

Artwork Excellence

Bear was installed in conjunction with the grand opening of Stuart Park, which was named for former long-term Kelowna Mayor James Stuart. Designed by Rhode Island, USA–based artist Brower Hatcher, Bear is a tribute to Kelowna's settlement on the shores of Okanagan Lake. The theme of the art work is a grizzly bear: "Kelowna" is the English translation of the Okanagan/Syilx first nation word for "female grizzly bear." Within Bear's body are objects that represent periods of Kelowna's history. The boat-like shape is a reference to the role the lake has played in the evolution of the community and the shaping of its identity. The sculpture sits on a promontory overlooking a public space that hosts a skating rink in the winter and programmed events in the summer. Along with other elements of the site, Bear creates a sense of place at the park. It is the most high-profile piece in the City of Kelowna's collection, and is enthusiastically embraced by the community.

在我死之前
Before I Die

艺术家：凯蒂·张（华裔加拿大籍）
地点：美国新奥尔良圣马里尼大街 900 号
推荐人：彼得·莫拉雷斯
Artist：Candy Chang (USA)
Location：900 Marigny St. New Orleans, USA
Researcher：Peter Morales

作品描述

华裔加拿大籍艺术家凯蒂·张将自家附近一所被废弃的房子侧面漆成一幅大黑板，在上面写了一个再简单不过的填充题，重复了几十遍："在我死之前，我想要 ＿＿＿"，向每个路人提出心灵的追问，并留下若干支粉笔，以供过路人直接在上面写出自己的回答。做该项目的念头是张在短短几个月的时间里，失去深爱的朋友后，思考了许多关于死亡的问题。为了提醒周围朋友在有限的人生里什么是真正重要的，"那几个月使我深刻地意识到人生是短暂而脆弱的，不该拖延着"。

这个项目意在创造一个供沉思的公共空间，能够让许多人产生共鸣，显示了如果大家有机会能够写出自己的心声并且与其他人分享，那么我们思想的公共空间就能够变得更加强大。以一种全新的、富有启发性的方式来理解街坊邻居。凯蒂·张和她的同事创建了 beforeidie.cc 网站和 Before I Die 工具箱，帮助人们筑墙并在线分享他们的墙。

在世界各地热心人的参与下，"在我死之前"墙壁已经风靡全球。凯蒂·张收到来自全世界范围成千上万份留言和请求，人们希望在自己的社区也开展这个活动。墙壁成了世界性参与的艺术项目，这个简单的问题使人们更了解自己的真实需要，增强彼此的真正交流，将人们团结到一起，意识到虽然各自有分歧但也有很多共同的东西，能更好地促进大家的生活。

解读

当你看到或听到"在我死之前，我想要____"的时候，你会想到什么？或许你会陷入沉思，或许你会困惑，也或许你会感到唏嘘。但真当你认真思考这个问题的时候，你会感到你想要的还很多，有很多事情还没做完，有很多人或物要珍惜，也许从这一刻起，你才会更加呵护与珍爱你已经拥有的，得到的、才懂得什么是最重要的，才深刻理解身体比金钱重要，驻足比匆忙重要，知足比欲望重要。艺术家凯蒂·张就是在自己房子的侧面漆成一块黑板，在上面为全世界的人们出了这道令人深思的题目。在这个公共空间里，人们都会拿起笔把自己的心声与思考写在上面，不仅使社区居民之间的关系更加友好与融洽，而且也打开了人们心中那堵看不见摸不着的心墙，在社会发展的今天，这个题目或公共空间可能留给我们更多的是启发与深思。（冯正龙）

Artwork Description

Chinese Canadian artist Candy Chang painted a large blackboard on the side of an abandoned house in her neighborhood, and wrote out a half-finished phrase: "Before I die I want to ____." Copying this simple but evocative phrase a few times on the board, she left out some chalk as an invitation for passers-by to fill in the blank. Chang had recently lost a close friend, and with the problem of death weighing on her mind, she was inspired to create a work that might help others recall those things in life truly important to them. "Those months made me very aware that life is brief and tender and cannot be postponed," she explains.

This project is intended to create a public space in which collective reflection is possible. It is a moving demonstration of how expression coming from deep in the heart that is openly shared can nourish public life. Further, it offers a totally new and inspiring way to learn about a local community. Candy Chang and her colleagues created the beforeidie.cc website and the Before I Die toolbox in order to give more people access to the wall and share their writing online.

The "before I die" wall has gained international attention, with enthusiastic participation from people around the world. Tens of thousands of followers have sent messages to Candy Chang inquiring about carrying out this activity in their communities. The wall has become a world art project. This simple phrase awakens our awareness of our real needs, enhances the possibility for meaningful exchange, and brings people together. By inviting people to recognize what we all share in common, the project helps people lay the groundwork for living more fruitful lives.

Artwork Excellence

Because of Chang's original New Orleans project, she has received hundreds of kind messages and requests from people around the world who want to make walls in their communities. This wall is turning into a global participatory art project and expanding to cities around the world, including Amsterdam, Portsmouth, Querétaro, Almaty, San Diego, Lisbon, Brooklyn, and beyond. The project brings people together to realise that they have differences but also share many things. In communities where it is illegal to post any information—including notices of community gatherings—and where only advertisings signs are permitted, Chang's work serves a vital human function that is otherwise denied many people.

布拉多克马赛克公园
Braddock Mosaic Park

艺术家：詹姆斯·赛门（美国）
地点：美国布拉多克维罗纳大街和布拉多克大道
推荐人：绍娜·迪伊
Artist: James Simon（USA）
Location: Verona St. and Braddock Ave. Braddock, United States
Researcher: Shauna Dee

作品描述

在宾夕法尼亚的布拉多克小镇，莫侬加希拉山谷，这里是钢铁制造业的中心，曾经是一个繁华的商业目的地，从 20 世纪 80 年代的后工业时期起遭受到了严重的衰退。当这个巨大的经济引擎崩溃了，布拉多克失去了 90% 的东西——建筑、商业和居民。

然而布拉多克仍然是一个精彩丰富和弹性的人类居住的历史悠久的地区。在留存下来的房屋里的社区基层人民努力振兴这座小镇，这股风气正在形成。

阿勒格尼县经济发展部在维罗纳大街上建造了一个公交车遮挡棚。建造的位置位于新房屋发展的入口处附近，但是该地相当地空旷，地上散落着垃圾，长满了野草，正好位于遮挡棚和新房屋之间，变成了很碍眼的地段。"布拉多克马赛克公园"项目把垃圾清理了出去，变成了一座美丽的社区马赛克和绿色公园/空间。

与布拉多克经济发展部，布拉多克市镇和阿勒格尼县经济发展部，布拉多克终极版担保资金合作，在一个闲置地块上，一起联手布置安装了这个马赛克公园。凯西·卡斯特纳是一名项目经理参与这个县的部门工作，她邀请了雕塑家詹姆斯·赛门为该地区创作作品。布拉多克终极版，向赛门先生提供了免租金的工作室。七个当地的年轻人受雇于当地的布拉多克青年项目，该项目是亚美利组织发起的，由阿勒格尼县人民服务部提供劳力。

2008年，在布拉多克大道的维罗纳大街公交车站，新设计的"池塘"马赛克完成了，并且被安置在布拉多克公共空间里。布拉多克终极版获得了一个著名的大西洋中部艺术基金会的赠款，用以设计建造这个马赛克公共空间。公共空间和房屋马赛克图腾柱由赛门和周边学校的学生共同设计。

在布拉多克，赛门的另一个公共艺术项目和马赛克公园很像，那是一个欢迎标志。当你穿过蓝晶大桥进入布拉多克市镇，迎接你的将是一座惊人的10—12英尺的马赛克壁画，上面拼写了几个单词，"欢迎来到历史悠久的布拉多克，"壁画上使用了一些闪闪发光的镜子、砖瓦、石头和彩色玻璃。2007年夏天，作为布拉多克社区日庆祝活动的一部分詹姆斯·赛门为马赛克壁画揭幕，壁画持续地启发着居民。

更多关于布拉多克终极版的信息：

哪里有人看见被遗弃的建筑，空旷的场地，无聊的、无所事事的青年，布拉多克终极版就看到了机会。2003年由约翰·费特曼（2005年被选举为市长）创立，布拉多克终极版促进了当地的环境管理，艺术项目，就业，培训，为当地青年提供辅导。他们寻求创意、参与社区复兴项目。布拉多克终极版与阿勒格尼县最大的青年就业项目合作，参与布拉多克青年计划。夏季，BYP通过美化、宣传和基层来雇佣一批希望自己所在的社区更美好的年轻人。

解读

点石能成金，化腐朽为神奇。政府出资鼓励，利用社区废弃的场地改建文化产业区，"布拉多克马赛克公园"很容易使人联想起国内的798、M50等创意园区。这项公共艺术作品，不仅是社区改建的文化项目，更带动了当地的区域文化产业，用艺术让社区恢复活力，为生活增添情调。而这项公共艺术最出色之处在于激励了公众的共同参与热情。社区参与者在漫长的制作过程中，蹲伏在大型壁画上，刷洗碎瓷玻璃的污垢尘埃，仔细镶嵌每一块马赛克，完成项目的同时，也在不断发现自我。每块单片的马赛克就像单独的个人，需要从群体的镜面中映射出自身的全新价值，当整片马赛克镶嵌完成，那绚丽的惊鸿一瞥无法不令人动容。无水池塘，也能一池璀璨。（吴昉）

Artwork Description

Once a thriving commercial destination amidst the industrial center of steel production in the Monongahela Valley, the small town of Braddock, Pennsylvania suffered through a brutal postindustrial decline in the 1980s. When its massive economic engine sputtered out, Braddock lost ninety percent of everything—its buildings, its businesses, its people.

Yet Braddock remains a richly historic area populated by wonderful and resilient people. Amidst the remains of housing and commerce a community grassroots effort to revitalize the town is taking shape.

Allegheny County Economic Development constructed a bus shelter on Verona Street. It was built near the entrance to a new housing development, but a vacant lot, strewn with trash and overgrown with weeds, lay between the shelter and the new homes, creating something of an eyesore. The Braddock Mosaic Park project turned the trashed out abandoned lot into a beautiful community mosaic and green park/space.

In partnership with the Braddock Economic Development Corporation, Braddock Borough, and the Allegheny County Department of Economic Development, Braddock Redux secured funding and coordinated the installation of a mosaic park on an unused lot. Kathy Castner, a project manager who works with the county department, invited sculptor James Simon to create something for the lot. Braddock Redux, contributed rent-free studio space for Mr. Simon to work. Seven local teenagers, employed through the local Braddock Youth Project that is run by AmeriCorp through Allegheny County Department of Human Services, provided labor.

In 2008 the "Pond" mosaic was completed in the newly designed and constructed Braddock parklet at the Verona Street bus stop on Braddock Avenue. Braddock Redux won a prestigious Mid Atlantic Arts Foundation grant to create and build the mosaic parklet. The parklet also houses mosaic totem poles designed by Simon and students surrounding school districts.

One of Simon's other public art projects in Braddock similar to the mosaic park is a welcome sign. As you cross the Rankin Bridge to enter the borough of Braddock and are greeted by a stunning 10-by-12-foot mosaic mural, which spells out the words, "Welcome to Historic Braddock," in bits of sparkling mirror, tile, stone and stained glass. James Simon unveiled the mosaic in the summer of 2007 as part of Braddock's Community Day festivities, and it continues to inspire residents.

Artwork Excellence

Experts say the most effective public art projects, as in the case of the Braddock mosaic, use a range of local community resources in some way, helping to broaden involvement and facilitate success. Karen Newell, the spokeswoman for the Mid-Atlantic Arts Foundation, which contributed $18,000 to the Braddock project, said, "The goals are to stimulate public participation in the artistic process... It has a place of pride in that community because they worked so hard to bring it to fruition." In this case, Braddock Youth Project participants performed most of the tedious labor of carefully breaking ceramic tiles and gluing them into place, and then filling in grout between the pieces. The numerous partners in the project included the Braddock Economic Development Corporation, Braddock Borough, the Allegheny County Department of Economic Development, and the nonprofit organization Braddock Redux.

数字统计
Census

艺术家：安妮塔·格雷斯塔（美国）
地点：美国马里兰州
推荐人：绍娜·迪伊
Artist: Anita Glesta（USA）
Location: Suitland Maryland, United States
Researcher: Shauna Dee

作品描述

随着联邦统计局大楼周围面积超过 7 英亩面积土地的覆盖，安妮塔·格雷斯塔的大型景观干预永久建立。此公共艺术方案由建筑项目中的 GSA 艺术委任，是对统计想法以及全球化视野要求下的深思熟虑的成果。

就像格雷斯塔其他的公共艺术作品一样，"数字"计划探索了自然环境与社会环境的有机结合的课题，并通过雕塑和城市景观的设计来强调人与大地间的一种联系。乔治·艾尔法的著作《数字的通史》激发了她的灵感：将数字从抽象的符号代表回归其"最初的作为改进、激励以及记录人类活动的功能"。在 Turf 景观建筑公司和艺术家诺尔·坎普拉的帮助下，格雷斯塔创造了一系列作品，它们充分利用这条被称为"数字长廊"的小径周围能利用的空间，因而我们看到，这条长廊由不计其数的砖头所砌成，蜿蜒盘旋，并且成了人们坐下来闲聊的胜地。

在这个作品中，格雷斯塔跳出数字符号的传统用法，引出了这些简易数字背后的神秘元素。而最终营造的气氛则是既神秘又平易近人的，并加强了参观者个体与周围物理环境的交流。

作为她项目的一部分，格雷斯塔也将 Skidmore，Owings & Merril-designed 统计大楼这个百万平方英尺面积里工作的 1 万名员工考虑到项目中了。她将整个空间拟人化，包括添加长凳、景观元素，还有小小旮旯之地可用来捉迷藏——这些都使得这块地儿显得更有趣更有探索意味。"我所能做的就是希望通过我这个艺术家的角色发挥作用，在一定程度上改变他人对于数字的思考，甚至从那些工作人员身上可以看到改变：他们每天都看着这些符号，可能有天会想：'呀，这事实上不仅仅是数据，而是关于人类传承的历史'，或者'这些国家里的都是这么些人啊'"她说。

图绘瓷砖四方板坐落在绿植和草坪之中。瓷砖板上的方格图案是丝网印刷的照片，都是一些无脸、无名氏的一群穿 T 恤和牛仔裤的普通人，又腰围对着观者。而数字是成整数套印的，且从数字 0 到 9 无穷尽地、依次排列。一个数字"1"形状的砖砌长方凳坐落在工程项目的正中间，这个形状借鉴了美国土著的墓碑形式。而每一个土堆上都有手绘的图表或符号，其灵感来自于美国土著的纹身图案，或是像手写的数字符号等。环绕着整个长廊的有阿拉伯数字、汉字数字或其他的一些数字符号的瓷砖板。这些瓷砖板个个都是图片、语言和文本的图解式组合，能快速地传播信息。

其实美国的人口普查自始至终贯穿着异议：其核心主题的不顺从正反映了美国历史上大杂烩式的凌乱状态，而在官僚机制的主宰下处置的方法又是强势而一以概之的。在"数字"这个项目中，格雷斯塔抓住了公共艺术的这个颇费心思的目的，并积极参与在下议院或绿色山庄（Village Green 有无其他特殊意思？）的会议。格雷斯塔暗示，和平集会的权利应和午间闲谈的机会一样多，而社团的责任是讨论谁是活动领导。最终，格雷斯塔的坚韧和俏皮并存的方式激发了游客齐聚讨论、休憩以及围观之所。

解读

安妮塔·格雷斯塔的这项公共艺术作品不凭借数据资料，而依托于对数字历史的记忆；不选择广为人知的阿拉伯、玛雅、波斯、闪族等计数形式，却表现连美国公众都还陌生的美国本土计数传统。一位活跃的女艺术家，一个灿如夏花的笑靥，注定安妮塔·格雷斯塔的"数字统计"不会亦步亦趋于数字逻辑的摆布，她用六年的时间，去思考数字统计这一概念对人类的影响与意义，在数字抽象化了的符号中觅得美国——这个变幻无穷国度的崭新轮廓线。

如果说独到的视野和表现力还不足以诠释公共艺术的全部内涵，那么"数字统计"这个项目有意识的"未完成性"则最大可能满足了作品与公众间交流的需求。不论是位于人口普查总部大楼前庭的实体作品，还是富于童趣的手绘俯瞰草图，都能看出明显的数字"缺口"，这些缺失的、尚不圆满的数字符号，等待着公众的涉足，当移动的人流到达某一位置时，数字被精确地显示。同时，缺失的本身也寓意了消失的人流和在美国未被统计在内的群体。

途经，显现，完善，消失。安妮塔·格雷斯塔的"数字统计"游戏于数字和意义抽象的可能性中，公众则游戏于作品空间的轨道上，从起点到终点，生命本质当空划过、落地生根。（吴昉）

Artwork Description

Covering over 7 acres around the Federal Census Building, Anita Glesta's large-scale landscape intervention is permanently installed. Commissioned by the GSA Art in Architecture Program, the public art project is a meditation on the notion of counting and order with a global perspective.

Like Glesta's other public works, the "Census" project explores the integration of the physical and social, using sculpture and landscaping to foment connection between people and the land. Georges Ifrah's book The Universal History of Numbers inspired her to transform numerals from abstract representations back into their "original function as a system to promote, stimulate, and/or record human interaction." With the assistance of Turf Landscape Architects and artist Noel Copeland, Glesta created a series of occupiable forms and spaces along a path called the Census Walk. Oversized integers built of bricks, stretching in, out, and up from the walk, become meeting places for people to sit and chat.

In "Census," Glesta has transcended the traditional use of numerical symbols, eliciting mythic elements from these usually straightforward symbols. The resulting atmosphere is both mystical and accessible, intensifying individuals' connection to the physical environment surrounding them.

As part of her project, she tried to take into account the 10,000 employees who work in the million square foot Skidmore, Owings & Merrill-designed Census building. Her humanizing of the space included adding benches, landscaping, and small nooks that make the area appear fun to explore. "All I can do is hope that my role as an artist can effect some kind of other thinking about census, even from the employees, who look at these symbols everyday and might think 'Gee, this is actually more than data, this is about the history of counting humans' and 'who are the people of this country,' " she says.

Illustrated, tiled slabs nestle among green plantings and grass. These grids depict screen-printed photographs of hundreds of faceless, nameless, generic human figures dressed in T-shirts and jeans, arms akimbo, confronting the viewer. The figures are overprinted with integers running endlessly, repetitively, from 0 to 9. A brick bench in the shape of the numeral 1 sits between two halves of an earthwork referencing Native American burial mounds. On each earthen hemisphere, hand-painted tiles chart numerals from Native American corporeal, or hand-based, numerical systems. Arabic, Chinese, and other numerical systems cover walls throughout the walk. The synthesis of image, language, and text is graphic and immediate in its ability to communicate diversity.

An element of dissent runs through Census: the core of its noncompliance reclaims the truly messy hodgepodge that is American history and places it directly in view of a hefty bureaucratic juggernaut. In "Census," Glesta seizes on the well-meaning intentions of public art and clarifies them into active participation in the Commons, or the Village Green. The right to peaceable assembly, she implies, applies as much to a lunchtime schmooze as it does to a community's responsibility to discuss who it is and where it is headed. Ultimately, her stoic and playful approach encourages viewers to sit and convene in conversation and people watching.

Artwork Excellence

Glesta turned a commission for a visual art piece at a government building into a participatory, placemaking work of art. The timing of the project was significant; there was an effort by the government to raise public awareness of the latest Census and to reach out to the American people to not be fearful of being counted. To add an educational component, Glesta incorporated a history lesson into the piece by addressing census methods in the past, and she showcased different cultures' numerical systems. In particular she found inspiration in Native American counting systems because historically the indigenous Americans have been underrepresented, and they express counting in sentences that she found incredibly poetic. It wasn't the commissioners' intention to fill all 7 acres, but Glesta took into account the 10,000 employees who work in the Census building and included benches, landscaping, and small nooks in the project to make it look fun to explore and to provide a nice walking space. She also wanted to challenge the employees, who might think "Gee, this is actually more than data, this is about the history of counting humans and who are the people of this country," she says.

鱼
F.I.S.H.

艺术家：唐纳德·利斯基（美国）
地点：美国圣安东尼奥
推荐人：绍娜·迪伊
Artist: Donald Lipski（USA）
Location: San Antonio, TX, United States
Researcher: Shauna Dee

作品描述

圣安东尼奥镇的一个显著特征就是穿过城市的圣安东尼奥河。在这座城市的整个历史中，人们利用这条河来抵挡洪水，并且还建造了散步小径和天桥来增加河岸的美观度。河畔步道，这条于1940年代初就落成如今样貌的小径，是来到德克萨斯州的游客们的第一目的地。

最近，技术方面的改造已被列为圣安东尼奥河道改造计划之一，多年来，已有3 845亿美元的资金不断地投入改造项目，通过创建一个直线型穿越城市中心地区的公园来促进城市环境、经济、休闲娱乐以及文化等方面的改善。项目的目的是恢复当地的栖息地以及圣安东尼奥河的支流小涧（或者是帮助直流储备水源作为支流）。成果将集中在这条河的13英里长的范围内，包括小镇南部的8英里长的密森站（Mission Reach），1英里长的鹰岛，还有4英里长的位于小镇北部的博物馆站。

博物馆站的初衷就是把文化设施和商业中心连接起来，以达到经济发展的普及化。圣安东尼奥艺术博物馆周围的地标景观，从博物馆起绵延到摇滚酒吧和工业基地。项目还改进了河道的公共入口，能让城镇里坐着游艇观光的游客们方便上岸，而从路上观赏河景的景色也因新的景观而更令人赏心悦目。

"F.I.S.H."这个项目是由圣安东尼奥河基金会委任的 11 个公共艺术项目中的一件作品。利斯基设计了 7 英尺长、玻璃纤维材料的大耳太阳鱼——这是一种和圣安东尼奥当地河道里的鱼差不多的品种——用钢丝绳把它们挂在高架下面，每一条鱼体内都有 1 000 瓦的 LED 灯，到了晚上这些"鱼"们就会被点亮。

高架桥上来往车辆发出的声音会因这些"鱼"而显得像是从桥下水中鱼儿们游动而发出的声音。桥下的水上巴士搭载着乘客穿梭往来，从河道一头到另一头；桥边有一条可以散步或慢跑的小径，从博物馆新开的平台上可以看到这些水上巴士和游人。一般来说，免费开放的高架桥并不会让人联想到散步、游船或者坐在它下面聊天，但"F.I.S.H."这个景观项目改变了人们的看法，把原有模式打造为一个充满活力的空间与色彩的地方，也反映了圣安东尼奥镇居民的愿望：希望有朝一日他们的母亲河能再次变成一条生机勃勃的水路。

解读

25 条 7 英尺长的手工绘制大鱼，腾空漂浮。这是公共艺术家唐纳德·利斯基一手缔造的美国版"鱼乐"。超越现实的场景营造，利用河面投影的空间复制，公共艺术在充分使用空间维度的实践中不断膨胀自身的气场，让观者被引入，让游者入境。

"鱼"的原形脱胎自圣安东尼奥当地河流中的长耳太阳鱼，阳光下的鱼群泛着金光，欢腾活跃，引人侧目，但只有到了夜晚，这项公共艺术作品所蕴含的温暖、体恤、包容而明亮的力量才不疾不徐展现在世人眼前。圣安东尼奥 8 英里的河廊走道，是赏心乐事去处，可美中不足，其中一段位于洲际高架 I-35 之下的商业区，一入夜，就变成被人遗忘又令人生怖的暗角。人是追求光明的生物，上帝说要有光，就有了光，就有了优游在洲际高架下黑暗空间里的发光鱼群。放置在每条鱼肚中的 1 000 盏 LED 灯管，透过彩绘的玻璃纤维树脂，散发着华彩，又因河面的折射和投影，使光晕徐徐，静且美。"鱼"成为圣安东尼奥灯塔般的新地标，远望有超越梦境般的绚烂。

据说第 26 条鱼被安放在圣安东尼奥艺术博物馆的当代馆，供人近赏。不知这第 26 条鱼，可曾也在某一刻，觊觎过自己水中朦胧的倒影？（吴昉）

Artwork Description

One of San Antonio's most distinguishing features is the San Antonio River that flows through the city. Throughout San Antonio's history, humans have controlled the river to minimize flooding and they have enhanced it aesthetically with walkways and footbridges. The River

Walk, the top tourist destination in the state of Texas, was born out of such improvements in the early 1940's.

More recently, technological advancements led to the San Antonio River Improvements Project, a multi-year, $384.5 million on-going investment to add environmental, economic, recreational, and cultural attributes of the city by creating a linear park through the heart of the city. The project aims to restore native habitat and the natural meander of the river. The efforts are concentrated on 13 miles of the river, including the eight-mile Mission Reach south of downtown, 1-mile long Eagleland, and the four-mile Museum Reach north of downtown.

The primary goal of the Museum Reach is to connect the cultural institutions and commercial centers for economic development at street level. The extension flows past the San Antonio Museum of Art, rock clubs, and industrial sites. The project improves public access to the river, allowing passenger barge traffic up the river from downtown, and new landscaping enhances views from locations overlooking the river.

"F.I.S.H." is one of 11 public artworks commissioned by the San Antonio River Foundation to enhance the Museum Reach of the river. Lipski's 7' long fiberglass long-eared sunfish, a familiar species native to the San Antonio River, hang from the I-35 overpass with cables. They are each filled with around 1,000 L.E.D. lights that illuminate the sculptures at night.

The sound from the traffic overhead adds to the underwater feeling that comes from being surrounded by fish. River taxis shuttle passengers under the bridge as they go from one part of the river to another, there is a walking/jogging path under the bridge from which to view the work, and it can be viewed from a new landing at the museum. Normally a freeway overpass might not be a desirable place to walk, boat, or sit under, but the "F.I.S.H." installation transforms the space with its exuberant scale and colors, and reflects San Antonians' hope that their river will be a living waterway again.

Artwork Excellence

Images of the artwork are featured in nearly every article about the new stretch of San Antonio's River Walk. The work that has added light and lightness to the underside of a major highway overpass, and the public reactions have been very positive-appealing to arts critics and children alike. The sculptures are large-scale and very detailed, as well as technologically advanced with the use of fiberglass and L.E.D. lights.

她的秘密是耐心
Her secret is patience

艺术家：珍妮特·艾科尔曼（美国）
地点：美国凤凰城
推荐人：绍娜·迪伊
Artist: Janet Echelman(USA)
Location: Phoenix, United States
Researcher: Shauna Dee

作品描述

"她的秘密是耐心"是一座坐落于亚利桑那凤凰城的高达 145 英尺的空中雕塑，这是一件全新的公共坐标，人们赞誉它为振兴市中心作出了贡献。悬挂在新双城街区公民空间公园的上空，这座雕塑非常巨大，但质感很软，被固定在一个位置上，但它却不停地在运动。它在空中微微地跳着舞，通过沙漠阵风的吹拂而编排着自己的舞蹈。

这件作品有着大型三维多层的造型，具有手工编织和机器制造的组合效果。它是国际获奖团队共同合作努力的成果，团队汇聚了航空和机械工程师、建筑师、灯光设计师、景观建筑师和制造者。这件作品通过指引观众的视线朝向天空专注于一个新型天体，从而重新定义了"艺术空间"。

白天，这件雕塑高高地悬在头顶、树顶和建筑上方。雕塑营造出了艺术家所称之为的"阴影图画"的效果，艺术家说自己深受凤凰城云朵阴影的启发，在她第一次实地考察时就为此着迷。

夜里，雕塑上的照明随着四季变化逐渐地变换着色彩。选择不同的色彩目的是给居民对气候的感觉做一些调剂，比如夏天给灯光加入凉爽的色调，而冬天就添些暖色调。灯光设计随着雕塑的哪一部分被照到而变换，让其余部分昏暗模糊，从而营造出神秘的朦胧感，就如同月照的变化一样。

艺术家受到当地独特的季风云层的形成和它们投射的阴影启发，此外雕塑的造型是受了沙漠植物群和当地化石的影响。该雕塑的名字引自美国诗人爱默生的诗句，他写道："依照大自然的步伐，她的秘密就是耐心。"

这件作品坐落于公民空间公园内，公园位于商业区的北部边缘以及亚利桑那州立大学的西部。亚利桑那州立大学的学生、当地商务人士和年长的居民在空闲时会从附近的 HUD 房屋（都市房屋发展计划）出门，聚集在公园里休憩。

这件作品对于凤凰城地标的重要意义就如同芝加哥著名的"豆子"，甚至可以和巴黎的埃菲尔铁塔相提并论。

解读

每当有微风吹拂在凤凰城的公民空间公园，这个漂浮在空中的网状作品就开始起舞。赤金色内核在膨胀，靛青色外围在收缩，"她的秘密是耐心"像火焰在跳跃。对这件"张扬又神秘"公共艺术作品的争议从未停止过，水母或子宫？惊艳大胆或蠢笨无用？可是，优秀的公共艺术，正是把人群聚集在一处，让言论自由。

凤凰城，历来是与速度、机械相提并论的汽车城，刚性的气息紧跟高速上飞转的车轮植入城市的容貌，催促行人的脚步不知不觉快那么一丁点儿、再一丁点儿。公共艺术常常映射出城市独有的气质，好比凤凰城以往最常见的公共艺术作品，总是放置在高速公路的两侧，庞大而有规律，车辆以每小时 65 英里的速度飞驰而过，随时准备好捕捉那惊鸿一瞥，惊心动魄。"她的秘密是耐心"似乎疏离于这种都市的主流姿态，以一种奇幻的舒缓节奏，催眠式曼舞，让那些停不了的速度、来不及的脚步，在神秘弧线的摇摆中，慢慢放弃原有的立场，凝视，出神。

在这里的市中心，固有的匆忙与自然的节奏不期而遇，自然的秘密是耐心，自然是"她"。珍妮特·艾科尔曼用作品看似的无序，呈现东方式的无为，悠悠晃晃间，熨帖人心。（吴昉）

Artwork Description

"Her Secret Is Patience," a 145-foot-tall aerial sculpture in Phoenix, Arizona, is a new civic icon hailed for contributing to the revitalization of downtown. Suspended above the new 2-city block Civic Space Park, the sculpture is monumental yet soft, fixed in place but constantly in motion. It dances gently in the air, choreographed by the flux of desert winds.

The large 3-dimensional multi-layered form is created by a combination of hand-baiting and machine-loomed knotting, and is the result of a collaborative effort with an international team of award-winning aeronautical and mechanical engineers, architects, lighting designers, landscape architects, and fabricators. This work redefines "art space" by bringing viewers' eyes upwards to the sky to focus on a new celestial object.

During the day, the sculpture hovers high above heads, treetops, and buildings. The sculpture creates what the artist calls "shadow drawings," which she says are inspired by Phoenix's cloud shadows that captivated her from the first site visit.

At night, the illumination changes color gradually through the seasons. The goal in selecting the colors was to provide residents some small climate relief, adding cool hues in summer and warm tones in winter. The lighting design also changes what portion of the sculpture is illuminated, leaving parts obscured in mystery, much like the phases of the moon.

The artist was inspired by the region's distinctive monsoon cloud formations and the shadows they cast, in addition to forms found in desert flora and the local fossil record. The title quotes American poet Ralph Waldo Emerson, who wrote, "Adopt the pace of nature; her secret is patience."

The Civic Space Park, where the artwork resides, is located at the northern edge of the business district and just west of the Arizona State University campus. ASU students, local businesspeople, and senior citizens from a nearby HUD housing development gather in the park in their downtime.

The significance of the piece to the identity of Phoenix has drawn comparisons to that of Chicago's famous "Bean" and even the Eiffel Tower in Paris.

Artwork Excellence

The excellence of Her Secret Is Patience has been recognized both locally and nationally, receiving the Public Art Network's Year in Review Award in 2010 and the Reader's Choice Award by the Phoenix New Times. Technologically it is as challenging as its large scale is striking, and the size of the team behind the work underscores its complexity. It also received the Excellence in Structural Engineering Award from the Arizona Structural Engineers Association. The placemaking quality is apparent in a number of ways. It is the most striking part of a new public space in downtown Phoenix, the Civic Space Park, which is 2 blocks long by 1 block wide, located at the northern edge of the business district and just west of the ASU campus. ASU students, local businesspeople, and senior citizens from a nearby HUD housing development gather in the park in their downtime. The artwork enhances the comfort level of the park at night by changing color tones to become cooler when the temperature is hot and warmer when it is cold. The design references the local climate patterns, flora, and geologic history. The significance of the piece to the identity of Phoenix has drawn comparisons to that of Chicago's famous "Bean" and even the Eiffel Tower in Paris.

光电瀑布
Light Showers

艺术家：吉尔·安霍尔特（加拿大）
地点：加拿大多伦多
推荐人：凯伦·奥尔森
Artist: Jill Anholt（Canada）
Location: Toronto, Canada
Researcher: Karen Olson

作品描述

"光电瀑布"坐落于多伦多海滨东湾地区新谢尔伯恩公园内，是一组 9 米高的雕塑，共有三尊。

这组雕塑是与公园融为一体的具有美学基础和创新型水系统的功能组件，这将安大略湖以前的工业地区转化成了急需的绿地。

谢尔伯恩地区有着独特的功能，在园亭内有一个综合的紫外线（UV）水处理设施。东湾周边的暴雨径流和湖水被收集起来，送入紫外线处理设施。

紫外线处理设施把水净化，通过地下管道把水输送到雕塑，水作为独立的元素通过了小管道。作为每个雕塑的基础，水都收集好了，然后用机械装置把水输送到顶部，水均衡地分布着，通过悬挂的不锈钢网网布或纱布，用来捕捉水在冬天形成的独特冰形态。然后水流入一个 240 米长的水管道（或称都市的河流）中，流经公园，流入安大略湖。

这些雕塑在水处理工艺中发挥着作用，通气通风，作为导水管将净化过的水排入公园内的饲养池和流入湖中。

该雕塑的形式也是一种视觉表达，传达出了社区周边对于可持续发展的愿望；他们提升这些水，为收集社区的水而庆祝，之前的人工瀑布像有纹理的面纱一样，水流入了管道内，再次流入安大略湖。

夜晚，一组灯光成为了这件艺术作品的附加元素。每当人们在雕塑周围走动，集成运动传感器就被触动了，落下的水帘开始转换着光线，以此强调当地人的行为和长远效应之间的联系。

9米高的雕塑由混凝土和镶嵌在雕塑两边的不锈钢长条制成。这个元素里使用了大型玻璃纤维模具而建成，并用一个环氧化物包裹的钢筋加固，里面注满了轻盈的混凝土。

可持续发展在安霍尔特作品的材料表达、概念的发展、形成方面起到了生成的作用，因此每个元素构建使用到的模具被设计成可以重复使用的。当混凝土固化后，雕塑被小心地从模具里移了出来，运到了现场，安装在混凝土地面上。

解读

"自然"一词在现代汉语中，包含两种不同义项，其一指与"人为"相对，是各种自然而然的事物与状态，其二指人类文明以外的世界，即"大自然""自然界"这两个义项的关联在于："自然界"是最"自然"的。

吉尔·安霍尔特的"光电瀑布"妙在以"非自然"的手法去实现"最自然"的念想，用"人为"的科技力量，化前工业区为公共景观。从雨水和湖水中来，回到湖水和空气中去，三组雕塑，成就了水流一次又一次的华丽演出。尤其夜晚，"光电瀑布"能凭借对人群迢远的感应，变幻五彩灯花，流淌出的霓虹，穿过极光，穿过行人，水流有了与人类生命相连的脉搏和呼吸。是融入了自然元素的科技感、未来感？还是用科技和想象去还原自然本真的面貌？沿着光电瀑布飞流直下，自然与科技，平行于银河的两岸。（吴昉）

Artwork Description

Light Showers is a series of three 9-meter sculptures located in the new Sherbourne Common Park along the East Bay section of Toronto's waterfront. The sculptures provide both an aesthetic foundation for the park and a functional component of an innovative water system that is integrated into the park, which has transformed a former industrial area into a much-needed greenspace on Lake Ontario.

A unique feature of the Sherbourne Common is an integrated Ultraviolet (UV) water treatment facility that is located in the park pavilion. Storm runoff and lake water from the surrounding East Bay neighborhood collects and enters the UV treatment facility. The UV treatment purifies the water and sends it through an underground conduit to the sculptures, where it is drawn to individual elements through small channels. The water then collects at the base of each structure, where it's mechanically elevated to the top and distributed down hanging stainless steel mesh scrims, or veils, designed to capture water in the winter to form unique ice patterns. The water then spills into a 240-meter long water channel—or urban river—that runs through the park and empties into Lake Ontario.

The sculptures play a role in the water treatment process by providing aeration and acting as a conduit to bring treated water to other raised pools in the park and to the lake. The forms are also a visual expression of the surrounding community's aspirations to sustainability; they celebrate collected community water by elevating it before the water falls as textured veils into the channel that returns it to Lake Ontario.

In the evening, a series of lights become an added element of the artwork. As people move around the area of the sculptures, integrated motion sensors trigger shifting light patterns in the falling sheets of water to emphasize the connection between local actions and distant effects.

The 9-meter tall structures are made of concrete with stainless steel strips placed in insets on the sides of the sculptures. The elements were constructed using large fiberglass molds that were reinforced with an epoxy-covered rebar and filled with agila concrete.

Sustainability plays a generative role in conceptual development, form, and material expression of Anholt's work, so the molds were designed to be reusable for the forming of each element. After the concrete was cured, the sculptures were carefully removed from the molds, craned on the site and installed on concrete floorings.

a collaborative effort with an international team of award-winning aeronautical and mechanical engineers, architects, lighting designers, landscape architects, and fabricators. This work redefines "art space"

Artwork Excellence

Light Showers is the central feature of Sherbourne Common Park, the first park in Canada to integrate an ultraviolet facility for neighbourhood-wide stormwater treatmentinto its design. Anholt's three sculptures provide a stunning aesthetic sequence as the focal point of the park, and at the same time they reveal and contribute to the water treatment process. After the water is treated, the sculptures lift the water from ground to sky, aerate it as it falls down scuptural veils, then release it to a water channel that runs through the park and delivers the water to Lake Ontario. The addition of the motion-sensor lights enables people to interact with the artwork at night and alter the imagery on the falling water.

纽约市空中步道公园
NYC High Line Park & Art

艺术家：詹姆斯·科纳景观设计公司、迪勒·科菲迪奥 + 兰弗罗（美国）、
植物设计师皮埃特·欧多尔夫（荷兰）
地点：美国纽约曼哈顿西区
推荐人：米娜·曼噶尔弗德赫卡
Artist：James Corner Field Operations, Diller Scofidio + Renfro（USA），
and planting designer Piet Oudolf (Netherland)
Location：Manhattan's West Side, NYC, USA
Researcher：Meena Mangalvedhekar

作品描述

空中步道利用工业基础设施作为公共绿色空间。这里原先是一段废弃的高架铁路，在利用和改造铁路结构的基础上，创造出凌驾街道上空的立体公园。这是一个体现历时性保护的更新项目，融合了三个历史时期的特征，并将园林与建筑相结合，使其成为一条历史与现代共融、建筑与景观交织的空中绿廊。

自 2009 年 6 月空中步道第一部分开放起，就成为城市的主要观光景点之一，是有别于普通公园的独特公共空间。这个公共空间把生机勃勃的植物和狭长的木板融为一体，构成一条平滑流畅、几乎没有空隙的行走路面。铁轨依然是主要元素，引导着参观流线。公园还保留了大约二十种自生植物，一年四季花卉盛开。还包含水景、观景平台和日光浴平台，还有一块聚集区可用于表演、艺术展览和教育项目。多孔的步道含有敞开式接口以便水流在木板之间排出，灌溉周边的花坛，减少从步道流入下水道系统的雨水量。

空中步道是纽约西区工业化历史的一座纪念碑。它提供了机会创造一个崭新的极具创意的公共空间，与传统的政府和开发商主导的模式不同，空中步道主要源自一个以社区为基础的非盈利性组织——"空中步道之友"，在保护、再开发及后期管理过程中发挥着积极作用，为其他城市的工业废弃地的重新使用提供了一种新模式。

解读

"纽约市空中步道公园"又是一个典型的公共艺术介入空间再造的案例。一段废弃的高架铁路，以前是无人问津，现在是人流冲冲，建筑与位置的独特，历史与现代的融合，园林与景观的交织，创造了其特色的公共空间。这条空中步道公园，设计师在它特定的位置和功能的基础上，以铁轨为参观流线，周边设置水景、观景平台和日光浴平台；为丰富行人的日常生活，步道公园还会不定期的举办一些艺术表演和展览。（冯正龙）

Artwork Description

The High Line is an expanse of public green space built on pre-existing industrial infrastructure. Previously an abandoned elevated railway, the structure provided the basic framework for the resulting continuous stretch of park above the street. The project weds together historic protection with neighborhood renewal, architectural form with landscape, and the built environment with the resuscitating air.

From the day the first section of the High Line opened in June 2009 it has been one of the city's major tourist destinations-a unique public space different from any other park. Plant life fills the spaces between the narrow railroad ties, forming a smooth, linear, virtually seamless walking surface. The railroad tracks themselves are a prominent visual feature ushering visitors forward. Approximately twenty varieties of perennials bloom throughout the year. Public amenities on the High Line include a water fountain, viewing platforms, a sundeck, and gathering areas available for performances, art exhibitions and educational programs. Rain water that would otherwise spill off the pavement is channeled through porous pathways and open joints into the plant beds, thereby reducing strain on local storm-sewers.

The High Line is a monument to the industrial history of New York's West Side. Compared with the routine development practices of government and private developers, it exemplifies an innovative way to create public space. The High Line was realized mainly through the efforts of a community-based nonprofit organization. "Friends of the High Line" , which continues to play an active role in the protection, development and management of the project. It also serves as a model of civic action for others who see opportunity in abandoned industrial sites in other cities.

Artwork Excellence

The public space blends plant life (reminiscent of the quiet contemplative nature of the self-seeded landscape and wild plants that once grew on the unused High Line) with long, narrow "planks," forming a smooth, linear, virtually seamless walking surface. The public environment on the High Line contain special features, including a water feature, viewing platforms, a sundeck, and gathering areas to be used for performances, art exhibitions and educational programs.

The High Line is inherently a green structure. It re-purposes of a piece of industrial infrastructure as public green space. The High Line landscape functions essentially like a green roof; porous pathways contain open joints, so water can drain between planks and water adjacent planting beds, cutting down on the amount of storm-water than runs off the site into the sewer system.

开渠之流
Open Channel Flow

艺术家：马修·格勒（美国）
地点：美国德克萨斯州休斯敦
推荐人：绍娜·迪伊
Artist: Matthew Geller（USA）
Location: Houston, TX77007, United States
Researcher: Shauna Dee

作品描述

"开渠之流"是常驻纽约的艺术家马修·格勒的雕塑作品，由休斯敦艺术联盟委托制作，成为休斯敦城市公共艺术的永久收藏之一。该雕塑坐落在水牛河口公园旁边的萨宾水泵站附近，从公园地面耸立而起，总共 60 英尺的高度。受管道基础设施奇形怪状的启发，作品呈现出勾连复杂的管道状，具有室外公共淋浴头的功能，由手压泵操作。附近的哲梅尔滑板公园（Jamail Skate Park）保证了滑板青年们和过路行人在休斯敦湿闷的下午会前来这里汲取一丝清凉。

萨宾水泵站不是那种典型意义上放置公共艺术作品的地点：它不是集市广场，不是某建筑物的入口通道，也不是一个对公众开放的公园。该地点看起来名不见经传，尤其是邻近不远有着貌似更加吸引人的滑板公园。水泵站的重要性在于提供人类赖以生存的水资源，公众对此习以为常，因此问起它在哪里的话，大多数人都答不上来。

马修·格勒通过这件作品巧妙地连接了互相接壤的李和乔哲梅尔滑板公园、水牛河口公园和周边其他设施，直至休斯敦市中心地带，动足了脑筋把人们的注意力吸引到此处。他细心寻找并整合这个地点本身所具有的各种活力元素：水蓝色的各种水管看似手忙脚乱一般从水泵站冒出来，神秘的小丘覆盖住了地面下浸没水中的水箱。所有这些视觉元素整合出一种有趣、荒诞却又兼具实用功能的都市公共设施。

这个 60 英尺高的装置作品由各种 12 英寸、8 英寸和 4 英寸的钢管构成，顶部装有琥珀色和蓝色的信号灯，置放在被严加维护的萨宾水泵站区域内。这件作品比萨宾水泵站内的其他管道结构要大许多，也要稍稍复杂些。其中有一根水管像树枝一样高耸，跨过栅栏伸展到水坝范围以外的水牛河口公园内，以一个 25 英尺高的莲蓬头在那里休止。

莲蓬头下面的公园地面上是一个直径为 8 英尺的不锈钢排水管盖，和一个复古样式的现代版人工井泵。人们只需按下井泵，25 英尺高的莲蓬头就会往下洒水。与此同时，水流在管道中流动引发琥珀色和蓝色信号灯闪动起来，远在城市另一端的人们便会知道此处莲蓬头下，有人正在享受一个片刻的清凉痛快的"淋浴"。

"开渠之流"仅用了很少的费用，连接起三种不同的环境：水泵站、滑板公园和水牛河口公园。这件装置作品在休斯敦下城区通过从地理上整合不同性质的地点，邀请公园嬉戏的人们和滑板爱好者一起来和作品产生积极互动。这件装置艺术在"不被使用"时看上去像一个古怪的建筑物，正如滑板公园在没有滑板客的时间里既美丽又怪诞，作品旨在为酷热的休斯敦增添一个精神清凉站，其信号灯滑稽的信号弹，使这块植根于周边环境的休闲地成为公众惬意休闲的目的地。

解读

马修·格勒搭建的"开渠之流"很容易让人联想起大友克洋的"机器人嘉年华"，有着后现代的荒诞和奇趣，好像水管在狂欢，而与其相对应的，是这项作品毫不荒诞，甚而有些严肃的创作意图。超现实的管道森林，室外的公共淋浴头，报警器一样的信号灯，所有这些"无厘头"的元素，都为一个目标：将人们的注意力吸引到提供生命之源的水泵站。

热闹的滑板公园一侧有个水泵站，无人知晓它的存在，一如喧嚣的日常生活中无法离开的水资源，而人们总是无暇关注。"开渠之流"希望用公共艺术的形式，唤醒人们对水资源的意识，从而由意识到重视。开渠之流的淋浴者，就是滑板公园的滑板客，使超现实的水管迷宫变得现实起来，人类需求水源，人类也终将保护水源。

按下井泵—蓝色信号灯亮起—按下井泵—黄色信号灯亮起—蓝色黄色—管道运转正常—蓝色黄色—地球运转正常。（吴昉）

Artwork Description

Open Channel Flow is a sculpture by New York–based artist Matthew Geller that was commissioned by the Houston Arts Alliance as a permanent piece for the City of Houston Art Collection. Located at the Sabine Water Pump Station next to Buffalo Bayou Park, the structure emerges from the landscape of to a height of 60'. Inspired by the strange protrusions of plumbing infrastructure, the colossal pipe works features a public outdoor shower activated by a hand pump. The nearby Jamail Skate Park ensures that a steady flow of skaters and passersby will indulge in a refreshing spritz on Houston's infamously humid afternoons.

The Sabine Water Pump Station is a highly unusual site for an artwork: it's not a plaza, an entryway, or a public park. The site can seem particularly mundane, especially given its proximity to an ostensibly more glamorous Skatepark. The Pump Station's importance—providing water, a vital human necessity—is generally taken for granted by the public, most of whom wouldn't be able to identify it if asked.

Geller set out to call attention to the site by subtly integrating it with both the bordering Lee and Joe Jamail Skate Park, Buffalo Bayou Park and its larger environs, reaching as far as downtown Houston. He sought to augment and activate existing elements of the site: the aquamarine water pipes that emerge seemingly helter-skelter from the Pump Station grounds, the mysterious mound that covers the submerged water tank, and water. All these elements combine to create a kind of urban earthwork that is playful, absurd and as entertaining as it is functional.

The sixty-foot tall structure of 12-, 8-, and 4-inch steel pipes, with amber and blue beacons on top, stands inside the restricted grounds of the Sabine Water Pump Station. It is a much bigger, slightly more elaborate version of other pipe structures located on the Pump Station grounds. One pipe, like the branch of a large tree, leaves the pump station property, passes over the pump station fence and ends with a showerhead twenty-five feet overhead in Buffalo Bayou Park.

On the ground below the showerhead is an eight-foot diameter stainless steel drain cover and a contemporary version of an old-fashioned manual well pump. As one pushes down on the pump handle, water rains down from the shower head twenty-five feet above. Simultaneously, and as a result of pumping water through the pipes, the orange and blue twin beacons on top of the structure flash, signaling people as far away as downtown that another person has doused themselves with a refreshing, albeit brief, shower.

With an economy of means, Open Channel Flow reconciles three disparate environments: the Pump Station, the Skatepark and Buffalo Bayou Park. It integrates the sites and downtown Houston as it invites park visitors and skateboarders to play an active role in animating the work. While it is an appealing architectural oddity when not "in use" (much as the Skatepark is a beautiful and strange structure even when devoid of skaters), it offers an added draw as a spirited cooling station in the Houston heat, with its beacon functioning as a wry smoke signal, locating the recreation area as a lively destination with real roots in its environs.

Artwork Excellence

The 60-foot tall Open Channel Flow is a delightfully complex configuration of pipes, valves and elbows that seems to grow out of Houston's Sabine Water Pump Station. Geller's intent with it was to call attention to the city's water delivery system—it's a little known fact that Houston has 7,000 miles of underground drinking water pipe—and how ease of access to clean water is sometimes taken for granted. It does this through a witty illustration of water delivery: it offers showers to passersby and beacons on top the work's highest element 60 feet flash when water courses through the pipes. All these elements combine to create a kind of urban earthwork that is playful, absurd, and as entertaining as it is functional. Open Channel Flow also integrates three environments that come together at the site—the Pump Station, the Jamail Skate Park and Buffalo Bayou Park—and highlights the backdrop of the Houston skyline.

通道
Passage

艺术家：麦格斯·哈里斯（英国），拉约什·赫德尔（匈牙利）
地点：美国凤凰城
推荐人：绍娜·迪伊
Artist: Mags Harries（UK），Lajos Héder (Hungary)
Location: Phoenix, United States
Researcher: Shauna Dee

作品描述

　　"通道"是一个多方位合作的公共艺术作品，位于南部山脉社区图书馆，由凤凰城公共图书馆和南部山脉社区学院共同操作。对使用这座图书馆的学生和社区居民而言，它是体现社区身份和振奋学习精神的一块磁铁。为了反映这种精神，艺术家把视觉元素与文字糅合在一起。这个项目包括景观设计，尤其是在图书馆入口处外面设计了几把音响座椅和四个棚架上的亚利桑那著名诗人阿尔伯托·里奥斯的诗歌中的摘句。

三把音响座椅排列在图书馆正门前。他们将图书馆建筑延伸为景观，使其视觉上和南部山脉连接起来。座椅座位的表面是用当地的华拉派石制成的。椅子两侧的材料使用彩色混凝土，表面嵌有钢质字母。周围的路面上也嵌有字母，仿佛从座椅上倾泻下来。每个字母都混杂在一起，这些字母可组成和当地景观有关的单词比如"沙漠"，"石头"，"山脉"和"水"。这些散乱的字母让观众可以尝试自己来组词和作诗。

座椅内的扬声器受到运动传感器的激活后能够播放诗歌的录音。舒缓播放着的诗歌营造了一种亲近的氛围。里奥斯策划了这些诗歌的挑选，其中有19部诗歌描绘的是凤凰城南部和当地的景致。

此外，为参加"2012年全国诗歌月"，里奥斯将编辑一部社区诗集，这些诗也打算放进音响座椅内。图书馆将征集描绘凤凰城南部的两行体诗句，里奥斯将会按照日本传统"连歌"的形式构成一部诗集。

在经过图书馆通向西部运河的小径上四个诗歌棚架一路横跨。字母焊接在棚架的顶棚上，透过阳光照射把字母组成的诗句阴影投射到小径上。阴影随着太阳照射位置的变化，提供了不断变化的感觉体验。

阿尔伯特·里奥斯为这个项目特意写了四个对联。四个对联可以独立阅读，也可以作为整篇诗歌来欣赏：

1. 闪亮的蛇和细针般的植物 / 这想象力千变万化的山谷

2. 阴影——无数小片段构成的凉爽夜晚 / 你脖颈后的梦幻气息

3. 当第一个雨点打在干燥的豆荚上 / 古老的歌就充满了沙漠

4. 南部的山脉啊 / 这沙漠海洋中的缓缓巨浪

解读

对于诗歌而言，最宝贵的特质恰恰是其私密性，直呈诗人内心最真实的情绪感受，而美好的诗歌往往又最能感化他人，这是千古以来，人们热爱诗歌、代代唱诵的缘由。当环境建设与公共场所设计成为城市及区域规划的重要主题，公共艺术就肩负起"成教化，助人伦"的社会职责。将诗的隐秘优美用某种温和而非强迫的形式传递给参与者，是"通道"所要面对的难题与挑战。

诗歌与朗诵的关系亲密无间，于是有了三把雕刻意象词汇的音响座椅；诗歌的美妙在于任由不同心境、语境、环境，释义微妙的千差万别，于是有了朝夕四季变幻无穷的光影投地；诗歌的神奇是与人类存在的往还酬答，不同情感不同人，造就了诗歌的丰富与完满，于是有了延展文字的诗的通道。麦格斯和拉约什的作品，无疑洞悉诗中三昧，更将公共艺术与人为善的特质发挥淋漓。

不用行至水穷，亦可坐看云起，多好。（吴昉）

Artwork Description

Passage is a multi-faceted, collaborative public artwork at the South Mountain Community Library, operated jointly by The Phoenix Public Library and South Mountain Community College. Used by students and the community at large, the library is a magnet for community identity and the spirit of learning. To reflect this energy, the artists integrated visual elements with words. The project includes landscaping, specially designed acoustic seating outside the library entrance, and four trellises featuring excerpts of poetry by distinguished Arizona poet Alberto Rios.

Three acoustic chairs are grouped in front of the library's main entrance. They extend the architecture of the library into the landscape, making a connection with South Mountain. The seat surface of the chairs is made of local Hualapai stone. The sides of the chairs are made of colored concrete with steel letters cast into the surface. Letters are also embedded into the surrounding pavement, as though cascading from the chairs. Each letter of the alphabet is represented in the jumble, as are letters that make words that reference the landscape, such as "desert," "stone," "mountain," and "water." The scattered letters encourage visitors to make their own words and poetry.

Speakers inside the chairs play recordings of poetry when activated by motion sensors. The poems play softly to create an intimate experience. Ríos curated the collection of poems, featuring 19 poets writing about South Phoenix and the landscape of the area.

In addition, for National Poetry Month 2012, Ríos will edit a community poem to be included in the acoustic chairs. The library will solicit submissions of two lines about South Phoenix, which Ríos will construct into a poem in the tradition of the Japanese form called the renga, or "linked elegance."

Four poetry trellises span the path as it runs past the library and toward the Western Canal. Letters welded to the canopy project shadow lines of poetry onto the path. The shadows shift with the sun to offer a constantly changing experience.

Alberto Ríos wrote these four couplets specifically for the project. The four couplets can be read individually or as a whole poem.

1.Diamond-clad snakes and plants made of needles,

This valley of constant imagination

2.Shade— small fragment of night, cool

Dream's breath on the back of your neck

3.When first raindrops hit dry mesquite pods,

  Ancient songs fill the desert

4.South Mountains

  –Slow waves in the ocean of this desert

Artwork Excellence

Passage is a multi-faceted, collaborative public artwork at the South Mountain Community Library in Phoenix. The library is a magnet for community identity and the spirit of learning. To reflect this energy, the artists wanted to integrate visual elements with words, with the help of noted local poet, Alberto Ríos. Before this project, the artists had been involved with the rapidly changing environment of the South Mountain area of Phoenix for over eight years. During that time, this area of citrus orchards and flower farms became a built-up residential community. The artists have been able to contribute to the pedestrian, bicycle and equestrian connections, and to the public image of this community through a series of projects: Arbors & Ghost Trees (2004) for Baseline Road and The Zanjero's Line (2009) for the Highline Canal Trail that have been completed, and the soon to be built Western Canal Bridge. Passage creates another link in these trails, one that relates to the others and adds new artistic ideas particularly fitting for the new South Mountain Community Library.

铁路站公园
Railyard Park and Plaza

艺术家：肯·史密斯，弗雷德·斯沃兹，玛丽·密斯（美国）
地点：美国新墨西哥州圣达菲
推荐人：绍娜·迪伊
Artist: Ken Smith, Fred Schwartz, Mary Miss（USA）
Location: Santa Fe,NM,United States
Researcher: Shauna Dee

作品描述

"铁路站场公园"的设想就是能成为圣达菲社区的一片令人愉悦的、绿洲般的地块，但其目的也是提供给人们这样的一个地方：能让他们逐渐了解到自己与周围环境的一种新型关系。

这个建立在废弃铁路站场的、面积为 13 英亩的公园是由社区发起的城市重建计划的成果，而这个项目已持续长达 20 年的时间。在公共土地信托管理的带领下，整个团队与公园管理员密切合作。项目由大厦、林荫道与自行车道以及一个 7 英亩的公园组成。

一个以前通常是用在旧铁路沿线的水塔，很快地就体现了它在其中的重要性。水是从屋顶或项目场地北段的人工设施处收集来，并储存在地上或地下，用以维持公园的园艺。从北端向南流动的路径由电线杆顶部的蓝色灯光来点缀，而这些电线杆则形成了连接整个项目的中枢。公园里有一片小树林，里面排列着 400 年历史的"马德里灌溉沟"——它是当地的一种传统灌溉方法，由马德里灌溉沟输送给公园南端的规模更小的"妮娜灌溉沟"。其使用过程类似于西南部城镇村庄或一座传统西班牙殖民式花园所用的可持续农业模式。整个公园遍布着旱生植物，并生长着 400 棵浓荫大树。

此地区的历史对参观者来说是显而易见的，他们可以观察灌溉渠的功能，沿着小径追寻旧铁路和干线的历史，并坐在 circular ramada ——能回忆起附近曾经存在过的蒸汽机修理处的旧圆屋。公园里这些元素的种种直线性设计，都反映了此地过去与铁路的种种印记。公园里所用的材料——ramadas 的电线杆、铁路铺就的花园小道、由木头堆起的木头长椅——唤起人们对公园工业时代的回忆。

参观者也被考虑到项目的设计之中。用来照明的电线杆、林荫道和长 ramada 串联起了项目中的所有部分，而直线型铁路花园在你走近公园北端时就能看见并吸引路人进来看看。大圆 ramada 被松树环绕并被回廊环抱着，这就是公园的入口处。此地的坡度也很平缓，周围一圈石笼网，隔开了来自外界交通的噪声。这个斜坡草坪类似于一个可容纳两到三千人的开放平台。

公园西边的景观设施：由石头、木墙或棚架子划分开来的一系列环形区域，使人想起了空中能见到的那些水田，或者美国当地建筑的那种地下室。这些都是用来划分公园不同集聚场所的——从草地午餐圈，到公园主入口的一个环形玫瑰花园。公园中心地带是由巨石圈划出的空地，同时用来做儿童玩耍场地，那里有一个弧形堤，一面可以爬的斜坡，另一面则是攀岩墙。而另一边则是婴儿区域，由沙包圈、草坪和树木环绕，一个毗邻的岩丘可用来俯瞰整个区域，也用来展示水从山顶流下的路径。毗邻沙丘还有一个用来收集和表演的小型的探险区域以及石头迷宫。而占据了两个边缘地区之间中心地带的有石柱和攀爬网的广场，这是一个更大的娱乐区域。

公园的这片中心地带，目的是为了给孩子创造一个能集聚玩耍的场所，能把孩子们都吸引到中心地带来。那里可以爬、捉迷藏、激发挑战以及闲逛，与此同时还可以午餐、表演或阅读。公园北端的林荫道和大厦是另一个集聚场所，那里因各种演出、戏剧和农贸市场而氛围活跃。在交易日，可以在小货车或搭起的帐篷下买卖，并安排尽可能多的场所供人们彼此会面和互动。

铁路站场管理处（The Railyard Stewards）是一个提供社区管理并提供资金维护运营的草根组织，并给铁路站场公园做规划，同时也与艺术家们合作策划了一些铁路站场艺术项目。

解读

"铁路站公园"是回忆与回忆的叠加。400 年历史的"马德里灌溉沟"，保留了历史久远的生活方式，20 年持续建设的项目进程，同样也作为一种记忆的痕迹，永久地留在孩子们的欢声笑语中。

这个对公众开放的公园项目，不仅是孩童的乐园，更是所有人回到过去、回到童真的时间飞船。项目的公共性在于：安排尽可能多的场所供人们彼此会面与互动。城市生活的节奏是以牺牲人们驻足闲谈为代价的，擦肩而过的一瞬，多少令人怀念往昔恬淡闲适的田园生活，这也吸引参观者们更热衷于"铁路站公园"。历史爱好者们，在这里成为生活的爱好者；孤独自扰者们，在这里放松心情，随着自主添加入公园的小绿化越来越多，慢慢把内心也建造成一座花园。

"铁路站公园"是一项景观工程，灌溉沟和旧铁路，把老城生活与摩登时代前后贯通，而孩童的嬉笑与成人的会心，成为联系过去与未来的承载物，鲜活跳跃。（吴昉）

Artwork Description

Railyard Park and Plaza was envisioned as a place of pleasure, an oasis for the community of Santa Fe. But the intent was also to create a place where people come to understand a new relationship between themselves and their environment.

This thirteen-acre park built on the site of the abandoned railyards is the result of a community-initiated urban redevelopment project that took place over a twenty-year period. The collaborative team worked closely with a group of park stewards led by the Trust for Public Land. The project consists of the plaza, an alameda and bike path and a seven-acre park.

An elevated water tank like those traditionally used along the old rail lines immediately announces the importance of this element. Water collected from the roofs and hardscape at the north end of the project is stored above and below ground and then used in the maintenance of the plantings of the park. The path of the water moving from the north end to the south is noted by blue lights atop the utility poles that form the connecting spine of the project. In the park a bosque of trees lines the 400-year-old Acequia Madre, an example of the traditional means of irrigation in this region. The acequia feeds the smaller Acequia Nina which runs to the south end of the park; along the way it sustains an example of the permaculture farming used in the pueblos in the southwest and a traditional Spanish Colonial garden. Xeric planting is used throughout the park and over four hundred trees have been planted to provide shade.

The history of the site and the area becomes apparent as visitors are able to observe a functioning acequia, follow paths that mark the old rail lines and spurs, and sit in a circular ramada that recalls the old round house for engine repairs that once existed nearby. The linear layout of the park elements reflects the site's past connection with the railroads. The materials used in the park—utility poles for the ramadas, steel rails lining garden walks, wooden benches in the form of stacks of wood—recall the industrial history of the park.

Visitors are engaged by the layout of the park. The spine of utility poles used for the lighting, alameda and long ramada connect all the parts of the project while the linear rail gardens visible as you approach the park from the north draw the visitor in. The large circular ramada surrounded by pine trees and hung with porch swings acts as a front porch to the park. The grade of the site has been lowered which, along with stone-filled gabions at the edge of the site, buffer visitors from the noise of the traffic. A sloped performance lawn adjacent to an open field accommodates two to three thousand visitors.

Another aspect of the western landscape is reflected in the park: a series of circular areas defined by stones, wooden walls or trellises recall the circular irrigated fields seen from the air as well as the underground chambers of Native American architecture. These all define gathering places in the park—from the grass picnic circles to the circular rose garden at one of the main entrances. At the center of the park is a gathering place defined by a ring of boulders that is also a children's playground. There is a curved embankment that has a series of slides on one side with a climbing wall at the back. On the opposite edge there is a toddler's area made up of circles of sand, grass and trees. An adjacent rock mound configured for an overview of the area also shows the water runoff pattern of a hilltop. A small stepped area for gatherings and performances as well as a stone labyrinth adjoin the mound. A field of posts and climbing nets occupy the area between the two edges of the larger play circle.

In this central area of the park, the intention was to create a gathering area that integrates children into the heart of the park. There are places to climb, hide, be challenged and hang out, while others may picnic, perform or read. The alameda and plaza at the north end of the park is another gathering area. It is activated by performances, festivals and the presence of the farmer's market. On market days produce is sold from pickup trucks and under awnings. In as many places as possible situations have been set up for people to meet and engage with each other.

The Railyard Stewards is a grassroots organization that provides community stewardship and advocacy for the care and programming of the Railyard Park and Plaza. There is also a Railyard Art Project for collaboration with artists.

Artwork Excellence

Railyard Park and Plaza is a community-initiated redevelopment of an abandoned railyard that involved much collaboration over 20 years. The project incorporates the history of the location, including its railroad traditions and the native habitants of the region, in the design and function of the park and plaza. The intent to have an impact on the community into the future is clear; from the artist's website: "Through repeated trips, there is the possibility that visitors to the park will discover new ways to think about planting their yards or handling the water they use; there will be the opportunity to encounter other members of the community outside their usual family or social groups; the previous uses of this land will become apparent. This public space in Santa Fe provides the opportunity for new bonds to be formed between the residents of this city and their surroundings." The Railyard Park and Plaza is an urban art environment and is itself a work of public art. Artists, organizations and collaborative, creative teams are encouraged to respond to this intentionally designed landscape with temporary art and performance projects that emphasize engagement, participation and education. Collaborations across disciplines—artists, designers, scientists, historians, sociologists, etc—are of particular interest. Additionally, projects that respond to specific locations in the Park and Plaza through community and historic context are encouraged.

反省缺失
Reflecting Absence

艺术家：米歇尔·阿拉德（美国）
地点：美国纽约市世贸中心遗址
推荐人：绍娜·迪伊
Artist: Michael Arad（USA）
Location: WorldTradeCenterSite, New York, United States
Researcher: Shauna Dee

作品描述

国家 9·11 纪念馆是为了纪念那些在 2001 年 9 月 11 日世贸中心、五角大楼和宾夕法尼亚尚克斯维尔三处的恐怖袭击中丧生的近 3 000 人，还有在 1993 年 2 月在五角大楼的 6 位因炸弹袭击而丧生的人，并向他们表达诚挚敬意。

纪念馆的"双子瀑布"各占 1 英亩地，其最大特色是：它是北美最大的人工瀑布。这个池塘被称为"反思的缺席"，坐落在曾经是双子楼竖立的地方。建筑师米歇尔·阿拉德和景观建筑师彼得·沃克创建的纪念馆，是从全球超过 63 个国家 5 200 多种设计方案中脱颖而出的。

纪念馆水池四周围绕着一圈铜嵌板，上面刻着 2001 年和 1993 年两次袭击中每一位遇难者的名字，触目惊心地提醒着人们，这是美利坚土地上有史以来损失最惨重的来自境外力量的袭击，以及美国历史上单次人员伤亡最大的一次救援。

建筑师的设计方案：

这座纪念馆提供给了人们一个空间，得以寄托对这次发生在世贸中心的两次袭击而离去的千万条生命的哀思之情。它坐落在一片树林之中，位于树林中由两个大凹槽隔出的空隙之处。两个凹槽（也就是水池）的大小正好是沿着原本双子楼的面积大小（也就是水池形状是依据双子楼留下的痕迹建造的）。水池广场是由四面不断循环的水流瀑布围出来的，这是一个大型的真空之地，是人们"缺失"的开放式、透明式的纪念。

纪念馆广场的地面看上去有着由一排排的落叶乔木组成的点线性排列韵律，从而围出了一个随性的小树林圈。同时这块地方还有铺路石、绿植和矮树丛等；而这座充满生机的公园会通过其一年一度的新陈代谢而不断扩充并延伸纪念馆的存在意义。

水塘四周的铜铭碑上刻着遇难者的名字，它们所占空间之大以及一个个平民名字的罗列，都提醒着这场灾难。站在瀑布之旁，看着汩汩流水流向深渊，来此地的人都能感受到弥漫在这些铭碑四周的一种难以名状的感伤。

纪念碑广场在设计之初就被设定为一个斡旋之地，为政府和纪念馆共有。而它临街而建也正因其为城市规划整合的一部分，纪念馆广场的存在意义也希望能融入纽约市民的日常生活，它们是城市鲜活的一部分，而不是与城市隔离而独立存在的。

解读

城市的公共环境空间，为人们提供小憩、放松、欢笑的场合，很多的情绪在都市规划中欢腾释放。然而，有没有一种公共场合，可以让行人过客默然凭吊呢？悼亡，是人类最自然的情感之一，太多时候，公共性的存在抑制住悼亡的心愿，强行否定了消极情绪的存在意义。911事件之后，巨大的哀恸和反思，超出了私人小空间可以承载的范围，所幸，世贸中心遗址及时转生出公共作品——"反省缺失"。

大型的真空之地，直指事件核心现场，为公众提供了开放式的纪念处。"缺失"的是什么呢？摩天高楼？生命迹象？一种信仰？各种纷争？这也是"反省缺失"对公众开放式的提问。

循环流转的人工瀑布，用动态的景观安抚世人躁动不平的心。水声潺潺，为悼念与缅怀平添了一份安全感，滤去外界的窥探和打扰。由此，公共艺术不仅是公众情绪的发散地，更是人们情绪自然流露的保护伞。悼念是为了记忆，凹陷的水池上方，某些东西正超越物质曾经的存在，变得永恒。（吴昉）

Artwork Description

The National September 11 Memorial is a tribute of remembrance and honor to the nearly 3,000 people killed in the terror attacks of September 11, 2001 at the World Trade Center site, near Shanksville, Pa., and at the Pentagon, as well as the six people killed in the World Trade Center bombing in February 1993.

The Memorial's twin reflecting pools are each nearly an acre in size and feature the largest manmade waterfalls in the North America. The pools, called "Reflecting Absence," sit within the footprints where the Twin Towers once stood. Architect Michael Arad and landscape architect Peter Walker created the Memorial design selected from a global design competition that included more than 5,200 entries from 63 nations.

The names of every person who died in the 2001 and 1993 attacks are inscribed into bronze panels edging the Memorial pools, a powerful reminder of the largest loss of life resulting from a foreign attack on American soil and the greatest single loss of rescue personnel in American history.

Architects' design statement:

This memorial proposes a space that resonates with the feelings of loss and absence that were generated by the destruction of the World Trade Center and the taking of thousands of lives on September 11, 2001 and February 26, 1993. It is located in a field of trees that is interrupted by two large voids containing recessed pools. The pools are set within the footprints of the Twin Towers. A cascade of water that describes the perimeter of each square feeds the pools with a continuous stream. They are large voids, open and visible reminders of the absence.

The surface of the memorial plaza is punctuated by the linear rhythms of rows of deciduous trees, forming informal clusters, clearings and groves. This surface consists of a composition of stone pavers, plantings and low ground cover. Through its annual cycle of rebirth, the living park extends and deepens the experience of the memorial.

Surrounding the pools on bronze parapets are the names. The enormity of this space and the multitude of names underscore the vast scope of the destruction. Standing there at the water's edge, looking at a pool of water that is flowing away into an abyss, a visitor to the site can sense that what is beyond this parapet edge is inaccessible.

The memorial plaza is designed to be a mediating space; it belongs both to the city and to the memorial. Located at street level to allow for its integration into the fabric of the city, the plaza encourages the use of this space by New Yorkers on a daily basis. The memorial grounds will not be isolated from the rest of the city; they will be a living part of it.

Artwork Excellence

Michael Arad's design was chosen in an extremely competitive pool of applicants for the 9/11 Memorial to be built in place of the void in lower Manhattan left after the September 11th attacks on the World Trade Center towers. The panel decided that Arad's design most successfully embodied the desire for healing. But that was only the beginning of the challenges as the work progressed, with financial and political obstacles presenting themselves along the way. What resulted was an impressive installation of two square fountains with water that cascades 60 feet down two levels, each occupying one of the exact footprints of the Twin Towers. Around the rim of each is a long bronze strip perforated with the names of victims: of the 2001 attacks on the Twin Towers and the Pentagon, of the hijack of Flight 93, which crashed in a field in Pennsylvania, and of the 1993 bombing of the World Trade Centre. The names, after years of agonising, are grouped by the location of each victim at the time of the attacks, modified by "adjacency requests" whereby relatives could ask for individual names to be by others with whom they were close. The 415 trees that occupy the surrounding 8-acre plaza are all the same size, which required an exceptional effort of selecting and nurturing. Overall, Michael Arad, and the landscape architect Peter Walker, made the space a place of both death and life—where victims can be properly remembered, but where office workers can come to eat their lunch. As of December 29, 2011, over 1 million people have visited the memorial since it opened on September 12, 2011.

索拉尼桥和广场
Soleri Bridge and Plaza

艺术家：保罗·索来利（意大利）
地点：美国斯科茨代尔
推荐人：米娜·曼噶尔韦德赫卡尔
Artist: Paolo Soleri (Italy)
Location: Scottsdale, United States
Researcher: Meena Mangalvedhekar

作品描述

斯科茨代尔的"索拉尼桥和广场"——由著名的艺术家、建筑师、哲学家保罗·索来利创作——成为了人行通道，阳历，也成为了斯科茨代尔海滨的聚集场所。这个期待已久的公共空间位于斯科茨代尔市中心，吸引着多样化的观众：从海滨的休闲游客和当地居民，到学生、游客、建筑师和艺术爱好者。通过庆祝太阳活动，这个标志性的桥和广场统一着过去和现在。水路站点有着丰富的历史意味，将当今文化生动地融合起来。整个站点将人和自然凝聚起来。这个动态项目的元素参考了保罗·索来利的其他作品：桥标志着太阳活动，连接着人类跨越时间的概念；单片铁板、地球铸板反映了艺术家的审美；经典的铜制大钟，支撑了索来利的项目，受到了国际公认。桥由两个 64 英尺的塔锚定，南部 27 英尺宽，北部缩短为 18 英尺。坐落于北部轴心线，桥会标记下太阳倒影而产生的太阳活动。两座塔中间有 6 英寸的空隙，当地球运转时使太阳产生一种光芒。每当太阳在正午时——它可以把 12 点正午的时钟改变 40 分钟——光线透过空隙产生了阴影。这个光芒的长度会有所不同，取决于一年的时间。每当夏至（6 月 21

日）太阳位于空中最高的位置，投射不了任何影子；每当冬至（12月21日）太阳在空中的位置最低，阴影是最长的，直达桥梁结构。沿着桥面长度的红色条纹遵循着光线，感知地带领观众跨越桥梁。每年的9月21日、3月22日，桥和广场庆祝着一年一度的秋分和春分。22 000平方英尺的广场包括了图腾面板，复制铸铁壁图案，代表着可桑迪和阿尔可桑迪，那是保罗·索来利的主要建筑项目。每个3 500磅重的面板单独由保罗·索来利博士和罗杰·托马尔蒂设计，在亚利桑那州斯科茨代尔的可桑迪制造。

10个地铸板构成了广场的南部边界，使用了伊利诺伊的形状，使用大地色不同地迭代，突出了各种类型的设计。更大的，11个面板构架了桥梁的北部。地球的铸造工艺使用了沙漠大地、水和水泥，在混凝土浇筑前，进行雕刻，然后就能成型了。这些面板的制作历时近8个月。面板上浓重的建筑雕刻与金水钟会一起形成了一种完美的对位，反映了保罗·索来利在建筑和生态学里的生活工作。1969年，钟完全由保罗·索来利制造，钟是艺术家在华盛顿的科克兰画廊举办的第一次美国回顾展的一部分，包裹着22英尺高的塔，位于广场南部边缘，钟和面板制造出了一种亲密的画廊，设置于开放的广场中。

解读

工业化的材质与造型，后现代的科技感，表现的却是与人类生存休戚相关的宇宙奥秘。索拉尼桥和广场，让人忽略了它的钢铁外壳，乐于融入艺术家创造的宇宙观中。

"主呵，是时候了。夏天盛极一时。把你的阴影置于日晷上，让风吹过牧场。"

里尔克的诗句在保罗·索来利的城市项目中得以实现。宇宙之浩渺，让微小如人类常常忘记头顶上的那片星云，就同一台精密准确的机器，保障着世间万物的日夜相继。熙熙攘攘的人群，在桥和广场构架出的宇宙空间里，投下小小的阴影，与天地奥秘相融合。或许，连欢欣鼓舞的行人都未曾意识到，自身已然处于跨越时间的维度中，那么，公共艺术的终极意义，是激发公众的意识还是融于公众的无意识呢？（吴昉）

Artwork Description

Scottsdale's Soleri Bridge and Plaza—by renowned artist, architect, and philosopher Paolo Soleri—is at once a pedestrian passage, solar calendar, and gathering place along the Scottsdale Waterfront. The long-awaited public space in downtown Scottsdale appeals to a diverse audience ranging from casual waterfront visitors and local residents, to students, tourists, architects and art lovers. By celebrating solar events, the signature bridge and plaza unify the past and the present. The site of the waterway is rich with historic undertones and serves a lively mix of present day cultures. The overall site provides a coherence between man and nature.

The dynamic project elements reference the range of Soleri's work: A bridge marks solar events and connects humans conceptually across time; monolithic, earth-cast panels reflect his aesthetic; and the classic bronze bells, which have supported Soleri's projects, are recognized internationally.

The bridge is anchored by two 64-foot pylons and is twenty-seven feet wide on the south side narrowing to eighteen feet on the north. Situated at a true north axis, the bridge is intended to mark solar events produced by the sun's shadow. The six-inch gap between both sets of pylons allows the sun to create a shaft of light as the earth moves. At each solar noon—which can vary up to 40 minutes from twelve o'clock noon—light coming through the gap produces a shadow. The length of this shaft of light varies depending upon the time of year.

At each summer solstice (June 21st) when the sun is highest in the sky, no shadow is cast; while at each winter solstice (December 21st) when the sun is lowest in the sky, the shadow is the longest, reaching to the bridge structure. A red stripe along the length of the bridge deck follows the light and perceptually leads the viewer across the bridge. The bridge and plaza also celebrate the annual equinox events that are approximately on Setpember 21st and March 22nd.

The 22,000 square foot plaza includes totemic panels replicating the cast wall motif representative of Cosanti and Arcosanti, Soleri's major architectural projects. Each 3500-pound panel was individually designed by Dr. Soleri with Roger Tomalty and produced at Cosanti in Scottsdale, Arizona. The ten earth cast panels framing the southern boundary of the plaza use the illinoidal shape in various iterations with earth color pigments accentuating the various designs. A larger, eleventh panel frames the north side of the bridge. The earth cast process uses the desert earth with water and cement, which is then carved, before being cast in concrete and allowed to cure. The fabrication of these panels took nearly eight months.

The strong architectural carving on the panels creates a perfect counterpoint for the Goldwater Bell assembly, which reflects Soleri's life work in architecture and ecology. Fabricated entirely by Paolo Soleri in 1969, the bell was part of the artist's first U.S. retrospective at the Corcoran Gallery in Washington, DC. Encased within the 22-foot tall pylons near the south edge of the plaza, the bell and the panels create an intimate gallery setting amidst the open plaza.

Artwork Excellence

The bridge is designed to bring awareness of our human connection to the sun and the natural world. The earth's rotation each day and the sun's location with relation to the earth are both keyed to the bridge's true north axis location and the 80 degree angle of the pylons. It is this symmetry that allows the solar geometry of the design to function. A 22,000-square-foot plaza on the south side of the canal creates a pedestrian-friendly gathering environment with connections from Scottsdale Road to the Waterfront pathways. The public space is pedestrian and bicycle friendly as well as completely ADA accessible. The bridge and plaza provide a destination for passive recreation such as walking and jogging and epitomizes the invention and innovation for which Soleri has been internationally recognized. The plaza also includes a portion of the regional, 141-mile Sun Circle Trail.

卡戎兄弟广场
Square des Frères-Charon

艺术家：拉菲埃尔·德·古鲁特（加拿大）
地点：加拿大魁北克蒙特利尔
推荐人：绍娜·迪伊
Artist: Raphaëlle de Groot（Canada）
Location: Montreal, Quebec Canada
Researcher: Shauna Dee

作品描述

本作品由建筑师德拉·里瓦·阿弗雷克，景观建筑师罗伯特·德雅尔丹和艺术家拉菲埃尔·德·古鲁特共同设计，于 2008 年完工，"卡戎兄弟广场"坐落于蒙特利尔一个老城区两条历史街道的十字交叉口。

"卡戎兄弟广场"是公共艺术空间网络的一部分，沿着马克吉尔大街主轴线而展开。那里是一条历史通道，连接着旧港与当代市中心。

广场提供了一种当代都市景观的体验，受到了最初该区域使用功能的启发，那里原来是草原湿地，17 世纪卡戎兄弟在那里建造了风车房。该广场犹如一种对照般的体验，连接着原始的草原湿地，如今被城市包围着，产生了新的维度，唤起了公共对于该地区历史和地理的意识。

这个项目使用了一种简单、雅致和极简主义的建筑语言来创造环形与圆柱形形式之间的对话，包括一座长着野草的花园、残留的风车房和一座以观景楼的形式出现的园亭。与这些姿态相配，使用照明方案营造出了一座多色彩的花园，寓意四季的变化。

这个项目的建造是对不满的工业区域都市复兴的一种反应，"卡戎兄弟广场"是一个全新的公共设施，而其所在的地域已经有150多年的历史。这个新广场提供了身份认同感、公民的自豪感及一年四季可供公众使用的开阔的户外场地。

该市组织了一支跨学科设计团队，其中包括一位艺术家、一位建筑师和一位景观建筑师，目的在于跨越传统的学科界限，在大家的共同点上进行合作和融汇，而不是根据习惯的专业领域进行划分和分配任务。这个团队运用了各种沟通工具以鼓励和便于公众参与整个设计的过程。

通过聚焦当代都市和都市生活方式的体验为重点，设计团队以一位使用者的观点出发去探讨概念，与周围的环境发生联系。

"卡戎兄弟广场"街道公共领域经过了精心的设计以确保其舒适性和安全性，可供轮椅通行。可持续发展的倡议包括了种植本地的野草，这对市政灌溉系统起到了重要的减负作用，园景工程和公园凉亭的贴面采用了耐用的魁北克花岗岩。

就其提供的生活质量而言，"卡戎兄弟广场"220万加拿大元的建造成本是一项获益的市政投资。

这个负担得起的便利设施，比如它的解说方案、大面积的植被和可回收利用的公园凉亭，使项目设计师控制了成本，给当地居民和旅游业带来了巨大的收益。有适度的规模和预算，这座广场的规模和预算都适中，它是麦克吉尔大街大型历史空间网的一个必不可少的组成部分，也是蒙特利尔文化旅游业品牌战略的关键要素。

解读

"卡戎兄弟广场"的中心主体是一个标准的圆形湿草埔，中间被一条修葺整齐的石路两边分开，摆上座椅，供游人休息交谈。通往圆弧两端的石路，连接起不同的地貌特征和时代象征，在旧港与新城中开辟忆古怀今的好去处。

现代城市建筑风格的熔炉，将地球打造成一个世界村，很像巴别塔的困境，不同无以沟通，大同又迷失自我。公共景观建设在有限的环境条件下，去唤醒人们的地域认同感，夯实文化逐渐被淡漠的根基，实在是一桩功德。从设计者和策划人的角度，"卡戎兄弟广场"跨学科的团队整合，也是未来公共艺术创作的发展趋势，公共艺术不再囿于环境和景观的范畴，更被赋予了社会使命与文化职责，积极参与到公众的日常生活中。（吴昉）

Artwork Description

Designed by Affleck + de la Riva architects, Robert Desjardins, landscape architect, and Raphaëlle de Groot, artist, and completed in 2008, Square des Frères-Charon is located at the crossroads of two historic streets in one of the oldest sectors of Montreal.

Square des Frères-Charon is part of a network of public spaces organized along the axis of McGill Street, a historic thoroughfare that links the Old Port to the contemporary city center. The square offers the experience of a contemporary urban landscape inspired by the original vocation of the site, a prairie wetland where the Charon brothers built a windmill in the seventeenth century. The square is an experience in contrast and connection where the prairie wetland, surrounded by the city, takes on new dimensions and raises public awareness of the history and geography of the site.

The project uses a simple, refined, and minimalist architectural language to create a dialogue between circular and cylindrical forms including a garden of wild grasses, the vestiges of the windmill and a park pavilion in the form of a belvedere-folly. Complementing these gestures, the lighting scheme introduces a chromatic garden that alludes to the changing seasons.

Built as a response to the urban revitalization of a disaffected industrial sector, Square des Frères-Charon is an entirely new public amenity in a space that is more than 150 years old. The new square provides identity, civic pride, and generous outdoor areas for all-season public use.

The city organized a trans-disciplinary design team including an artist, an architect and a landscape architect, whose objective was to cross traditional disciplinary boundaries, collaborate, and converge on common ground rather than divide and distribute tasks according to habitual professional domains. The team employed a variety of communications tools to encourage and facilitate citizen involvement in the design process.

By focusing on the experience of the contemporary city and urban lifestyles, the design team explored concepts from a user's point of view and initiated a connection with the immediate surroundings. Square des Frères-Charon's street level public domain was carefully designed to insure it is comfortable, safe, and wheel chair accessible. Sustainable initiatives include the planting of local species of wild grasses, which take a significant load off the municipal irrigation system, and the use of durable Quebec granite for hard landscaping and the cladding of the park pavilion.

Considering the quality of life it provides, Square des Frères-Charon's $2,2 M construction cost was a profitable municipal investment. Affordable amenities such as an interpretive program, extensive vegetation and a recycled park pavilion allowed the project designers to control costs while providing significant benefits to residents and the local tourist industry. While modest in scale and budget, the square is an essential component of McGill Street's larger network of historic spaces and a key element in Montreal's cultural tourism branding strategy.

Artwork Excellence

Built as a response to the urban revitalization of a disaffected industrial sector, Square des Frères-Charon is an entirely new public amenity in a public space that is more than 150 years old. The new square provides identity, civic pride, and generous outdoor areas for all-season public use. Square des Frères-Charon is the result of a rich interdisciplinary collaboration and an innovative consultation process. The project innovates both in a team approach that encourages members to cross professional boundaries and in the guarantee of public input resulting from the use of novel citizen-friendly communication technologies. The city organized a transdisciplinary design team whose collaborate and converge on common ground rather than divide and distribute tasks according to habitual professional domains. The team employed a variety of communications tools to encourage and facilitate citizen involvement in the design process. Citizens became complicit in the creation of a new identity for Square des Frères-Charon.

献给堪萨斯城的歌舞女神
Terpsichore for Kansas City

艺术家：麦格斯·哈里斯（英国）和拉约尔·赫德（匈牙利）
地点：美国密苏里州堪萨斯城
推荐人：凯伦·奥尔森
Artist: Mags Harries（UK）& Lajos Héder（Hungary）
Location: Kansas City, Missouri United States
Researcher: Karen Olson

作品描述

"献给堪萨斯城的歌舞女神"为新考夫曼表演艺术中心而创作，声光结合成了一件仪器，由车库改建而成，成为了与该表演中心附加的场地。这件作品由麦格斯·哈里斯和拉约尔·赫德创作而成，延伸贯穿了艺术区停车库的整个空间，那个车库能停 1 000 辆车，四层楼高，能够服务堪萨斯城的艺术区。

"献给堪萨斯城的歌舞女神"的主要特点是——英文发音为 turp-SICK-uh-ree ，由希腊神话里的舞蹈歌唱女神的名字命名，这是一件"发光的管风琴"，一座垂直的声光雕塑穿透了四层楼高车库的楼梯间。

参观者们靠近这座发光的管风琴能够看到七个四层楼高的亚克力管子里面有编过程序的 LED 灯，上上下下地移动着，散发着白色和蓝色的光，鸣起了编排好的音乐。

当参观者们从车库的任何角落停好车后出来，他们激活了运动传感器触发了来自 112 个扬声器的音乐。当他们穿过车库，走向一个出口或主楼梯，他们注意到对于该空间来说，音乐是如此的特别，三首原创曲子特意为该作品所创，由作曲家大卫·莫尔顿、罗贝塔·瓦卡和来自堪萨斯城的波比·华生作曲。

不仅仅是作曲家创作音乐，他们也为发光的管风琴创作视觉曲子。罗贝塔瓦卡在她的音乐中使用网络摄像头，准时地在不同物体和手势之间移动，为发光的曲子制造出一种动画效果，用来伴随她的音乐。大卫·莫尔顿使用了完全不同的手法，构成光线来展示他的音乐；他在黑白图像上缓慢地移动网络摄像头。

这个软件格式由马特·哈特尔编写。不仅仅是发光管子和扬声器的排列需要被可靠地驱动，对于未来的艺术家长远地发展这项内容也是一种友好的方式。因此哈特尔创作了一种内容编辑，当作曲者实时地编写他们的项目时，这种编辑功能能够使作曲者看到安装模式，以及内容调度和专门的软件来驱动传感器、LED 灯、20 轨音频和其他硬件。

除了布线和气候控制系统，传感器阵列检测速度和运动方向贯穿整座建筑。该系统已经内置诊断和故障恢复例程，旨在保持它运行正常可靠。

哈里斯和赫德希望这个车库的体验将给过路人一种享受和一种温柔的过渡，无论对于刚刚来到这儿的人，还是刚刚在大厅里欣赏过演出的人。

同时他们希望为这个城市建立一种规划，考夫曼中心制定未来的布局安排，应该由其他艺术家们加入这个原创文库中。

解读

传统的演出总是分为开始、发展、高潮、结束，也就是所谓的起承转合。通常表演最具张力的阶段过后，就慢慢缓和直至结束，为的是使观众能在剧院中完成整出表演的心理感受，而不将这种情绪感受带出剧场。因而，作为进入表演情境与脱离表演情境的剧院外部公共空间，就承担了过渡观者情绪的使命，从现实投入演出，再从演出的虚幻中全身而退，还原真实。"献给堪萨斯城的歌舞女神"正是这种过渡的载体。

新考夫曼表演艺术中心的车库，在发光管风琴的装点下，声光电色，几乎成为第二个舞台。观众和来访者从进入车库的那一刻起，就开始酝酿起舞台的情绪，又从离开车库的过程中，舒缓激荡的情绪，逐渐平复，回到从前。观众通过公共艺术空间搭建起的桥梁，在幻境与真实间往返，最大限度地享受艺术之美。（吴昉）

Artwork Description

Terpsichore for Kansas City transforms the parking garage for the new Kauffman Center for the Performing Arts into an instrument of light and sound, a complementary venue to the performances at the Center. The artwork, by Mags Harries and Lajor Héder, extends throughout the entire space of the Arts District Parking Garage, a 1000-car, four-story garage that serves Kansas City's Arts District.

The main feature of Terpsichore—pronounced turp-SICK-uh-ree and named after the Muse of Dance and the Chorus in Greek mythology—is the "light organ," a vertical sculpture of light and sound that penetrates the four-story stairwell of the garage. Visitors approaching the light organ see white and blue light, choreographed with the music, moving up and down programmable LEDs inside seven, four-story acrylic tubes. When visitors exit their vehicles anywhere in the garage, they activate motion sensors that trigger music coming from 112 speakers. As they pass through the garage toward one of the exits or the main stairway, they notice the music is unique to the space, one of three original compositions created for the artwork by composers David Moulton, Roberta Vacca and Kansas City's Bobby Watson.

Not only did the composers create music, they also created visual compositions for the light organ. Roberta Vacca used a web camera and moved different objects and gestures in time to her music to make the animation for a light composition to accompany her music. David Moulton took a different approach to composing the light display for his music; he moved a web camera slowly over black and white graphics.

Specifications for the software, programmed by Matt Harter, were demanding. Not only did the light tubes and speaker arrays need to be driven reliably, there also had to be a user-friendly way for future artists to develop content remotely. So Harter created a content editor that allows composers to see a simulation of the installation while they compose their projects in real time, as well as a content scheduler and specialized software to drive sensors, LEDs, 20 track audio, and other hardware.

In addition to wiring and climate control systems, sensor arrays detect speed and direction of movement throughout the building. The system has built-in diagnostics and trouble-recovery routines designed to keep it running reliably.

Harries and Héder hope that the experience in the garage will provide enjoyment for passersby and a gentle transition for those coming to or who have just enjoyed a performance inside the hall. They also hope to establish a program with the city and the Kaufman Center to commission future compositions by other artists to be added to this original library.

Artwork Excellence

Terpsichore for Kansas City turned a 1000-car parking garage into a musical instrument. Terpsichore for Kansas City embraces the notion that a garage might serve a more innovative purpose than merely storing cars. Properly outfitted, a garage can become an inventive and inspiring musical and visual instrument; an experiential device for the pleasure of people moving through the garage, particularly those going to and coming from performances at the Kauffman Center for the Performing Arts; and, in its way an inspiration to draw the arts into one's life after leaving a performance.

鸟
The Birds

艺术家：麦芳维・麦克劳德（加拿大）
地点：加拿大温哥华
推荐人：绍娜・迪伊
Artist: Myfanwy Macleod（Canada）
Location: Vancouver, Canada
Researcher: Shauna Dee

作品描述

温哥华艺术家麦芳维・麦克劳德为位于东南部福溪的温哥华奥运村创作了"鸟"，该地区曾经是工业区，主要用作停车场。后来又来建造运动员居住的房屋，奥运村拥有 56 000 平方米的住宿中心，里面有 600 多间房间。整个基地按照低能量电子衍射（LEED）黄金和白金标准而设计。因此该奥运村被确认为世界上最环保的居住地。

"鸟"成型于这种对可持续发展的关注。这件作品的亮点是同时关注更轻微和更严重的方面：当一个外来物种被引入当地会造成什么样的影响，鸟类的美丽外表有时掩饰了它们对生物多样性可能造成的危害。两尊比实体更大的麻雀雕塑组成了"鸟"，鸟身上涂着小鸟的真实颜色，那个品种的鸟在 1800 年代从英格兰引入到了北美洲。从那时起，麻雀变得非常普遍，它们把很多本地物种赶了出去。麦克劳德那两尊巨型麻雀（一只雄性，另一只雌性）有 18 英尺高，当游客走近时，相形之下就显得渺小了。艺术家说把小型鸟类变得更大这个观念，是为了强调一种信息：当一个外来物种被引入，它能够严重地破坏该地区的生态系统，因此我们应该注意自然界的相互依存关系。

2010 年奥运会之后，负责 2010 年奥运会和残奥会（VANOC）的温哥华组委会把这个基地移交给了温哥华市。奥运村更名为"千年水"，并转换成了住宅，一个社区中心，小学，托儿所和零售、服务的空间。"千年水"的海滨区域是福溪海堤长廊和自行车道的一部分，能够清楚地看到科学世界和温哥华市中心。但是这个开发计划困难重重：

《加拿大艺术》（2011 年春季刊）这样评论道：

背景是一座混凝土广场，周围是昂贵公寓，能够看到繁忙市中心的天际线和连绵隐现的山脉。两座巨型鸟雕塑，一雄一雌，被建造成纪念碑式的塑像，卸掉了瓷器的外壳，降落在了温哥华。"鸟"被安放于广场的两侧，铜制爪子卷曲着，就像正要起飞，开始俯冲轰炸，或是从它们的板凳座上走下来，左右蹬脚。当广场几乎无人时，显得很荒凉，你会有种异乎寻常的感觉，好像发生了奇怪或可怕的事，鸟儿们回家要栖息过夜了。场景虽然热闹，但预示了一种不确定的未来，它们随时准备着发起攻击。

麦克劳德通过"鸟"这个雕塑对我们提出了一个警告，让我们对过去的传说和做过的蠢事以及对未来的看法进行反思。本来应该很平常的都市环境中的这一幽灵现场刺激我们把这一情绪体验深深地烙进了记忆中。

"鸟"是这个时代和地点的荒谬的、令人诧异的、危险的、带有威胁性的看守者。它们还是把我们引入童话般景观中，把现实缩小的原型。像郊外的某一片曾经显得如此巨大和恐怖的树林，儿童也许重回那里，发现它们比记忆中的小了很多。或者由于奥运村刚刚被接管，呈现出城市中的一座鬼镇，里面有空置的公寓和实体公园——理想的麦克劳德地形——他们或许会发现鸟全都飞走了，仅留下了主宰着它们生存的不可靠的记忆。

解读

王尔德的《快乐王子》中，人们仰视"像风标一样漂亮"的巨型王子雕像，小燕子依附于王子脚边，快乐王子和小燕子在寒冷中互相陪伴，不断消耗自身以救助他人。加拿大艺术家麦芳维·麦克劳德在其作品"鸟"中讲述的却是另一个故事。

两尊麻雀雕像高 18 英尺，寓意外来物种引入对当地生态可能的威胁。更为深层次的，作品探讨了人与自然界的相互依存关系，对于人类盲目的、错误的生态意识予以警示。但仅从外表而言，"鸟"并未给人以凌厉的指控感，相反，仿真的色彩和细腻的工艺，都让市民更愿意亲近它，在它脚边休息闲谈，被乌托邦的超现实幻象所吸引。这是不是意味着公共艺术在承担社会教化功能、履行公益职责的宏图目标之前，首先还是要给予人温暖的融入感和归属性，减少与当地社会及文化环境的违和感呢？

麻雀当然不是燕子，然而快乐王子也并不快乐。如果麦芳维·麦克劳德的"鸟"能让人们快乐，那么这件公共艺术作品已经成功了一大半。（吴昉）

Artwork Description

Vancouver artist Myfanwy MacLeod created The Birds for Vancouver's Olympic Village in Southeast False Creek, which previously had been a former industrial area that was once mostly parking lots. When built to house athletes, Olympic Village was a 56,000m squared accommodation centre with over 600 units. The whole site was designed to follow LEED Gold and Platinum standards. As a result, the Olympic Village was recognized as the World's Greenest Neighborhood.

The Birds was shaped by this focus on sustainability. The work highlights both the lighter and graver sides of what can happen when a non-native species is introduced to an environment, how the beauty of birds can sometimes mask their threat to biodiversity. Two larger-than-life sparrow sculptures that make up The Birds are painted with the true colors of the small bird that was introduced to North America from England in the 1800s. Since then, sparrows have become so commonplace that they have driven out many other native species. MacLeod's two giant sparrows (one male and one female) are 18 ft. tall and they dwarf visitors who come near them. The artist said the idea of making the small bird bigger was to underline the message that when a foreign species is introduced, it can adversely upset the eco-system of an area and thus we should pay attention to the interdependence of nature.

After the 2010 Olympics, the Vancouver Organizing Committee for the 2010 Olympic and Paralympic Games (VANOC) turned over the site to the City of Vancouver. The Village was then renamed "Millennium Water" and was converted into residential housing, a community centre, elementary school, daycare centres, and retail and service spaces. The Millennium Water's waterfront area is part of the False Creek seawall promenade and bike route, with a clear view of Science World and downtown Vancouver. But the development has struggled:

From Canadian Art, Spring 2011:

The backdrop is a concrete plaza surrounded by generic stacks of expensive condos with a view of the busy downtown skyline and the ever-looming mountains. The two massive birds, a male and a female, are fabricated to look like monumental figurines that have busted out of a china cabinet and descended on Vancouver. The Birds are positioned on opposite sides of the square, bronze claws curled, as if poised to take flight and start dive-bombing, or to step down from their bench pedestals and stomp around. When the square is unpopulated...and appears desolate, you get this uncanny feeling that something bizarre or terrible has happened, that the birds have come home to roost. When the scene is bustling, they're harbingers of an uncertain future, keeping watch over their charges.

With The Birds, MacLeod has created a portent for us to reflect on our past fables and follies and on our visions of the future. This spectral scene in an otherwise homogenous urban environment incites us to fix the experience to memory. The Birds are preposterous, surprising, precarious and menacing keepers of this time and place. They're also archetypes, drawing us into a fairy-tale landscape that dwarfs reality. And like the suburban patch of woods that once seemed so vast and spooky, children might return to find them so much smaller than they remembered. Or, given that the Olympic village just went into receivership, rendering the empty condos and its concrete park a ghost town in the middle of the city—ideal MacLeod terrain—they might find that the birds are gone altogether, leaving unreliable memory charged with their survival.

Artwork Excellence

Vancouver artist Myfanwy MacLeod has always been interested in the ability of art to convey political and historical meaning. The Birds, her work for Southeast False Creek Olympic Plaza, has been shaped by this new community's focus on sustainability. The work highlights both the lighter and graver sides of what can happen when a non-native species is introduced to an environment, how the beauty of birds can sometimes mask their threat to biodiversity. From the artist: "Locating this artwork in an urban plaza not only highlights what has become the 'natural' environment of the sparrow, it also reinforces the 'small' problem of introducing a foreign species and the subsequent havoc wreaked upon our ecosystems.... The Birds reminds us of our past, but it aspires to challenge the future. It is my hope that the work stimulates understanding that will lead to a greater sense of shared responsibility and caring."

何塞·福斯特的陶瓷
The Ceramics of José Fuster

艺术家：何塞·福斯特（古巴）
地点：古巴哈瓦那海玛尼塔斯居民区
推荐人：卡利·英格布里特森
Artist：José Fuster（Cuba）
Location：the Jaimanitas neighborhood, Habana, Cuba
Researcher：Cally Ingebritson

作品描述

古巴陶艺通常被称为"加勒比海的毕加索"。在哈瓦那，可以很轻易地辨别出福斯特所在的居民区，他的房子别出心裁地被装饰得好像巴塞罗那的古埃尔公园。福斯特曾在艺术教师学院学习，后来在哈瓦那的一所陶瓷工作室工作，深受西班牙艺术家毕加索和建筑家高迪作品的影响。1987年，福斯特在哈瓦那美术馆举办了自己的陶瓷展"我的城市"，获得巨大成功。

1996年，何塞·福斯特决定把自己艺术带到社区，让人们参与其中。当年他的家只是一所小木屋。福斯特开始用陶瓷马赛克覆盖每个墙面，由此开始了他的创梦之旅。随着时间的推移，他的房子被彻底改变：每堵墙，每扇门甚至房顶都被装饰成鲜亮的颜色，或被改装成了雕塑。福斯特的邻居非常热衷于他的装修活动，他们一个接一个的要求他以独特的方式用陶瓷装修自己的家。此后，福斯特开始雇佣他的邻居作为他的助手装修民居。

何塞·福斯特已在国际上获得关注，被称为"有着成熟男孩丰富多彩的想象力"的艺术家。他的杰出之处在于其作品植根于草根阶层，在社区项目中，邻居们的想法和意见成为福斯特作品的重要来源之一。在他的努力下，社区居民的审美情趣得到了提升。他说："这个项目不只是关于我一个人的房子，而是一个以社区为基础的整个地区的项目。我所有的邻居都是我的合伙人。"

解读

我们的生活是根植于传统的，陶瓷作为传统工艺，在过去的人类日常生活中发挥了至关重要的作用。但随着科技的一步步发展，很多新兴材料以其更加实用的特性和方便的制作工艺逐渐将陶瓷工艺品取代。

但是我们的生活却并没有在美感上得到多大的改变，反而精神的空虚逐渐注入日常的生活，反过头开始怀念起那千百年来就一直伴随着我们生活的传统工艺。美丽的纹饰，清脆的声音，和绝对取自自然的原材料。当房子和城市被陶瓷的记忆碎片铺满，我们的记忆似乎也被重拾。福斯特的陶瓷梦起源于哈瓦那的某所陶瓷工作室，然后就这样在长达数年的时间里，伴随着他的生活轨迹和创作轨迹逐渐传播开来。时间流逝，理想依旧，那是关于陶瓷的一个色彩斑斓的梦境，也是现代人类对于传统生活方式的回顾。（高浅）

Artwork Description

The Cuban ceramist José Fuster is often called the "Picasso of the Caribbean." It's easy to tell when you've reached his neighborhood outside La Havana, Cuba. Fuster's home workshop itself looks like it's been transported from Barcelona's Parc Güell. Pablo Picasso and the Spanish architect Antoni Gaudi have always been strong influences for him, whether during his studies at the School of Arts Education , or later while working at a ceramics studio in Havana. In 1987 Fuster held a ceramics exhibition entitled "My City" at the Havana Art Museum, which brought him great acclaim.

In 1996 Fuster decided to take his art into the community involving everyday people in his process. His creative journey began when he started covering the walls of his log cabin home with ceramic mosaic, over time completely transforming the space. Every surface including the roof became decorated in bright colors or converted into sculpture. Fuster's neighbors became interested in his renovation activities and one by one requested that he redecorate their homes in his unique style. Thereafter, Fuster began employing neighbors as assistants to help decorate homes.

Fuster has received international attention, and has been likened to "a grown up kid with a wild imagination ". His success can be attributed partly to the grassroots nature of his artwork. His neighbors' thoughts and ideas have been integral to the creative process of these community projects. Through his efforts, he has provoked greater aesthetic curiosity and sensibility in the whole community. Fuster comments, " This project isn't just for my house, it's a community-based project for the whole area. All my neighbors are my partners."

Artwork Excellence

Fuster's excellence rests in his grassroots approach to improving the aesthetics of his entire neighborhood. Through creating his "dream world" first at home, his fame grew over time and he has employed many of his neighbors in his work. He has reinvested money in Jamanitas, the neighborhood where he lives in Cuba.

无声的进化
The Silent Evolution

艺术家：杰森·德凯尔·泰勒（英国）
地点：墨西哥坎昆国家海洋公园
推荐人：凯伦·奥尔森
Artist：Jason deCaires Taylor（UK）
location：The National Marine Park of Cancun,Mexico
Researcher：Karen Olson

作品描述

这个项目是世界上规模最大的水下人工构筑物，由四百多个与真人等大的人物雕塑组成。雕塑原形来自墨西哥社会各阶层的人物，还包括全球的各种人物形象，其精致的细节具有强烈吸引力。雕塑由中性酸碱度的水泥制成，这种生态混凝土为海洋生物营造了一个盘踞和栖居的环境，从而为自然珊瑚礁减轻生存压力。

据墨西哥旅游局统计，每年有近75万名游客来这里潜水。因此，作品在壮观地把公共艺术和生态主题融合起来之际，也吸引附近给自然珊瑚礁带来巨大压力的游客。雕塑的形状和颜色因水深和观察者的位置而有所差别，时刻变动的水流也会改变人们观察到的景象。水下的雕塑没有墙壁的阻隔，没有博物馆的灯光，参观者可以通过潜水和乘坐水下游船来欣赏。在浮力与失重的感觉下，从不同的观赏角度观看，都会带来特殊的感觉，并且驱使人们进行理性的认识。

雕塑在沉入海底之后孕育了自己的生命力量，传递出人类可以通过意志和观念使自然更加强大的愿望。随着珊瑚逐渐生长和海洋生物建立起生态环境，雕塑群在外观上会发生改变，体现出艺术与自然环境之间奇妙的错综复杂的关系。提醒每个人直面各种生存问题和我们给赖以生存的环境带来的影响，也提醒人类自身和大自然互相依存的事实以及对环境的尊重。

解读

这是一场海底艺术展，占地 420 平方米的展场，就设在国家海洋公园一处光秃秃的海床上。这场名为"无声的进化"的展览是由英国雕刻家兼潜水员泰勒创作的。泰勒磨铸出 403 具真人大小的雕像，这些雕像的原形有来自墨西哥社会各阶层的人物，还有全球的各种人物，并使用特殊混凝土作为素材，希望能刺激珊瑚生长。泰勒说，他希望能"创造一个不朽的人工堡礁，一个鼓励鱼群聚居、栖息的空间，并且期望这片人工礁床能吸引游客注意、远离自然珊瑚，让自然珊瑚有足够时间修复、生长"。这次艺术展的目的与用意也在于凸显全球珊瑚礁数量的锐减。艺术家使用公共艺术与生态主题完美融合的做法，给游客提供的特色体验欣赏方式，以及给游客留下的深刻感悟和影响，吸引众多游客的关注和喜爱。（冯正龙）

Artwork Description

This art work is one of the largest man-made underwater attractions in the world, consisting of over 400 permanent life-sized sculptures. The sculptures portray people engaged in a wide variety of occupations, each rendered in exquisite detail. Casts for the sculptures were made from people representing a broad cross-section of Mexican society as well as from other parts of the world. Each sculpture is made from pH-neutral concrete, an ecologically non-disruptive material appropriate as a dwelling environment for marine organisms which otherwise depend on the highly stressed natural coral reef.

According to the Mexico Tourism Bureau statistics, the Cancun Marine Park attracts over 750,000 tourists each year. The project dramatically blends public art together with an ecological objective, drawing tourists who would otherwise bring enormous pressure to the natural coral reef. The sculptures' shape and color differs depending on the depth and position of the observer, and the ocean current is constantly producing changes in the underwater scene. The Underwater Sculpture Museum has no barrier walls, special lighting or other fixtures; visitors can freely scuba dive around the artworks or take an underwater sightseeing boat. Buoyant and weightless, visitors experience a very different way of viewing sculptural works, and as a result come away with very strong impressions and thoughts about the meaning of this artwork.

After settling on the sea floor the sculptures began to take on a life of their own. This is an example of human ingenuity and will abetting desired processes of nature. Furthermore, the project conveys the perplexing relationship between art and nature, for as coral and marine life flourishes within this ecosystem the appearance of the sculpture group will be transformed. It is a reminder of the impact we have on the environment which everyone depends upon for survival, and of the respect for the environment which our interdependent relationship demands.

boat. Buoyant and weightless, visitors experience a very different way of viewing sculptural works, and as a result come away with very strong impressions and thoughts about the meaning of this artwork.

After settling on the sea floor the sculptures began to take on a life of their own. This is an example of human ingenuity and will abetting desired processes of nature. Furthermore, the project conveys the perplexing relationship between art and nature, for as coral and marine life flourishes within this ecosystem the appearance of the sculpture group will be transformed. It is a reminder of the impact we have on the environment which everyone depends upon for survival, and of the respect for the environment which our interdependent relationship demands.

Artwork Excellence

The Silent Evolution is unique in fusing public art with an environmental conservation objective on a grand scale. British sculptor Jason deCaires Taylor placed over 400 life-size figurative works on the bottom of the sea. As one of four projects by deCaires that make up the Museo Subacuático de Arte (MUSA), which opened in 2010, The Silent Evolution is not only a human-made reef that attracts natural aquatic growth, it also draws visitors away from more fragile natural reefs nearby. Visitors can view the works, made from pH-neutral concrete, via scuba diving, snorkeling, or riding in glass-bottomed boats. The completed work is one of the largest and most ambitious underwater artificial attractions in the world, occupying an area of over 420 sq meters and with a total weight of over 180 tons.

水上空中花园
WaterSkyGarden

艺术家：珍妮特·恩特尔曼（美国）
地点：加拿大不列颠哥伦比亚省
推荐人：绍娜·迪伊
Artist: Janet Echelman（USA）
Location: RichmondBritish Columbia, Canada
Researcher: Shauna Dee

作品描述

　　"水上空中花园"是在里士满奥林匹克椭圆速滑馆（2010 年温哥华冬奥会速滑比赛的官方滑道所在地）周围的广场上改建的一个永久性的社区艺术园区。里士满市想要这个椭圆速滑馆成为城市的一道新型、再生的湖滨景观的焦点。恩科尔曼的设计考虑了体育馆周围所有的视觉元素——水、天空和散步小径——用石块、木头、水、空气泡泡喷泉、钢筋、网兜和照明设备等，建立起了一个大型的整体景观艺术。

　　红染雪松木铺就的木板路带领观者从整个艺术作品中穿过。水净化增氧器在池塘上方"吹泡泡"——实际是在收集从椭圆速滑馆 5 英亩面积的屋顶上溢出的水——而这个"泡泡"其实就是高悬的网兜，能随风飘动，在夜晚则变成空中照明灯笼。

红色木板路和灯笼式屋顶的构思来自这座城市的文化底蕴。里士满是加拿大移民人口比例最高的城市，而其中大部分又是亚裔。所以这些曲曲折折的红色木板路看上去就像是传统中国春节的"舞龙"的化身。温哥华的日本新渡户花园（Nitobe Japanese garden）和中国中山公园（Sun Yat Sen Chinese garden）是重要的代表例子，尤其是他们因地制宜的材料、交叉的小径和有倒影的池塘以及田园风光。"水上空中花园"则是一个想吸引观者更多逗留的一个深思熟虑的艺术园区。顶头上的网状物提供了一种全新的视觉体验，是将艺术放到天空中；而在晚上它们就像灯笼一样飘着。网兜的构想与此地有一种特殊的联系，此地区也有着源远流长的渔业和罐头制造业历史，当地的玛斯昆族人（Musqueam Band）至今都在教他们的后代在菲沙河（Fraser River）用这种特殊弧度的网兜来捕鱼，并雇佣了许多当地少数民族的人。

这个项目由恩科尔曼的团队共同完成，包括曾在国际上获奖的建筑师、工程师、照明与治水顾问、景观建筑师以及操作师等。

解读

东西方文化大冲撞的近几十年中，文化多数时候被用做一种符号去提升公众认知度，很多"混血"的文化形态得以展现在世界各地的艺术作品中，产生的新符号及新文化现象，完全符合格式塔心理学的1+1＞2模式，其中，珍妮特·恩特尔曼的"水上空中花园"就是一个典型案例。

在这座人为的空中花园里，红色、灯笼、之字形的红色木板桥象征舞龙，历历昭告天下，中国来了，中国气象来了。但"水上空中花园"的成功之处，恰是其不仅仅停留在对异域文化元素的铺排上。红色网兜的构想代表了当地历史悠久的渔业和罐头制造业历史，当地人时至今日，仍然用这种有着独特而优美弧线的网兜来捕鱼。异文化符号与本地文化元素相融合，互相借鉴，借形共形，从而产生新一种有别于任何母体文化的视觉形态，这才是文化衍生的奥秘，也是类似公共艺术作品的意趣所在。"水上空中花园"，无疑切中肯綮。（吴昉）

Artwork Description

"Water Sky Garden" transforms the plaza surrounding the Richmond Olympic Oval, official venue for the 2010 Vancouver Olympic Winter Games speed-skating events, into a permanent art environment for the community. The City of Richmond intends for the Oval to become the centerpiece of the new, revitalized urban waterfront section of the city. Echelman's design engages the space all around the viewer—water, sky, and pedestrian pathways—to create an immersive whole using rock, wood, water, air bubble fountains, steel, netting, and light.

Red-stained cedar boardwalks lead visitors through the artwork. Water purifying aerators draw shapes with bubbles on the surface of a pond that collects runoff water from the Ovals' 5-acre roof, while suspended net-forms undulate overhead in the wind, becoming sky lanterns during nighttime illumination.

The red boardwalk and sky lanterns are inspired by the city's cultural communities. Richmond has the largest immigrant population by proportion of any city in Canada with the majority of those immigrants being of Asian descent. The wooden boardwalk follows a curving path similar to the choreography of the Dragon Dance, a performance frequently seen in local Chinese festivals. The Nitobe Japanese garden and the Sun Yat Sen Chinese garden of the Vancouver region are important references, especially their material presence, intersecting paths and reflective ponds, and their framing of views. "Water Sky Garden" is a contemplative art environment that encourages participants to linger. The overhead netted forms provide a new visual experience, putting art in the sky; at night they glow like lanterns. Nets have a special relationship with the site, as the native Musqueam Band continue to teach their children to fish using nets at this particular bend in the Fraser River to this day, and this area has a history of the fishing/canning industry which employed many ethnic groups.

This project was achieved through Echelman's collaboration with a team of international award-winning architects, engineers, lighting and water consultants, landscape architects, and fabricators.

Artwork Excellence

The Water Sky Garden is situated at the northeast corner of the Richmond Olympic Oval. The purpose of the work is to encourage people to spend time in the shared public space and connect the main roads to the architecture and the waterfront as the city is re-establishing its connection to the Fraser River. The designs of the boardwalk and the artwork reflect the diverse cultural communities of Richmond, which has the largest immigrant population by proportion of any city in Canada, the majority being of Asian descent. Not only does the artwork reflect Asian cultures, the netting component evokes the tradition of fishing the Fraser River using nets that the native Musqeum Band continues to pass on to their children. The water gardens collect the runoff from the building's 5-acre roof, providing water quality treatment and water storage for irrigation. It provides a natural wetland habitat for birds, mammals, and aquatic life and creates an authentic native wetland garden experience for visitors.

沃兹房屋计划
Watts House Project

艺术家：里克·罗威（美国）
地点：美国洛杉矶
推荐人：绍娜·迪伊
Artist: Rick Lowe（USA）
Location: Los AngelesCA, United States
Researcher: Shauna Dee

作品描述

这是一件通过邻里重建项目而合作成型的艺术作品，沃兹房屋项目与艺术、建筑、可持续发展的社区联姻，随着大量艺术观光客的涌入参观传说中的沃兹塔，而重新定义了邻里的意义。

频繁地有观光客们参观赛门·罗地亚那奇怪、不可思议的钢塔，而社区并没有获得很多收益，做这个项目的意义在于弥合塔、艺术中心和社区之间的间隙。

里克·罗威把沃兹房屋项目基于他休斯敦的项目上，他是这样形容这个一排房子的项目：

从事社会或社区工作从来就不是呆滞的。这和绝大多数艺术品相比都是完全不同的，该作品的物质和社会关系都被恰当地定义过。有时你建造一个结构时有机会可以长期可持续下去。

作品将不再成为一次性产品：一群艺术家在这个消沉的邻里间做一些装置，博得大众的关注，并且可以走出去。

但是民众们感兴趣于贡献自己的特长发展这个项目，使之成为一个正在进行的活动。因此形成了一个组织。

通过反思和对话，沃兹房屋项目植根于罗威的信念。作为一名艺术家置身于居住方案中，能够提升生活的品质和扩大地方的意义。"艺术家，"罗威说，"在发展中加入一种不同的分层。"因此这个项目始于艺术家的住宅。

但是罗威开始理解到有人需要在洛杉矶掌舵项目，因为他本身就是当地人，能够更好地理解它。那个人就是洛杉矶艺术家埃德加·阿森诺。罗威在洛杉矶奠定好了基础工作后回到了休斯敦。在沃兹，他把主控权交接给了阿森诺，阿森诺留下来成为了项目的执行主管。

实际上，沃兹房屋项目（WHP）寻求制定邻里家庭共享的愿景，通过以下几项来实现：

关注于现有住房的翻新，包括通过艺术家和建筑师来改善小区景观。

利用现有城市范围内的资源促进伙伴关系，在项目区满足家庭的需要。

减少生态足迹，区域物业的维护和运营成本；以及在沃兹塔近邻地区吸引再投资。

但是这个项目主要是受细微之处和概观中得到、失去了什么而定义。一天之间洛杉矶部分的故事，很少有机会为自身而定义自身——更不用说世界了。

为此，阿森诺把邻里街坊周边居民的故事等历史一起合拢起来，将艺术家与建筑师一起配对，体现了创意与实用性。"与使用黏土相反，我们用时间和空间建构了一个邻里和关系。"

在 2011 年 9 月，WHP 收到了"艺术地方"的补助金，在附近的一个原始结构上翻新平台。这个平台起着集线器一样的作用，也是一系列项目和社区活动运作的基地。这次全面的翻修将会提供给沃兹塔社区丰富的服务。

目前的日程安排包括家庭文化艺术和教育暑期系列，沃兹艺术讲座教室系列和基层就业培训。这同样是 WHP 即将迎来的房屋装修的管理和操作基地：爱心之家、马德里加尔楼和加西亚楼。在交付方案和计划操作该项目时，这个翻新的平台将确保一个安全、卫生和审美的氛围。

解读

"邻国相望，鸡犬之声相闻，民至老死，不相往来。"老子的"小国寡民"意在民风淳朴，而今时今日，人类在时代激进的发展变化中倍感孤独，交流、沟通、整合，成为新世纪文化的重点，赋予了设计"以人为本"与时俱进的新释义。

里克·罗威的"沃兹房屋项目"，通过邻里间同心协力去完成一项公共作品，而深得人心。这个项目的持续性使作品不再是一次性的终结，不是完成时而是永远的进行时态，具备了无限发展的可能性。公众和邻里也愿意在这项公开的活动中，奉献自己的特长和才智，在此过程中，重新找到自身的归属和认同感。

"沃兹房屋项目"在某种意义上更像"黏合剂"，黏合了旧宅与新屋、民居与社区、生活与艺术，以及最为关键的——人与人。（吴昉）

Artwork Description

A collaborative artwork in the shape of a neighborhood redevelopment project, the Watts House Project marries art, architecture, and sustainable community development to redefine a neighborhood that receives an influx of art tourists visiting the fabled Watts Towers. Because the community hasn't benefited much from these frequent visits to Simon Rodia's strange and incredible steel towers, the idea was to bridge the gap between the Towers, the Art Center, and the community.

Rick Lowe based the Watts House Project on his Houston project, Project Row Houses, which he describes this way: Socially engaged or community work is never stagnant. This is very different than most traditional artwork, where physical and social relationships are pretty defined. Sometimes there are opportunities for long-term sustainability in which you build structures... It could have stopped as a one-shot thing: a group of artists doing some installation stuff in this depressed neighborhood, calling attention to it, and walking away. But folks got interested in contributing their expertise to develop the project into an ongoing activity. It became an organization.

The Watts House Project was rooted in Lowe's belief that through reflection and dialogue, an artist-in-residency program can elevate the quality of life and amplify the meaning of place. "Artists," says Lowe, "add a different kind of layering to 'development.' " So the program started with artist residencies.

But Lowe came to understand that someone needed to be in L.A. to anchor the project, someone from there who understood it. That person was L. A. artist Edgar Arceneaux. Lowe went back to Houston after having laid the groundwork for the L.A. effort. He handed the reins in Watts to Arceneaux, who remains the project's executive director.

Practically, The Watts House Project (WHP) seeks to enact the shared visions of neighborhood families by:-Focusing on the renovation of existing housing, including the improvement of residential landscaping by partnering families with artists and architects-Facilitating partnerships with existing city-wide resources to meet the needs of families in the project area-Reducing the ecological footprint, maintenance and operating costs of area properties; and-Attracting reinvestment in the Watts Towers neighborhood area

But the project is mostly defined by nuances, by what gets lost in the overview: the day-in, day-out stories of a section of L. A. that has seldom had a chance to define itself for itself—let alone for the world. To this end, Arceneaux folds together the history of the neighborhood alongside the stories of its residents, pairing artists with architects, creativity with practicality. "Instead of using clay, we're using time and space to sculpt a neighborhood and relationships."

In September 2011, WHP received a $370,000 ArtPlace grant to renovate The Platform, one of the original structures in the neighborhood. The Platform serves as the hub and base of operations for a number of projects and community programs. Its full renovation will provide a wealth of services to the Watts Tower community. The current programs include a Family Arts and Education Summer Series, Watts Arts Lecture Classroom Series, and Grassroots Job Training. It is also the management and operation base for WHP's upcoming housing renovations: the Love House, Madrigal House, and the Garcia House. The Platform's renovation will assure a safe, sanitary and aesthetic atmosphere for the delivery of programs and the planning and operations of projects.

Artwork Excellence

A collaborative artwork in the shape of a neighborhood redevelopment project, the Watts House Project marries art, architecture, and sustainable community development to redefine a neighborhood. A testament to the excellence of the Watts House Project is the way it became a nonprofit organization to carry on the legacy of its artist founder, Rick Lowe. Lowe understood that it needed a leader who understood the Watts neighborhood and who could be there to facilitate the community development. Edgar Arceneaux was that person, and he remains the executive director. WHP does not dictate a vision for the neighborhood, but sees itself as a facilitator, mediating conversation among creative professionals, families, and local stakeholders. WHP is a process for working with many disparate groups, and is primarily a connector and facilitator. Arceneaux's essential tools are volunteers, designers, residents and people who can be described only as neighborhood griots who have witnessed Watts' changes—people like Felix Madrigal, who has lived in the neighborhood for more than 20 years, and artist John Outerbridge, who served as the director of the Watts Towers Art Center for more than 30 years. What Outerbridge finds most noteworthy is that this endeavor honors both the history of the people and that of these hand-crafted houses. "Up until now, the story of gentrification and architectural philosophy [in Watts] didn't include this." Arceneaux's project broadens the concept of both art and its process. "It demonstrates that art has the audacity to be anything that it needs to be."

波形
Wave Forms

艺术家：丹尼斯·欧本海姆（美国）
地点：美国宾夕法尼亚州费城
推荐人：米娜·曼噶尔韦德海克尔
Artist: Dennis Oppenheim（USA）
Location: Philadelphia, Pennsylvania, United States
Researcher: Meena Mangalvedhekar

作品描述

"波形"是丹尼斯·欧本海姆对于费城的著名地标——自由钟的俏皮性诠释。在这个案例中，六只房屋大小的钟铃聚集在一起，钟铃均由铝管制造，把这些铝管穿孔在一起，放置在露天庭院里，毗邻宾夕法尼亚大学附近的一幢新建公寓楼。在这件作品的构思过程中，艺术家与景观建筑师莎拉·佩斯切尔一起进行了咨询讨论。该作品的设置包括了一个沉重的花岗岩基座和生长在雕塑上的藤蔓。

"以声波中的视觉配置构建了广场花园，我可以避开由"自由钟"——费城最有名的形象——所引起的很明显的联想。通过增大钟铃的尺寸以近似于建筑物，我能够干扰作为'物体'的钟铃和作为'住宅'的钟铃之间的联系。"欧本海姆告诉《美国雕塑家》杂志。

这个作品向公众开放，观众可以在钟铃下面、四周或中间散步（但禁止攀爬）。钟铃下面的花岗岩上有着声波图案，在钟铃下面交错碰撞。雕塑的造型很像那口著名大钟的形状，但欧本海姆重新设计了这个标志性的符号，他加入了建筑元素，包括多加了窗户，提高了透明度。钟铃在夜晚有照明。

欧本海姆钟铃的概念使用了现代材料和理念，将观众引入一种当代的宣言，关注如何最终解释和使用历史。钟铃的造型和声波模式能引起对历史与现代、自由和围栏、沉默和演讲的沉思。艺术家带来了有关住所和有机的、不可预测的自然变迁的观念。

该项目由费城"% 为艺术计划"负责建造，毗邻价值 7 100 万美元的高层豪华公寓。这个花费 120 万美元的项目由汉诺威公司出资，在费城的重建公共艺术计划的指导下进行。斯劳特基金会于 2006 年春季展出了一个名为"波形：费城栗子大街 3401 号计划书"的公共展览。该作品由加州太阳谷拉帕洛玛美术进行施工、运送和安装。

解读

美国费城的自由钟，上刻《圣经》名言："向世界所有的人们宣告自由。"不仅是费城的象征，更是美国精神不朽的传奇。当艺术家丹尼斯·欧本海姆选择钟形为其"波形"的主体形象时，可想而知，成功与压力将如影随形。

借形于自由钟，本身就是费城的地标象征，很轻易就能让当地的市民产生认同与好感。然而对于公众知名形象物的套用，也很可能直接导致人们的轻视，从而掩盖或忽略作品的一些良苦用心。丹尼斯的解决之道在于将建筑元素引入主体的设计，通透的空间，消弭彼此的距离与隔膜；微启的天窗，为金属骨架增添人情的趣味。"波形"让费城的自由钟不再是遥远落日下的精神符号，转而成为人们可以穿越其中、乐享生活的自由地。（吴昉）

Artwork Description

Wave Forms is Dennis Oppenheim's playful reinterpretation of Philadelphia's famous landmark, the Liberty Bell. In this case, the six house-sized bells are grouped in clusters and constructed from aluminum tubes and perforated aluminum in open-air courtyards adjacent to a new apartment complex at the University of Pennsylvania. The work was conceived in consultation with landscape architect Sara Peschel. The setting includes a heavy granite base and vines that grow over the sculpture.

"By taking the visual configuration found in sound waves to structure the plaza garden, I could detour from the more obvious associations brought forth by the Liberty Bell, Philadelphia's most famous image.

By increasing the scale of the bells to approximate architecture, I was able to disturb the association of the bell as 'object' to bell as 'dwelling,' " Oppenheim told American Sculptor. The works are accessible to the public, and spectators can walk under, around, and through the bell forms. (Climbing, however, is not allowed.) The granite is patterned after sound waves that crisscross and collide under the bells. The sculptural forms resemble the famous bell's shape but Oppenheim reinvented the iconic symbol with inserted architectural elements including windows, promoting transparency. The bells are illuminated at night.

The conceptual bells of Oppenheim, using modern materials and ideas, brings to the viewer a contemporary statement regarding how one interprets and ultimately uses history. The bell forms and sound wave patterns evoke history and modernity, freedom and enclosure, silence and speech. The artist brings forth the ideas of dwelling place and the organic, unpredictable nature of change.

The project was built under Philadelphia's percent-for-art scheme, connected to the $71 million adjacent luxury apartment high rise. The $1.2 million project was sponsored by the Hanover Company under the guidelines of Philadelphia's Redevelopment Authority Public Art Program. The Slought Foundation presented a public exhibit titled: "Wave Forms: Proposal for 3401 Chestnut St., Philadelphia" in the spring of 2006. The work was engineered, transported and installed by La Paloma Fine Art of Sun Valley, California.

Artwork Excellence

Oppenheim's piece is a reworking of the familiar Liberty Bell theme, ubiquitous to Philadelphians. In this case, he's reinvested the symbol with new meaning, literally etching the sounds of liberty in stone, while building a "house of liberty," integrated with nature, and scaled for human interaction.

维恩伍德艺术区协会
Wynwood Arts District Association

艺术家：大卫·隆巴迪（美国）
地点：美国佛罗里达州迈阿密
推荐人：绍娜·迪伊
Artist: David Lombardi and Tony Goldman (USA)
Location: Miami , FL , United States
Researcher: Shauna Dee

作品描述

维恩伍德艺术区是大迈阿密的艺术中心，那里有 50 多家画廊、4 家博物馆和主要收藏，以及更多的机构。它坐落于设计区南部，I-95 和比斯坎大道，从第 20 至第 36 街。

维恩伍德艺术区协会的任务是促进维恩伍德成为大迈阿密的艺术和创意行业的中心，同时加强工厂和仓库区的特色。此外，维恩伍德艺术区协会致力于给行人、住户和访客提供一个清洁的、安全的、丰富的环境——并且继续改造邻里。

维恩伍德，一个最容易被忽视的内陆城市，街坊由仓库工厂和特卖店组成，其中很多店和仓库都是长期关闭的，里面没有留下什么东西，除了幽灵之外。20 世纪七八十年代纽约的 SOHO 和东村，或是如今洛杉矶市中心，维恩伍德是一个可供艺术家和画廊做事的地方，这个空间非常适合他们，那里曾经是历史性扩张的地区。他们甚少知道关于他们所做的事不仅仅会重建整个区域，更会重塑迈阿密的面貌。

在这种精神的驱使下，维恩伍德艺术区协会 (WADA) 应运而生了。由画廊、收藏家策展人，博物馆头领和艺术家组成，WADA 事实上充当着这个地区的管家角色，如今驻扎了 50 多个画廊，四个博物馆，数量众多的复杂艺术和艺术家工作室，以及逐年增加的卫星般的展会，隶属于巴塞尔艺博会。来自这个协会每月的第二个周六的艺术漫步到繁华的咖啡文化，这是一个活跃的街区，众多活动都是由维恩伍德艺术区协会发起的。

街道壁画由知名街头艺术家创作，从 2009 年起壁画开始覆盖满维恩伍德的墙壁，这些墙壁犹如画布一样，使更多的艺术家参与进该项目中，托尼古德曼在 2010 年打开了维恩伍德的大门，卷起了店面那 176 英尺高的大门。维恩伍德墙引来了世界上最棒的艺术家们来到迈阿密创作涂鸦和各类街头艺术。杰弗里戴奇共同策划了 2009 年的项目，这是该项目开展的第一年，非常成功。

解读

老厂区改造利用，国内外都已有比较成熟的经验，德国的鲁尔工业区、美国的 LOFT 风格、北京的 789、上海的 1933 等，都是相对成功的范例。对于仓储式办公区、工场艺术园区的规划和建设，以及其经营之路，一直都是城市建设课题中值得探索的话题。

大卫·隆巴迪的维恩伍德艺术区协会项目，期图通过老场区的改建和艺术家的驻扎，重塑迈阿密的面貌。在艺术家的努力下，老墙面成为新画布，废弃的场地重焕光彩，不断有新锐杰出的艺术家涌入艺术区。维恩伍德艺术区协会的改建在规划中注重了综合性、可持续性和广泛性，为前来此处的艺术家提供良好的人文环境和创作空间，为进驻的商家提供可靠的商业运营机制和政府支持，更为当地居民创造出良好的公共空间，使其成为人群聚拢的时尚之地。这种"双赢"乃至"多赢"的建造模式，值得我们借鉴。（冯正龙）

Artwork Description

The Wynwood Arts District is the epicenter of the arts in Greater Miami, home to over 50 art galleries, 4 museums and major collections, and more. It's nestled between I-95 and Biscayne Boulevard, from 20 to 36th street, south of the Design District. The Wynwood Arts District Association's mission is to promote Wynwood as the epicenter of the arts and creative businesses in Greater Miami while enhancing the character of the factory and warehouse district. Furthermore, the Wynwood Arts District Association is committed to providing a clean, safe and enriching environment for pedestrians, occupants and visitors—and continuing to transform the neighborhood.

Wynwood, a mostly forgotten inner city neighborhood made up of warehouses and factories and wholesale outlets, many of which had been long closed and left in their wake nothing but ghosts. Like New York's SOHO and East Village back in the '70s and '80s, or the Downtown L.A. of today, Wynwood was a place where artists and gallerists could do what they are driven to do, as they saw fit, in spaces that were at once historical and expansive. Little did these driven folks know though that what they did would not only remake an entire district; it would help to remake the image of Miami itself.

It is in that spirit that the Wynwood Arts District Association (WADA) was formed. Comprised of gallerists, collection curators, museum heads and artists, WADA is the de facto steward of a District that is now home to over 50 galleries, four museums, numerous arts complexes and artist studios, as well as an increasing array of satellite fairs affiliated with Art Basel. From its monthly Second Saturday Art Walks to its bustling café culture, this is a neighborhood alive with action, and much of that action is guided by the Wynwood Arts District Association.

Murals by renowned street artists have covered the walls of the Wynwood Walls complex since 2009, and to create more canvases and bring more artists to the project, Tony Goldman opened the Wynwood Doors in 2010 with 176 feet of roll-up storefront gates. The Wynwood Walls has brought the world's greatest artists working in the graffiti and street art genre to Miami. Jeffrey Deitch co-curated the first successful year of the project in 2009.

Artwork Excellence

Before it became an arts district, Wynwood was a mostly forgotten inner city neighborhood made up of warehouses and factories and wholesale outlets, many of which had long been closed and left in their wake nothing but ghosts. Like New York's SOHO and East Village back in the '70s and '80s, or the Downtown L.A. of today, Wynwood was a place where artists and gallerists could do what they are driven to do, as they saw fit, in spaces that were at once historical and expansive. Not only did the artists remake an entire district; they helped to remake the image of Miami itself. From that platform the Wynwood Arts District Association was formed to become the stewards of the district.

西佩·托堤克
Xipe Totec

艺术家：托马斯·格拉斯福德（美国）
地点：墨西哥墨西哥城
推荐人：莱恩·柏格森
Artist: Thomas Glassford（USA）
Location: Mexico City, Mexico
Researcher: Laine Bergeson

作品描述

白天，格拉斯福德的纪念碑式的作品"西佩·托堤克"是一件相对而言无形的 LED 灯组成的网络，编织在特拉特洛尔科广场高楼的大理石外墙上——这幢大楼是一件现代的里程碑，于 1963 年由佩德罗·拉米雷斯·巴斯科斯设计，这幢房子便是墨西哥外交部所在地。夜晚，它看上去是一幢视觉错综复杂的，由蓝色和红色的灯光组成的一层层铺在大楼上的肌肤构成的建筑；LED 灯散发的光辉横贯特拉特洛尔科广场，从视觉上增强了整个墨西哥城的天际线。

LED 灯基于类晶体、非周期性的平铺和配置，遵循了几何图案，按照定义来说缺乏对称性。虽然它超出了这个复杂的几何美感，但格拉斯福德着迷于文化推定，成型于最近对于这些造型的研究。1970 年代，作为一种抽象的数学难题，西方科学家探索了这些造型，但随后类晶体的结构被发现存在于自然界中，正是原子的晶体结构。但是该故事更加的复杂：在 2007 年，一位哈佛大学物理学博士生在去现代伊朗的旅行中发现了 15 世纪波斯建筑也有相同的图案。中世纪使用如此复杂的，西方要很久以后才知道的造型，这一事实困扰着科学界。文化偏见阻止他们承认在五百年前波斯人就已经知道了这种图案。如今《西佩·托堤克》加入了这个古老的数学和审美传统；这件作品是跨越悠久年代之后对于数学、哲学和美的持续对话的一部分。

"西佩·托堤克"的创作是为纪念墨西哥国立大学（如今简称为：UNAM）一百周年，特拉特洛尔科广场上的外交部大楼如今属于该大学所有。在阿兹特克时期，该广场是一座重要的城市和市场中心，后来成为墨西哥山谷中最重要的考古遗址之一。阿兹特克人建造的庙宇遍布整个广场。西班牙人 1530 年代到达这里把大量庙宇夷为平地，他们在此地建造了圣地亚哥·阿波斯托尔教堂和修道院。圣地亚哥·阿波斯托尔至今耸立在广场上，毗邻残存的两座阿兹特克庙宇旁。特拉特洛尔科是 1968 年 44 名抗议学生被狙击手杀害的地方，时间就在墨西哥城举办夏季奥运会前夕。

阿兹特克神灵"西佩·托堤克"与本件艺术品同名是很贴切的。这位神灵被人熟知为"剥皮之王"或"夜晚的饮酒者"，"西佩·托堤克"把自己的皮剥掉以喂养人类，这种行为类似于把玉米的外壳剥掉让其发芽。

格拉斯福德希望他的现代"西佩·托堤克"为特拉特洛尔科广场注入新活力，使整个墨西哥城的天际线发生了变革性的效果。

解读

阿兹特克是 14 世纪至 16 世纪的墨西哥古文明，阿兹特克文明与印加文明、玛雅文明并称为中南美三大文明，托马斯·格拉斯福德以阿兹特克神灵西佩·托堤克命名其灯光装置艺术，在传统的古文明与现代的光艺术间游移。建筑直线条的俊朗和色彩跳脱的不对称几何灯管，在视觉心理上产生某种极具内在张力的美。

白天的"西佩·托堤克"是具有隐藏性的，正如神灵西佩·托堤克"夜晚饮酒者"的身份，只有到了夜间，才散发出直逼人心的夺目妖娆。有趣的是，"西佩·托堤克"的发光几何形状来源于自然界的晶体结构，那种无懈可击数学般的精确之美，竟然与光艺术完美结合，在现代感的耀眼光芒中，闪耀出古老传承的神秘与优雅。这似乎也是另一种形式的纪念，纪念当地的悠久历史与重大事件，告诉人们：从哪里来，到哪里去。（吴昉）

Artwork Description

By day, Glassford's monumental work Xipe Totec is a relatively invisible web of LED lights woven across the marble façade of a tall building in Tlatelolco plaza—a modernist landmark that was designed in 1963 by Pedro Ramírez Vásquez to house Mexico's Ministry of Foreign Affairs. By night, it is a visually intricate architectural "skin" of blue and red lights layered over the building; the LEDs glow across Tlatelolco plaza and enhance the whole Mexico-city skyline.

The LEDs follow a geometric pattern that is based on quasicrystals and aperiodic tiling, configurations that by definition lack symmetry. Beyond the aesthetic beauty of this intricate geometry, however, Glassford was fascinated by the cultural presumptions that shaped recent studies of these forms. In the 1970s, Western scientists "discovered" these forms as an abstract mathematical conundrum, but then quasicrystal formations were found to exist in nature as the crystalline substructures of atoms. But the story is more complicated: in 2007, a Harvard doctoral student specializing in this field of physics recognized the same patterns in fifteenth-century Persian architecture while traveling through modern Iran. The medieval use of such complex forms, long unknown in the West, baffled the scientific community. Cultural biases prevented them from admitting that the Persians had figured it all out half a millennium before. Now Xipe Totec joins this ancient mathematical and aesthetic tradition; the work is part of a continuing conversation through the ages about math, philosophy, and beauty.

Xipe Totec was commissioned to commemorate the centennial of the National University of Mexico (today known as the UNAM), which now owns the Foreign Ministry building in Tlatelolco plaza. The plaza was an important city and market center during the Aztec period, and remains one of the most important archeological sites in the Valley of Mexico. The Aztec built temples throughout the Plaza. When the Spanish arrived in the 1530s, they razed many of the temples and built the church and monastery of Santiago Apóstol. Santiago Apóstol still stands in the square adjacent to two of the existing Aztec temples. Tlatelolco was also where 44 student protestors were killed buy a sniper in 1968, just before Mexico City hosted the summer Olympics.

Appropriately, the Aztec deity Xipe Totec is artwork's namesake. Also known as "Our Lord the Flayed One" or "Drinker of the Night," Xipe Totec skinned himself in order to feed humanity, an act akin to maize shedding its outer layer in order to germinate. Glassford hopes that his modern-day Xipe Totec rejuvenates Tlatelolco plaza and has a transformative effect on the whole Mexico City skyline.

Artwork Excellence

In this monumental work, Glassford has helped to reframe as beautiful a part of the city that has been defined and defiled by conquerors and rocked by senseless violence. The patterned lights across the building invoke a "skin" that seems to make the building—and the plaza that shines in its reflected light—glow with new and positive energy. Glassford is using public art to bring new life to an area with a complicated history. The title of the work—Xipe Totec—refers to a deity in Aztec mythology who represents fertility and sacrifices, and who sheds his skin to feed humanity. Glassford hopes that, like the Aztec god, his monumental work will help feed the soul of new generation. Glassford has sheathed Tlatelolco in a new skin: its capillaries glow to commemorate a new life as a cultural center—a beacon visible from any vantage point of the Valley of Mexico.

中南美洲地区提名作品
CENTRAL AND SOUTH AMERICA NOMINATED CASES

16 吨
16 Tons

奥摩罗（小山）
O Morro (The Hill)

汽车主题公园
AutoTheme Park

潘恩纪念馆
Paine Memorial

钻石小种子
Diamante de Semillitas

大豆
Que Soy

家具装置
Furniture Installation

萨尔瓦多壁画项目
Salvador Mural Project

雅尔米里亚姆艺术俱乐部
JAMAC (Jardim Miriam Art Club)

考卡山谷省议会代表
The Deputies of the Assembly of
Valle del Cauca

Inhotim 研究所
Instituto Inhotim

提乌纳的堡垒文化公园
Tiuna el Fuerte Cultural Park

活着的历史
Living History

回忆·壁画
Memory Murals

占用氧气
Occupation Oxygen

16 吨
16 Tons

艺术家：赛斯·乌尔辛（美国）
地点：阿根廷布宜诺斯艾里斯
推荐人：莱恩·柏格森
Artist: Seth Wulsin (USA)
Location: Buenos Aires, Argentina
Researcher: Laine Bergeson

作品描述

　　"16 吨"位于阿根廷不宜诺斯艾里斯，艺术家赛斯·乌尔辛和一个团队花了 5 周时间打破了 22 层卡塞罗斯监狱大楼的窗户。他选择用窗户制造了脸部图像（的出现），剩下的窗格在阳光的照射下成像。这件作品包含了 48 张脸，跨越了 18 层，在每天不同时段从监狱窗口电网上闪耀着光芒。在未来一年半的时间内，这幢大楼连同这些图像一起一层又一层地被拆毁。监狱的窗户电网——每个 17 英尺长（5.2 米），9 英尺（2.7 米）宽，由 209 个圆形，每个直径 8 英寸（203 毫米）的半透明窗户组成，用作像素画屏幕。脸部完全体现了空间和光的功能——监狱内漆黑的内部空间，闪耀的光芒透过光的反射显示出了剩余窗户的空间感。图像设计成从大街上不同的位置看过去都是可见的，根据地球中太阳的相对位置导致的不同时间段和季节的变化而定。

全年太阳上升在天空中变化，观看的视角就发生了改变。在一个基本的水平上，布宜诺斯艾利斯政府下令拆除监狱，由阿根廷军队执行，这是本件艺术品的萌芽。这幢大楼预计在 2001 年拆迁，但是实施的进程受到了不同法律、环境和官僚的阻碍。原先计划对该大楼实行三个阶段的爆破。但是爆破工程在最后一秒被停了下来，因为一群居民担心这可能对环境造成破坏，包括石棉中毒，监狱地下通道内可能会逃出来上百万只老鼠。

相反地，卡塞罗斯监狱被实施了机械拆除，从楼顶到楼底逐层地拆除。这个项目联系着阿根廷的历史。"出现"是西班牙动词"出现"的过去分词。这个词的第二层意思是幽灵或鬼。它用倾斜的方式涉及阿根廷肮脏的战争，政治犯在飞机上，飞机驶向大西洋，接着犯人们就被抛出了机舱。这些政治犯便是所谓的"失踪者"。

"16 吨"是创作于 1940 年代晚期的流行歌曲的名字。它指的是一天内煤矿工人的预计煤产量，但在本项目的背景下涉及的是通过装置，或解除装置，这些过程下，打破玻璃的数量。

制造这类"出现"的能见度依赖于每日、每月月亮和太阳的年度周期，地面上观众的位置的不同，根据更大的宇宙周期运动，乌尔辛连接着这个监狱丑陋的历史。通过拆除工作，大楼消失了，向观众暴露出了隐藏维度的地方和自然历史。

解读

"16 吨"指向五周时间内，打破 22 层卡塞罗斯监狱大楼的窗户，浮现的人脸在一层层地拆除后，终将消失在布宜诺斯艾利斯上方蔚蓝的天空中，成为一个阴暗潮湿的旧梦，慢慢在阳光下散尽所有的罪恶。

赛斯·乌尔辛的公共艺术，在一种不断推进的完成过程中，让公众直面洗涤丑恶的艰辛，一如分娩的阵痛，每一层的拆毁，都杂糅着疼痛与希望。布宜诺斯艾利斯有监狱历史文物馆，卡塞罗斯监狱的一切都将成为过往，是为了忘却的纪念，也是为了更和平的明天。"16 吨"的沉重包含着历史的沉重，是以公共艺术的形式发出对战争的谴责，而作品的本身却并不令人生厌。这也正是赛斯·乌尔辛的杰出之处，以暴易暴、以恶制恶，或许适用于统治和战场，却终究不适用于艺术。（吴昉）

Artwork Description

For 16 Tons and Aparecidos, artist Seth Wulsin and a team spent five weeks breaking out windows in the 22-story Caseros Prison building in Buenos Aires, Argentina. His selection of windows created the images of faces (aparecidos), which appeared when sun reflected in the remaining panes. The work included 48 faces spanning 18 stories, which shined out from the prison's window grids at varying times of day. Over the course of the next year and a half, the building was demolished floor by floor, and with it the images.

The prison's window grids—each one 17-feet (5.2 m) tall and 9-feet (2.7 m) wide, and comprised of 209 circular, semi-opaque windows each eight inches (203 mm) in diameter—served as pixelated screens. The faces were completely a function of space and light—the dark interior space of the prison, and the light shining through the optically reflective space of the remaining windows. The images were engineered to be visible from varying positions in the street depending on time of day and according to seasonal changes in the sun's relative position to the earth. The viewing angles change throughout the year as the sun's elevation in the sky changes.

On a basic level, the demolition of the prison, contracted out by the city government of Buenos Aires to the Argentine military, is the seed for the artwork. The building was slated for demolition in 2001, but the process has been subject to various legal, environmental and bureaucratic roadblocks. The original plan was to implode the building in three steps. But the implosion was stopped at the last minute by a group of neighbors concerned about the possibility of damaging environmental effects, including asbestos poisoning and the possibility of driving millions of rats out of the tunnels they occupy underneath the prison. Instead, Caseros was demolished by mechanical means, floor by floor, from the top down.

The project connects to Argentinian history. Aparecido is the past participle for the Spanish verb aparecer, to appear. Its second meaning is apparition or ghost. It may also refer in an oblique way to Argentina's Dirty War, in which political prisoners were thrown out of airplanes over the Atlantic Ocean. These political prisoners were known as "los desaperecidos." "Sixteen Tons" is the name of a popular song written in the late 1940s. It referred to the amount of coal a miner was expected to load in a day, but in this context may refer to the amount of glass broken out through the installation, or de-installation, process.

By making the visibility of the aparecidos dependent on the daily, monthly, and yearly lunar and solar cycles, and the position of the viewer on the ground, Wulsin connects the ugly history of the prison to the larger cycle of cosmic movement. As the building disappeared through the demolition processes, viewers were exposed to hidden dimensions of place and physical history.

Artwork Excellence

By using nothing but the sun and the Caseros Prison's demolition as the raw material for 16 Tons and Aparecidos, Wulsin created a sustainable and self-contained public artwork that transformed a crumbling building into a shining beacon of beauty.

汽车主题公园
AutoTheme Park

艺术家：多位艺术家
地点：秘鲁利马
推荐人：彼得·莫拉雷斯
Artist: Multiple artists
Location: Surquillo, LimaPeru
Researcher: Peter Morales

作品描述

汽车主题公园位于秘鲁利马，它是艺术团队巴苏拉玛在美洲城市里实施的八个"都市固体废物"项目中的一项。参与的艺术家对这个项目提出建议并建造了一系列游乐设施，将一座被废弃的基础设施改造成一座小型娱乐公园，用明确的行动庆祝这片公共空间。

巴苏拉玛是一支基地在西班牙马德里的团队，致力于对可回收材料的研究和创新性使用。"都市固体废物项目，所有种类垃圾的进口／出口"是个多平台的公共艺术项目，致力于回收使用各类废弃物，包括固体废物、家庭废物、工业废物或是无形废物。在世界各地的城市里，该项目取得的成果有明确的地点，并取决于当地的背景及合作者。各项活动在破败的公共空间里展开，最终把它们改造成了充满活力的可进行社会和文化交流的新环境。

在利马，艺术家着力回收汽车零部件和被废弃的公共交通基础设施并将其改造成为休闲场所，呼吁并质疑都市空间的管理是否迎合行人的需求。20世纪的大多数城市的改变方向都在朝着为最大限度地提高汽车的畅通，而公共空间和行人的需求，甚至在小镇中心，都被放到了次要的位置。秘鲁的利马提供了一个极端的例子：这个城市缺乏城市规划，毫无计划地不断扩大，交通网络承载能力不足，有些地方甚至根本就没有，这种情况在城市里那些公共空间处于破败或被遗弃状态的地区格外严重。

1986年开始在一个平台上建造一条穿越苏尔基玛洛利马街区的高架轻轨线，此后不久该项目被突然暂停了。一年又一年，这个街区的居民等待着这个项目的完工。与此同时这广阔的都市空间未被充分利用。汽车主题公园利用了废弃的建造基地沿线，这条宽9米长达几公里的以前很少有人去的高架混凝土平台的上下空间。在短时期内，各类当地艺术家和艺术团队与巴苏拉玛一起工作，将这个空间转换成一座小型娱乐公园。每个艺术家或组织集中于公园的一种特定元素上。工艺坊构成了公园的组成部分，在创造装置之前已经在附近创办了工艺坊。汽车零部件和轮胎的广泛使用是为了激起大家对于公共交通和私人交通运输问题的反思。

"汽车主题公园"提出如果可以开放的话，这几公里长的废弃混凝土平台可以作为行人散步的高架公园，它那长长的漫步走廊为行人提供了娱乐和休闲的空间，而又躲开了街道上来往交通的危险。从艺术家的观点看，那些矗立在未完工平台前面的未完工的柱子伸向天空，它们可说是一种制度化的标志，象征着这座900万人口的城市年复一年对进步和持续增长的期待。在2010年该公园开放了15天。

解读

交通是城市发展的核心枢纽，越来越错综密集的高架和轨道交通或漂浮或沉潜在城市的各个角落，有没有幻想过在夜晚空旷的高架上遛弯吹风？在地铁的轨道槽里注水激流勇进呢？抽离出白天喧嚣的人群后，城市就像一座空荡荡的妖魔游乐城。在发展跃进的加速度里，人们对公共空间的需求，缺席了。

在秘鲁的利马，"都市固体废物"项目鼓励公共艺术家们将废弃物和场地改造成一系列的娱乐公园，"汽车主题公园"是其中一项已经落实了的方案。残缺的高架立柱被绑上了五彩的色条，无数废弃的橡胶轮胎在艺术家的巧思下，化身为秋千和攀爬网，"汽车主题公园"的多姿多彩与一旁灰蒙蒙的车道、拥堵的马路，形成鲜明的对峙状态。对于嬉笑的孩童而言，这是一个肆意玩耍的好去处，对于依附于城市发展的人们而言，却能激起一种反思：公共空间的建设是否要以人的需求为首位？那么，当需求由城市经济急速发展转向人群心理健康与提高生活质量时，公共环境的建设是否也该调整方向？以那个小小的幸福指数为圭臬呢？（吴昉）

Artwork Description

Auto Theme Park (RUS. Lima, autoparque de diversiones) in Lima, Perú, was one of the artist collective Basurama's eight "Urban Solid Waste" projects held in cities throughout the Americas. For it, participating artists proposed and built a series of rides and games, transforming a neglected infrastructure into a small amusement park in a clear act of celebration of public space.

Basurama is a collective based in Madrid, Spain, dedicated to the research and creative uses of recyclable materials. "Urban Solid Waste

Project, Import/Export of all kinds of garbage" is its multiplatform public art project focused on working with any kind of waste, whether solid, domestic, industrial or intangible. In cities around the world, the outcome is site specific and depends on the local context and collaborators. Activities take place in degraded public spaces, converting them into lively new contexts for social and cultural exchange.

In Lima, artists focused on reclaiming car parts and abandoned public transportation infrastructure to create a recreational place that calls into question the management of urban spaces to meet the needs of pedestrians.

Most modern cities in the 20th century have evolved to maximize the circulation of automobiles while public spaces and the needs of pedestrians, even in the centers of town, become a secondary preoccupation at best. Lima, Perú offers an extreme example of this: the city suffers from a lack of urban planning, indiscriminate informal growth and a transportation network that is insufficient and often nonexistent, especially in those parts of the city where public spaces are either in a state of decay or abandonment.

In 1986, construction began on a platform for an elevated light rail line through the Lima neighborhood of Surquillo. Shortly thereafter the project was abruptly suspended. Year after year the neighborhood awaits the completion of the project. In the mean time this wide swath of urban space goes underutilized. Auto Theme Park took up space along the abandoned construction site, under and just forward of an inaccessible, elevated concrete platform 9 meters wide and several kilometers long. Within a short period of time, various local artists and artist collectives working with Basurama converted the space into a small amusement park. Each artist or group focusing on a particular element of the park. Workshops to create the components of the park were held in workshops nearby prior to installation. The extensive use of car parts and tires was meant to provoke reflection on issues of public and private transportation.

Auto Theme Park suggests that if made accessible, the kilometers-long abandoned concrete platform might serve pedestrians as an elevated park with a long promenade that offers space for recreation and respite from the dangers of traffic at street level. From the artists' perspective, the unfinished columns that stand ahead of the incomplete platform, reach up to the sky as an institutionalized symbol of the perennial waiting for progress to sustain growth in a city of 9 million. The park was open for 15 days in 2010.

Artwork Excellence

The excellence of Auto Theme Park as a placemaking public art project resides in a powerful confluence of various elements: it reclaims public space, recycles car parts to create a functional playground, and rewrites a surreal narrative of missed opportunities into an uplifting story that is fun, provocative and informative. Auto Theme Park is a brilliant example of collaborative, community based placemaking on a large scale with a modest budget.

钻石小种子
Diamante de Semillitas

艺术家：杰姆·吉利（委内瑞拉）
地点：委内瑞拉加拉加斯的何塞·菲利克斯·里巴斯大道
形式和材料：金属框上的不锈钢薄片，标准瓷漆
推荐人：彼得·莫拉雷斯
Artist：Jaime Gili（Venezuela）
Location：Calle principal José Félix Ribas, Caracas, Venezuela
Researcher：Peter Morales

作品描述

在陡峭山坡上建造着许多贫民区的房子——比如在委内瑞拉加拉加斯佩塔雷的巴里奥·何塞·菲利克斯·里巴斯——由内而外的建造，使用下面建筑的平屋顶作为平台向上建造。居住在这些区域的人经常把最突出的外墙空出来，不涂上油漆，因为这块墙人不容易上去。即使当脚手架，这块墙也是危险的，因为山坡陡峭不平整。这导致了墙面蓬头垢面的、未完成的外观。

艺术家杰姆·吉利收到城市的邀请创作艺术，周边的社区将得以改观。这个地区能俯瞰到地铁的入口。除了外观以外，巴里奥·何塞·菲利克斯·里巴斯在附近建造地铁期间遭遇了意外的山体滑坡。被留下的是一个火山口状的空间，最终城市创建了一个露天剧场，在那里建造了一个小型练习场。

吉利在这里创作了"钻石小种子",作品中的外语名 Semillitas 在西班牙语里意为"小种子",这个名字其实是当地一支棒球队的名字。作品中的外语名 Diamante 在西班牙语里是钻石的意思,在英语里这个词也是一样的意思:它既是一种宝贵的石头,也是一个比赛、玩耍的区域。

当艺术家和当地孩子谈话时,他把自己想象成是一个孩子,安打,努力地击打涂板。他想象这些颜色具有不同的价值。蓝色的板是本垒打,黑色的是双击,等等。他把附近的居民召集起来,一起商讨最终的项目该是怎么个外观。根据一些业主的要求,艺术家粉刷了部分房屋光秃秃的墙壁。他还修改了颜色设置,因为业主在自己的房屋里生活了多年也未粉刷过。

艺术家使用脚手架和专业人员粉刷墙面,并且安装颜色鲜艳的金属板。艺术家的金属板色彩配色方案用来处理未完成的砖墙。最高的墙,最难爬上去的一部分仍旧保留了原先的红砖。艺术家创作的几何形状表现了一块切割过的钻石闪闪发光,突出山腰居住区的多面形象——居住区犹如一颗钻石。由于本作品,艺术家获得了市长办公室颁发的优秀奖。

解读

独特的地形特点,造就了其别样的建筑形式。陡峭山坡、险峻地形,形成了建筑外观杂乱无章。为改变原貌,回归自然美感,艺术家受到委派开始了一段寻找设计方案的旅程。

根据实际情况,因地制宜,在保留了原先的红墙的基础上,采用鲜艳色彩的金属板做遮挡,形成如"钻石"般闪耀的墙面,并与附近的棒球场形成呼应,相得益彰,最后以一支棒球队的名字"小种子"来命名,另有一番韵味。这个公共艺术项目不仅改善了社区的形象,而且也为当地儿童创造了一个游戏玩耍的乐园。

夜晚的社区如梦幻般美丽,到处闪烁着钻石般的光芒,以前的冷清感已经散去,杂乱无章的外观也已不见。白天这里充满欢乐的笑声和运动的气息,晚上这里是美的海洋,到处展现着公共艺术带给社区的魅力。(冯正龙)

Artwork Description

Most dwellings built on steep mountainsides in deprived areas—such as in the Barrio José Félix Ribas in Petare, Caracas, Venezuela—are built from the inside, using the flat roof of the building below as a platform to build upward. Those who live in these neighborhoods often leave the most prominent outer walls unfinished and unpainted because they are inaccessible. Even when scaffolding is available, it is dangerous to use due to the unevenness and steepness of the hills. This gives the walls an unkempt, unfinished appearance.

Artist Jaime Gili was invited by the city to create art that would address the appearance of the neighborhood. This site overlooks the entrance to the metro. For in addition to appearnace, the Barrio José Félix Ribas experienced an accidental landslide during construction of the metro line in the neighborhood. What was left behind was a crater-like space the city eventually used to create an amphitheatre where a small practice field was built.

Here Gili created Diamante de Semillitas. Semillitas means "little seeds" in Spanish, and it is the name of the local baseball team. Diamante is the Spanish word for "diamond" and the word has the same meaning as it does in English: it is both a precious stone and a field of play.

As the artist spoke with the local kids, he imagined himself as a child, batting and trying to hit the painted panels. He imagined the colors having different values. A blue panel is home-run, black one is a double, etc. He also brought residents of the neighborhood together to determine what the project would ultimately look like. At some of the homeowners' request, the artist painted some of the houses' bare walls. He also modified color arrangements because many of the owners had been living in their homes for years without being able to paint them.

The artist used scaffolding and professionals to paint some of the walls and install the brightly painted metal panels. The artist color scheme for the metal panels dealt with unfinished brick walls. The highest walls, the most difficult ones to access remain in the original red brick.

The artist's geometric shapes suggest the sparkle of a cut diamond and highlight the multi-faceted aspect of the mountainside neighborhood— the neighborhood as diamond. The artist received a Medal of Merit from the Mayor's office for his work.

Artwork Excellence

The excellence of Diamante de las Semillitas is evident in the scale of the project. The work covers a large area, incorporating several private dwellings and serves to frame and identify a neighborhood baseball playing field. The brightly painted panels invite young baseball players to imagine hitting the ball out of the park and onto the target. The project brought neighbors together to work out the details of what the walls of the neighborhood would look like, and it brought the possibility of painting walls that, given their condition, had been difficult to paint. The brightly painted and sharp panels shine like diamonds fallen from the sky, sparking the imaginations of future artists and participants from the municipality.

家具装置
Furniture Installation

艺术家：加布里埃尔、蒂亚戈 • 普利莫（巴西）
地点：巴西里约热内卢
推荐人：莱恩 • 柏格森
Artist: Gabriel & Tiago Primo (Brazil)
Location: Rio de Janeiro, Brazil
Researcher: Laine Bergeson

作品描述

每四个月，位于巴西里约热内卢的 Gentil Carioca 画廊的外墙就会被不同的艺术作品所装饰。2009 年的 5 月至 8 月，那面墙又被占满了，作品的作者是加布里埃尔和蒂亚戈 • 普利莫兄弟。兄弟俩把床、餐桌、吊床、椅子、橱柜、凳子、茶几等都钉在了红黄亮色的外墙上。然后每天 12—14 个小时用攀岩工具在这独特的室外房间跨来跨去，他们大多数时间就在吊床或床上闲躺着，并和下面过往的行人交谈。整个的"户外公寓"位于街道上方 33 英尺。要上厕所的话他们会爬过画廊的窗，去里面的盥洗室。

普利莫兄弟的"家具装置"的重要性在于：它提出了有关城市生活的所有问题，通过在里约热内卢街头持续四个月的表演引导公众参与了进来，并对城市生活的传统观点提出了挑战。

通过淡化公众与私人的界限，兄弟俩尖锐地提出了一组关于艺术本质、家和隐私的相关问题。例如，既然这个项目是他们临时四个月的私家住宅，它更多是否作为一种生活方式的试验，而非单纯是一件公共艺术作品？大城市里公寓和住宅之间彼此靠得那么近，那么居住在城市里从某种意义上讲是否始终是在进行公开表演呢？

"家具装置"也促进了一种对城市居住体验的新思考方式：住在一个几乎是垂直的环境里意味着什么？怎样利用城市的垂直空间——有没有荒废或者误用？城市里这些不计其数的垂直空间是否应该更好、更有效地加以使用？而人们真的能住在这样的环境下吗？植物能种在哪里吗？如果真的有必要居住在垂直空间的话，那城市居民如何利用这样的一个垂直的都市环境？

通过与行人的直接交谈，兄弟俩鼓励公众成为他们作品的一部分。他们会挑战路人对家及安全感的传统观念，促使他们对都市生活深入思考。许多行人会问兄弟俩："租金多少？"这一问就暴露出了城市生活高成本这个幽灵一样萦绕不去的问题。然而，当你的屋子只是一面外墙时，都市生活的种种方便和大都市的活力还值得用这么高昂的价格去换取吗？居住在建筑的一侧，人们又愿意付多少租金？

这件引人思考的装置作品将长期留在参观者的记忆之中。

解读

从弯腰四肢攀爬的猿猴进化成今天的人类，历经了千万年的演变，而今再要由直立行走的人类转变为攀爬为生的状态，犹如逆生长一般不切实际。"家具装置"的实验性决定了其定位于行为艺术的一种，意在启发人们对于公众与私人界限的某种思考，而与实际生活无关。

从视觉外观的感受而言，"家具装置"侧身于公共建筑的一墙，艳丽的色彩与紧邻画廊的庄严白墙形成强烈的视觉冲击，与此同时，有悖于生存常态的生活方式、摆脱地心引力的空间布局、真人上演的活体秀，以及艺术家与行人的对话，都让人落入超现实的意境，并由超越现实的虚无去激发思考现实的勇气，这是加布里埃尔和蒂亚戈·普利莫兄弟艺术家组合期望通过他们的作品传递给行人的信息，即：

"大城市里公寓和住宅之间彼此靠得那么近，那么居住在城市里从某种意义上讲是否始终是在进行公开表演呢？"

然而，我们不禁也要发问：没有观众参与的表演，是否也算是公开的呢？（吴昉）

Artwork Description

Every four months, the exterior wall of the Gentil Carioca gallery in Rio de Janeiro, Brazil, is adorned with a different work of art. From May to August 2009, that wall was occupied—literally—by brothers Gabriel and Tiago Primo. The brothers bolted a bed, dining table, hammock, chair, dresser, stool, and side table to the bright red and yellow exterior wall. Then, using mountain climbing gear, they navigated the unique outdoor room for 12-to 14-hours a day, spending most of their days lounging the hammock or bed and talking to the passerby below. The whole "exterior apartment" was located 33-feet above street level.

When nature called, they climbed through the window of the gallery to use an indoor bathroom.

The Primo brother's Furniture Installation is important because of all the questions it raises about life in the city, and because it brought performance art to the streets of Rio de Janeiro for four consecutive months, engaging the public and challenging people's views of life in the city.

By blurring the distinction between public and private, the brothers raised pointed and relevant questions about the nature of art, home, and privacy. For example, since the project was their private home for four months, is it more of a lifestyle experiment than a public art project? Is living in the city always somewhat an exercise in public performance, since apartments and homes are so close to each other?

Furniture Installation also prompted new ways of thinking about the experience of living in a city: what does it mean to live in a mostly vertical environment? How is vertical space used—or unused or misused—in cities? Could the endless amount of vertical space in cities be put to better, more effective use? Could people really live there? Could plants grow there? How should city dwellers utilize the vertical urban environment, if at all?

By engaging in direct dialogue with passerby, the brothers encouraged the public to become part of their piece. The brothers challenged the passerby's perception of home and safety, and made them think about urban life. Many pedestrians asked the brothers: "How much is rent?" —a question that raises the specter of the (often) steep cost of city living. Is proximity to urban amenities and big-city vitality still worth a premium price when your house is an exterior wall? How much would people pay to live on the side of a building?

This thought-provoking installation will live on in viewers' memories for much time to come.

Artwork Excellence

The Primo brother's Furniture Installation is important because of all the questions it raises about life in the city, and because it brought performance art to the streets of Rio de Janeiro for four consecutive months, engaging the public and challenging people's views of life in the city.

雅尔米里亚姆艺术俱乐部（JAMAC）
JAMAC (Jardim Miriam Art Club)

艺术家：莫妮卡·纳多尔（巴西）
地点：巴西大圣保罗附近雅尔米里亚姆
推荐人：卡提亚·坎顿
Artist: Monica Nado（Brazilian）
Location: Jardim Miriam, a neighborhood in the Greater Sao Paulo,
Brazil
Researcher: Katia Canton

作品描述

1996 年，莫妮卡·纳多尔决定舍弃有画廊代理的当代艺术家的舒适地位，与巴西的贫困社区合作，创建一个公共性的工作。她从访问恶化和贫困社区，创作壁画的图案（其中许多涉及家庭生活的模式，如厨房抹布）开始。在这些模式中，她刻画出模板并教当地人如何创作和应用它们，使用不同颜色喷涂房屋的外墙和被遗弃的空间。她帮助居民创造一个有凝聚力的环境，提高他们的自尊。这个想法在许多不同的社区被证明是成功的。

2000 年，纳多尔开始在靠近圣保罗的雅尔米里亚姆工作，这里曾经是一个充满暴力和危险的地方。纳多尔与这个近邻关联如此密切，以至于她还是搬到了那里。最初，在当地居民的帮助下，她在当地的一处房屋里成立了"艺术俱乐部"，在那里她开始邀请艺术家，来自许多领域的研究人员和知识分子为当地居民讲课，并推广一种免费的创作室。该俱乐部获得了投影机，并为社区免费放映电影。

纳多尔的首创引起了公众的关注，其中包括巴西政府和圣保罗州，他们后来委托她进一步发展俱乐部。今天，"雅尔米里亚姆艺术俱乐部"（JAMAC）已经成为其他街区的典范。很多人去那里参加讲习班和学习空间制造的可能性和技术，以使它们能够复制和适应于他们自己的社区。2006 年，"雅尔米里亚姆艺术俱乐部"成为圣保罗艺术双年展的一部分，并在国际艺术界得到公众的认可。

解读

"艺术到底是不是有钱人的玩意儿？"这是一个值得大书特书的问题。在很长的一段时间里，我们都以为艺术是只属于高阶层人士的休闲玩具，他们掌握这个世界上 80% 的财富和几乎同等比例的艺术品。但是，伴随着信息的传播，人们的创造力被不断地超越，艺术这个词的语境也发生了翻天覆地的变化，我们甚至开始不能用简单的一句话，几个专有名词去定义到底什么是艺术，到底什么是艺术品。

那么，艺术又为了什么存在，为什么而创作，又是谁创作出来的？伴随着哲学的思考，我们开始考虑艺术是否是为了生活而存在。美好的生活包括着我们生活在舒适的房间，美丽的环境，拥有自尊，对未来充满期望。如果艺术的目的真的就是这个，那么莫妮卡·纳多尔开办的雅尔米里亚姆艺术俱乐部已然大获成功。（高浅）

Artwork Description

In 1996, Monica Nador left her confortable position as a contemporary artist represented by galleries and decided to create a public work in partnership with poor communities in Brazil. She started by visiting deteriorated and poor neighborhoods and creating drawings and painting patterns on walls (many of the patterns pertained to domestic life, such as kitchen towels). From these patterns she would cut stencils and teach locals how to create and apply them, using spray paint in different colors, on the facades of houses and abandoned spaces. She helped residents create a cohesive environment and improve their self-esteem. The idea proved successful in many different communities.

In 2000, Nador started working with the Jardim Mirian neighborhood in Sao Paulo, which used to be a violent and dangerous place. Nador got so involved with the neighborhood that she moved there. With the initial help of local inhabitants she founded an "Art Club" in a local house, where she began inviting artists, researchers and intellectuals from many fields to lecture and promote free workshops for the residents. The club secured a projector and presented free film screenings for the community.

Artwork Excellence

Her initiative drew public attention, including the Brazilian government and the Sao Paulo State, which later commissioned her to grow the club. Today the Jardim Mirian Art Club (JAMAC) has transformed itself into a model for other neighborhoods. Many people go there to participate in workshops and learn placemaking possibilities and techniques to be able to reproduce and adapt them to their own communities. In 2006, JAMAC was part of the Sao Paulo Art Biennial, and received public recognition from the international art community.

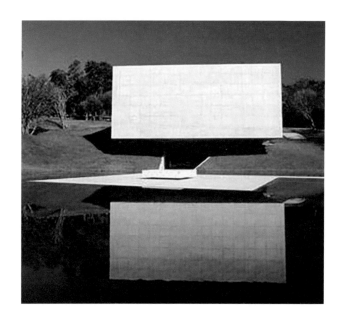

Inhotim 研究所
Instituto Inhotim

艺术家：博纳多·帕兹（巴西）
地点：巴西布鲁马蒂诺
推荐人：彼得·莫拉雷斯
Artist：Bernardo Paz （Brazil）
Location：Brumadinho, Brazil
Researcher：Peter Morales

作品描述

位于巴西第三大城市贝罗·奥里藏特 60 公里远，Inhotim 是一座面向于巴西和国际艺术家的乌托邦式私人画廊和博物馆，就像一座 21 世纪当代艺术的凡尔赛宫，里面涵盖了来自 30 多个不同国家的 100 多名艺术家创作的 500 多件艺术品，以户外公园的形式呈现，里面有景观雕塑和建筑展馆。Inhotim 里也涵盖着珍稀动物和一座热带稀树草原的植物园，其中有 3 500 种物种，全都放在一座 178 英亩的有着森林、湖和小山的景观内，设计构想是由景观设计师罗贝·伯乐·马克斯提出的。

于 2004 年 9 月开业，2006 年面向普通观众开放，Inhotim 获得了巴西商人博纳多·帕兹的资助而创立，他致力于慈善事业，他通过采矿业获得财富从而开始了艺术收藏。

帕兹在 1980 年代开始购买艺术品，但是 Inhotim 项目酝酿于他第一次购买当代艺术品时；"真正的红"（1997 年），那是一件由屯嘎创作的装置作品，具有一系列的红色元素——悬挂着玻璃瓶、网线和液体。这次购买这件艺术品使帕兹开始关注大型装置艺术，所以需要建设专门的画廊来展示这些收藏品。

这个画廊发展成为了一个机构，如今拥有 500 多名雇员，包括馆长部门有艾伦·施瓦兹曼，约亨·福尔兹和罗德里格·穆拉，这几位都经常主办艺术展览会，画廊和展会保持 Inhotim 展品的国际化背景。着重于创作于1960 年代到现在的当代艺术品，藏品包括雕塑、装置、绘画、素描、摄影、电影和录像。

Inhotim 扩大自己的藏品的一个主要策略是邀请艺术家来此访问，为这里创作新的艺术品。艺术家经常创作特定地点的艺术项目，与具体环境中的自然和文化特质对话。Inhotim 同时寻找奇特的艺术品纳入藏品中，然后创建设施在永久基础上展示这些藏品。因此这里收藏了新生代艺术家的作品，将他们极具意义的作品汇聚起来。这些展品里包括克里斯·伯顿、珍妮特·卡迪夫、艾里欧·奥义提希卡和内维尔·达尔美达、马修·巴尼、约尔格·马奇、加巴斯·洛佩兹、道格·艾特肯、阿德里阿娜·瓦拉约、奥拉夫·埃里阿森、多利斯·萨尔赛多、齐尔多·美乐莱斯、维克·穆尼兹、恩内斯多·耐托、约翰·阿罕恩和里格贝托·托雷斯，这些都是国际知名的艺术家。

自从 1990 年代以后，藏品持续地扩充着，发展成为目前世界性的规模，藏品中有一些是当今最知名的艺术家的作品，Inhotim 里的艺术家有着各自不同的文化背景，在巴西，面向公众展出这些真正的国际当代艺术精品。

解读

大型装置艺术的奇思异想与彪悍大胆，加上稀有动物群与热带珍稀草原，博纳多·帕兹的“研究所”，无愧于乌托邦式的私人收藏博物馆。如果这个“研究所”是对公众开放的，而不仅仅面向前来创作的艺术家群体，那么，它的公共性就在于让艺术创作顺从于环境的限定，使人为性强烈的装置艺术与外在大环境相处融洽。

艺术家们在帕兹的“研究所”，创作特定地点的艺术项目，与有限且具体的自然文化特质相对话。这是公共艺术的一个本质属性，作为独立个体的艺术作品本身是未完成的，只有与外在大环境及人群心理相契合，才一次又一次完善自我，成为一项成熟的作品。通过帕兹这个项目的动机和进展，我们似乎也看到了公共艺术创作中“自由”与“逍遥”的区别。任何生活实践中的自由都有其限度，而逍遥则指向彻底的解脱，可是这种解脱，在实际的环境中，却是以委顺、配合的姿态得以实现的。（吴昉）

Artwork Description

Located 60 kilometers from Brazi's third city Belo Horizonte, Inhotim is a utopian private gallery and museum for Brazilian and international artists. A 21st-century Versailles for contemporary art, it includes more than 500 artworks by 100 artists of over 30 different nationalities presented in an outdoor art park with sculpture in the landscape and architectural pavilions. Inhotim also includes rare fauna and a tropical savanna botanical garden containing 3,500 species, all set in a 178-acre landscape of forest, lakes and hills developed from proposals by landscape designer Roberto Burle-Marx.

Inaugurated in September 2004 and opened to the general public in 2006, Inhotim was created through the philanthropy of Brazilian businessman Bernardo Paz, who funded the art collection through a fortune derived from mining. Paz started buying art in the 1980s, but the gestation of Inhotim stems from a purchase of his first contemporary piece; True Rouge (1997), an installation by Tunga, is a series of red elements—suspended glass bottles, netting and liquids. This purchase led to Paz's focus on large-scale installations that necessitated the construction of purpose-built galleries to exhibit the collection. The galleries developed into an institution that now employs over 500 staff, including the curatorial board of Allan Schwartzman, Jochen Volz and Rodrigo Moura, who constantly monitor art fairs, galleries and exhibitions to maintain the international context of the Inhotim collection. Focusing on contemporary art produced from the 1960s to the present, this collection includes sculpture, installation, painting, drawing, photography, film and video.

One of the main strategies adopted by Inhotim for the enlargement of its collection was to offer visiting artists the opportunity to create new artworks expressly for the collection. Artists often create site-specific art projects in dialogue with the natural and cultural characteristics of the environment. Inhotim also seeks to identify singular artworks to incorporate in the collection, and then to create facilities to exhibit them on a permanent basis. It has thus collected artists of the new generations, gathering significant sets of their works.

The collection includes work from artists Chris Burden, Janet Cardiff, Hélio Oiticica and Neville d'Almeida, Matthew Barney. Jorge Macchi, Jarbas Lopes, Doug Aitken, Adriana Varajão, Olafur Eliasson, Doris Salcedo, Cildo Meireles, Vik Muniz, Ernesto Neto, and John Ahearn and Rigoberto Torress, among other important international artists.

Since the late 1990s the collection has been continuously expanded and developed to achieve its current worldwide relevance, with artworks by some of today's most powerful artistic voices. Inhotim, with artists from different cultural contexts, offers the only truly international collection of contemporary art accessible to the public in Brazil.

Artwork Excellence

Instituto Ihotim offers the only truly international collection of contemporary art accessible to the public in Brazil. It integrates a botanical park that focuses on native flora and fauna with a first class contemporary art collection that includes artists of different nationalities all within a 178- acre site. Both the State Government of Minas Gerais and the Federal Government of Brazil recognized the Inhotim as a Civil Society Organization for Public Interest (OSCIP). In 2010 the gardens received the official title of Botanical Garden from the National Committee of Botanical Gardens (CNJB).

Inhotim participates fully in the social, economic and cultural life of the region and works to contribute toward the development of the people who live and work there. Its commitment to education is also evident in its programs. Inhotim, which attracts 160,000 visitors a year, has established a credo of education and social development with the Department of Inclusion and Citizenship and the Labatorio Inhotim, which works with the adjacent town of Brumadinho. It has also created a program involving music as well as art for teenagers designed to raise awareness of their cultural heritage.

The ever-present culture of Brazil is music and dance, but as the country's contemporary art scene is being discovered through events such as the Biennial de São Paulo, Semana de Arte do Rio in 2011 and ArtRio in 2012, Inhotim is an impressive and specifically Brazilian philanthropic model for the pedagogical and theatrical presentation of artists' works in an enchanted art wonderland.

活着的历史
Living History

艺术家：阿莱娜·乌尔（美国）
地点：危地马拉拉维纳尔公墓
推荐人：卡利·英格布里特森
Artist: Alayna Wool（USA）
Location: Rabinal, Guatemala
Researcher: Cally Ingebritson

作品描述

长期以来危地马拉面临着许多挑战，只举数例：殖民地化、贫困、食品短缺和内战。当艺术家阿莱娜·乌尔第一次拜访这个国家时，她的身份是ArtCorps的驻场艺术家。这个和平队式的项目把艺术家派到中美洲，以完成"通过推广艺术和文化来支持社会变化主动性"的使命——艺术和文化是促成发展组织与其服务社群之间的合作性的、可持续的工作的有力工具。

在访问处于这个国家中心地区的一个有着 36 000 居民的小镇拉维纳尔时，乌尔被了解到的事实所震动：这里有六千多人在危地马拉内战的规模最大的几次屠杀中丧生。她解释说，这个项目最艰难的部分是"使我感到心情消沉，同时听到可怕的故事一再讲述"。在确保一个纪念死者的公共艺术项目的资金筹备到位之后，乌尔带着为幸存者服务和为冲突中的那些死难者留下记录的决心，返回这一地区。

乌尔主持了一系列前期会议，并邀请 1982 年屠杀中受祸最深的七个社区的居民参加。随后她单独会见人们，倾听他们的故事，并为归档的照片制作文件以确保不丢失。受害者的家人制作了地图，以说明事件在 1982 年的武装冲突中是如何展开的。在共同开发埋葬着许多受害者的城市公墓邻近的大型装置艺术时，乌尔组织了性别工作室，并为受害者家庭协调寻求法律支持。她一共收集和修复了两千多张照片，并将它们用于创作一幅 400 英尺的壁画。

在"活着的历史"的开幕式上，当地的社区领袖出席现场并为装置艺术作品祝福，那些失去家庭成员的人都领到了尺寸为 8×10 英寸的带框的亲人照片。随着这一项目的完成，乌尔期待着与其他有着类似需求的社区合作，她说："同样的概念可以用于任何一个曾经面临这种暴力的社区，不管是危地马拉还是其他地方。"乌尔希望这个项目"努力让这些痛苦的事实不再重现，为拉维纳尔和危地马拉的人民带来和解"。

女性和平力量基金会把 2010 年的妇女和平奖授予她，基金会声称"阿莱娜致力于运用她的艺术家天赋为世界带来和平，为此甚至不顾个人的危难。她最近的志愿者项目在危地马拉的村庄已经运作三年了，她在这里为拉维纳尔的武装冲突的死难者创建了一个影像档案库。她努力帮助幸存者把那些没有得到惩罚的罪行广为传播，并为他们的法庭案件提供协助。

解读

苏珊桑塔格在其《自信的梅普尔索普》一文中，称梅普尔索普的照片并不奢望给人以醍醐灌顶的启示，存在于拍摄者与被拍者之间的并非是一种掠夺式的关系。影像在很多时候，并不是某种人事物的真相，而是其最强烈的表现形式，阿莱娜·乌尔在"活着的历史"中，用布满人物肖像的墙面与生存者对视，这种表现形式带来震撼人心的仪式感，生者因而与死者有了平等交流的机会。

当然，一切历史都是当代史，一切都会过去，一切又都不会过去。阿莱娜乌尔以这种公共艺术的手段，去昭示一段时间上已成为历史的事件，用激起痛苦回忆的方式给未来些许警醒。作为一个公共艺术家，乌尔已尽其所能，只是，有暴力的表述者，有暴力的受害者，似乎暴力的施与者，也不该缺席。

时间是单向的轨道，向死而生，而情感、文化、信仰却可以超越时间之上，所幸，历史所涵盖的，不仅仅是时间。（吴昉）

Artwork Description

Guatemala has long faced many challenges: colonization, poverty, food shortages and civil war to name just a few. Artist Alayna Wool first visited the country during her time as an artist in residence with ArtCorps. The Peace Corps-esque program sends artists to Central America with the mission to support "social change initiatives by promoting art and culture as powerful tools to generate cooperative and sustainable work between development organizations and the communities they serve."

While visiting the town of Rabinal, a small town of 36,000 inhabitants in the central area of the country, Wool was impacted by learning of the more than 6,000 individuals killed there during some of the largest massacres of the Guatemalan civil war. The hardest part of this project was "feeling my heart sink, while hearing horrific stories retold," she explains. After securing funding for a public art project in remembrance of those who died, Wool returned to the area determined to serve the survivors and document those lost to the conflict.

Wool hosted initial meetings inviting residents to participate in the project in seven of the communities most affected by the 1982 massacre. She later met individually with people to hear their stories and to document archived photos to ensure they would not be lost. Victims' families made maps to show how events unfolded during the armed conflict of 1982. In tandem with developing large installations hung around the city cemetery where many of the victims were buried, Wool organized gender workshops and coordinated legal support for the victims' families. In total, she gathered and restored over 2000 photos, which she used to create a 400-foot mural.

At the inauguration of Living History, indigenous community leaders attended to bless the installation and those who had lost family members were each given a framed 8x10 photo of their loved ones. With this project under her belt, Wool looks forward to working with other communities with similar needs, saying, "This same concept could be used in any community that's faced this kind of violence, whether in Guatemala or elsewhere." Wool hopes this project "will bring reconciliation to the people of Rabinal and Guatemala in an effort for these torturous acts not to be repeated."

Awarded the 2010 Women of Peace Award by the Women's Peacepower Foundation, the organization has said, "Alayna is dedicated to using her talent as an artist to bring peace around the world even at her own personal risk. Her current volunteer project has been living in a rural Guatemalan village for three years, where she has been creating a photographic archive of the victims of the armed conflict in the town of Rabinal. Her efforts are helping the survivors disseminate impunity and assists with their court cases."

Artwork Excellence

The artist was determined to help heal wounds caused during Guatemala's Civil War during which 6,000 people were killed in one of the bloodiest massacres of this era. The artist included the victims' families in the entire process and gathered 2,000 photos that she made into a public mural.

回忆·壁画
Memory Murals

艺术家：爱德华多·考博拉（巴西）
地点：巴西圣保罗
推荐人：莱恩·柏格森
Artist: Eduardo Kobra（Brazil）
Location: Sao Paulo, Brazil
Researcher: Laine Bergeson

作品描述

爱德华多·考博拉的"回忆·壁画"是一组 21 幅大型壁画的组合，画在巴西圣保罗市区和市郊的建筑物墙壁上和其他一些外墙上。壁画的尺寸由于内容和地点不同而大小不一，壁画的内容都是描绘圣保罗以前的街景，委托项目的主题是为了纪念这座城市丰富的历史积淀。壁画中的画面跨越了 20 世纪，由考博拉和他工作室的 10 位助手组成的团队共同创作绘制。壁画中的许多场景都与街道平行，创造了一种引人注目的，几乎是互动的画面，把过去与现在融为一体。其中有一处与街道平行的壁画，画中内容比实物更大，它营造出历史幻觉的上升，于是把现代汽车和行人矮化了。

考博拉使用喷枪绘制壁画，他以浓重的黑色阴影，另一部分涂以明亮颜色的手法营造出现实主义风格。其效果看起来很具立体感，像要从墙上跃出冲进圣保罗熙熙攘攘的街道似的。

考博拉这位涂鸦艺术家自从 11 岁起就开始涂鸦。2005 年，他发现了一本 1920 年代描述桑托斯港口的图画书，自此开始对绘制旧时的场景发生了兴趣。他把其中的一幅照片作为创作灵感，决定在苏马瑞大道的墙壁上创作一幅"实验性"壁画。第一幅壁画描绘了一群工人把袋子装运上船。他脱离了自己习惯的明亮涂鸦风格，选用黑白两色创作壁画（虽然后期一些壁画使用了彩色颜料）。

受到这些经验的启发，考博拉着手寻找更多的 20 年代早期的电车、凉亭、撑伞的妇女、穿西服的男人之类的老照片，把他们创作成壁画的内容。任何路过这些壁画的过路人现在沿着地铁或从始终拥挤不堪的高速公路边的高处看到这样的几片历史场景。他选择了加ська山教堂附近的一堵墙，目的是想突出安静的电车和永不停息的当代街道之间的对比。他说道："我想要人们去感受一种宁静感，一种更人性化的生活节奏，哪怕这感受只有片刻。"加萨山教堂很喜欢这幅壁画，他们把一块更有价值的、面积更大的教堂前 200 平方米的一面墙让给了考博拉的项目。

在寻找更多圣保罗老照片的过程中，他对保利斯塔马路特别着迷，那里的汽车、有轨电车和人行道都装饰着金色喇叭。他得到许可在保利斯塔路现代购物中心的一块特定空间里绘制壁画。考博拉说他感觉自己就像一位历史老师，唯一的区别是他的教室不在室内，没有四面墙和桌子，他的教室就是这挤满了行人和驾车人的城市。

解读

与街道平行或延伸、和现实等比或放大了的场景形象，带领行人穿过现实的境况，行走在往昔与今日的临界线上。巴西艺术家爱德华多·考博拉的"回忆·壁画"，用不同于以往涂鸦的现实主义风格，将圣保罗的街头延展至无穷远。

考博拉的作品，拥有 3D 绘画的立体效果，巧妙地镶嵌于城市的墙面中，打破了横亘在历史与今天之间的认知障碍，壁画与真实错位交叉，不知不觉中，迈开的步伐就跟上了古老的节奏。"于是我们奋力向前划，逆流而上的小舟，不停地倒退，进入过去。"菲茨杰拉德笔下的盖茨比相信人可以通过自身的意愿和努力，回到过去。我们愿意相信巴西圣保罗的街头，巨大的壁画可以唤起人们对往日追忆的种种温情，至少，"回忆·壁画"已成为圣保罗美丽的妆点，让周围的空间变得富有弹性，人情味十足。（吴昉）

Artwork Description

Eduardo Kobra's Memory Murals (Muros da Memoria) is a collection of 21 large-scale murals on the sides of buildings and other exterior walls throughout and around Sao Paulo, Brazil. The murals, which vary in size depending on where (and on what) they're painted, all depict street scenes from Sao Paulo's past and were commissioned to celebrate the rich history of the city. The images depicted in the murals span the 20th Century and were created by Kobra and a team of 10 assistants from his studio. Many of the scenes are painted at street level, creating a compelling, almost interactive tableau where past and present mix together. When one of the street-level murals is larger than life, it creates the illusion of history rising up and dwarfing the modern day cars and passersby.

Kobra created the murals with an airbrush, and his almost hyper-realistic style relies on deep graphite-rich shadows and bright rays of painted light. The results are pieces that feel three-dimensional and seem to bounce write off the wall and into the bustling, modern-day streets of Sao Paulo.

A graffiti artist since the age of 11, Kobra became interested in painting scenes of the past in 2005 when he discovered a book with pictures Santos port from the 1920s. Using one of those photos as inspiration, he decided to do a "test mural" on a wall on Sumaré Avenue. That first mural portrays a group of workers with bags filling a ship. In a departure from his usual bright-colored graffiti style, he chose to do the mural in black-and-white (though some of the later murals are in color).

Inspired, Kobra went looking for more old photos of early 20th century trams, gazebos, and women with umbrellas and men dressed in suits, and he recreated them as murals. Anyone passing by these murals now will see a little of the past along the subway or rising above the always-congested motorways. When he chose a wall that runs alongside the Church of Calvary, his goal was specifically to highlight the contrast between the tranquility of the tram and the never-ending movement of the modern-day street, saying, "I wanted people to feel, even for a moment, a sense of calm, a more human pace," he says. Calvary Church was so enamored of the work that they donated another, more coveted wall space—the 200 square meter wall in front of the church—to Kobra's project.

As he began to search out even more photos of old Sao Paulo, he became particularly enchanted with the Avenida Paulista, in which the streetcars, trams and sidewalks were decorated with golden trumpets, and he obtained permission to paint in a privileged space in a modern-day shopping center on Avenida Paulista. Kobra has said that he feels like a "history teacher" with the only difference being that his classroom is not an interior room with four walls and desks, but the city itself crammed with pedestrians and motorists.

Artwork Excellence

Kobra's urban intervention, Memory Murals (Muros de Memoria), seeks to transform the urban landscape with large-scale murals that resuscitate memories of the city. The project weaves together modernity and nostalgia, creating scenes of tranquility and beauty for modern-day Sao Paulo residents, and raises awareness about the city's rich history. The murals are rendered in a realistic style that makes the pieces—especially those painted at street level—eel almost interactive, as if present-day residents could literally step back in time 100 years. It also allows the figures from the past to "comment" on the present day. Many of the painted figures in the murals wear expressions that seem to question—and even perhaps indict—contemporary residents on some unknown charge. Whatever is being said, the murals allow for a vibrant conversation between past and present. Kobra has said that his goal was to capture the tension between romantic nostalgia and contemporary bustle.

Memory Murals is also significant for the way it shows an evolution in the world of graffiti and street art. Kobra abandons graffiti's trademark bright colors in many of the murals (a few are rendered in color) and anthropomorphic caricatures to create sumptuous, realistic portraits of people and places. It is as if Kobra has brought the world of the gallery to everyday life on the busy urban streets.

占用氧气
Occupation Oxygen

艺术家：多位艺术家
地点：巴西圣保罗
推荐人：莱恩·柏格森
Artist: Multiple artists
Location: Sao Paulo, Brazil
Researcher: Laine Bergeson

作品描述

"占用氧气"是一年一度的艺术装置／干预项目，在巴西圣保罗的布宜诺斯艾利斯公园里持续展出好几周。在策展指导达奇奥·比库多的策划下该项目始于 2009 年，已经举办第三届了，最近这届的装置作品由比库多和卡地亚·坎顿联合策展。2011 年装置开展于 2011 年 9 月的下半年，其中包括超过 25 名独立的艺术家参与创作了公园内的临时装置作品。

公园里的每件作品都与公园里的自然元素（树木、水、植物、草地、风和空气）在某种程度上产生互动。许多艺术家都把自然元素用作"画布"，在这画布上面"构建"他们的装置作品。举个例子来说，艺术家谷托·拉查兹在公园水景（一片小型喷泉池）里安装了一根 1 米高的银棒。棒垂直地漂浮于表面，同时像一束光线穿透浅浅的水面。附加的银棒可以随着风和水流朝任何方向移动几英寸，这样使得每次看它时银棒的布局都会发生变化。

艺术家费尔南达·伊娃逼真地创作了当地野生动物的复制品，似乎这些动物在公园内逍遥漫游，完全忘掉了还有人类和当代生活的陷阱。这一装置作品使观众对这座公园及周围区域产生了远古时候环境状况的想象。艺术家桑德拉·马提耐利在公园里竖立起一块巨大而空旷的框架。从一个方向观看，这个框架包围了公园里的绿色树梢。美丽的枝叶犹如被框在画廊的一幅画中，可以毫不夸张地说它提供给了观众一个新的框架，通过它来对自然界进行沉思。从另一方向来观看，该框架包围了公园里的那座永久性雕塑。通过框住这座雕塑，作品强调了自己的存在，让进入这座公园的观众从一个全新的角度来欣赏这座雕塑，并赋予它全新的意义和诠释。费尔南多·伊娃创作的"卡米尼奥斯天堂"装置涉及到把很轻的羽毛安装在细长的垂直杆子上。于是伊娃把羽毛杆子放置在公园小道上排成一列直线，使观众们路过时让这些羽毛杆子随着空气的波动微微舞动起来。

好几位艺术家都用树木作画布。联合策展人卡地亚·坎顿创作了一件名为"莴苣"的作品，用一棵大树当做假发支架。从下面看上去，观众能够看到十几个鲜艳的假发在树枝上摇曳着。雷吉纳·德·巴罗斯在一棵树上挂着巨大的透明活动塑料管，雷吉纳·卡莫纳把薄薄的红布悬挂在树枝上。完成后的作品使人联想起西藏人祈祷用的旗帜和一种城市里的晾衣绳。

这座有 25 名以上独立艺术家共同合作的作品营造了一座户外画廊，使得参观公园的人——这些人原本从不踏入博物馆或画廊——有机会参与了艺术品观赏的体验。该项目同时也对地貌艺术提出了很多问题，寻求改造户外空间，赋予展示地点特别的感觉，它还为经常机械地不加思索地经过室外都市空间的城市居民一个停歇一下，重新思考周围空间的机会。

解读

"占用氧气"的名字起得很生动，拟人化使人倍感与自然的亲近。如同建筑存在的真正意义是为了让空间昭然可睹那样，氧气本无形，空气抓不住，所以巴西圣保罗的艺术家们就借用各种人间的有形物态，作为自然界的参照物，去验证自然的存在。

翩翩起舞的布幔，看到的是熏人的微风；漂浮水面的银棒，察觉出水波的回旋；垂直的羽毛杆子，让空气的舞动变得真切可感。诸如此类，达成人与自然的互为参照。按本体论的说法，人是借助于"非我"来认识自我的，那么如果自然有灵，它也需要借助一些参照物，来显示自身。人和自然，彼此认识的过程同样也是自我不断发掘的进程。圣保罗的 25 位艺术家，用公众行为的方式，召唤公众意识的自我觉醒。（吴昉）

Artwork Description

Occupation Oxygen (Ocupação Oxigênio in Portuguese) is an annual art installation/intervention that lasts for several weeks and is staged in Buenos Aires Park in Sao Paulo, Brazil. The project, which began in 2009 under the curatorial direction of Dacio Bicudo, is in its third and most recent installation, co-curated by Bicudo and Katia Canton. The 2011 installation took place during the latter half of September 2011 and included over 25 individual artists who installed temporary constructed works in the park.

Every piece in the park interacts with the park's natural elements (the trees, water, plants, grasses, wind and air) in some way. Many of the artists used the natural elements as a "canvas" upon which they "built" their installations. Artist Guto Lacaz, for example, mounted 1-meter high silver rods in the park's water feature (a small fountain and pond). The rods appear to float vertically on the surface, yet at the same, like rays of light, penetrate its shallow water. The rods were affixed in such a way that they were allowed to move a few inches in any direction with the wind and flow of water, changing the composition of the rods each time it was viewed.

Artist Fernanda Eva created lifelike replicas of wild, native animals that appear to roam the park, oblivious to the people and the trappings of contemporary life. This installation allows viewers to imagine the park and the surrounding area as it might have been thousands of years ago. Artist Sandra Martinelli erected a huge, empty frame in the park. Viewed from one direction, the frame encloses the green treetops of the park. The beautiful foliage is framed as a painting in a gallery might be, which gives viewers a new frame—quite literally—through which to contemplate the natural world. Viewed from the other direction, the frame encloses a permanent sculpture that stands in the park year round. By framing the sculpture, the work highlights its existence and lets visitors to the park see it anew and give it new meanings and interpretations. The installation Caminos do Paraiso by Fernanda Eva involved lightweight features mounted to thin vertical rods. Eva then lined the paved paths in the park with the feather rods, allowing them to catch the movement of the air as viewers passed by.

Several artists used trees as their canvas. Co-curator, Katia Canton, contributed a piece called Rapunzel, which used a large tree as an oversized wig stand. From below, viewers could see a dozen or more brightly colored wigs dangling from tree branches. Regina de Barros hung a giant mobile of clear plastic tubing in a tree, and Regina Carmona strung thin sheets of red fabric from tree branches. The finished work evokes both Tibetan prayer flags and an urban clothesline

Taken together, the 25-plus works create an outdoor gallery that allows park goers—who may never otherwise step foot in a museum or gallery—to participate in the art-viewing experience. The project also raises many of the questions and ideas of the Land Art movement, which sought to transform outdoor space and give it a specific sense of place. It also gives urban dwellers, who often move through outdoor urban space in a rote and inquiry-less fashion, a chance to pause and reconsider their surroundings.

Artwork Excellence

Occupation Oxygen is more than just an exhibition and installation. It is a large-scale, easily accessible group of public works that explores the relationship between an indoor gallery space and an outdoor "gallery" space. In an outdoor gallery, the entire city is invited through its "doors" to explore art and nature. Art is available for everyone.

The use of the word "oxygen" in the title is specific and intentional, and it highlights other aspects of excellence in this work. Oxygen is the basis of all life—it is the air we breathe and a guarantee of existence. To oxygenate something means to fill it with life—to expand, in this case, the city and the park with ideas, feelings and thoughts. Much of the intention of this installation is to refresh and renew the park and city by helping all parkgoers see their environment in new ways. The works of art in the park give passerby a counterpoint to the homogenization and rote nature of daily life in the city.

Occupation Oxygen acts as kind of forum where artists come annually to celebrate and reinvigorate life in the city, to explore the relationship between art and nature, and to question the role of the art museum—or lack thereof—in the contemporary art world. Occupation Oxygen engages the public of Sao Paolo and asks them to consider questions about art, nature, and city life that they might not otherwise ask themselves.

奥摩罗（小山）
O Morro (The Hill)

艺术家：吉荣·库哈斯、德鲁·乌尔哈恩（荷兰）
地点：巴西里约热内卢
推荐人：泰德·德克尔
Artist: Jeroen Koolhaas and Dre Urhahn（Netherland）
Location: Rio de Janeiro, RJ Brazil
Researcher: Ted Decker

作品描述

2010 年在里约热内卢的桑塔·玛尔塔社区，"奥摩罗（小山）"创作完成，该作品展示了公共领域创新的、私人出资的、艺术家领导的区域营造项目。

这个公共艺术项目外观上很美观，给人一种愉悦感，催生出希望，激发人们对自己社区的自豪感和主人翁意识，从而对当地产生久远的影响力。

哈斯 & 哈恩是一个艺术团队的名字，由吉荣·库哈斯（鹿特丹人，生于 1977 年）和德鲁·乌尔哈恩（阿姆斯特丹人，生于 1973 年）组成。自从 2005 年起，他们为音乐电视拍摄了一部描述圣保罗和里约热内卢贫民窟嘻哈文化的纪录片，由此经历而受到启发，他们制定了一个计划，要在别人意想不到的地方制作"离谱"的艺术介入项目。他们与当地青年以及周边居民一起在巴西贫民窟绘制巨型壁画。他们的第一个项目是在维拉·克鲁塞罗社区，那是一个臭名昭著的贫民区，因贩毒导致暴力犯罪案发率很高。2007年艺术家在那里创作了一件巨幅壁画后，又在一座大型混凝土防洪装置上作画，那座装置的设计目的是在雨季到来时抵挡山洪使当地社区免遭洪灾的。

这些项目最初在维拉·克鲁塞罗获得的成功启发了艺术家在里约热内卢一个非常大的贫民窟桑塔·玛尔塔进行"奥摩罗"项目，这个坐落于山坡上的社区非常突出，在该城市的多个地点能够看到，因为它的位置在著名的地标和旅游目的地面包山和科尔科瓦之间，那里最高处有巨型基督救世主纪念碑。从附录的"被改造前"的图片来看，艺术家选择了桑塔·玛尔塔中心一个普通的广场作为项目执行地。

乌尔哈恩形容"奥摩罗"是一项自由的艺术。它不是由政府机构委托的，而是由坚定基金会独立资助的。（起初，里约珊瑚涂料油漆公司是合作伙伴之一，但项目启动不久该公司就退出了。珊瑚涂料的员工们指导当地居民如何在脚手架上安全工作，以及怎样使用涂料）贫民窟的居民受雇按照事先设计好的图案彩绘他们自家的房子（见附录的项目绘画图）。艺术家的目标是把居民社区改造成一件"具有史诗规模的艺术品"，使它产生一种爆炸般的色彩绽放，欢乐地辐射到世界各地。

这个项目是有机的，而且变通性很强，随着募集到越来越多的资金，以后能够继续扩展。桑塔·玛尔塔地区的转变取得了显著的成功。游客们被鼓励搭乘缆车到陡峭的山坡，到该社区直接俯瞰这片彩绘的区域。根据一位里约热内卢居民所述，该项目吸引她参观这个社区，如今她经常到这座山上玩，并且去桑塔·玛尔塔一家很棒的餐厅用餐。

解读

里约热内卢是巴西贫民窟数量最多的城市之一，这个城市600万人口中曾经有三分之一都住在贫民窟内，当人们猎奇于贫民窟百万富翁的传奇故事之余，肮脏混乱、穷困饥饿，一直就是贫民窟的现状。号召当地居民集体参与完成的公共艺术作品"奥摩罗"，用艺术美化生活的积极心态，去修复当地居民严重受损的自豪感与主人翁意识，收到了良好的社会反馈。

相比改建前贫民窟的灰暗色调，完成改建的"奥摩罗"色彩奔放，像是七彩石惊爆了里约热内卢的街头。艺术家们前期绘稿，当地居民的后期介入，使作品有了与公众良好互动的基础，正是在精心涂刷色彩的日子里，当地居民的自豪感被一点一滴调动起来，可以想见，最终完成的一刻，心情就如墙面彩绘般靓丽欢快。

"奥摩罗"项目的完成，不仅仅是抚慰人心的一次艺术尝试，当地政府更用这项公共艺术拓展地区的旅游业，为贫民窟争取实实在在的生存机会，用实例证实了公共艺术体现在社会功能上的价值。（吴昉）

Artwork Description

The O Morro (The Hill) project in the Santa Marta Community of Rio de Janeiro was completed in 2010 and demonstrates excellence in innovative, privately-funded, artist-led placemaking in the public realm. The contemporary design, while being aesthetically pleasing and cheerful, also encourages hope, community pride and a sense of ownership, resulting in long-term impact.

Haas & Hahn is the working name of artist team Jeroen Koolhaas (Rotterdam, 1977) and Dre Urhahn (Amsterdam, 1973). They started working together in 2005, when they filmed a documentary about hip hop in the favelas of Rio de Janeiro and São Paolo for MTV. Inspired by this visit, they developed a plan to make "outrageous" art interventions in unexpected places, starting with painting enormous murals in the slums of Brazil together with the local youth and other residents. Their first project was in the Community of Vila Cruzeiro, a notoriously impoverished favela with high levels of violence due to drug trafficking. After completing a large mural painting there in 2007, they painted a massive concrete flood control device designed to protect the community from flash floods during the rainy season.

The successes of these initial projects in Vila Cruzeiro inspired the O Morro project in Santa Marta, a very large favela community in Rio de Janeiro. The hillside community is prominent and visible from many viewpoints in the city because of its location between the famous geographic landmarks and tourist destinations Pão do Açucar (Sugar Loaf Mountain) and Corcovado, which has the giant Cristo Redentor (Christ the Redeemer) monument on top. As seen from the attached before image, the artists selected an ordinary, rather nondescript plaza in the heart of Santa Marta.

Urhahn describes O Morro as "free art." It was not commissioned by a governmental agency, but rather funded privately through the Firmeza Foundation. (Initially, the Rio paint company Tintas Coral was a partner, but that company pulled out shortly after the project started. Tintas Coral's employees did, however, train local residents how to use and apply paint with instructions about how to safely work on scaffolding.) Inhabitants of the favela were employed to paint their own houses according to a pre-arranged pattern (see attached drawing for the project). The vision is to turn their community into an "artwork of epic scale" which will "produce an explosion of color, joyfully radiating into the world."

The project is organic and flexible enough to allow future expansion as more funds are raised. The transformation of this area of Santa Marta is visibly successful. Visitors are encouraged to ride a funicular up the steep mountain slope to the community to observe the area first-hand. According to one Rio de Janeiro resident, the project enticed her to visit the community, and now she frequently makes the trip up the mountain to dine at a marvelous Santa Marta restaurant.

Artwork Excellence

O Morro is an excellent example of response to place. It is a site-specific public art work with far reaching ramifications for the Santa Marta community and the larger metropolitan area of Rio de Janeiro, Brazil's second largest city. The project's unique combination of private fundraising coupled with training and employing people to paint their own homes makes it a model for future projects in the public realm. Its entrepreneurial approach to funding and collaboration is noteworthy. And, the overall effect of the 34 homes and the samba school which were painted by residents represents a remarkable transformation of an area that was considered blighted and an eyesore. It is a point of pride for the residents of the community of Santa Marta. Other favelas in the Rio de Janeiro metropolitan area are employing this model currently. While there are other similar noteworthy projects, this is a top-notch project that has been completed and has transformed a community.

潘恩纪念馆
Paine Memorial

艺术家：多人参与，潘恩社区居民三代居民（智利）
地点：智利潘恩
推荐人：卡利·英格布里特森
Artist: Various, three generations of Paine community members (Chile)
Location: Paine, Chile
Researcher: Cally Ingebritson

作品描述

"潘恩纪念馆"坐落于智利首都圣地亚哥南边约 20 英里的潘恩小镇，那里大概共有 50 000 居民，在马普切语中佩为"蓝天"，这是在智利和阿根廷通用的一种主要的当地语。潘恩在圣地亚哥闻名遐迩，因为它有植被茂盛动物成群的露营地，另外，它还因举办佩恩田园博览会而知名，那是一个每年一月举行的庆祝当地农业的盛会。

"潘恩纪念馆"是在对 70 名左翼人士表示敬意，他们皆是在 1973 年的皮诺切特独裁统治中被处决或是强行消失的激进的潘恩公民。这是颁给潘恩的苦痛勋章，因它在那个皮诺切特时代拥有全智利最高的人口失踪率。这个纪念馆自称"一个缅怀的地方"（"un lugar para la memoria"），并试图承认那些在智利政府暴力下失败的人。

"潘恩纪念馆"包含一个由单独的原木柱子组成的木头森林，总共有 1000 少 70 棵。每一棵柱子代表那些死者遗留下的朋友和亲人，而每一棵缺失的柱子则象征那些在 1973 惨案中失去生命的人。每一名受害者的家人受邀设计一个个性化的陶瓷锦砖取代缺失的柱子，以此来纪念他们挚爱的亲人。这些木柱子被排在一块方方整整、覆盖着灰色和浅褐色鹅卵石的地上。

2000 年，受害者亲属们组成了 AFDD——潘恩组织，即被处决与强行消失者家人联盟，作为一种途径让他们可以大声说出曾经发生的惨案并哀悼他们痛失的亲人。三代潘恩社区的成员参与了潘恩纪念馆的创立，让它成为一个真正的集合艺术品项目。纪念馆的布局可以让人们来反思过去，同时又可以举办活动和集会。最初的设计是想建造一个永久性的建筑，但智利政府所提供的资金仅够让它由一个放置在纪念馆旁的半永久拖车勉强对付和维持。

解读

又是一项关于纪念的公共作品，原木柱子的木头森林，是坚守的象征，也是无言的受难。由潘恩三代居民和其他公众集体参与的"潘恩纪念馆"，为潘恩家族也为其他过客，提供一个缅怀的场所。被迫害的 70 人，并不是孤单的受害者，留下的亲朋要在这一世界吞咽下苦果；缅怀的哀恸，也并非不会痊愈，而缺乏自由悼亡的压抑，才是永久的阴影。

亲属和居民联合实施这个项目，为的是能有一种公然的途径供他们大声讲出各自的遭遇和伤痛。只要公开哀悼的自由仍被限定，迫害的余孽就一日未曾消除，公共艺术项目在这里，充当了民意自由的催化剂，使压抑的沉痛得以缓慢纾解。

成功的公共艺术作品，绝不局限于纪念碑式的意义，它联系过往与未来，让公众从过去中抓住些什么，用于当下，寄望明天。这是"潘恩纪念馆"不单纯为悼亡，也为集会公众，成为公众受益项目的缘由，也是那 70 块陶瓷锦砖存在的意义。（吴昉）

Artwork Description

Paine Memorial ("Memorial de Paine" in Spanish) is located in the small town of Paine, approximately 20 miles south of Santiago, the Chilean capital. Home to roughly 50,000 inhabitants, Paine means "blue sky" in Mapudungun, one of the main indigenous languages spoken in both Chile and Argentina. Paine is popular with Santiago residents for its camping sites covered in beautiful flora and fauna and is well known for hosting Expo Paine Rural, a festival celebrating local agriculture held each January.

Paine Memorial pays homage to 70 leftist, active Paine citizens, who were either executed or forcefully disappeared during the Pinochet dictatorship in 1973. This gives Paine the grievous distinction of having the highest disappearance rate per capita in all of Chile during the Pinochet era. The memorial calls itself "a place of remembrance" ("un lugar para la memoria") and attempts to recognize those lost to the violence caused by the Chilean government.

Paine Memorial consists of a forest made of individual timber logs totaling 1000, minus 70. Each log represents the friends and family left behind by those who died, while each missing log symbolizes those who lost their lives in the 1973 tragedy. In place of a log, family members of each victim were invited to design a personalized ceramic mosaic honoring their loved ones. The wooden poles are arranged in a square plot of land, covered with gray and beige pebbles.

In 2000 family members of the victims formed the group AFDD-Paine, Association of Families of the Detained-Disappeared (Agrupación de Familiares de Detenidos Desaparecidos y Ejecutados de Paine) as a means to speak out about what happened and also grief their loss. Three generations of Paine community members were involved in creating the Paine Memorial, making this a truly collective art project. The layout of the memorial invites individuals to reflect on the past as well as host events and meetings. The original design intended to make the memorial a permanent structure, but the Chilean government only provided enough funding to manage and maintain the memorial from a semi-permanent trailer placed alongside the memorial.

Artwork Excellence

The Paine Memorial's excellence not only lies in its tribute honoring activists who lost their lives during the Pinochet dictatorship as well as acknowledging their friends and families, but in the unique process by which the memorial was created. In 2000, family members of the victims formed the group AFDD-Paine (Association of Families of the Detained-Disappeared of Paine, Agrupación de Familiares de Detenidos Desaparecidos y Ejecutados de Paine) as a means to speak out about what happened and also grief their loss. Three generations of Paine community members were involved in creating the Paine Memorial, making this a truly collective art project.

大豆
Que Soy

艺术家：多洛雷斯·卡塞雷斯（阿根廷）
地点：阿根廷科尔多瓦
推荐人：莱恩·柏格森
Artist: Dolores Caceres（Argentina）
Location: Cordobam, Argentina
Researcher: Laine Bergeson

作品描述

从 2008 年 12 月到 2009 年 4 月，艺术家多洛雷斯·卡塞雷斯接管了阿根廷科尔多瓦的卡拉法博物馆前面的花园，并且将它转变成了一片大豆农田。这个都市农业干预项目名为"大豆"，是对 2008 年阿根廷政府和农业部门之间冲突的回应。

2008 年 3 月，阿根廷政府在克里斯蒂娜总统领导下推出一个新的农业出口税制，把大豆出口的征税从 35% 提高到 44%。目标在于两方面：一是提高由于世界粮食价格上涨而增加的政府份额回报，从而筹集资金投资于社会；二是通过鼓励农民转变种植的主食比如小麦和玉米而不是出口诸如大豆一类的作物来降低国内食品价格。但是农民感觉税增加太多于是开始抗议，导致了全国范围长期的政治动荡和动乱。抗议了几个月后，对增税措施进行了几轮紧张的投票，结果是否决了这一措施，把出口税恢复到了三个月前的 35% 的水平。这一逆转是农业部门的一次重大胜利，也是阿根廷政治历史上的一个重大时刻。

卡塞雷斯创作了"大豆"回应了这段政治动荡的时期。这件作品西班牙语（Que Soy）的意思是"我是"，同时也有英文单词大豆的意思。这件作品是暂时的装置，使用土壤如同画布，用手种植大豆犹如作画。按照卡塞雷斯的说法，这件作品的部分目标是要建造一件与城市互动的公共艺术作品，让观众参与进来对政治局势进行反思。这件作品还要唤起观众思索永恒与无常、都市生活与农业生活以及阿根廷种植粮食的人和吃粮食的人之间的社会经济问题。作品坐落于博物馆门前，在大街上就清晰可见，这件作品迫使那些一般不进博物馆的人们面对了一件高度政治化的公共艺术作品。这件装置也激起了环保活动家们很强的反应，他们对这件作品以及阿根廷和全球的大规模农业生产有着反感情绪。他们反对这件装置，呼吁大家意识到单一种植农作物正对全球变暖产生影响以及毁掉原始森林用作农耕地的做法。但是这件作品背后的主要动机及其回应则是政治方面的，所涉及的相关问题是与农业、食物、社会、营养诸问题上的公正性以及抗议的伟大力量。

解读

艺术家多洛雷斯·卡塞雷斯在阿根廷科尔多瓦的卡拉法博物馆门前花园，种起了大豆，并将花圃改建为农田。这个略带无厘头趣味的公共项目，实则回应了阿根廷政府与农业部之间的决策矛盾。小小的大豆，在这场具有政治意味的公共项目中，荷枪实弹，饶有生趣。

"大豆"的寓意在于召唤公众对政治局势与经济发展进行思考。政策不再是当局者孤芳自赏的一拍脑袋，人们也不必再扮演无知的良民，民主的决策是以公众利益为主旨，借政府窗口发出的行为指导，如此而已。艺术家敏感到人们对当局政府决策的态度，很多时候就像对博物馆的观望，或许从未亲身涉足博物馆的内部。那么，将这项公共艺术移植到博物馆门前的公开场地，其实也是提醒公众有义务参与到政府行为中，公众是政府的第二决策力，犹如"大豆"已经成为了第二个公开的卡拉法博物馆。（吴昉）

Artwork Description

Que Soy—made of a simple, single soybean field—is notable for the manifold and complex ways it raised awareness of ongoing political and social justice issues, environmental concerns, and agricultural and nutritional issues in Argentina, and for how it blurred the line between city and country, art and public space.

Installed in front of the museum immediately after the political upheaval in Argentina, the piece prompted viewers—many of whom would not have otherwise entered a museum of gallery—to contemplate land use in Argentina (should it be used to grow food for export or local consumption?); issues around and global food production (should the land be used to grow soybeans—in greater demand for export—or Argentine staple crops such as wheat and corn?); government regulatory issues (to what degree should the government determine how the land is used and the crops are grown? How much autonomy should farmers have?); social justice issues (who has access to food? What food do they have access to?); environmental issues (how does farming and the large-scale monoculture crops impact global climate change and biodiversity?); issues around the nature of the urban environment (what belongs in a city? A soybean field? Or an ornamental garden? Can cities be reimagined to include food crops? What are the benefits and drawbacks of reimagining the urban environment?) and, not least, conceptual issues of what constitutes art (Is a soybean field art? Is it art if its not in a museum? What makes it art and not just a commercial crop, subject to its own export and tax levy?)

When one crop of soybeans raises all these complex questions and transgresses all these boundaries between the expected and unexpected, it is transformed—just as part of the city was transformed into a field—into a singularly excellent public art installation.

Artwork Excellence

From December 2008 to April 2009, artist Dolores Caceres took over the garden in front of the Museo Caraffa in Cordoba, Argentina, and turned it into a soybean field. The urban agricultural intervention, called Que Soy, was in response to the 2008 conflict between the Argentine government and the agricultural sector.

In March 2008, the government of Argentina, led by president Cristina Fernandez de Kirchner, introduced a new taxation system for agricultural exports, raising levies on soybean exports to 44 percent from 35 percent. The goal was two-fold: one, raise funds for social investment by increasing the government's share of returns from rising world grain prices, and two, reduce domestic food prices by encouraging farmers to switch to growing staples like wheat and corn, rather than export crops such as soybeans. Farmers, however, felt that the tax increase was too high and began to protest, leading to a prolonged period of turbulent politics and upheaval all across the country. After several months of protests, the measure was put to several tense votes and reversed, returning the export tax to its pre-March level of 35 percent. The reversal was a major victory for the agricultural sector and a significant moment in the history of Argentine politics.

Caceres created Que Soy in response to this period of political upheaval. The name Que Soy literally translates as "I am;" it also refers to the English word for "soybeans." The piece was an ephemeral installation that used the soil as its canvas and hand-planted soybeans as its paint. Part of the goal, according to Caceres, was to build a piece of public art that interacted with the city and engaged viewers in thoughtful reflection about the political situation. The piece was also meant to provoke viewers to think about permanence versus impermanence, urban life versus agricultural life, and the socio-economic concerns of both the people who grow food in Argentina and the people who eat it. Visible from the street in front of the museum, the piece forced people who would not normally go into a museum to confront public, and highly political, art.

The installation also provoked strong reactions from environmental activists, who had negative feelings about the piece and about the state of large-scale agricultural production in Argentina and across the globe. They protested the installation to raise awareness about the effects of monoculture crops on global warming and the destruction of wild forests to make room for agricultural land.

But the main motivation behind the piece—and reaction to it—was political, addressing relevant issues related to the politics of agriculture, food, social and nutritional justice, and the power of protest.

萨尔瓦多壁画项目
Salvador Mural Project

艺术家：朱迪·巴卡（美国）
地点：萨尔瓦多阿塔哥市街头
推荐人：彼得·莫拉雷斯
Artist：Judy Baca（USA）
Location：Street of Ataco, Salvador
Researcher: Peter Morales

作品描述

这个壁画项目的初衷是描绘联合国的千禧年发展目标。在开始时，艺术家朱迪对她的作品做了公开演示，并且意识到这个项目已经吸引了整个中南美洲的壁画艺术家，他们从未有过自己的壁画工作室，更没有机会参与大型合作项目。

该壁画全长 350 英尺，其中 6 座壁画有 8 英尺高，画在城市的多面墙上，在巴卡的团队以及周边社区的热心成员的协助下共同完成。社区的居民最大限度地参与了该项目的方方面面，他们也由此得到一次宝贵的自由表达意见的机会。壁画体现了对社会问题的关注，没有任何明显的政治派别意识，最终的设计展示给整个社区以获得认可。艺术家力图包容各种观念，通过共同协商达成一致。

壁画表达了对诸多社会问题的关注，其中一个重要主题是环境问题，包括附近咖啡种植园喷洒农药导致的水质污染、烹饪用的木材也饱受农药污染、产生的烟雾导致心肺疾病等等；另一个问题是儿童保护，许多儿童因被迫到咖啡园打工而辍学；关于土著身份的问题也有大量篇幅的描绘；另一个有争议的主题是男女平等，该地区 80% 的女性遭受家庭暴力，对于妇女及其家庭的正当和不正当行为的画面也穿插在壁画中。因此，这个项目吸引了中南美洲的许多壁画艺术家，引发了中南美洲和萨尔瓦多公共艺术状况的一场大众讨论。

解读

"萨尔瓦多壁画项目" 共 350 英尺长，贯穿整个城市，其中 8 英尺高壁画有 6 座，展示在城市的多面墙面上，由巴卡的 LA 团队以及在周边社区的热心成员的协助下，在交流互动基础上共同完成。社区充分地参与了该项目的整个过程，参与该项目的乡民也得到了一次自由表达意见的特殊机会。

壁画表达了对社会问题的关注，环境问题、儿童保护问题、土著身份和根的问题贯穿了整个壁画。另一个争议主题是性别平等。这个项目吸引了整个中南美洲的壁画艺术家，成为中南美洲和萨尔瓦多公共艺术状况的一场即兴会议。（冯正龙）

Artwork Description

The original intention of this mural project was to illustrate the United Nations Millennium Development Goals. At the outset the artist Judy Baca made a public presentation of her initial work, and discovered that the project had already attracted interest from mural artists throughout Central and South America who had no studios of their own nor opportunity to participate in large-scale cooperative projects.

The total work measures 350 feet long, comprised of six murals each 8 feet high, painted on various walls of the city. The artwork was completed by Baca's L.A. team with enthusiastic assistance from members of the community, who participated in all aspects of the project. For the townspeople this project represented an unusual opportunity for free expression about a range of concerns and issues in the community without the intrusion of partisan politics. Final designs were presented to the whole community for approval. The artists strived to include all the ideas brought forward through consensus.

The murals gave expression to the concerns and issues of the community, notably the health of the environment. Environmental concerns included excessive water pollution due to pesticide runoff from neighboring coffee fields, and smoke from pesticide-laden wood used for cooking indoors leading to enlarged hearts and lung disease. Another important theme was the protection of children. Many children are faced with the need to work in the coffee fields and are therefore unable to attend school. The theme of indigenous identity and cultural roots was also woven throughout the murals. Another controversial theme was gender equality. Up to eighty percent of the women in the region are thought to be victims of violence. Images of proper and improper behavior toward women and families were depicted in the paintings. As a result this project attracted the attention of many mural artists throughout Central and South America, and has become a catalyst for public discussion about the status of public art in El Salvador and throughout the region.

Artwork Excellence

The excellence of Judy Baca' s El Salvador Mural Project is evident in her capacity to respond to substantial changes in the dynamics and conditions of the social environment. It is a large-scale work that is channeling the energies of a large number of people in a collective communal fashion. This project created an impressive opportunity for self-expression in this otherwise marginalized community under politically fraught conditions. The work was completed in a mere ten days. Eight months later the mayor, who had originally supported the project, defaced and repainted more than half of it.

An organization has emerged as a result of the defacement of the murals: El Comité de Apoyo al Frente is promoting a woman to run against the mayor in the upcoming elections later this year (2012). The project has given the people of Ataco a sense of agency against powerful interests and even as the work was defaced the images that were expunged are now used to promote issues that are important to women and the community.

考卡山谷省议会代表
The Deputies of the Assembly of Valle del Cauca

艺术家：多利斯·萨尔赛多（哥伦比亚）
地点：哥伦比亚波哥大
推荐人：卡利·英格布里特森
Artist: Doris Salcedo（Colombia）
Location: Bogotá, Colombia
Researcher: Cally Ingebritson

作品描述

玻利瓦尔广场是哥伦比亚波哥大的主要广场，是哥伦比亚司法宫、国立国会、波哥大市长大楼和波哥大主教堂的所在地。

近 200 年来这里成了哥伦比亚政治和历史的中心，如今游客们涌向这里在西蒙·玻利瓦尔雕像前拍照，在风景如画的蒙塞拉特山俯瞰全城。

2007 年的一个夜晚，玻利瓦尔广场被艺术家多利斯·萨尔赛多转换了。创作了 "考卡山谷省议会代表"，她在广场上铺满了蜡烛以纪念 11 位被 FARC（哥伦比亚革命武装部队）杀害的哥伦比亚政府代表。立法委员在卡利被绑架为人质，这是哥伦比亚第三大城市也是考卡山谷省的首府，在这些人质被押五年后遭杀害。FARC 经常实施政治绑架给哥伦比亚政府施压迫使政府释放被关押的游击队员以换取人质。政府一直对这种战术对策感到反感。游击队员把这些代表关押在丛林里，后来他们觉得遭到了政府军的袭击，于是杀害了 12 位代表中的 11 位。

听到这个消息后，多利斯·萨尔赛多到波哥大中心和众多志愿者一起，悄悄地逐一点燃蜡烛，直到 25 000 支蜡烛覆盖满玻利瓦尔广场。"纪念是我作品的真髓，"多利斯·萨尔赛多说道。"如果我们不了解自己的过去，我们不可能正确地生活在现时，也就无法去面对未来。"这件作品是一种哀悼行为，它制造了一种空间，在那里人们可以去缅怀这些逝去的代表。

多利斯·萨尔赛多利用她的这种经验，作为"另一种来自错误地点"的艺术行为来表达她对于哥伦比亚政治事件的诠释。她批判性地看待自己作为一名艺术家的角色任务："我不认为艺术能够解决问题。我没有为这些家庭做任何事；我没有为这些受害者做任何事，当我们谈论起这种类型的试图解决政治问题的艺术时，我们必须面对这一现实。艺术没有能力挽回一切。"

将玻利瓦尔广场铺满蜡烛的行为默默地显示着哥伦比亚的尊严，这是一个经常在举行哀悼的国家，没有遗体，常常连遗体都没有。"如果要让人的生命保持住尊严，那么就必须要回归美丽，"多利斯·萨尔赛多说，"我们就在这里找到的尊严，几乎把它变成了一个神圣的空间。"

玻利瓦尔广场三面由历史悠久的政府大楼构成，建筑高高地矗立在广场上，政治家在那里做出过许多高层决议。那些蜡烛小而静穆，时时在提醒着高层人士不要忘记自己的失职。

解读

哥伦比亚艺术家多利斯萨尔赛多清醒地认识到："艺术没有能力挽回一切。"她所做的，只是用公共艺术的形式，开放性地，为那些值得纪念的生命，挽回些许应有的尊严。而对于尊严的获取，多利斯将其定义为回归艺术之美。

"夜正长，路也正长，我不如忘却，不说的好罢。但我知道，即使不是我，将来总会有记起他们，再说他们的时候的……"《为了忘却的记念》，鲁迅也深谙文艺救不了国难，但这不影响其文风其思想的感染力。艺术的纯粹能够涤荡罪恶带来的破坏，这是艺术自身所具备的特质，公共艺术家所要做的，就是将此特质自然地展现出来。"考卡山谷省议会代表"万朵烛花闪烁，照亮了天际，身临现场，那份庄严与宏阔的气象，让人不得不重新认识到生命的肃穆、壮阔，同时也是脆弱。"用一个神圣的空间，去找回生命的尊严。"就此而言，多利斯做到了。（吴昉）

Artwork Description

Plaza de Bolívar, the main square of Bogotá, Colombia, is home to Colombia's Palace of Justice, National Capitol, Bogotá's Mayor's building and the Primary Cathedral of Bogotá. For nearly 200 years it has been the political and historical heart of Colombia. Tourists now flock here to take pictures of the Simón Bolívar statue and the picturesque Monserrate Mountain overlooking the city.

One evening in July 2007, Plaza de Bolívar was transformed by artist Doris Salcedo. For The Deputies of the Assembly of Valle del Cauca (Los Diputados de la Asamblea del Valle del Cauca), she filled the plaza with candles honoring 11 Colombian government representatives killed by the FARC (Fuerzas Armadas Revolucionarias de Colombia, or Revolutionary Armed Forces of Colombia). The legislators had been taken hostage in Cali, the third largest city in Colombia and capital of the department Valle del Cauca, and held for five years before they lost their lives.

The FARC used political kidnappings in an effort to pressure the Colombian government: they would release kidnapees in exchange for imprisoned guerrillas. Until this point the government had been opposed to this tactic. After holding the representatives in the jungle, the guerrillas thought they were being attacked by army forces; the FARC reacted by killing 11 of the 12 representatives they had sequestered five years prior.

Upon hearing this news, Doris Salcedo took to the center of Bogotá accompanied volunteers and quietly lit candles one by one until 25,000 of them covered Plaza de Bolívar. "Memory is the essence of my work," Doris says. "If we don't know our past, there is no way we can live the present properly and there no way we can face the future." This piece was an act of mourning that created a space where people could remember the deceased representatives.

Doris uses her experience as an "other from the 'wrong' place" to express her interpretation of political events in Colombia through her work. She is critical of her role as an artist: "I don't think art can solve problems. I'm not doing anything for these families; I'm not doing anything for these victims. That's a reality we have to face when we talk about this type of art that's trying to address political issues. Art does not have the ability to redeem."

The act of filling Plaza de Bolívar with candles silently honored Colombia, a country that constantly mourns—oftentimes grieving death without bodies. "If you want to dignify a human life then you have to come back to beauty," Doris says, "That's where we find dignity and almost turn (it) into a sacred space." Plaza de Bolívar is flanked on three sides by historic government buildings rising high above the square where politicians make high level decisions. The candles, small and quiet, reminded them of the work they have failed to do.

Artwork Excellence

The Deputies created a space of mourning for 11 Colombian representatives killed by the FARC, turning Bogota's Plaza de Bolivar into a memorial of light.

提乌纳的堡垒文化公园
Tiuna el Fuerte Cultural Park

艺术家：亚历杭德罗·海埃克·科尔、艾莉娜·卡达尔索、
米歇尔·桑切斯·莱昂·布拉伊科维奇（委内瑞拉）
地点：委内瑞拉加拉加斯隆加雷大街
推荐人：彼得·莫拉雷斯
Artist：Alejandro Haiek Coll, Eleanna Cadalso and
Michelle Sánchez de León Brajkovich (Venezuela)
Location：Calle Longaray, Caracas, Venezuela
Researcher：Peter Morales

作品描述

该项目选择了一座被废弃的停车场用来开发公园，这是一场由建筑师和艺术家主导的新兴集体文化运动。为了节约经费和能源，基础设施均以非主流施工技术建造。回收来的船运集装箱被组合在一起，转换成多元的模块元素，配置成各种功能的空间并且能够逐步增长。这个项目的另一个目的是为了改善当地的环境，提高绿化覆盖率。目前，加拉加斯首都区域的绿化覆盖率人均只有 0.26 平方米，远低于世界卫生组织规定的每个城市居民应享有 10—12 平方米公园空间的标准。

几年来，该项目的占地面积不断扩大，逐渐成为一座城市文化公园，不仅建立起办公室、教室、餐饮、绿色和体育区域，还计划通过举办工作坊和开展活动来促进各类艺术和科学的发展。目前，计划中的五个项目已经完成了两项。公园现有一家商店、一座自助餐厅、一个广播站和一个音乐编辑台。每天有超过500名儿童和青少年前来参加文化和艺术活动。

公园的名字引用自附近一个军事基地的称谓："提乌纳堡垒"。提乌纳是当地一位英雄的名字，以他的名字命名是为了让团队回归这个名字本身的自然和社会意义。艺术家们没有将改变社区面貌的希望寄托于政府体系，而是依靠民间的力量来推动公共艺术，体现出创造性、连贯性及其改善当前社会和人类生存环境的能力。

解读

"提乌纳的堡垒文化公园"这个项目选择在一个被废弃的停车场，这是在20世纪80年代一个很危险的街区，它处在高速公路当中，也没有市政监管，等于处在交叉区，艺术家认为这是一个对这个社区进行改造的很好的机会。他们要重新打造这个地区的环境形态和社会形态，同时对坐牢的、监狱里的孩子通过艺术的吸引力对他们进行改造，能够让他们把手里的那些作案刀具武器变成他们唱歌的话筒。另外，他们希望通过废弃空间以及废弃物的重新回收利用，重新把物、地形、人三种因素平衡起来，致力于打造一种所谓的微城市主义，也就是说它在一个比较小的规模上进行社会化，并且在一定程度上影响法律，能够让目前的局面实现再平衡。

社会责任心的体现，多方的支持与赞助，棚户区人才济济，使这个项目受到居民的热情参与与合作，这个地方可以说已经成为新兴集体文化运动场所、居民的精神堡垒，民间的创造力可以在这里尽情发挥，生存环境得到积极的改善，从而提升了这个公共社群的精神性生活覆盖率。（冯正龙）

Artwork Description

The project encompasses an emerging collective cultural movement led by architects and artists, which is focused around the development of a park on an abandoned parking lot. The park infrastructure was built using cost-effective, low-energy technologies. Recycled shipping containers, for instance, were grouped together as modular elements in expandable multi-use spaces. Increasing green coverage in the city was another project priority. Currently the Capitol District of Caracas contains an average of .26 square meters of park space per inhabitant, compared to the World Health Organization's recommended standard of 10 to 12 square meters of park space per city inhabitant.

In recent years, the project area has continued to expand, becoming a city cultural park containing offices, classrooms, dining spaces, green spaces and sports areas, and with plans for organized workshops and other activities promoting development in the arts and sciences. At present, of a total of five planned projects two have been completed.

There currently exist a store, a cafeteria, administrative offices, a radio station, and a music-editing studio. On a daily basis more than 500 children and adolescents participate in cultural and artistic events in the park.

The name of the park is a reference to a nearby military base, Fuerte Tiuna. However, "Tiuna" was originally the name of a native warrior from the region. Thus the name has been re-appropriated to restore its natural and social connotations. The artists did not put their hopes in government assistance, but rather relied on civic involvement to promote public art. The project reflects the creativity, cohesiveness and the capacity that is possible when hands join together to improve the social and human environment.

Artwork Excellence

Haiek and Cadalso are spearheading an emergent cultural movement of collective action led by architects and artists who do not wait for a failed system of government to respond to the needs of the community but make things happen by informal means. Haiek and Cadalso's Tiuna el Fuerte Cultural Park received 1st Place in the International Festival for Architecture eme3 in Barcelona, Spain, in 2011. This award was based on the creativity of the project, its coherence, and its capacity to improve its immediate social and human environment. Their collective work has also received mention in Harvard Design Magazine blog post in the same year. The project has received a National Prize for Culture in Venezuela.

大洋洲地区提名作品
OCEANIA NOMINATED CASES

21 海滩牢笼
21 Beach cells

艺术家：格雷戈尔·施耐德（德国）
地点：澳大利亚悉尼邦迪海滩
推荐人：凯利·卡迈克尔
Artist: Gregor Schneider (Germany)
Location: Bondi Beach of Sydney, Australia
Researcher: Kelly Carmichael

作品描述

作品设置在澳大利亚悉尼最著名的旅游景点——邦迪海滩，每个4米见方的网状笼子与周边环境形成强烈反差，形成了令人不寒而栗的对应物。休闲的美丽海滩的一部分被转换成转换军事区域般的景象，破坏了海滩原本具有的娱乐休闲、公共开放的空间概念。作者擅长驾驭不同形态的空间表现力，他认识到建筑空间能够影响人的深层意识并产生异化感，作品以政治见解透彻著称。

这个装置利用人们对海滩的日常感知，让参与者在建筑空间中经历一次心理上的逆向体验。这些网状笼子里除了蓝色气垫、海滩遮阳伞之外，还有令人不安的黑色塑料垃圾袋。虽然这里的阳光依旧明媚，参与者透过网格和垫子能够听到海浪拍打沙滩的声音，但在貌似舒适的听觉里，却感到被这个网状结构束缚住，由此产生某种被囚禁的心理暗示。作品正是将快乐与不安融合在一起，通过颠覆现实来揭示在平凡时的不安感。

装置通过营造人为构筑空间与个人心理空间的冲突，不仅使置身其中的参观者游移于自由和被监控、隐私和曝光、内部和外部的错觉之间，而且从更深层次看，作品还意在影射当时澳大利亚的政治气氛，例如难民被拘留在国外的中转站、附近的克罗纳拉海滩上爆发的种族骚乱以及政府在移民问题上的僵化立场等，体现出强烈的批判现实的寓意。

解读

这个是邦迪海滩项目。2007 年在澳大利亚邦迪海滩的项目"21 海滩牢笼"，这里面是 20 个大小一致的牢笼，把它安置在澳大利亚东海岸的邦迪海滩。这实际上是一个展出空间，一种非常休闲的平等社会当中的一个装置。在这里，海滩游客会和这个装置产生互动与交流。施耐德先生在访问了澳大利亚后，制作了这个作品，并把这个装置安置在邦迪海滩，长度是 44 米，21 个海滩牢笼。他通过这种方式，将人们的安全、威胁进行对照。人们一进入笼子里之后，这个时候他们看外面的景观就有了一些变化。有些牢笼是关闭的，要通过一个阶梯才能进去。而有些是开放的，通过开放的走廊在里面穿行。这个项目也造成了很多争议，不仅创造一种休闲以及和项目之间的对照，而且也充分体现了澳大利亚这个国家的身份。

施耐德先生讲到，2005 年 12 月 11 号出现在澳大利亚这样一个看似非常平静的海滩的种族暴动，对这个作品产生影响。施耐德先生可以说是空间的心理学家，所以说他的整个海滩牢房的摆放充分利用了人们的心理。它没有屋顶，所以就向空中开放。每个牢笼里都会有一些供人使用的设施，包括垫子和遮阳伞。有人问施耐德，说你是不是一个政治艺术家，他说不是，我是一个当代艺术家。我作为艺术家也关注世界，包括这个作品也关注澳大利亚的政治问题。

这个海滩创造了一种既有自由、又有限制的景象。这是充满矛盾、充满冲突的景象。一些评论者说，关塔那摩以及澳大利亚自己的拘留营都是投射在这上面的阴影，"21 个海滩牢笼"反映了在恐怖主义大的背景之中，社会里的一些紧张局势包括旁边海滩曾经发生的种族暴乱等。（冯正龙）

Artwork Description

This art installation was mounted at Bondi Beach, a major tourist destination in Sydney, Australia. Large mesh cages, each four meters square, were erected on the beach in striking contrast to their surrounding environment. The normal recreational appearance of the public beach was arrestingly interrupted by the impression of a military installation. Gregor Schneider is adept at manipulating the expressive quality of a space, and at using architectural elements in particular to elicit a more profound awareness of space, often creating a sense of alienation. His artworks are renowned for their strongly political views.

This installation plays on conventional assumptions about beaches, and through construction of space elicits a psychologically tense experience. The placement of blue air mattresses and beach umbrellas inside the mesh cages alongside several disconcerting black plastic rubbish bags contributes to this effect. Although the sun shines brightly and participants can hear the crashing surf on the beach, the audience still feels weighed down by the mesh structure which conveys the psychological suggestion of imprisonment. The artwork juxtaposes artifacts of happiness and anxiety, and in its subversive portrayal of reality exposes a persistent sense of unease.

This installation creates a tension between physical space and psychological space. Visitors wander at liberty and yet feel monitored; there is privacy yet exposure; there is both an internal illusion and an external one. On a deeper level, this artwork also makes allusions to the Australian political climate. The incidence of refugees detained at foreign ports, race riots erupting on the beaches of nearby Cronulla, and the government's unabashadely tough stance on immigration and refugees, are all subjects of strong critical reflection.

Artwork Excellence

In 21 Beach Cells, Schneider took an everyday Sydney beach experience and provided a chilling counterpart. Blended pleasure and trepidation in a disquieting and very public art work, Schneider allowed beach goers to see and understand a different physical and psychological space. They were no longer observing but participating.

钢铁和草坪的干预项目
Breather/Tenefüs

Breather / Tenefüs
艺术家：穆拉特和弗阿特·萨英勒（土耳其）
地点：新西兰基督城美术馆前院
推荐人：阿尔苏·雅英塔斯
Artists: Murat and Fuat Sahinler (Turkey)
Location: Forecourt of Christchurch Art Gallery, New Zealand
Researcher: Arzu Yayıntaş

作品描述

穆拉特和弗阿特·萨英勒对公共空间的干预方式通常是挑战公共空间的限制，以此建立新的空间概念，同时质疑现有的公共空间。他们的项目主要是为寻找新的机遇，扩大和繁殖公共空间，以及社会和空间的可能性。他们曾在伊斯坦布尔通过使用照片和谷歌地球图像开展项目。他们从宏观层面上观察到画廊前面的空间，包括前院，可以统一作为一个整体，形成一个新的集会广场，因为在现有情况下，这个空间只用于街道和人行道。

在位于美术馆的建筑和街道之间的前院，在现有绿地之间，那里是空置的，有可能被转换为同一目的。简单地带走一个石头框架周围的绿化面积的部分，它被塑造成一个圆形剧场形式，创建了草地之间的连续性，使民众更容易进入前院，对大众产生吸引力。他们对这个露天形式的偏爱，主要源于其古典结构，提供了一个聚会、社交和休憩的地方，同时对比美术馆门面的有机形式。这种圆形剧场提供了休息的时间和空间。

这个项目打破了我们在都市环境中的日常体验，在美术馆的大型建筑前，打破了现有的空间秩序与被动接受公共空间的现状，增进了公众之间分享的多样可能性。

解读

公共艺术对公共空间的干预有多种方式，干预方式的成功与否关键在于敢于打破限制，质疑现有或已经存在的公共空间，建立新的空间概念与内涵。

这个项目的目的就是寻找新的机遇，扩大和繁殖公共空间，以及社会和空间的可能性。他们在现有的绿地上，进行了空间转换，把原先的一部分绿地面积，塑造成了一个圆形剧场的形式，这样不仅使绿地之间更具联系性，而且还创建了一个既有互动性又具休憩性的露天形式的地方。

原先的这块绿地仅仅为了观赏与划分功能区，现今却成为公众驻足聊天互动的场所。这个项目打破了公众在都市环境中的日常体验，在质疑现有空间秩序与公众被动接受固定的公共空间现状的情况下，打破空间限制，增进公众在公共空间互动、分享、交流的多种可能性。（冯正龙）

Artwork Description

Murat and Fuat Sahinler's interventions usually push the limits of public space, seeking opportunities to expand and multiply the possibilities for social meaning in public space. For this project, they started work from their base in Istanbul, first looking at photographs, plans and Google Earth images of the site. From this birds-eye perspective they discerned that the spaces in front of Christchurch Art Gallery, including the forecourt, street and the sidewalk, could be integrated to form a new gathering place, in contrast to their existing function as transportation corridors.

An existing vacant green patch located between the Gallery building and the street suggested a possible approach. Simply by removing a section of masonry barrier and shaping the green area into an amphitheater form, the duo created continuity among the various spaces and made them more accessible and appealing. They chose an amphitheater form for its suggestion of a place for gathering, social encounter and leisure while offering a contrast with the organic form of the gallery façade. The grass amphitheater introduces a break in space and time for the pedestrians passing by.

Artwork Excellence

The project constitutes a challenge to passive acceptance of the existing urban environment, and to the massiveness of the gallery building. It indicates the diverse possibilities for expanding the public sphere.

角柱和怀佩罗沼泽人行道
Corner Post and WaiperoSwamp Walk

艺术家：比利·爱普尔（新西兰）
地点：新西兰奥克兰
推荐人：凯利·卡迈克尔
Artist: Billy Apple（New Zealand）
Location: Auckland, New Zealand
Researcher: Kelly Carmichael

作品描述

伊甸公园是新西兰最大城市奥克兰的橄榄球故乡。当时正准备主办 2011年橄榄球世界杯，公园需要升级，要对这块体育场地进行小规模的关键性改造。比利·爱普尔非常擅长改造任务，他是一位新西兰艺术家，出生名是巴里·贝茨，在 1962 年改姓了爱普尔，因为他得到了伊卡璐速效头发漂色膏女士的一点帮助——他组建了一支核心设计团队，负责"怀佩罗沼泽漫步"项目。他们在两条主干道中间进行分割，伊甸公园有一条捷径：怀佩罗沼泽人行道是新西兰第一条作为艺术品而设计的道路。这是行人专用的人行道——除了在非比赛日，道路面向车辆开放一条单向通道，成为了奥克兰第一个官方的大众分享的空间，行人和车辆可以共享同样的区域。

"怀佩罗沼泽漫步"由大型黑色和白色的砖块和浇灌的混凝土所构成。道路两旁左右各围着火山干石墙，使用的材料进行了严格地几何对称设置，材料与奥克兰当地的景观相对应。临近"怀佩罗沼泽漫步"的是爱普尔的作品"角柱",是一个实地的具体项目,是体育场角柱的缩小版,矗立在一片充满活力的绿色回收玻璃制成的树脂场地上,越过了白色边线,让人联想起了体育场上的边界线。毗邻格子路径,黑色与白色对于体育爱好者来说如同起到了助记符设备的作用,触发与橄榄球的联系(让人联想起新西兰全黑队的黑色队服),板球队的白色,或是赛车格子旗。超大号的角柱顶部有一面旗帜与几何设计相吻合,同时也用作导航设施引导体育爱好者走对路,也可用作比赛前和比赛结束后的集合地点。

这件作品得到了爱普尔喜欢的神圣的分割或称为黄金比例的数学概念的强调,定义数值比是 1:1.618,这一比例在自然界很常见,最早在古代被发现。本件作品艺术家采用了黄金比例来决定整体尺寸以及几何设计元素的位置。

艺术家的职业生涯历时 50 年,地点跨度有三处(伦敦、纽约和新西兰),比利·爱普尔制作过物品、文学作品、摄影作品和装置,他采取过行动来验证艺术的定义和挑战对艺术身份的构建假想,揭示艺术系统的运作,展示了艺术对于大型社会、政治和经济力量的渗透性影响力。

解读

新西兰艺术家比利·爱普尔注重细节处理,"角柱和怀佩罗沼泽人行道"处处体现出奥克兰当地值得骄傲的橄榄球运动象征,从黑白几何格纹地面,到绿色树脂场地,跨越白色的边界线,视线停落在角柱上方格状的迷你旗帜上,所有的设计元素,都鲜明围绕着体育精神,为当地民众增添一个行走休闲新场所的同时,也激发了人们的荣誉感与地方归属感。

评论称比利·爱普尔善于揭示艺术系统的运作,展示了艺术对大型社会、政治及经济力量的渗透性影响力。其中,渗透性可谓公共艺术建设的一个关键词。一个项目,如果在周遭环境中过于突出,则未必能自然地融入所在氛围,难免牵强;如果过于默默,又多半导致忽略乃至遗忘,不能将作品的公众性激发出来,最终和光同尘,失去存在的价值。而成功的公共艺术项目,能把控住"出世"与"入世"的度,用潜移默化的力量,渗透公众的日常生活和精神世界。这也是"角柱和怀佩罗沼泽人行道"给予我们的实践经验。(吴昉)

Artwork Description

Eden Park is the home of rugby in Auckland, New Zealand's largest city. When it needed upgrading to host the 2011 Rugby World Cup, a small but important access to the sporting venue was transformed. No stranger to transformation himself, Billy Apple—a New Zealand artist born Barrie Bates conceived his Apple identity in 1962 with a little help from Lady Clairol Instant Crème Whip hair bleach—formed an integral part of the design team charged with the Wairepo Swamp Walk project.

Cutting between two main roads and a much used shortcut to Eden Park, Wairepo Swamp Walk is the first road in New Zealand designed in its entirety as artwork. It is a pedestrian-only lane—except on nonmatch days when it is open one-way to vehicles, making it Auckland's first official "shared space" where pedestrians and vehicles use the same area. Wairepo Swamp Walk consists of large black and white slabs of poured concrete. It is framed on either side by volcanic dry stone walls, its strict geometric nature juxtaposed with the use of materials that are part of the local Auckland landscape.

Adjacent to the Wairepo Swamp Walk is Apple's The Corner Post, a site-specific, scaled-up version of a sports field's corner post standing on a vibrant green "field" of recycled glass imbedded in resin and crossed with white markings recalling boundary lines on a sports pitch. Bordered by a chequered path, the black and white acts as mnemonic devices for sports fans, triggering associations with rugby (the black uniform of New Zealand's All Blacks), cricket whites or the checkered flag of motor racing. The oversize corner post is topped by a flag echoing the geometric design and also serves as a navigational device to orientate fans on their journey, or as meeting point before and after the game. The work is underlined by Apple's beloved mathematical concept of the divine proportion or "golden section," a numerical ratio defined as 1:1.618 that frequently occurs in nature and was first recognized in ancient times. For this work, the artist called upon the golden section to dictate the overall dimensions as well as the location of elements within the geometric design.

In a career spanning 50 years and three locations (London, New York and New Zealand) Billy Apple has produced objects, text pieces, photographs, and installations, and undertaken actions, that test definitions of art, challenge the structuring suppositions of artistic identity, expose the workings of the art system and demonstrate art's permeability to larger social, political and economic forces.

Artwork Excellence

Fitting for an artist known for his rigorous investigation of the sites, systems and social relationships that structure the art world, Billy Apple has created not only two major installations but also a new public space, a forum for discussion of the sporting match and a work that is ignited by those who use the space to meet, traverse and interact with the site. Just as a sports venue is only a field without its players to ignite the game, so too does the intervention of artist and public bring this formally quotidian suburban environment to life and create sense of place.

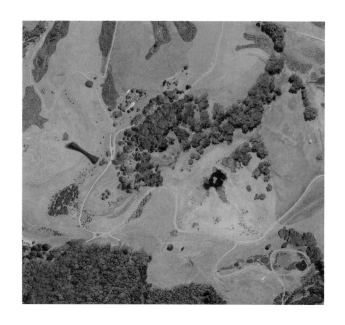

吉布斯农场
Gibbs Farm

艺术家：多位艺术家
地点：新西兰奥克兰凯普拉港北部雕塑公园
推荐人：凯利·卡迈克尔
Artist: Multiple artists
Location: Sculpture Park in Kaipara Harbour of North Island, New Zealand
Researcher: Kelly Carmichael

作品描述

"吉卜斯农场"，一位名为艾伦·吉卜斯的商人精心设计和创办的产物。该农场是一片绵延一千英亩的雕塑公园，坐落于新西兰最大城市奥克兰凯普拉港北部，驱车大概一小时左右。作为新西兰几位头号艺术赞助人之一，吉卜斯于 1991 年购买了此块地产并开始邀请和资助艺术家前往那里创作雕塑艺术品，直接从农场和周边的环境中激发出创作灵感。

按面积看，凯普拉港算得上是世界上最大的港口之一。涨潮时港口面积覆盖 947 平方公里，在退潮时露出大约 400 平方公里的泥滩和沙地。该港口面积巨大，它占据了整个地区面向西方的地平线。同时该港口面向西风盛行的方向，有时风力巨大，横扫整个区域。吉布斯农场面朝大海，并最终延绵进海边水中，每一件作品依次放置，仿如与海争辩。新西兰著名的沿海地形和空阔的天海之势，在吉布斯农场与艺术作品一起被独特地保留和呈现。

吉布斯农场的地形为雕塑艺术品的收藏塑造了新的美学，使得农场本身成为有关在地极简主义最有趣的收藏案例之一。为了呼应农场宏大的规模和地理环境苛刻的自然属性，艺术家们不得不在前有的艺术创作基础之上更加努力一筹。河谷深处，山坡沿线，水流之中，平原低地之间，山丘自上而下，当代艺术中最知名艺术家们的雕塑作品臣服于如此开阔的凯普拉海港，并与轮廓线粗旷的古老地质景观完美融合。

理查德·塞拉（Richard Serra）的"钢制风景线装置"，从体量上分割了地景；肯尼思·斯内尔森（Kenneth Snelson）的"简单 K 系列"拒绝地心引力的"漂浮压缩"；还有托尼·奥斯勒（Tony Oursler）的"泥巴歌剧"，给以原始沼泽泥地的丰厚的致敬。形式上串联整合这些作品的是丹尼尔·布瑞经典的"绿白相间的栅栏"，和整个地形纵横相交。艺术家安尼施·卡普尔（AnishKapoor）"分割，地点一"装置嵌在山脊之间，朝着农场的内陆和山坡另一端的大海进行视觉和听觉的延伸。在潮汐区域，新西兰本土艺术家鲁塞尔·莫斯（Russell Moses）的名为"凯普拉洼卡"的大地艺术作品和艺术家安迪·戈兹沃西（Andy Goldsworthy）的"拱门"都拥抱当地现有沉淀砂粒、岩石和泥土，甚至企图超越其上，升华出一种在别处的境界，或似一种精神之旅，或如一种迁徙模式，抑或一种向遥不可及的地平线步步踏去的执着。

吉卜斯农场以平均每 12 个月的速率收藏作品，但是很多作品需要三到五年的时间去计划和实施，一些国际上著名的艺术家如艾瑞克·欧（Eric Orr），乔治·瑞奇（George Rickey），马奇克·德·高伊（Marijke de Goey)，索尔勒维（Sol LeWitt）和中国艺术家展望联合新西兰本土的艺术家如比尔·卡尔波特（Bill Culbert），克里斯·布斯（Chris Booth ），格瑞汉·班奈特（Graham Bennett），莱耶（Len Lye），里欧·凡·戴·欤克尔（Leon van den Eijkel），内尔·道森（Neil Dawson），彼得·尼尔克斯（Peter Nicholls），彼得·瑞奇（Peter Roche），拉斐尔·霍特热（Ralph Hotere）和理查德·汤姆森（Richard Thompson）。吉卜斯农场聚集了当今最为出色的一批艺术家的重要代表作，超越了任何画廊和美术馆在新西兰的收藏实力。

解读

看到新西兰的雕塑公园"吉布斯农场"，不禁让人联想起 20 世纪 60 年代末出现在欧美的"大地艺术"，由极少主义艺术的简单和无细节形式发展而来，艺术家主张返回自然，以大地作为艺术创作的对象。"吉布斯农场"坐落于新西兰最大城市奥克兰凯普拉港北部，绵延 1 000 英亩，艺术家们从周边的农场和自然风光中汲取创作灵感，是一项将"大地艺术"创作有规划地聚拢在一处的景观项目。

目前"吉布斯农场"收藏的艺术家作品，已经超过了新西兰任何一家画廊和博物馆的收藏实力，成为室内装置或雕塑在室外空间的无限延展区。"吉布斯农场"内置的所有公共作品，都无一例外，格外关注作品的"场地感"，艺术家们勘探农场内的不同地形地貌特征，选择与自己创意匹配度最佳的区域进行创作。所有的公共艺术作品不仅与环境有机结合，更通过设计来加强或削弱农场本身的地形、地质、季节变化等特性，从而引导人们更为深入地感受自然。

作为固定的主题公园，当地的人们有了一个可以随时拜访的公开景区，人为的景观雕塑与大自然的万千气象交融复合，呈现层次丰富的光影变幻。天、地、人，三者在此汇聚，广袤无垠中，一派勃勃生机。（吴昉）

Artwork Description

Gibbs Farm, the brainchild of businessman Alan Gibbs, is a sprawling 1,000-acre sculpture park located on Kaipara Harbour about an hour north of Auckland, New Zealand's largest city. One of New Zealand's leading art patrons, Gibbs bought the property in 1991 and began commissioning sculpture for it soon after, inviting artists to respond to the directly to the site and surroundings.

By area, the Kaipara Harbour is one of the largest harbours in the world. It covers 947 square km at high tide, with some 400 square km exposed as mudflats and sand-flats at low tide. The harbour is so vast it occupies the whole western horizon of the site. It is also home to a prevailing westerly wind that sweeps, sometimes with extreme force, across the land. Gibbs Farm flows towards and eventually into the sea and every work contends in some way with the slide seaward. Famous for its coastal terrain, big skies and wide-open spaces, New Zealand's topography is captured at Gibbs Farm in a nutshell.

The terrain of Gibbs Farm has shaped the aesthetics of the resulting sculptures, making the Farm one of the most interesting collections of site-specific minimalism. In response to the sheer scale and demanding nature of the landscape, artists have pushed beyond what they have previously attempted or achieved. Tucked into valleys, rolling down slopes, bridging bodies of water, peaking between clefts in the land or rising ethereally from hill tops, major sculptures by some of the most well known names in contemporary art bend to a landscape contoured by a turbulent geological past and dominance of the vast Kaipara Harbour.

There is Richard Serra's Te Tuhirangi Contour, collecting the volume of the land above and below it; the gravity-defying 'floating compression' of Kenneth Snelson's Easy K; and Tony Oursler's Mud Opera grasping the final return of all matter to the primordial ooze. Linking these forms is the classical formality of Daniel Buren's Green and White Fence, which runs both with and against the land; and the ridge-hugging Dismemberment, Site 1 by Anish Kapoor, which extends a red eye-ear telescopically out to sea and landward, bridging the inland and coastal aspects of the farm. Finally, in the tidal zone itself, New Zeleander Russell Moses' Kaipara Waka and Andy Goldsworthy's Arches both embrace the settling sand, rock and mud as well as some sense of a beyond, somewhere else, whether a spirit path, a migratory pattern or the drive to march in loping steps towards a distant horizon.

Collected at a rate of about one a year, though many take three to five years to develop, other works by international artists Eric Orr, George

Rickey, Marijke de Goey, Sol LeWitt, and Zhan Wang accompany other well known New Zealand names such as Bill Culbert, Chris Booth, Graham Bennett, Len Lye, Leon van den Eijkel, Neil Dawson, Peter Nicholls, Peter Roche, Ralph Hotere, and Richard Thompson. Gibbs Farm brings together major works by some of the world's most recognised artists, surpassing any gallery or museum collection in the country.

Artwork Excellence

The establishment of Gibbs Farm, its audacious commissioning policy, history of bringing major international artists to New Zealand and opening to the public is quite literally a landmark in public art for this region. Some of the art works may be large enough to be seen from satellite images at high magnification, but it is their impact on a small country where art and culture often come second to sport that will cement their legacy. The scope, ambition and artistic quality of the Gibbs Farm collection now rivals other major collections and sculpture parks—and in many cases the works surpass major works by the same artists elsewhere. For instance, the Goldsworthy exceeds the scale of his Cairnhead arches; and each of the Serra, Buren and large Rickey are more significant works than their counterparts at Storm King Art Center near New York City. For a small country like New Zealand, this is a remarkable achievement. Gibbs Farm has made a total commitment to an open-brief of commissioning and building major site-specific works from key artists that, without its existence, would be off-limits to the local population so far from international major art centres.

酒店
Hotel

艺术家：卡勒姆·莫顿（澳大利亚）
地点：澳大利亚墨尔本
推荐人：凯利·卡迈克尔
Artist: Callum Morton（Australia）
Location: Melbourne, Australia
Researcher: Kelly Carmichael

作品描述

卡勒姆·莫顿"酒店"（2008）矗立在澳大利亚维多利亚州舒展开阔的乡村。"酒店"仿佛被困在高速公路边上的空旷田野。因为周边没有其他建筑，酒店看起来像一块为疲惫旅客准备的沙漠绿洲。但是，这栋楼看起来似乎不像在营业，有一点古怪。从远距离看，酒店像任何其他在西方世界的酒店一样使用常见的混凝土和玻璃搭建。但是近看似乎有些不对了：这家酒店的建筑结构出了问题：它没有通路进出，人无法进到里面。理性和非理性、看见的和未曾预料的，此时此地这种种感觉在路过的驾车人脑海里一起闪过。

这个委托项目属于 Greens Road 路，作为 ConnectEast 的公共艺术项目的一部分。这个酒店是澳大利亚在主干道上呈现的最大委托公共艺术项目之一。酒店位于 Greens Road 路边上，Greens Road 路和 Bangholme 路的中间，酒店高 20 米，长 12 米，宽 5 米。它的一些窗户都在夜间由太阳能发电而亮。

像许多艺术家卡勒姆·莫顿的作品一样，酒店被处理成建筑和雕塑的共同体。莫顿的作品探索了人造环境和人的情感及其社会影响与我们的相互作用关系。他的作品研究我们如何遇到、感知或体验私人和公共空间。他的标志性作品以改造现代主义建筑而闻名，以前的作品曾将现有的建筑通过破坏、伪装、音响、改变比例、置换地点或材料而使其改头换面。最后在这件酒店作品里，莫顿尝试用这种古怪的欢迎仪式和终极的呈现效果，以一种挖苦的幽默方式嘲弄了现代主义建筑自以为是的傲慢态度，它揭示了艺术与生活、理想与现状之间的矛盾。这家酒店的展出意图是希望人们从新评估建筑的失败之处：同时要满足建造者的理念与居住者的期待会建造出多么失败的建筑。毫不夸张地说，正是这种空虚的酒店生活造就了今天这种根本不适合人居住的空洞人造建筑，它就竖立我们面前。

卡勒姆·莫顿是一个拥有细致入微的洞察力的艺术家，他清晰地表达出一个大概念，这个概念不仅仅包括"空间"如何变成"场所"，也包括多少当代人是怎样住在这种"非场所"中的。这些空间包括体育馆、机场、电影院和赌场这些都已经被转变过的地带。它们的存在是为了方便我们的活动，但它们不会鼓励我们在某个时间长度内占有这个场地。现代生活中，最为贴合这个概念的地点就是生活中司空见惯的汽车。从这个出发点来看，这家酒店开始不仅仅被看作是一个有趣的噱头或巧妙的双关语。艺术家把这栋建筑描述成一辆汽车和公路空间共同创造出来的丰碑。这家奇怪的酒店与周围的环境和经过它的人都格格不入。它让人觉得熟悉却又陌生，坚实而又空洞，真实却又不真实。这栋建筑虽然在我们的世界存在，却又显得不是为这个世界而生。

解读

当惯有的生活逻辑被打破之后，就会引起疑问，而疑问带来反思，这是澳大利亚艺术家卡勒姆·莫顿的"酒店"给观者设定的思维轨迹。

从外观来看，卡勒姆·莫顿的"酒店"完全形似于一般的公路酒店，当汽车驶在公路上远远遥望，几乎不会对其产生任何怀疑，直到汽车经过"酒店"的那一瞬，真相以极迅速的画面闪过，可以想象司机或乘客在疑惑中回望的有趣场面。"酒店"所制造的心理效果是一种滞后的，或者说是持续性的，而这一段持续的记忆痕迹，会促使观者去思索作品的意味，从而实现卡勒姆的创作意图。

无论观看者有没有猜中卡勒姆·莫顿作品深奥的内在寓意，还是仅仅作为一段车程的笑谈，"酒店"存在的本身就已经是一个话题，公共艺术项目的众多形式正是通过制造某个公共的话题，从而达成公众与环境的一种协调。（吴昉）

Artwork Description

Marooned in the middle of an open field along a rural stretch of motorway in the state of Victoria, Australia, stands Callum Morton's Hotel (2008). With no other buildings in site for miles around, it looks like an oasis for tired travellers. But there is something slightly off and a little bit odd about this building. From a distance Hotel looks like any other generic concrete-and-glass budget hotel in the western world, instantly recognizable and offering an average night's sleep and mediocre shower pressure. Up close, however, something's not quite right: the scale of the building is wrong, there's no access road and no way of getting inside. Here the rational and irrational, the visible and the expected collide in one uneasy, fleeting drive-by.

Commissioned as part of ConnectEast's public artwork program alongside the EastLink highway, Hotel is one of a series of works representing the largest commitment to public art on a major road in Australia. Located next to EastLink between Greens Road and Bangholme road, Hotel stands 20 meters high, 12 meters long and 5 meters wide. Some of its windows are lit at night by solar power.

Like much of the artist Callum Morton's work, Hotel sits somewhere between architecture and sculpture. Morton's practice explores our interaction and relationship with the built environment and the emotional and social impact it has on people. His works examine how we encounter, perceive or experience private and public space. Known for his transformation of iconic modernist architectural forms, previous works have taken existing structures and altered them through destruction, camouflage, sound, and changes in scale, location or material. Culminating in Hotel, with its oddly welcoming and ultimately destabilizing effect, Callum Morton's practice undermines the lofty aspirations of modernist architecture with a satirical humor, revealing the tension between art and life, the idea and the present. Hotel exhibits a desire to re-evaluate the failure of architecture to satisfy both its creator's idealism and its inhabitants' expectations. The emptiness of hotel life is, quite literally, reflected in a hollow, faux building erected in a scale that could never be habitable.

Callum Morton is an artist who possesses a nuanced ability to articulate large ideas—not only how "space" becomes "place," but also how much of contemporary life is lived in "non-places." These spaces—including sports stadiums, airports, cinemas, and casinos among others—are transit zones, built to facilitate our movement through them rather than encourage us to occupy them for any length of time. The most accessible and quotidian of all the "non-places" in contemporary life has to be the car, and it is from this starting point that Hotel begins to be understood as more than an amusing gimmick or clever pun. Described by the artist as "a type of monument or mirror to the effect produced by both the car and the space of the freeway itself," Hotel is strangely disconnected from the landscape it exists in and those who pass by. Familiar yet strange, solid yet hollow, real yet not real, the building is in our world but not quite of it.

Artwork Excellence

Hotel is one of a series of works representing the largest commitment to public art on a major road in Australia. It exhibits a desire to re-evaluate the failure of architecture to satisfy both its creator's idealism and its inhabitants' expectations. The emptiness of hotel life is, quite literally, reflected in a hollow, faux building erected in a scale that could never be habitable. Here the rational and irrational, the visible and the expected collide in one uneasy, fleeting drive-by.

壁画项目
Mural projects

艺术家：塞缪尔·英德拉特玛（印度尼西亚）
地点：新西兰奥克兰
推荐人：凯利·卡迈克尔
Artist: Samuel Indratma（Indonesia）
Location: Auckland, New Zealand
Researcher: Kelly Carmichael

作品描述

1998 年 5 月印尼总统苏哈托的戏剧性辞职标志着近代史上东南亚地区最长独裁统治的终结。在其执政的 30 年中，军人出身的苏哈托保持着对政治的独断统治，整整一代印尼人是在他的统治下成长起来的。在他的"新秩序"统治下，建立起了高度集权的中央政府，由军队掌控政府。到 20 世纪 90 年代，随着年轻人追求政治自由和自由表达，苏哈托的独裁统治、到处蔓延的贪污腐败及和反对者的打压都遭到了前所未有的挑战。苏哈托下台后，印尼实现了言论自由，与此前高压统治时期形成鲜明对比。与此同时，文化表达繁荣发展。

塞缪尔·英德拉特玛以壁画创作闻名，在艺术领域和把印尼还给人民的斗争中都是中坚力量，他的创作属于公共艺术，紧密联系现实社会。他的作品围绕着身份构建展开，这对一个在新的政治体制下正重新定位的国家是十分重要的。同时，在公共空间这一概念已被政府破坏殆尽的情况下，他的作品对于促进与公共空间的联系也发挥着重要作用。同许多从压制中解放出来的国家人民一样，印尼人把街道作为探讨意义和认知身份的空间。英德拉特玛的发展壁画项目将第三空间转化为具体的地点，将公共空间的所有权切实地交还给了印尼人民。

英德拉特玛同时进行着独立创作和集体创作。作为日惹（雅加达）年轻艺术家街头艺术运动的先锋，英德拉特玛的个人创作试图探讨当代印尼社会面临的问题，特别是年轻人所遇到的问题，印尼政坛的动荡起伏也是他的灵感来源之一。

英德拉特玛是艺术团体"漫画药剂学"（Apotik Komik，1997—2005）的活跃分子，这个团体旨在"为公共艺术发声，为印尼公共空间的理解进行表达"。团体以漫画为灵感的集体创作受到了瞩目，这不仅因为公共艺术在印尼尚属新奇，还因为艺术家们使用的都是简单且经济的材料。

日惹壁画论坛（Jogja Mural Forum）始于 2005 年，英德拉特玛是成立者之一。他的壁画是文化身份的标志，他常与团体、年轻人或者传统工匠合作。在近期的"纪念物壁画项目"（2007—2008）中，英德拉特玛在雅加达的六个贫民区进行了联合创作。

从 20 世纪开始，日常生活的范围逐渐向室内发展。评论家理查德·伯顿在《论支持公共艺术的发展》一文中提出，当前社会正经历着"令人震惊的无人性……感觉普通人对日常公共空间的使用毫无权利"。而事实上，日常公共场所正是街头艺术及壁上涂鸦这一普遍的形式之一赖以存在的空间。壁画将空间转化成地点，通过对公共空间的再造对人产生直接影响。这样的创作对于印尼这样遭受过独裁统治、自由曾受到束缚的国家来说无疑是最有效的。历经十余年，塞缪尔·英德拉特玛的创作已经成为了雅加达社区人们争取公共空间权利的不可分割的一部分。

解读

随着独裁统治的倒台，原总统所谓的"新秩序"统治也就不名一文，街头艺术家们迎向自由之风，用涂鸦艺术去建造全新的文化秩序，还街头空间于公众。

街头涂鸦，本身就伴随着随性、自由、年轻、张扬、热情等关键词，常以绚丽大胆的色彩和破除陈规的构图，夺人眼球，塞缪尔·英德拉特玛的"壁画项目"，在特定的政治环境和文化格局中，尤其具有革命性的开拓精神。

"壁画项目"是奥克兰的一种文化标志，也是一种政治新风向，塞缪尔召集涂鸦爱好者、当地年轻人，一起群体创作，小心翼翼地将"公共空间"——这个遗落了近半个世纪的概念重新塑造起来。当街道、公共设施、广场空间都不赋予民众自由权时，这些空间本身的意义也就被架空了。随心所欲地在街头刷下自己喜爱的图案与色彩，突破公共空间与公众的敌对或漠视状态，这是左脚迈开的第一步；用构思精良、有品质感的壁画涂鸦，努力带给陈旧的街道不一样的风景，企盼居住的环境能有美好的明天，这是右脚走出的第二步。

在为民众争取公共空间权利的道路上，一步一个脚印，步履坚实。（吴昉）

Artwork Description

The dramatic resignation of Indonesia's President Suharto in May 1998 brought to an end the longest period of one-man rule in modern South-East Asian history. The former general had been in continuous political command of Indonesia for more than 30 years, and a generation had grown up which knew no other leader. Under his "New Order" administration, Suharto constructed a strong, centralized and military-dominated government. By the 1990s, challenges to the Suharto government's authoritarianism, widespread corruption and repression of opponents gathered pace as the younger generation began to challenge the stifling of political and personal expression. After the President's fall from grace, the Indonesia of the last decade saw a freedom of speech in marked contrast to the censorship of the New Order era and also a flowering of cultural expression.

A key player in both the art scene and the reclaiming of Indonesia for its citizens, Samuel Indratma is known for an ongoing mural project of socially engaged public art. Urban in focus and feel, Indratma's murals centre around the construction of identity, significant not only for a nation redefining itself under a new political regime, but also because they seek to encourage connection to public space, a concept formally corrupted by government. Like many countries liberated from repressive governments, the streets of Indonesia are very consciously used as a space for negotiating meaning and understanding identity. Indratma's ongoing mural project takes what was once a third space and helps it become a place, effectively returning the ownership of public space to the Indonesian people.

Indratma has worked both as a solo artist and in collectives. As one of the pioneers of the street art movement among young artists in the city of Yogyakarta (Jakarta), Samuel Indratma's solo practice explores problems facing contemporary Indonesian society, in particular youth, and is often inspired by his country's political upheavals.

Indratma was an active member of the Apotik Komik (1997 to 2005) a group of artists who aimed to "bring out the discourse of public art and an understanding of public space to our community in Indonesia," the collective's comic-influenced practice attracted much attention, not only because of the novelty of public art in Indonesia but also for the artists' use of simple and inexpensive materials.

Co-founder of the Jogja Mural Forum established in 2005, Indratma's murals act as markers of cultural identity, often created in collaboration with communities, youth and traditional craftspeople. A recent project, Tanda Mata Mural Art Projects (2007-2008) saw Indratma co-ordinate a mural project for six slum areas in Yogyakarta.

As the business of daily living has progressively moved indoors over the last century, many commentators—like Richard Burton in his essay "The Arguments for Public Art" —have observed that the modern city suffers from "an alarming inhumanity...a feeling that ordinary people have no claim to the spaces of daily public living." This is where street art and perhaps its most ubiquitous form of the mural comes into play. Murals offer the transformation of spaces into places in an attempt to directly affect members of the community by reclaiming public spaces. This is never more effective than in areas and nations that have suffered authoritarian rule or restrictions on freedom of expression. Indonesia is such a place, and for over a decade the works of Samuel Indratma has been an integral element in creating rights to public space for communities.

Artwork Excellence

A key player in both the art scene and the reclaiming of Indonesia for its citizens, Samuel Indratma is known for an ongoing mural project of socially engaged public art. Urban in focus and feel, Indratma's murals centre around the construction of identity, significant not only for a nation redefining itself under a new political regime, but also because they seek to encourage connection to public space, a concept formally corrupted by government. For over a decade, the works of Indratma have been an integral element in creating rights to public space for communities in Yogyakarta (Jakarta). As one of the pioneers of the street art movement, Indratma and his murals have inspired a generation of new artists.

窄轨摆式列车公共艺术项目
Tilt Train Public Art Project

艺术家：多位艺术家
地点：新西兰奥克兰
推荐人：凯利·卡迈克尔
Artist: Multiple artists
Location: Queensland, Australia
Researcher: Kelly Carmichael

作品描述

由三十多位原住民的和托雷斯海峡岛的艺术家携手合作，"窄轨摆式列车公共艺术项目"创造了澳大利亚规格最长的当代艺术项目，整整覆盖了两列在布里斯班和凯恩斯这两座相距 1 700 公里的沿海城市之间运营载客的火车车厢——位于昆士兰州的东北部。这件创新性的项目对于澳大利亚土著艺术和艺术家来说代表着一个里程碑式的意义，如同电影般的速度，它见证了本土文化生产机制的视野和活力，跳脱了与西方市场相近的思维和审美范畴。

昆士兰铁力和昆士兰公共艺术基金的 art+place 签了一份联合协议，这个项目其实是昆士兰政府的和解行动计划之一。2008 年，澳大利亚政府为不公正的法律和过渡政府的政策，向原住民和托雷斯海峡岛的居民道歉，尤其是为曾经将孩子从他们的家庭中迁离的政策，这是政府公开承认这段可怕的历史及其对该地区居民带来的永久影响。政府的道歉是迈向和解的重要一步。在几位当地最著名的艺术家的努力下，"窄轨摆式列车公共艺术项目"是对这段被掩埋的历史及澳洲原住民的生活加以承认的重要举措。布里斯班和凯恩斯的火车路线穿过很多土著民监管的土地，而这与土地的联系给予了澳大利亚土著民认同感和归属感，这些土地是昆士兰沿海地区的原住民和少数民族与国家之间精神上和所属关系上的联系的活见证。火车上的艺术来自当地土著民的传统图腾和传说等原住民的风土人情，并深入浅出地展示出来。

艺术的图片和历史由 150 米长的 7 节车厢运载，它可携带 300 米长的土著艺术作品。在作品的"三步走"过程创建之后，要拍照并由平面设计师复制出来，最后丙烯包装材料做来了并用到火车上。2011 年 5 月，昆士兰铁路开始了"窄轨摆式列车公共艺术项目"的首发，艺术家朱迪·沃特森和埃里克·蒂波奇分别站在火车两头。现在这个项目已有第二列火车了。

朱迪·沃特森是昆士兰州东北部 Waanyi 人的后裔，她的"窄轨摆式列车"的灵感来自她在 2009 年做为昆士兰州苍鹭岛大学研究站的驻馆艺术家（AIR）时的经历，那时她做了一些蚀刻画和印刷品，这些作品融入了环境数据，记录了濒危物种和气候变化，用的是一种自然的方式来体现美感。这些具有挑战性的图像是循环利用贝壳、垃圾堆、化石、白蚁丘等做成的，反映了昆士兰沿岸自然风景的脆弱性。

埃里克·蒂波奇来自托雷斯海峡的巴杜岛，也是当地最有创新性和最受尊重的艺术家之一。他的独特而深奥的作品述说的是过去时代的勇士和传奇英雄的故事。他为"窄轨摆式列车"项目做的作品想法来自动物的迁徙。他从自己对托雷斯海峡文化的亲近了解中汲取设计灵感，的隐喻文化的知识，当地动物迁徙所组成的路线和踪迹构成了一个极为重要的环境地图。

北昆士兰卡德维尔 Girringun 原住民艺术中心的代表艺术家来自九个保持老传统团体：the Nywaigi, Gugu Badhun, Warrgamay, Warungnu, Bandjin, Girramay, Gulnay, Jirrbal and Djiru people. "窄轨摆式列车公共艺术项目"用了编织材料、雕刻容器、图文并茂的形式，Bagu 和黏土鸟类。一种被称为"Jawun"（音译）的东西是由编织或耐火土制做的带两个角的容器组成，是北昆士兰热带雨林地区的特产。

Josiah Omeenyoshi 是来自 Lockhart 河的 Umpila/ Kanthanampu 族，他的创作灵感取自他的土地和文化，这些作品展示了他对生命的热情以及他认为什么才是重要的。"窄轨摆式列车公共艺术项目"的灵感取自他的充满生机、色彩丰富的作品"我的散步与垂钓之地"和"何处是鲑鱼谷"，他的作品用大海展示了他的亲和力，以及他对光与色彩多变性的控制能力，而这些都源自其最爱的垂钓地点的真实景色之美。

解读

在奥兰多的"窄轨摆式列车公共艺术项目"中，我们看到了土地与文化、图腾与信仰、动物的迁徙、气候的变化、沿海的风景、热带雨林的生命，所有的元素和主题，都展现了当地本土文化的鲜明特征，以及这些特征散发出的艺术感染力：生命生生不息，新新相续。

无论在何种国际艺术主流的冲击下，本土文化都有着旺盛的生命力及不可撼动的独立性，因其在当下仍是活着的、发展着的文化，所以才具备源源不断的活力，供公共艺术家们采撷灵感。"窄轨摆式列车公共艺术项目"公共项目，由政府的和解项目引发，进而蜕变成一种公然展示地方文化特色的自豪之旅。火车从起点驶向终点，就像移动的画布闯入当地人的眼帘，燃起他们的好奇、惊叹、赞美以及文化自豪感。列车上的绘画是当地公众所熟悉的，因而也是喜闻乐见的，所谓艺术的公共性，正在于此。（吴昉）

Artwork Description

Bringing together over 30 Aboriginal and Torres Strait Island artists, the Tilt Train Public Art Project has created Australia's longest piece of contemporary art. Covering two passenger trains operating between the coastal cities of Brisbane and Cairns—a distance of some 1,700 km or 1,056 miles—in the North Eastern state of Queensland, the innovative project represents a major milestone for indigenous art and artists in Australia. Almost filmic as it passes at speed, the project is testimony to the breadth and vitality of indigenous cultural production outside of familiar Western market expectations and aesthetics.

A joint initiative between Queensland Rail and art+place, Queensland's public art fund, this project is part of the Queensland Government's Reconciliation Action Plan. In 2008 the Australian Government apologised to Aboriginal and Torres Strait Islander peoples for the unjust laws and policies of successive governments, and especially for the removal of children from their families. In acknowledging the terrible history and its enduring effect on Aboriginal and Torres Strait Islander peoples. The national apology was an important step towards reconciliation. Working with some of the region's most celebrated artists, the Tilt Train project represents an important initiative towards recognition of the buried histories and lives of Australia's indigenous population. The train's route between Brisbane and Cairns is a journey that crosses many Aboriginal custodial lands. The connection to land gives Australia's indigenous people their identity and sense of belonging. These lands are a living testimony of the physical, spiritual and custodial connections to country for the Aboriginal clan and language groups along with Queensland coastline. The artworks on the train are proclaiming the traditional imagery and stories of this area's indigenous people, quite literally reclaiming the metaphorical and literal ground as the train passes.

The images and history of the artists are conveyed on a seven-carriage train measuring 150 metres and thereby carrying a 300-metre canvas of indigenous art. In a three-step process artwork was created, then photographed and reproduced by a graphic designer, and finally a special vinyl wrap created and applied to the train. In May 2011 Queensland Rail launched the first tilt train project with artists Judy Watson and Alick Tipoti taking one side of the train each. The project now includes a second train.

Judy Watson is Aboriginal descendant of the Waanyi people of North-East Queensland. Her inspiration for the Tilt Train was based from set of etchings and screenprints she created when she was artist-in-residence at the University of Queensland's Heron Island Research Station in 2009. These prints fuse environmental data, signalling endangered species and climate change, with aesthetic renditions of natural forms. This challenging imagery, with recurring themes of shells, middens, fossils and termite mounds, reflects the fragile nature of Queensland's scenic coastline.

Alick Tipoti is from Badu Island in the Torres Strait and is one of the region's most innovative and widely respected artists. His distinctive and highly intricate works tell stories from the past about warriors and legendary heroes. His artwork for the Tilt Train explores the notion of travel by reflecting the wakes made by different animals as they move. To create the design, Alick drew on his intimate cultural knowledge of the Torres Strait, where paths and tracks of animal movements constitute vital environmental mapping.

Girringun Aboriginal Art Centre in Cardwell, North Queensland, represents artists from nine Traditional Owner groups: the Nywaigi, Gugu Badhun, Warrgamay, Warungnu, Bandjin, Girramay, Gulnay, Jirrbal and Djiru people. The Tilt Train artwork captures the woven and sculpted vessels, illustrated boomerangs, Bagu and clay birds. The woven and fired clay bicornual vessels, known as Jawun, are unique to the rainforest regions of North Queensland.

Josiah Omeenyo is an Umpila/ Kanthanampu man from Lockhart River who is inspired by his traditional lands and culture to create works that reference his enthusiasm for life and what he feels is important. Josiah's Tilt Train artwork was adapted from his vibrant and colourful painting Where I Walk and Fish and Where the Salmon Hide. This artwork demonstrates his affinity with the sea and his ability to celebrate its abundance through a spectrum of colours that are true to his favourite fishing spots.

Artwork Excellence

Bringing together over 30 Aboriginal and Torres Strait Island artists, the Tilt Train Public Art Project has created Australia's longest piece of contemporary art. In an official statement the Queensland Arts Minister commented, "This striking piece of mobile artwork is set to become one of our state's most recognisable tourism attractions." In this statement we find the true essence of what makes this particular public art work remarkable: its makers and their art forms have gone from government sanctioned racism and abuse to not only being recognised as a valuable conduit for dialogue and exchange, but also the face of Queensland for residents and tourists alike. This public art project—through the contributions of artists aimed at rebuilding cultural identity and initiating dialogue and social cohesion in this post-conflict situation—is forging new ground. Each train delivers not only travellers and holidaymakers to their destination but also the imagery and culture of Australia's first peoples back to the land, reaffirming their rightful sense of place, their belonging.

酒店
Tiwatawata

艺术家：约翰·雷诺兹（新西兰）
地点：新西兰奥克兰
推荐人：凯利·卡迈克尔
Artist: John Reynolds （New Zealand）
Location: Auckland, New Zealand
Researcher: Kelly Carmichael

作品描述

奥克兰的人类历史从 14 世纪早期毛利定居者延伸到 18 世纪晚期第一代欧洲探险者，目前它是全国最具商业主导地位，发展最快的一座大城市。奥克兰加速的都市发展和充裕的土地导致该市迅速扩张，二战前开始至今它早已成为世界上汽车拥有率最高的城市之一。

基于这些因素，奥克兰是一座大型郊区型，低密度，规划不足的城市，有着严重的交通问题，很糟的公共交通。

约翰·雷诺兹的作品"酒店"坐落于新霍布森维尔一个新住宅区，建造它是为如今有限的土地空间提供高密度的房屋，以满足奥克兰快速增长人口的需要。霍布森维尔是都市发展总体规划的一部分，它综合了公共交通，服务和社会基础设施。

作品以 19 世纪当地的实例为参照，在它正处于转型的时候，能够使观者同时回想起这块地方的历史，雷诺兹的雕塑回顾了毛利人历史上的标界杆，给新来霍布森维尔社区居民预先划标了地界，那地方以前曾是奥克兰的空军基地。108 根烧黑了的长杆排列着，不断变化地穿越过这个地形发生了变化的地区，"酒店"戏剧性地描述了这块土地的分割和标界的过程。

"酒店"融入了当地景观，意在把周边的地区连接起来。在该地区的深厚历史、坚强性格、当地环境生态和未来目的启发指引下，艺术家的意图是让自己的艺术作品扎根于这片风景，而不仅仅是放置在这片风景地。108 根烧黑的 / 或涂过色的木杆有着不同的直径，有着不同的高度，在随机的间距中紧密地摆放着，通过水塘和种植区域，用笔直的黑线把公园一分为二。

这件作品构成了一个扩展的图形标定，定义了视角，视觉途径和并列的都市和乡村。虽然回顾了旧的边界标示，本作品故意建造地不平整，不规则，以保留一种有机的感觉。在某些地方，烧黑的长杆的疏密度有变化，紧密地排列在一起，而其他地方则松散地排列着。

约翰·雷诺兹的作品大都充满了游戏感，浪漫感和天真的审美情趣，尽管他的实践属于知识分子行为的一种，带有丰富的寓言性质，把最终的诠释留给了观众。三十多年来，新西兰艺术界始终未能把雷诺兹的艺术加以归类，他的作品看起来都是些简单的符号和绘制文本，然而却含有无限的复杂性。

人们很容易把雷诺兹的作品理解为是简单的和游戏性的，但那些作品中所含的文学、宗教、艺术史和建筑方面的内容却表明了并非是那么回事。他的作品大多是关于如何以新的方式去理解事物，无论文字、丛生植物或标志图案，让观众直接体验，而不仅仅只看到他提出的新关系和新概念。

解读

新西兰艺术家约翰·雷诺兹创造"酒店"，选用了当地毛利人历史上使用的标界杆为主体元素，108 根高低粗细不同的标界杆，是历史符号化身的新地标，也是公众眼里的延绵景观；是土地分割和标界的过程，也是连接周边的友好标记。俯瞰"酒店"全景，就像构成中的点，在视觉上连结成线，贯穿起新霍布森维尔住宅区与周围的外环境，强调地域分割和区域整体的相互关系。

当然，木桩还暗含"扎根"于当地文化的寓意，约翰·雷诺兹不希望自己作品风格的"游戏性"分散了当地居民对于自身文化的关注，遥相呼应的木桩之间，承接了约翰·雷诺兹对新旧城市合理规划的理解，也借此项目，为当地的悠久文化留存可视化的记号。（吴昉）

Artwork Description

The human history of Auckland stretches from early Maori settlers in the 14th century to the first European explorers in the late 18th century, to its current position as the commercially dominating, and fastest growing, metropolis of the country. Auckland's intense urban growth and plentiful land has led to urban sprawl, a rapid decentralisation enhanced by the city boasting one of the highest car-ownership rates in the world, even before World War Ⅱ. Due to these factors, Auckland is a largely suburban, low-density and under-planned city with an acute traffic problem and poor public transport.

The new Hobsonville subdivision where John Reynolds' work

Tiwatawata is located is something new in this regard. It aims to offer higher-density housing to address Auckland's fast population growth within what is now limited land space. Hobsonville is a master-planned urban development with integrated public transport, services and social infrastructure.

Taking its cue from nineteenth century illustrations of the local area, Tiwatawata simultaneously recalls the history of its site while becoming part of its transformation. Reynolds' sculpture recalls the historical demarcation poles of the Maori people and pre-empts the fencing of boundaries with the arrival of more recent communities at Hobsonville, an ex-air force base in Auckland. A procession of 108 charred poles dynamically tracking across the shifting topology of the site, Tiwatawata dramatises the process of dividing and marking off the land.

Tiwatawata integrates into the landscape, intended to connect with the surrounding area. Guided by the site's rich history, strong character, local ecology and future purpose, the intention is an artwork in the landscape rather than on the landscape. The 108 charred and/or stained wooden poles of various diameters closely staged at irregular distances and differing heights, chart a straight dark line bisecting the park through a pond and planted areas. The work forms an extended graphic demarcating and defining views, visual pathways and juxtapositions of urban and rural. Though recalling formal boundary markers of old, this construction is deliberately uneven and irregular retaining an organic feel. In certain places the charred poles change in density, becoming more closely packed as they negotiate the land, loosening out in others. There's a playful, romantic, even naive aesthetic about much of John Reynolds' work, though his practice is an intellectual one, richly allegorical and leaving the final interpretation to the viewer.

Reynolds has eluded categorisation in the New Zealand art world for some thirty years, appearing to circumnavigate expectation with a practice replete with simple symbols and drawn text, yet one that is endlessly complex. It is possible to under-read Reynolds' work as simple and playful, but its literary, religious, art historical and architectural references tell otherwise. His work is most often about understanding things in a new way, whether words, tussock plants or iconic pattern, letting the viewer experience rather than simply see the new connections and concepts he offers up.

Artwork Excellence

The latest in a series of outdoor works in the landscape by John Reynolds, Tiwatawata both contributes to and comments upon the place-making practices of historical and contemporary communities, allowing us to reflect on the universal desire to make space into place.abuse

战争与和平及两者之间
War and Peace and In Between

艺术家：塔特祖·尼希
地点：澳大利亚悉尼
推荐人：凯利·卡迈克尔
Artist: Tatzu Nishi (Japan)
Location: Sydney, Australia
Researcher: Kelly Carmichael

作品描述

2009 年塔特祖·尼希将英国雕塑家吉尔伯特·贝叶斯的两尊古典骑马雕塑进行了转换，"和平祭"和"战争祭"都是铜制雕塑，坐落于澳大利亚新南威尔士的画廊外面。尼希将它们包裹住，变成很俏皮的结构——不避开公众，但却为参观者提供了一种完全不同而异常独特的体验。由此产生的"战争与和平及两者之间"同两个标志性的悉尼纪念碑和国家与帝国的标志产生了非常亲密而又明显的碰撞。"战争与和平及两者之间"的参观者通过建造在画廊顶端阶梯和尼希造的临时结构之间的楼梯和坡道进入高高架起的室内，那里看起来就像围绕在雕塑四周的家庭生活空间。雕塑"和平"位于精美装饰过的起居室内，雕塑"战争"位于卧室内，每样物品看上去都很普通，除了起居室内出现在一张咖啡桌上的骑手，其上半身像一座巨型胸像，马头却现身于一个橱柜里；在卧室内，马和骑手看上去像在跋涉，穿过床上的雪堆。

公众习惯于从一定的距离来欣赏这些位于高过我们头顶的基座上的经典纪念碑。在这里，塔特祖·尼希让观众以面对面的近距离来遇见贝叶斯的雕塑。国内的建筑和雕塑的位置和比例都不甚恰当，这为作品抹上了一层超现实主义和迷失的效果。该装置作品请求我们靠近后再次、深入地欣赏，来为新的观众创造了一种新鲜的亲密感。"战争与和平及两者之间"将观众和他们的思索替换了，唤起大众生动的回应，打破了他们与含有丰富内涵的装置结构之间的距离。塔特祖·尼希提供了一个能够遇见、触摸并近距离亲密分享"权利标志"的空间这样千载难逢的机会。

这位艺术家以许多不同的名字而闻名。在此工作时他的名字是塔特祖·尼希，他采取了一种转移和阻断个人身份的战略，在他的装置里反映了他将熟悉的物品进行替换或重置语境的手法。这位艺术家以在公共空间和世界各地创作摆脱常规尺寸和地点的作品而闻名。他的艺术实践的宗旨是要从传统和社会的层面上打破种种壁垒。

高级宾馆、昂贵的设计师商店，具有影响力的国家级形象一直被塔特祖·尼希所引用，他的装置允许普通人在生理与情感上密切地接近这些偶像和大型结构进行近距离接触，而在平时由于社会差距原因这些人是没有机会和这些东西接触的。他也转换街灯、停着的汽车和纪念碑等等，在它们周围建造新的空间，他不仅仅改变它们的布景，同时也改变我们日常的欣赏和观看方式。

解读

公众习惯于仰头注视公共雕塑的姿态，实际是对一种思维模式的习惯性遵循，所以，当"战争与和平及两者之间"引领公众进入移步换景式的超现实现场，推翻固有逻辑的强烈心理震撼，会使公众的情绪被作品牢牢牵引。与现实相矛盾所产生的内在张力，造就了这项公共艺术作品的成功。

瞻仰与平视、遥望与近观、室外与室内的转换，必然带来迥异的心理体验。艺术家通过减少公共雕塑与公众之间的物理距离，以及对崇高和平凡两者移花接木的错位布局，让公众获得一种崭新的公共体验。改变了日常观看方式的人们，在经过短暂的心理与视觉适应期后，对"战争与和平及两者之间"的态度是欢快的、喜爱的，因为这将是从未有过的首次经历：崇高的精神象征与琐碎的日常生活互为一体。普罗大众，终于也有了阐释经典的话语权。

战争与和平，历来是高高在上的两面旗帜，任凭历史时局的旋风扶摇而上，成为善与恶的堂皇理由，而在两者之间，是否也该预留一处空间，作为公众的立场呢？（吴昉）

Artwork Description

In 2009 Tatzu Nishi transformed two classical equestrian works by English sculptor Gilbert Bayes. The Offerings of Peace and The Offerings of War, both bronze sculptures, are situated outside the art gallery of New South Wales, Australia. Nishi encased them in playful constructions—not to shield them from the public, but to offer visitors an entirely different and unique experience. The resulting War and Peace and In Between offered a surprisingly intimate and stripped down encounter with two iconic Sydney monuments and emblems of state and empire.

Visitors of War and Peace and In Between entered elevated rooms—what appeared to be domestic living spaces around the statues—via stairs and ramps constructed between the top step of the Gallery and Nishi's temporary structures. In the beautifully decorated living room of "Peace" and the bedroom for "War," everything seemed completely normal except that, in the living room, the top of a rider appeared on a coffee table like a giant bust while a horse's head was revealed inside a cabinet; in the bedroom, the horse and rider were seemingly wading through a snowdrift of sheets atop the bed.

The public is accustomed to seeing such classical monuments from a distance and from plinths that raise them high above our heads. Here, Tatzu Nishi let visitors encounter Bayes' sculptures face to face at very close quarters. Both the domestic architecture and the sculpture were rendered impossibly out of place and out of scale, adding to the slightly surreal and disorientating effect of the work. Asking us to look again and look deeper, in close up, the installation created a fresh intimacy to a new audience. War and Peace and In Between displaced the viewers and their thinking, provoking fresh responses and allowing people to breach the distance between themselves and structures loaded with accumulated meaning. Nishi offered quite possibly a once-in-a-lifetime opportunity to encounter, touch and share close space with emblems of power.

Known by numerous different names, the artist working here under the name of Tatzu Nishi has a strategy of shifting and occluding his own identity, reflecting the way he displaces or re-contextualises the familiar in his installations. The artist is known for creating out-of-scale and out-of-place encounters in public spaces around the world. His practice is about breaking down barriers both literal and social. High class hotels, expensive designer stores, the powerful and figures of state have all been addressed by Tatzu Nishi, his installations allowing everyday people to connect in close physical and emotional proximity with icons and edifices social distance may have prevented. He has transformed street lights, parked cars and monuments amongst other things, building new spaces around them and not only altering their setting, but also the way we see the everyday.

Artwork Excellence

For the Sydney project War and Peace and iIn Between, Tatzu Nishi took two well-known public sculptures and rescued them from the invisibility that over-familiarity can bring. The nature of his structures allowed the viewer to approach and notice the size and detail of the classic sculptures, and in a private setting. There is an intense intimacy to what the installation offers. Nishi gave his audience an entirely new relationship to iconic symbols of place and status.

风树
Wind Tree

艺术家：井原道夫
地点：新西兰奥克兰
推荐人：凯利·卡迈克尔
Artist: Michio Ihara (USA)
Location: Auckland, New Zealand
Researcher: Kelly Carmichael

作品描述

"风树"就像一块用不锈钢管和桁架搭成的挂毯，设计安装于水面上方，能在风中摇摆。它是1971年奥克兰建城百年之际献给自己的百年纪念。五件作品中的一件制作于1971年奥克兰国际雕塑专题讨论会期间。这件雕塑被视为一种机会，能够进一步提升自己在泛太平洋地区的身份。这件作品由在法国出生的日本雕塑家道夫井原创作，在那个年代的新西兰具有革命性，但是作品本身充满了坎坷。很多人似乎都不想看"风树"，一开始其定位就不恰当。雕塑选址不佳，装在盒子里——使得其功能无法完全得到发挥——处于被密密麻麻的中层大楼所包围，这地方更像一个过渡性的地点或被遗忘的城市空间，而不像是一处市民广场。多年以来这件作品放置在那里遭人冷落，在奥克兰的这块中心商业区内无人注意，犹如蒙上了阴影。雕塑几乎变成隐形的了，为了给公共交通发展的一个大型翻新工程让道，雕塑被拆除了。在它从最初的地点被移除后，并没有计划要重新安装这件作品。它仍然被存放在炼狱中，被移除后已经静置了七年。

然后，到了 2009 年 2 月，一个公共艺术的意见小组提出计划要重新放置"风树"，使这件作品最初的创作意图得以复生。重新安置后，雕塑有了呼吸的空间，公共座椅两侧被原生树木和草包围着，2011 年"风树"重生了。这件雕塑与周围的工业化当代城市建筑非常谐调，它已回复了自我。它的新家位于温亚德季，该地区的重振也犹如经历了重新的转世投胎，实现了从工业散装石油化学品储存到面向公众开放的海滨公共空间转型，有着一些餐馆、一家活动中心、儿童的游乐场地和旅游景点。奥克兰的市中心以往没有优质休憩用地，如今有了一些亮点，温亚德季正争夺榜首位。如今重新注入了活力，风树犹如导电板一样激活了整个区域。当地的艺术圈和普通奥克兰居民都感到很迷惑，为什么这件出色的作品这么长时间里遭到了忽视。"风树"成了获得重生的一个象征，特别对于在艰难经济环境中的小城市而言。

作为一位艺术家，道夫井原使用不锈钢、黄铜和铜焊，但主要他的作品是运用空间和移动的手法。他的作品反映出了纯度、净度和信心。通过精心组织和安排比例，空气和光线通过透明的结构自由流动，他的雕塑作品是其周围空间的凝结，从而成为其中的组成部分。经常有着动力学性质，他的作品有着空气振动电流，创造出了生活和呼吸形式。道夫井原的作品有一种固有的清晰度，对于他血脉渊源中的日本传统审美以及伟大的建筑做出了呼应。

解读

从 1971 年到 2011 年，"风树"从不受欢迎到重见阳光，波折的命运其实明确了公共艺术的一个本质属性：公共空间的设计与规划只有与周围外环境相融合，才有机会展现自己的别具一格。

忽略"风树"前世的惨淡境况，仅从重新赢取公众青睐的这一刻来看，金属骨架构成的线条与紧邻的当代工业化城市气脉相通、节奏一致，仿生的创意立足点，又好像是工业城镇向海滨公共空间转换的信号，"风树"成为一种中介，让公众感受到钢筋铁树也有呼吸，从而向往自然与人类发展的的和谐关系。

另一方面，由"风树"坎坷的经历，或者也可以看到公共艺术家潜在具备的先验眼光，对于世界发展进程有种超越时间的远视，只是，在将此种超越投注于公共创作中时，仍需权衡公众与环境在当时当地的接受度与承受力，这也是每一个公共艺术家所应拥有的专业能力。（冯正龙）

Artwork Description

Wind Tree is a tapestry of stainless-steel tubes and trusses that are designed to swing in the wind above a water feature. It was Auckland's 1971 centennial present to itself. One of five works produced at the Auckland International Sculpture Symposium of 1971, it was seen as an opportunity to further cement Pacific Rim identity internationally. The work by French-born, Japanese sculptor Michio Ihara was revolutionary for its time within the New Zealand context, but it had a troubled life.

Many people just didn't seem to get Wind Tree and it was badly positioned from the start. Poorly sited and boxed in, the sculpture—which wasn't able to fully function—was surrounded by densely built medium-rise buildings in an area that functioned more as a transitory point or forgotten urban space than civic square. For years the work sat ignored, unnoticed and overshadowed in Auckland's central business district.

The sculpture became practically invisible and was dismantled as part of a major renovation to make way for a public transportation development. After removal from its original site there were no plans to reinstall the work. It remained hidden in storage purgatory, forgotten again, for more than seven years after being taken down.

Then, in February 2009, an Advisory Panel for Public Art endorsed a proposal to re-site Wind Tree in keeping with the original intentions of the work. Resituated with room to breathe and surrounded by public seating flanked by native trees and grasses, Wind Tree was reborn in 2011. In harmony with surrounding industrial and contemporary urban architecture, the sculpture has come into its own. Its new home in the Wynyard Quarter redevelopment has undergone a similar reincarnation, from industrial bulk petro-chemical storage to publically accessible waterfront social space with restaurants, an events centre, children's playground and tourism activities.

Auckland's city centre, formerly bereft of quality open spaces, now has a few jewels, and Wynyard Quarter is vying for top place. Now revitalized, Wind Tree is credited with galvanizing the entire area. The local art scene and ordinary Aucklanders alike are stumped as to why the remarkable work was so overlooked for so long. Wind Tree has become a symbol of rebirth, especially for a small city in a tough economic climate.

As an artist, Michio Ihara works with stainless steel, with brass and copper, but mainly he works with space and movement. His practice radiates purity, clarity and confidence. Carefully organized and proportioned, with air and light flowing freely through their transparent structures, his sculptures are a condensation of their surrounding spaces and thus become integral parts of them. Often kinetic in nature, his works vibrate with currents of air, creating living and breathing forms. Ihara's work has an inherent clarity, a nod to great architecture and the aesthetic of his Japanese heritage.

Artwork Excellence

Wind Tree was originally Auckland's 1971 centennial present to itself and seen as an opportunity to further cement Pacific Rim identity internationally. But poor siting didn't allow the moving sculpture to function. It was eventually taken down. The work has recently been re-sited. It is now credited with helping to revitalize a public space in downtown Auckland and has become a symbol of rebirth.

风棒
Wind Wand

艺术家：兰·莱耶（新西兰）
地点：新西兰新普利茅斯
推荐人：凯利·卡迈克尔
Artist: Len Lye (New Zealand)
Location: New Plymouth, New Zealand
Researcher: Kelly Carmichael

作品描述

和许多地方一样，海岸城市新普利茅斯也背朝大海。位于新西兰北岛崎岖的西海岸，这座过去的驻军小镇被四车道的高速公路和大多数废弃的铁路轨道与大海隔开。即使有些喜欢冒险的人尝试跨越这里，他们所能看到的也只有被常年汹涌的塔斯曼海侵蚀的海岸线，被大风刮秃的海滩，仅留下荒凉的巨石滩和一片破败的海堤。这是一项持续中的项目，新普利茅斯海岸步道目前有 10 公里，从东部的希克福德公园到西部的塔拉纳吉港，这一段路上有许多行人的步道接入点。这个获奖项目不仅把一系列小型公共储备区连接了起来，同时还把城市向海滨开放，将居民引入这片被遗忘的、无人关注的海岸线。

这是一项正在进行的项目，计划是再往东北方向延伸 15 公里，最终连接到海滨小镇怀塔拉。新普利茅斯的海岸步道是一个综合项目，它结合了社会空间、环境、地貌改建和当代艺术。其中一件主要的艺术品由新西兰艺术家兰·莱耶完成，另一件由新西兰艺术家约翰·雷诺兹来完成的作品已有了计划。

海岸步道的中央长廊有很多优点，而其中最为突出的是兰·莱耶创作的"风棒"，这是新西兰公共艺术中最大胆和最优秀的案例。

这件巍峨的 45 米高的动感雕塑用玻璃纤维柄制成，顶部有个球体，在夜里会散发红光。风棒会在风中弯曲摆动，幅度最大可达 20 米，就像一座摇曳的灯塔或是一座高科技的蒲公英。作品显示了艺术家对电影媒介和动感雕塑手法富有远见性的使用，以及他对于动感始终不渝的痴迷。莱耶以其独特的风格和实验技术而闻名，这位艺术家的艺术实践重申了他的信仰：动感和物理移情相比媒介更为重要。"风棒"的形式虽然简单，却包含了代表和反映人类形式的能量与活力。作品常常吵闹、有时危险，甚至公开表达性感。它摇摆着震动着对人生和环境作出回应。

当受到热烈讨论的"风棒"临时出现在海岸步道上来测试它的坚固程度时，它马上就成了全镇居民的话题。当顶部球体的半截被强风切断，作品被搬走做调整时，这种争议就越发强烈了。但新普利茅斯的居民们决不认输，他们在自家后院、在信箱上、有的甚至在帽子上建造自己的风棒。一家当地电台发起了一项"制作你自己的风棒"比赛，吸引了 141 名参赛者。

经过重新安装后，"风棒"成为了新普利茅斯的一个标志，当地旅游业的一项重要品牌，在壮丽的海岸线和海岸步道上高耸着。这座只有 7 万人口的与世隔绝的新普利茅斯小城市如今成了该地区主要经济活动的服务中心，其中包括集约化养殖和小规模的石油、天然气和石化生产。

在偏远的地方进行如此规模的项目，其卓越的眼光以及对当地社区和环境需求的灵敏反应都非同凡响。海岸步道在物质和隐喻上延伸了新普利茅斯市，依靠着世界级的设计和当代艺术，创造了一个全新的休闲和体育比赛场所。

解读

没有什么比未完成的作品更令人昂首期盼。沿着海岸线蜿蜒生长的公共艺术项目"海岸步道"，在持续不断的进展中，吸引公众的视线与参与热情，已成为新西兰北岛西海岸与东北海滨小镇怀塔拉之间的衔接纽带，为滨海城镇刻画出面貌一新的轮廓线。

在整个项目中，新西兰艺术家兰·莱耶的作品"风棒"，一直以其遗世独立的大胆构想，成为人们视觉乃至话题的中心。迎向海风可以大幅度弯曲的玻璃纤维柄，配合顶端红色的发光球体，像极了金蛇狂舞，又像大风中的蒲公英，将落未落。"风棒"以肆意的美感征服了当地居民的心，以至于当"风棒"因安全问题被挪走后，当地居民不惜自发制作风棒大赛，以唤回那份招摇的美。能够赢得当地公众的认可，是任何一件公共艺术作品最期待的掌声。

而整个项目的功德还远不止此，这项公共艺术行为逐渐成为当地的一个标志特征，吸引更多的人前来观看，沉寂多年的海岸线又再一次苏醒，充满活力与挑战。公共艺术项目的能量，使偏远变为中心，枯槁朽重焕新颜。"风棒"如一支随风摇曳的红色罂粟，北国海岸线的漫漫无边，从此不再孤独。
（吴昉）

Artwork Description

Like many places, the coastal city of New Plymouth had turned its back to the sea. Positioned along the rugged west coast of New Zealand's North Island, the former garrison town was separated from the sea by a 4-lane highway and mostly disused railway track. Even if the adventurous did manage to cross the obstacles little awaited them, coastal erosion by the often-turbulent Tasman Sea and extreme winds had taken away the beaches, leaving only inhospitable piles of boulders and a deteriorating sea wall. An ongoing project, New Plymouth's Coastal Walkway is currently a 10 km path from the Hickford Park in the east to Port Taranaki in the west with numerous pedestrian access points. The award-winning project has not only linked a series of small public reserves, but also opened up the city to the foreshore, connecting residents to a forgotten and previously unloved coastline.

An ongoing project with plans to eventually link to the small coastal town of Waitara, a further 15 km to the North East, New Plymouth's Coastal Walkway is a integrated project combining social space, the environment, placemaking and contemporary art. One major piece of art by New Zealand artist Len Lye has been completed and another by New Zealand artist John Reynolds is planned.

The Coastal Walkway's central promenade has many attributes but the jewel is Len Lye's Wind Wand, one of the most daring examples of public art in New Zealand, and one of the best. The towering 45 metre kinetic sculpture is a fibreglass stalk crowned with a sphere that glows red at night. Wind Wand bends and jiggles up to 20 metres in the wind, like a swaying lighthouse or a high-tech dandelion. The work demonstrates Lye's visionary approach to the medium of film and kinetic sculpture and a lifelong obsession with motion. Known for his distinctive style and experimental technique, Lye's practice reiterates his belief that motion and physical empathy are more fundamental than medium. Wind Wand is simple in form yet embodies an energy and dynamism that stands in for and reflects the human form. Often noisy, sometimes dangerous and even overtly sexual, the work sashays and vibrates in response to life and the environment.

The hotly debated Wind Wand was the talk of the town when it temporarily appeared on the Coastal Walkway to test its robustness. Controversy increased after the top half of the sphere sheared off in high winds and the work was removed for fine-tuning offsite. Not to be defeated, the people of New Plymouth built their own wind wands in backyards, on letterboxes and even on hats. A local radio station launched a make-your-own Wind Wand competition attracting 141 entries. Reinstalled, Wind Wand has become an icon of New Plymouth, a major part of tourism branding and soars over the dramatic coastline and Coastal Walkway.

With a population of 70,000, the small and somewhat isolated city of New Plymouth is the service centre for the region's principal economic activities, which include intensive farming as well as small-scale oil, natural gas and petrochemical production. A project of this magnitude in a rural environment is exceptional for its vision and responsiveness to the needs of community and environment. Coastal Walkway has extended the city of New Plymouth both physically and metaphorically, creating a new leisure and sporting events destination alongside world-class design and contemporary art.

Artwork Excellence

An ongoing project with plans to eventually link to the small coastal town of Waitara, a further 15 km to the North East, New Plymouth's Coastal Walkway is a integrated project combining social space, the environment, placemaking and contemporary art. The award-winning project has not only linked a series of small public reserves, but also opened up the city to the foreshore, connecting residents to a forgotten and previously unloved coastline. The Coastal Walkway's central promenade has many attributes but the jewel is Wind Wand, one of the most daring examples of public art in New Zealand, and one of the best. Coastal Walkway has extended the city of New Plymouth both physically and metaphorically, creating a new leisure and sporting events destination alongside world-class design and contemporary art. A project of this magnitude in a rural environment is exceptional for its vision and responsiveness to the needs of community and environment.

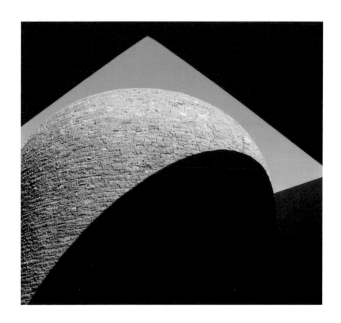

内无
Within without

艺术家：詹姆斯·特瑞尔（美国）
地点：澳大利亚堪培拉澳大利亚花园，澳大利亚 国家美术馆
推荐人：凯利·卡迈克尔
Artist : James Turrell（USA）
Location : Australia Garden, National Gallery of Australia (NGA)),
Canberra, Australia
Researcher : Kelly Carmichael

作品描述

詹姆斯·特瑞尔的作品被称为"光雕塑"，他把光和知觉作为创作的媒体和主题。这件作品外形像一座金字塔，正方形基座，赭红色墙壁，内部中央有一座绿松石水池，水流经佛塔中央内室周围。建筑有一部分深入地下，创造一种身临其境的观赏体验，使用空间、形状和照明来营造天空的感觉。作品引人注目之处并不是出自艺术家的人为制作，而是来自上方天体结构的自然美。

优雅简洁的圆顶室内有一条长凳围绕在边缘，中央内室的周围布置着玄武岩。房顶有一个圆形窗口，一块月光石设置在地板中央，与房顶的孔径相呼应。让观众在意识到周围自然环境的同时又感觉到自己与自然环境密切相关。该作品从黎明开放至深夜，使观众能够在里面体验到天体周期，可以静观天空由淡蓝至群青渐到漆黑的变化过程，习以为常的自然景象被巧妙地演化成作品的一部分，促使观众徘徊、观察、吸收并沉思，让人们以新的方式体验和重新认识生命中无处不在的光线。

詹姆斯·特瑞尔善于将传统与现代以及自然与人为等对立的概念完美地融合起来，变化莫测的光可以视为人类心灵的物化形态，对欣赏艺术时匆匆忙忙走过场的习惯提出了挑战。他的作品是对光线与空间的探索，提供了朴实、简洁和悠闲的感知过程，以唤醒精神的力量影响着视觉和心灵。

解读

"光雕塑" 是这件作品的名称，詹姆斯·特瑞尔以光和知觉作为创作媒体和主题，把建筑的结构美与人工美完美地结合在一起，让人们能够感受到光的变换和感悟时间的流逝，并且他也富有创意地解决了人与自然对立的难题，让观众在徘徊中观察、沉思并感悟自然。他也改变了观众 "走马观花" 的欣赏和感知过程，提供了一种朴实、简洁和悠闲的欣赏过程。这件作品用其内在的精神内涵影响着每一位观众的心灵和感受。（冯正龙）

Artwork Description

James Turrell's artworks are called "light sculpture". He takes light and perception of light as the medium and creative impulse of his work. This piece is shaped like a Pyramid, with a square base and red ochre walls. The internal courtyard contains a turquoise pool, water flowing around a stupa in the center. Part of the building is deep underground, creating a highly site-specific viewing experience. The perception of the sky is enhanced through the use of space, shape and lighting. Nevertheless it is the natural beauty of the sky itself which holds the audience's attention.

The elegant interior of the dome is ringed by a seating bench arranged around the central stupa. The circular window in the roof is echoed by the shape of the moon stone on the floor, creating both an awareness of the interior environment while sensing the mystery of the natural environment. The artwork is open from dawn til dusk, providing an opportunity for visitors to experience the changes in the sky throughout a celestial cycle— light blue gradually turning to dark. A familiar natural scene is skillfully re-envisioned as an artwork. Viewers can wander freely in observation, absorption and reflection, experiencing a new way to understand the life that radiates everywhere from light.

James Turrell is skilled at fusing contradictories— tradition and modernity, nature and culture. His works offer an alternative to spaces that perpetuate the usual rapid "sampling" of art, inviting visitors instead to linger. They stimulate basic perceptual faculties using light and space in order to animate the spirit and awaken the mind.

Artwork Excellence

The location for Within Without was chosen by the artist to complement the NGA's new Australia Garden and alert its audience to the clear and sometimes harsh natural light of Canberra, a quality unique to this setting. Within Without draws from and responds to its surroundings, allowing its audience to become both aware of and connected to their natural environment. It asks visitors to linger, to watch, absorb and reflect. It offers artlessness, simplicity, unhurried perception.

非洲地区提名作品
Afica Cases

拱门的记忆
Arches of Memory

艺术家：桑德琳・多尔（法国）
地点：喀麦隆杜阿拉
推荐人：维拉・托尔曼
Artist: Sandrine Dole (France)
Location: Douala, Cameroon
Researcher: Vera Tollmann

作品描述

本项目由玛丽琳・杜阿拉・贝尔和迪迪埃・绍布主管，当代艺术中心杜阿拉艺术于 1991 年在喀麦隆的经济和文化首都杜阿拉成立。该非盈利性组织致力于实验性都市干预活动，特别是在社区开展，这些社区是贫困社区，受影响严重，基础设施不足。杜阿拉艺术的都市项目总是在特定地点实施的，有时非常具有现实性，比如提供水泵系统或建造一座急需用的桥梁。从 2006 年以来，杜阿拉艺术聚焦于加强城市的建筑遗产。其长期的项目名为："杜阿拉，艺术和历史之城"，目的是要标注这个城市的历史景点。从而推广城市的历史和古代建筑，杜阿拉艺术委派法国设计师和当地居民桑德琳・多尔设计标志，用来为建筑旅游路线做标志点。

该项目被称为"拱门的记忆"。多尔决定旅途的每个地点都作为对过去的一种见证。她想要赋予一个确定的存在地一个标记，以使它们不会被人们遗忘。她使用的这种造型唤起了普通的棕榈树拱门，标志着喀麦隆传统节日举办地的入口。这些拱门，如今使用钢铁制作，象征性地打开着门，代表着早期抵抗殖民暴力统治。

拱门上的信息文本铭刻在有机玻璃版上，由历史学家瓦莱热·索德、曼噶和莱恩内尔·布莱兹·恩德耶霍亚亚撰写。文章是双语的，法语和英语（都是喀麦隆的官方语言），在杜阿拉市的处境下，诠释了特定地点的历史。除了让杜阿拉历史清晰可见外——1884 年喀麦隆被德国殖民了，1919 年被法国殖民，从 1940 到 1946 年该地成为了喀麦隆的首都——"拱门的记忆"项目作为一种宣传手段，用来保护城市的建筑遗产。项目的目的是标记 30 个地点和历史建筑，这些建筑的时间跨度是 16 世纪后期到 1960 年喀麦隆独立。杜阿拉艺术组织了导游带队观光游，出版了有每座拱门位置的城市地图。

目前为止有 18 座拱门：12 座坐落于波南约区，联通了司法机构、监狱、贸易和医疗机构的建筑，这些建筑均由先后的殖民政府（英国、德国和法国）和贝尔皇室建造。6 座拱门坐落于阿克瓦区。它们的功能是将基督教介绍到喀麦隆，观看杜阿拉市的城市建设基础，如同献给当地人的第一座公共医院。在这些建筑和纪念碑中，有贝尔国王的丧葬纪念碑、波纳库阿姆昂烟囱、曼德西·贝尔别墅、商会、贝尔国王的老宫殿、德国老邮局、正义之宫。

自从 1995 年起，杜阿拉艺术运营了一座展览空间，从此主持了一个驻场项目，也开展了一系列展览、表演、会议和研讨会活动。此外，杜阿拉艺术组织了杜阿拉城市沙龙三年展（2007 年和 2010 年），为城市制作了大量委派作品和永久性作品。

解读

罗斯福所言："建筑是石头写成的历史"。历史传承下的建筑是一座城市丰厚的遗产宝藏。建筑体现了城市的物质基础和意识形态，人们的物质生活方式和精神意念都会在建筑上得到反映与体现。它具有双重性而绝非片面的。城市建筑内涵包含了两方面的内容：一方面它反映了某一时代、社会、民族的整体形态，一方面是它自身形态的文化。

石头的历史即一个城市的建筑。城市建筑体现历史，衬托历史，并映射着历史，这是一条不可否定的规律。因为城市建筑既是历史的载体，又是历史的一种形态，是历史的有机组成部分。

聚焦于城市建筑遗产的保护与推广，关注于历史和过去的铭记与见证，"拱门的记忆"项目以标记历史建筑景点为出发点，以旅游线路标志点为宣传手段，以唤起人们对历史建筑遗产的关注与保护为目的。目前为止，矗立在杜阿拉的 18 座拱门，为人们打开了回忆历史的时间隧道，沧桑的历史、久违的过往，一幕幕，一幕幕的在述说着过去的故事。（冯正龙）

Artwork Description

Directed by Marilyn Douala Bell and Didier Schaub, the contemporary art center Doual'Art was founded in 1991 in Douala, the economic and cultural capital of Cameroon. The nonprofit organization's focus is on experimental urban interventions, especially in neighborhoods most affected by poverty and inadequate infrastructure. Doual'Art's urban projects are always site-specific and sometimes have a very practical nature, such as providing a water pump system or constructing a much-needed bridge.

Since 2006, Doual'Art has concentrated on enhancing the architectural heritage of the city. Its long-term project called "Douala,

City of Art and History" aims to mark the historic sights of the city. To popularize the history of the city and the ancient buildings, Doul'Art commissioned French designer and then-local resident Sandrine Dole to design signs that would define spots along an architectural tour. The project is called Arches of Memory (Arches de la mémoire).

Dole considered each place on the tour a witness of the past. She wanted to give the signs a certain presence, so that they would not remain unnoticed. The form she used evokes the common palm arches that mark the entrance of traditionally festive places in Cameroon. These arches, now made of metal, symbolically open doors to an earlier time of resistance to colonial violence.

The informational texts on the arches are inscribed on slabs of Plexiglas and written by historians Valère Sword, Manga and Lionel Blaise Ndjehoya. The texts are bilingual, French and English (both official languages of Cameroon) and explain the history of the specific sites in the context of the city of Douala.

Besides making visible the history of Douala—which was colonized by the Germans in 1884 (Kamerunstadt) and the French in 1919, and was the capital of Cameroon from 1940 to 1946—the Arches of Memory project has served as an advocate for the preservation of the architectural heritage for the municipality. The project's aim is to mark thirty sites and historic buildings from the late sixteenth century to the independence of Cameroon in 1960. Doual'Art organizes guided tours and publishes a city-map with the locations of each arch.

So far there are 18 arches: Twelve are located in the district Bonanjo and trace the development of judicial institutions, prisons, trade and health, by successive colonial governments (British, German and French), and the lineage of King Bell. Six arches are located in Akwa district. Those function as an introduction to Christianity in Cameroon and look at the basics of urban construction in the city of Douala, like the first public hospital dedicated to the natives. Among those buildings and monuments are Funerary Monument of King Bell, chimney of Bonakouamouang, Villa Mandessi Bell, Chamber of Commerce, old Palace of King Bell, old German post office or the Palace of Justice.

Since 1995, Doual'Art has also run an exhibition space that has since hosted a residency program, as well as a number of exhibitions, performances, conferences, and seminars. Additionally, Doual'Art organizes the triennial Salon Urbain de Douala (SUD) (2007 & 2010) which has produced a number of commissions and permanent works for the city.

Artwork Excellence

The project Arches of Memory is part of Doual'art's long-term effort to raise awareness for the (architectural) history of the city of Douala. Therefore, the project is partly an educational walk to point out the architectural heritage which represents the political history. By highlighting a selection of existing buildings, this urban intervention creates an urban choreography that locals or visitors can individually follow.

厄内斯特·奥本海默公园
公共环境升级：艺术品项目
Ernest Oppenheimer Park Public Environment
Upgrade :Artworks Programme

艺术家：多位艺术家、三位一体会议和图书馆合作
地点：南非约翰内斯堡
推荐人：吉乌希·切科拉
Artist: Multiple artists
Location: Johannesburg , South Africa
Researcher: Giusy Checola

作品描述

约翰内斯堡一个重要的景点：厄内斯特·奥本海默公园近期正在升级（2009年11月10日—2010年11月3日），艺术品项目中包括五个艺术介入的作品的实施，其中大多数是雕塑作品。艺术家西波·格瓦拉、姆芬多·凯特耶、斯东·马本达、马拉基亚·莫莎泊，三位一体会议和图书馆，从约翰内斯堡的当代文化和历史上得到了灵感启发，一起创作了五件兼具俏皮与严谨的作品，以此来纪念这座城市的兴起，同时纪念城市的更新再建。

厄内斯特·奥本海默公园位于约翰内斯堡的市中心。在1980—1990年之间经历了衰败期以后，社区以犯罪、污垢及其他弊病而成为了其标志特点。2007年约翰内斯堡和一群投资者制定了城市内部章程，聚焦于公共空间、公共艺术和促进遗产保护。厄内斯特·奥本海默公园艺术品也包括在这个章程内，目标是要建造一个场所能够使得附近的居民受益，他们可以在此

503

打篮球或者休闲娱乐，让市中心的商业重新焕发生机，让人们更愿意置身其中。

当地劳动力的加入参与，是本项目的一个重要组成部分。事实上在 2010 年 2 月，初步的概念由三位一体会议和图书馆共同构思，但是被负责选择项目的约翰内斯堡发展署否决了。约翰内斯堡发展署要求一份新的策划书，基于新的雕塑公园，有着大量的劳动力和详细的手工艺作品。

以下是五个主要艺术品的描述。

该公园的历史性复兴始于厄内斯特·奥本海默公园钻石（一种复制品）。创作这件令人印象深刻的雕塑是为了纪念厄内斯特·奥本海默的人生，他是一位钻石与金矿业的大亨，他掌控着戴比尔斯钻石公司，创立了英美资源集团。这件雕塑主导了厄内斯特·奥本海默公园，为了纪念他，在 4 月 15 日由执行市长阿莫斯·梅森多重新启动的项目。虽然很多人被唤起了记忆，开始重新使用该公园，然而很少人知道这座公园的名字，然而奥本海默的名字是享誉世界的。这颗粗切割的金刚石晶体不是最大的，然而被公认为是世界上最美的。在一种童心未泯的精神中，这件 236 克拉的厄内斯特·奥本海默钻石犹如注入了生机。米克·阿穆尔使用 3D 扫描这件作品，使用了华盛顿的约翰·哈特尔伯格的设备，放大了 50 倍，矗立起来，成为了约翰内斯堡中心及公园里一个闪亮的焦点。

这座公园因其有着跳跃的黑斑羚的水景在 1980 年代非常有名。该艺术品在遭受到严重的破坏之后，黑斑羚雕塑被重新安放在英美资源集团总部对面的主干道上。为了怀念这座著名的雕塑，艺术家创作了 14 件新的羚羊雕塑，让它们重返公园，在这儿牧草，休憩，享受新的景观。

约翰内斯堡市在自我重塑的过程中见证了大量建筑的诞生与消亡。从这个现象起，开始发展两项建筑干预项目：标准剧院和里斯克街邮局，按照以前建筑样式的照片而建造。公园的位置最初是标准剧院的地址，早期采矿先驱者在那里进行娱乐活动。这座剧院曾给予了 1900 年代早期脏兮兮的矿工们一些许必要的安抚。"鉴于约翰内斯堡不断地发展着，市内出现了大量建筑脚手架，建筑起重机上升到天际线，当代钢材支撑着早期约翰内斯堡那光谱般的外墙，"三位一体会议的斯蒂芬·霍布斯说道。

艺术家使用类似的手段实现了建于 1897 年的里斯克街邮局的外部风貌，"它经历了多次的修改，后来成为了大火的牺牲品。在城市保护自身建筑遗产的进程中，里斯克街邮局得到了再次翻新。这里的建筑装置被用来支撑里斯克街邮局的门面，在城市档案里这是一张蓝图，从没有建造过，"霍布斯说。

设计建造了文字雕塑，是为了启发公园的参观者。内容包括短语 EGOLI——你使我的梦想成真，姆赞斯基——全世界大团结，MJIPPA——你赋予我勇气和 JOZI——任何事都是有可能实现的。

解读

艺术提点着人类这个社会曾经的发展轨迹。历史与艺术，融合、交织、再创，无论在世界的什么地方，但凡被重建或复制的每一件艺术作品的背后追根溯源，都有着独一无二的故事，文化和历史意义。在这个项目中，历史重现了此区域内的辉煌历史，随着时间研磨消逝，埋藏在记忆背后的故事无论是可圈可点或羞于见人，但却证明了其曾经真实存在过。

约翰内斯堡市的过去被重建，经过重新设计后的雕塑作品虽没有按照 1:1 的尺寸放大，但几乎完全相同的细节已经具备唤醒了人们记忆的力量。艺术品的安置颇有讲究，这些新建的反映城市历史的艺术品被安置于公共空

间之内。寓教于乐，这是生活空间和艺术完美结合，市民们会永远记得这个城市，这个公共空间，有这样一段历史。（高浅）

Artwork Description

An important element Johannesburg's Ernest Oppenheimer Park recent upgrade (10th November 2009 – 3rd November 2010) is the implementation of five artistic interventions, mostly sculptures, included in the Artworks Program. Artists Sipho Gwala, Mfundo Ketye, Stone Mabunda, Malakia Mothapo, the Trinity Session and The Library, created five works that are both playful and serious, drawing their inspiration from the history and contemporary culture of Johannesburg to commemorate the city's early beginnings and celebrate its urban regeneration.

The Ernest Oppenheimer Park is in the center of Johannesburg. After a period of deterioration between 1980 and 1990, the neighborhood was characterized by crime and grime and other related ills. In 2007 the City of Johannesburg and a group of stakeholders developed the Inner City Charter, focused on public space, public art and the promotion of heritage. The Ernest Oppenheimer Park Artworks are included in the charter, with the aim of helping to create a place able to host citizens who live nearby, who want to play basketball or simply enjoy time, and revive both the vibrant business in the heart of the city so people will want to spend time there.

The involvement of the local labor was an important part of the project. In fact in February 2010 the first concept developed by Trinity Session and The Library was rejected by the Johannesburg Development Agency, which oversaw selection. The JDA requested a new proposal based on new sculpture garden brief with highly labor intensive and hand crafted artwork specification.

Here are descriptions of the five main artworks.

The historical revival of the park started with the Ernest Oppenheimer Diamond (a replica). This impressive sculpture was created in order to celebrate the life of Ernest Oppenheimer, a diamond and gold mining tycoon who controlled De Beers and founded Anglo-American. This sculpture presides over the Ernest Oppenheimer Park, which was relaunched in his honor by Executive Mayor Amos Masondo on 15 April. While many recall and continue to use this park, few know its name. Yet the Oppenheimer name is famous worldwide. The rough-cut diamond crystal is not the largest but is recognized as the most beautiful in the world. In a playful spirit, the 236-carat Ernest Oppenheimer diamond has been brought to life. It was 3D scanned by Mik Armour with the supplies of John Hatelberg in Washington, and scaled up 50 times to stand as a shining focal point of the park and central Johannesburg.

The park was perhaps most famous in the 1980s for its water feature with leaping impala. After serious vandalism, the impala were relocated to Main Street opposite the headquarters of Anglo American. In remembrance of the famous sculpture, the artists made fourteen new Bokkies that have made their way back to the park and are found here grazing, resting and enjoying the new landscape.

The City of Johannesburg has seen numerous buildings come and go during the process of reinventing itself. From this reflection started the development of two architectural interventions: The Standard Theatre and The Rissik Street Post Office, which were constructed from the photographs of the former facades.

The park was originally the site of the Standard Theatre, where early mining pioneers went to be entertained. The theatre brought much-needed light relief to the dusty gold diggers of the early 1900s. "In reference to Johannesburg's continual progress expressed in the ever-present scaffold and construction cranes punctuating the skyline, contemporary steel bracing supports a spectral façade of early modern Johannesburg," says Stephen Hobbs of The Trinity Session.

With a similar process the artists realized the facade of The Rissik Street Post Office, built in 1897. "It has undergone several revisions and has also been the victim of major fires. It will undergo yet another facelift when the City preserves its architectural heritage. Here construction rigging is used to support a façade of the Rissik Street Post Office that was never built and remains a blueprint in the City's archive," says Hobbs.

Text Sculptures was designed to inspire park visitors. It includes the phrases EGOLI-YOU MAKE MY DREAMS COME TRUE, MZANSKI - UNITING THE WORLD, MJIPPA-YOU GIVE ME COURAGE and JOZI - NOTHING IS IMPOSSIBLE.

Artwork Excellence

The success of this project can be measured on a number of levels:

1. The acceptance and interest from the users of the park and surrounding area, using the artworks as photo opportunities and portraiture props;

2. The lack of any significant vandalism;

3. Celebration of the innery city of Johannesburg in relation to its cultural heritage;

4. The interrelationship between this park upgrade and related successes of urban regeneration of other parts of the inner city;

5. The gratification of the client in pursuit of a sculpture garden for the inner city.

艺术之语
Freddy Sam (murals) - AWOA - A World Of Art

艺术家：里基·李·戈登（南非）
地点：南非开普敦
推荐人：吉乌希·切克拉
Artist: Ricky Lee Gordon (South Africa)
Location: Cape Town, South Africa
Researcher: Giusy Checola

作品描述

艺术家和艺术活动家里基·李·戈登 (Ricky Lee Gordon，又名 Freddy Sam) 从事街头壁画已创作超过 12 年。他的创作主要在南非和一些其他城市，如柏林、洛杉矶、伦敦、摩洛哥、亚特兰大、波特兰和纽约。他正在运作一个画廊／艺术空间项目，名为 /AWOA（艺术之语）。该项目为国际艺术家驻场计划，地点是在伍德斯托克——开普敦最古老的郊区之一。

戈登解释说："身为日常生活中的普通人，我觉得壁画创作过程具有任何其他艺术形式无法替代的体验和满足。"作为一名壁画家，他表示："我的意图是用公共艺术这一工具探索周边的社区，尝试与各界人士沟通和连接。因为相比较于结果，我更看重整个过程的体验。我相信，将灰色从城市的灵魂中抹去是艺术家、音乐家和诗人的工作。"

壁画是非常快速、廉价而有效的公共艺术形式，这就是为什么戈登有超过 70 幅壁画是永久性的作品，它们必须由社区拥有，当它们旧损后会被新的作品所替代。

过去的伍德斯托克地区主要是一个穆斯林社区，因为它是一个独特的地区，种族隔离时期这些家庭被允许住在这里。但是此后纺织行业迁入，居民不得不搬离并住到乡镇（离城外一个小时以上的路程）。因受到毒品和帮派团伙的滋扰，如今纺织业已迁离该地区。

戈登的项目建立了与其他学科的专家的合作，特别是与维罗·史密斯（一名调研公共艺术项目对社会变迁影响的人类学家）以及卡尔·伯恩斯（一名社会学社区战略家和大品牌和当地政府顾问）之间的联系。戈登的目的是用艺术提升南非贫困学校的层次，并创建一个 5 年的研究计划，旨在记录和研究艺术可能带来的影响。

他主要运作的项目是成立于 2008 年的 / AWOA。/ AWOA 使用创造性方法在开普敦伍德斯托克和卡雅利沙社区实施积极的社会变革，同时在斯威士兰、特兰斯斯凯和冈比亚开展其他项目。津巴布韦艺术家 Juma Mkwela 和 Willard Kambeva 领导艺术推广计划，到现在为止仍在非洲以"书写非洲"的名字运行。

这些方案包括儿童和青少年的艺术课程、工作坊以及儿童和青少年参与的壁画项目。为了使社区维持机构的整体积极性，项目往往用本地和来访艺术家当嘉宾导师及主持人，这样一来参与者便可以共享作品的所有权和完成作品的自豪感。

通过 / AWOA，戈登鼓励有责任感的企业进行社会投资，并与多个企业实体建立了合作关系，其中最引人注目的是和 OgilvyCT 的合作关系。这个组织正在经营一个名为"帕西•巴特利之屋"的项目——为流落街头的男孩们重新建立团体之家。2010 年以来，通过本地及国际艺术家的共同努力，这个项目已见证了许多房子的内部和外部是怎样被壁画装点得丰富多彩、令人振奋的过程。

解读

壁画自是无言的，可贵在艺术家成功地赋予了壁画说话的能力。里基•李•戈登绘制的头像轮廓清晰，虽然主基调是暗色，却很明显能够看出这是大写的非洲男性侧面像。洁净的白色石灰粉砌的墙上，如此大面积的反差对比，让周围的环境连带着陷入静谧无声。艺术家赐予这幅作品吸引市民和宣告信息的巨大力量，让欣赏者肃然静默，鸦雀无声，在凝神的瞬间，失去凡尘间的轻狂和浮华，霎那间拜跪虔诚，仰视庄重。（高浅）

Artwork Description

Artist and art activist Ricky Lee Gordon (aka Freddy Sam) painted murals on the streets for over 12 years in South Africa and other cities like Berlin, Los Angeles, London, Morocco, Atlanta, Portland, and New York. He runs /AWOA (A Word Of Art), a gallery/project space and a movement, which provides an international artist residency program in Woodstock, one of the oldest suburbs of Cape Town.

Gordon explains that "for the everyday man, murals are an experience and satisfaction that I feel can not be replicated by any other art form." As a muralist, his intention is " to explore my community and surrounding using public art as a tool to communicate and connect with people from all walks of life, as I am more interested in the experience than the result. I believe removing the greyness from the soul of the city is the job of artist, musicians, and poets."

Murals are a very quick, cheap, and effective form of public art; that's why Gordon' s more than 70 murals are created to be permanent, to be owned by the community and to be replaced when they deteriorate.

In the past the Woodstock area was predominantly a Muslim community, because it was a unique area during the apartheid's housing act and these families were allowed to live here. After a textile industry located in the community, however, the inhabitants were forcibly removed to the townships (over an hour outside the city). Industry has left the area, and today it is plagues by drugs and gangs.

Gordon' s projects are developed together with experts from other disciplines, especially with anthropologist Sydelle Willow Smith, for the investigation of the effects of public art projects for social change, and Cal Burns, who is a social and community strategist and consultant for big brands and local government. Gordon's aim is use art to uplift poor schools in South Africa and create a 5-year research program that aims to document and study the effects of art for change.

The project on which he's mostly focused is /AWOA, founded by him in 2008. /AWOA uses creative means to effect positive social change in the Cape Town communities of Woodstock and Khayelitsha, with additional projects located in Swaziland, Transkei and Gambia. Zimbabwean artists Juma Mkwela and Willard Kambeva lead the arts outreach programs, which until now have been run under the name Write on Africa.

These programs include art classes, workshops, and participatory mural projects with children and youth, and often involve local and visiting artists as guest instructors & facilitators in order to make the communities maintain a sense of agency within each activation, by sharing ownership and pride for the outcome.

Through /AWOA he encourages responsible corporate social investment and have built relationships with several corporate entities, most notably a partnership with OgilvyCT in the ongoing rejuvenation of Percy Bartley House, a group home for boys living on the streets. Since 2010, this project has seen the interior and exterior of the house transformed into a colorful and inspiring environment with murals by local and international artists.

The future plan of /AWOA is to create a one-room art school with the new name of Art in the City, providing a safe and supporting environment to create, share, and grow. The school will host a program of daytime classes for at-risk youth not attending school and promising adult artists from around Cape Town, as well as after school and weekend classes for youth from Woodstock and Khayelitsha. Students from outside the Woodstock area will be provided transport to and from their home communities.

Artwork Excellence

In the words of the artist: "We are building bridges. We host 5 artists from around the world every month, and we are working on big projects that effect entire communities. Until I can prove the science behind how my art and my projects effect social change, I can only claim that our work inspires people. I believe colour creates energy, and energy creates inspiration, and inspiration creates change. In addition we are making "doing good" cool-we are inspiring younger artists to use their talent to create the change they wish to see." (Ricky Lee Gordon, March 2012, interviewed by the researcher)

伊奥威尔酒店
Hotel Yeoville

艺术家：特里·库尔干（南非）
地点：南非约翰内斯堡
推荐人：吉乌希·切克拉
Artist: Terry Kurgan (South Africa)
Location: Johannesburg, South Africa
Researcher: Giusy Checola

作品描述

电视和印刷媒体无情地引导公众聚焦于南非人和非洲人之间的暴力和冲突——大多数是由于战争而移民，还有经济的原因——这些人从非洲各国来到了约翰内斯堡。"伊奥威尔酒店"（2007年3月—2010年12月）是一项合作性与参与性艺术项目，由约翰内斯堡的艺术家特里·库尔干负责策展，项目在伊奥威尔的一座新型公共图书馆里开展，伊奥威尔是约翰内斯堡市内东部边缘一个被忽视的陈旧郊区，是4万名非洲移民和难民的家。这些居民被南非主流社会和正规经济隔离和排除了。该项目的目标是通过揭示司空见惯的故事来提供观察城市的另一种途径。项目包括一个社区网站和一个由12个单独展位组成的展览，在那里观众受邀可以通过一系列的数字接口、互动媒体和在线申请来纪录自己。

"伊奥威尔酒店"的目的是制作一张伊奥威尔无形移民社区的社会地图，

开启新的对话 揭示非常普通的日常生活叙事 述说具体的个人故事——爱、失落、梦、欲望、孤独和地区的特质——通过这样的方式来表明移民们不能脱离现有社会的基础设施，当今每天的艺术实践都是基于这一观点意见，引用南非建筑师希尔顿·朱丹创作的一个术语，即"文化作为一种基础设施。"

"伊奥威尔酒店"包含了四个方面：研究、一个社区网站、一个互动展览和一本书。

这项研究是通过实地考察以了解社会邻里和空间基础设施。与居民和小型企业业主进行对话，关注于两方面：整个郊区的墙面空间被大量手写的社区广告所覆盖，广告内容有住宿、招工、汇款、寻人启事和其他服务和制品；网吧有着不寻常的密集度，

大多数是由有着特定的民族身份的外国人所并经营，为居住在南非的外国人提供连接的纽带。人们去网吧与遥远地方的熟人进行联系，分享消息与新闻。在广告与沟通的实践之间，工作小组决定制作一个特定的网站，以此作为项目的一部分。可以在伊奥威尔社区的网吧里浏览该网站，网站由大多数居住在同样郊区的南非和移民研究者负责研发，他们设计了一系列的研究工具，问卷面向网吧使用者、拥有者、经理、员工和街上的人群。

这一系列过程的结果便是设计"伊奥威尔酒店"网站，使用这类科技和普遍的社会媒介已经难以置信地被"有线"和创业社区使用了，它作为一种生存的手段来应对非常难得的机会。网站强调政治在个人日常生活细节上的重要性。通过对于主观和个人身份的强调，网站的结构与导航分成目录的形式，比如"家""爱""学习"和"工作"。从一方面来说，按战略构思的网站是一种有效的干预；从另一方面来说，也是出于文化上的目的：通过公共参与创造自身，同时作为项目范围的社会地图。

"伊奥威尔酒店"网站虚拟的空间被转至一个真实的空间内提供实时的展览体验，一年时间内会在图书馆一周开放 5 天，从 2010 年 1 月起至 2010 年 12 月为止。使用者同时塑造、制作了展览和网站的内容。新的伊奥威尔公共图书馆被选为装置的理想实施地点，因为这是一个密集的公共空间，是社会图像归档的最佳地点，故事也和图书馆本身的功能有关。提更·布里斯托设计建造了一系列自我纪录的装置，位于展位内。理念建筑师团队中的奥珀和利弗纳精致地设计了新图书馆的结构。

参观者走入展览空间之后，要在标着"你从哪里来"的墙上写下自己祖国的名字。这堵墙述说着众多不同身份背景的人、不同的个人和出生的地方。在"旅途展位"这一部分使用了改写过的交互式谷歌地图应用程序，他们标记着各自穿越非洲的旅途，经常在故事里加入图片。在"爱·照片展位"内，参观者们可以给自己拍照，可以单独拍或者和爱人一起拍，照片会上传至 Flickr 网站，能自动打印出来，留下空间，带回家作为礼物。在"录像展位"内，他们自己制作短片，能够上传到 Youtube 网站。在"故事展位"，他们编写故事，选择八个类别／主题之一的散文或诗歌。然后在"商务展位"，观众可以上传业务、住宿和其他分类广告和日益增长的目录清单，同时可以参与谈话论坛。

"伊奥威尔酒店"项目的最终成果／目标是写成一本书，目前正在编写中，将于 2012 年末由约翰内斯堡的第四堵墙图书公司出版。

解读

文化是一种基础设施，无论其最终的表现形式是什么，作为生活的一部分，无非关联情感，欲望，满足感和幸福感。美好的艺术展示方式给每天辛苦拼搏的人们继续存在在这个世界上的动力，赐予人们努力生活下去的欲望，市民们通过观察不同时期的公共艺术，掌握城市在不同阶段所要传达的信息，对比自己的生活，获得坚实的生活动力。

在电子信息技术支持下的公共区域是获取生活信息的最佳途径，伊奥威尔酒店项目将群众集合起来，提供一个可以信息交流，信息互换，汲取文化的场所。在这样的一个免费的类似于公益文化活动空间的大力支持下，市民的精神得到了升华，而城市公共设施也有了其继续存在并继续扩展的意义。（高浅）

Artwork Description

Television and print media relentlessly direct the public gaze towards the violence and conflict between South Africans and the Africans—largely forced migrants of war, conflict and economic reasons—who arrived in Johannesburg from other parts of the continent.

Hotel Yeoville (March 2007 – December 2010) was a collaborative and participatory art project curated by Terry Kurgan, an artist based in Johannesburg. It was run in a new public library in Yeoville, an old, neglected suburb on the eastern edge of the Johannesburg inner city, home to 40,000 African migrants and refugees. These inhabitants are isolated and excluded from the formal economy and mainstream South African society.

The project aims to provide an alternative way of seeing the city by revealing commonplace stories that might subvert the dominant discourse "out there" on migration. It includes a community website and an exhibition of 12 private booths in which audience members were invited to document themselves through a range of digital interfaces, interactive media and online applications.

The aim of Hotel Yeoville was to produce a social map of the invisible migrant community of Yeoville and start new conversations that reveal very ordinary everyday life narratives that speak of specific and individual people—love, loss, dreams, desire, loneliness and the idiosyncrasies of place—in such a way that they could not be separated from existing social infrastructure and contemporary everyday practices based on the notion, to quote a term innovated by South African architect Hilton Judin, of "Culture as Infrastructure."

Hotel Yeoville was had four aspects: research, a community website, an interactive exhibition, and a book.

The research was conducted through site visits to understand the social and spatial infrastructure of the neighborhood. Conversation with residents and small business owners was focused on two elements: the entire suburban block of wall space covered in bits and scraps of paper with hand-written community notices advertising accommodation, employment, money transfers, missing persons and other services and products; and the unusual density of Internet cafés, mostly owned and run by foreigners, with specific national identities providing the link between people living in South Africa and the places that they have come from. People use the cafés to be with familiar others, connect with long distance places, and share news and information. Most of the cafés are owned and run by foreigners and have specific national identities providing the link between people living in South Africa and the places that they have come from. Between the advertising and communication practices the working group decided to produce a customised website as part of the project, available at the online Internet Café community of Yeoville, by envolving South African and immigrant researchers mostly resident inthe same suburb who designed a series of research tools, like questionnaires orientated towards Internet café users, owners, managers, staff and people on the street.

The result of this process was the design of the Hotel Yeoville website, using the kind of technology and popular social media that are already being used by an incredibly "wired" and entrepreneurial community, as a means of survival against often quite difficult odds. The website underlines the political importance of the details of personal daily life. Through an emphasis on the subjective and personal identity, and the site's structure and navigation is divided into categories like "home," "love," "study," and "work." On the one hand the website was conceived as a strategic and useful intervention; on the other, as a cultural object that might produce itself through public participation and produce at the same time a social map of "the territory" of the project.

The virtual spaces of the Hotel Yeoville project's website were transformed into a real-space and real-time exhibition experience, which was open 5 days a week in the library for a whole year, from January 2010 to December 2010. The user shaped and produced both the exhibition and the website content. The new Yeoville public library was chosen as the appropriate venue for the installation, because it's a densely trafficked public space and it's the best place to locate the social archive of images and stories related to the function of the library itself. Tegan Bristow designed and built a series of self-documenting located in the booths. Opper and Livneh of Notion Architects worked sensitively on the structure of the new library.

As visitors entered the exhibition space, they wrote their home country on the "So where are you from Wall." This wall told the story of many different identities, individuals and places of origin. In the Journeys Booth section, using an adapted interactive Google maps application, they marked and mapped their journeys across Africa and beyond, often adding an image to the story. In the "Love Photo Booth" they made photographs of themselves, alone or with loved others, which were uploaded to Flickr, physically (automatically) printed out, left behind in the space, and taken home as a gift. In the "Video Booth" they made short movies, which were uploaded to Youtube. In the "Story Booth" they wrote stories, prose or poems choosing one of 8 categories/themes available. Then, in the Business Booth section they uploaded business, accommodation and other classifieds and listings for the growing directory and also participated in conversation forums.

The last product/object of the Hotel Yeoville project was a book, which is currently in production, and will be published by Fourthwall Books, Johannesburg in late 2012.

Artwork Excellence

Hotel Yeoville makes a difference because it has been inserted into a community whose agency and social capital are hugely devalued, in an extremely xenophobic times in South Africa. The participation in the production process is an attempt towards the validation of self by observing the world and being observed, and bringing people to find an alternative way to stimulate a lively and controversial public conversation about migrants and refugees living in South Africa.

It attracted a huge number of visitors/participants over the year that it ran, was very well received, broadly covered by local and International print and online media, television and radio stations and it appears even on a range of South African and African diaspora community organizations. Through the design and architecture of the exhibition space, the performative nature of the online interaction were often amplified in live performances, as answer to the claim step up to perform and show yourself!

On 2011, Terry Kurgan created a partnership with the University of Witwatersrand's The School of Arts and the African Centre for Migration and Society in order to develop a second phase of the project and raise enough funds to install Hotel Yeoville project into a shop-front on the suburb's high street, thus it can continue to make positive impact upon the lives of people living in the suburb, helping them in very practical ways to negotiate their way to live and move through the city, and at the same time to continue to make visible the most ordinary lives of a large, entrepreneurial and invisible immigrant community who can potentially contribute hugely to the social, economic and cultural life of Johannesburg city.

感染的城市公共艺术节
ITC Infecting The City Public Arts Festival

艺术家：非洲中心（南非）
地点：南非开普敦
推荐人：吉乌希·切科拉
Artist: The Africa Centre（South Africa）
Location: Cape Town, South Africa
Researcher: Giusy Checola

作品描述

"感染的城市"（ITC）是一个充满活力、富于创新的年度公共艺术节，在开普敦举办。艺术节由非洲中心组织策划，该中心是一个探索当代泛非文化实践的国际平台，是改变社会的催化剂，涉及的范围涵盖了音乐、视觉艺术、表演艺术、舞蹈、戏剧和诗歌。在"斯皮尔表演艺术节"结束以后，非洲中心决定在 2008 年举办"感染的城市公共艺术节"，该项活动的唯一目标是"用捕捉到我们日常生活复杂性的表演来感染城市"。在成立之时策展人杰·帕特尔和布莱特·贝利说，这一节日的愿景就是要搞活并重塑开普敦的中心商业区。自从那时起，ITC 开始成为一个得到充分认识和可持续发展的项目，拥有日益增多的观众和大家乐意接受的品牌，其名字基于慈善机构及合作模式。

2010 年，"感染的城市"这一名字进一步扩大了，增加了"公共艺术节"几个词，目的是把视觉艺术和公共干预融合进来，以便更确切地反映艺术节呈现的内容。

"在非洲，公共艺术始终是我们用来证明自我的一部分。在这个国家它又是我们的公共仪式、公共抗议和公共庆祝活动的历史写照。这个非洲的"我们"之间的相互联系在公众和社会的层面受到过一次又一次的挑战，这给艺术表达带来了充沛的活力。我们历史中的那一部分也同样阻碍了公众之间的相互联系，它把人们隔离开来，让大家彼此疏远，这种身体和精神上的疏离所造成的创伤仍有待愈合。'受感染的城市'是试图重新激发起这种内在联系的一个小小尝试。"

除了南非公共艺术的国际化推广外，ITC 的目标还要给公共空间带来一种别样的生活，鼓励促成民众团体收回城市的公共空间，通过自由地创作艺术和面向大众来发展艺术欣赏的文化。

基于这些理由，艺术节的一个重要方面是"艺术你好"学校项目，由私人赞助并与表演者马里卡·诺德罗弗合作。该项目每天吸引来 120 位学习者参与活动（共有 600 位学习者参与艺术周的课程），每天有 10 位来自市中心以外的教育工作者前来参加艺术节并讨论他们各自的体验。通过这些体验和对视觉语言、当代视觉和表演艺术、公共艺术和相关话题、社会和文化动力、环境方面、艺术行动和社会政治相关的艺术以及多媒体的讨论，该项目让高中生通过培养文化意识、体验艺术以及对艺术品表达自己的抒情性和批判性意见来提高自己。

如今由于受到财政或地域上的限制，居住在开普敦的 450 万人口中，大部分人很少能够体验舞蹈、戏剧、表演艺术、音乐和大型视觉艺术装置。ITC 作品旨在通过主办广泛的艺术表演形式来改变这一现实。艺术家来自南非以及国外，他们致力于找到办法让每个人都能够去看、去听、在艺术创作中发现自我。各种艺术团体、艺术家、该城市和普通公众在参与、受雇佣、创作新艺术作品、接触艺术、参与对公共空间的重申和娱乐的过程中以不同的方式受益良多。

这些年来，参加艺术节的艺术家和光临 ITC 的观众数量在不断增加。2011 年艺术节参与者超过了 25 000 人，有 314 名来自南非和国际的知名艺术家参与了干预项目，艺术家们来自舞蹈、表演、公共艺术和城市再生等领域。参加 2012 艺术节演出的还将有：阿西-帕特雷·路嘎、文森特·曼特索和尼古拉·汉那康姆，舞者兼编舞达达·马斯洛，艺术家维克托里纳·穆勒和欧勒·哈马，钢琴家贾斯汀·克拉维兹。

ITC 把公共艺术作品作为一种工具来增强社会凝聚力，同时它代表一种通过扩大和培养新观众来支持艺术行业的方式，而这些观众的任务是在未来支持艺术和艺术产业。

解读

每年一度的感染城市公共艺术节都会在南非开普敦如期举行。艺术节的举办主要是通过艺术家的作品为媒介，进行艺术家与民众的间接交流与互动。作品的风格呈现出多样化，主题具有针对性，艺术节中的各项活动的唯一目标是"用捕捉到我们日常生活复杂性的表演来感染城市"，深受民众的欢迎和喜爱。

南非开普敦这座城市中也充斥着各种社会矛盾和问题，生活问题、环境问题、工作问题以及精神压力问题等等，因此，艺术家通过作品来反映这些发生在市民身边的与他们有关的社会问题，从而引起他们对这个问题的反思与思考。同时，各种艺术团体、艺术家、普通公众在艺术节中参与、互动、交流、共享中受益匪浅。（冯正龙）

Artwork Description

Infecting The City (ITC) is a vibrant and innovative annual public arts festival that takes place in the city of Cape Town. It was conceived and organized by the Africa Centre, an international platform for exploring contemporary Pan-African cultural practice as a catalyst for social change across the disciplines of music, visual art, performance art, dance, theatre and poetry.

After the denouement of The Spier Performing Arts Festival, Africa Center decided to start ITC in 2008, with the singular aim to "infect the city with performance that captures the complexities of our daily life." At its inception, curators Jay Pather and Brett Bailey said the festival's vision was to enliven and reinvent the city of Cape Town's central business district. Since then, ITC has become a fully realized and sustainable project, with a growing audience and a readily recognized brand name based on a philanthropic and partnership model.

In 2010, Infecting the City expanded its name with the words Public Arts Festival to incorporate visual art and public interventions and to better reflect what the Festival represented.

Public art has always been part of who we are on this continent and in this country given our history of public ritual, public protest and celebration. The interconnectedness of the African "us" meets challenge after challenge in a public, social way, brought to vibrant life in artistic expression. There is too that part of our history that impeded this public interconnectedness, throwing people apart and far away from each other, a physical and psychic separation still waiting to be healed. Infecting the City is a small attempt at igniting this interconnectedness.

Besides the international promotion of South African public arts, the aims of ITC are to bring another kind of life to public spaces, to facilitate and encourage civil society to reclaim the common urban spaces, and to develop a culture of art appreciation by making art free and accessible to all.

For those reasons, an importance aspect of the festival is the Arts Aweh! school program, supported by private sponsors and coordinated by performer Malika Ndlovu. It facilitates the participation of 120 learners per day (600 learners over the course of the Festival week) and 10 educators per day, from schools outside of the city centre, to participate in the festival and have discussions about what they experienced. Through these experiences and discussions about visual language, contemporary visual and performance art, public art and related issues, social and cultural dynamics, environmental aspects, arts activism and socio-political relevance of the arts and multi-media, the program empowers high school students through fostering cultural awareness, experiencing art, creating their own artworks and giving their poetic or critical impressions of the artworks.

Today, due to financial or geographical constraints, most of the 4.5

million people living in Cape Town rarely experience dance, theatre, performance art, music and large-scale visual art installations. ITC works towards changing this reality by hosting a wide range of artistic forms, with artists from throughout South Africa and abroad, and focuses on finding ways for everyone to see, hear and find themselves in the art produced. Arts organizations, artists, the city and the general public have benefitted in various ways through participation, employment, creation of new works of art, exposure to the arts, and participation in the re-claiming and enjoyment of public spaces.

Over the years the number of artists participating and visitors of the ITC has steadily grown. The 2011 festival was attended by over 25000 people, with the interventions of 314 well-known South African and international artists from the fields of dance, performance, public art and urban regeneration.

Among others performing at the 2012 festival will be Athi-Patra Ruga, Vincent Mantsoe and Nicola Hannekom's Lot, the dancer and choreographer Dada Masilo, the artists Victorine Meuller and Ole Hamre, the pianist Justin Krawitz.

ITC works on public art as tool to build social cohesion and at the same time it represents a means to support the arts industry by developing and nurturing new audiences, whose task will be to support the arts and the arts industry in the future.

Artwork Excellence

Infecting The City was developed in response to the complete void of public art in the city of Cape Town. This festival is the only one of its kind in the entire country. It strives to bring socially-engaged performance and visual art out of theatres and galleries, and to transfer it into the communal spaces in the city of Cape Town, thus transforming the city for one week into an outdoor venue, art is free and accessible to everyone. ITC is also a way for citizens of Cape Town to own and engage with their collective space. Central to ITC is "Arts AWEH!" , a youth development project which aims to empower high school students through fostering cultural awareness. Arts organisations, artists, the city and the general public have benefitted in various ways through participation, employment, creation of new works of art, exposure to the arts, and participation in the re-claiming and enjoyment of public spaces. ITC has grown from an audience of more than 5,000 in 2008 to in excess of 25,000 in 2011.ITC works on public art as tool to build social cohesion and at the same time represents a means to sustain the arts industry by developing and nurturing new audiences, whose task it will be to sustain the arts and the arts industry in the future.

尼日尔建筑
Niger Buildings

艺术家：诺特·维塔尔（瑞士）
地点：尼日尔阿拉伯北部5公里的一片绿洲
推荐人：薇拉·托尔曼
Artist：Not Vital, Switzerland
Location：An Oasis 5 km north of Agadab, Niger
Researcher：Vera Tollmann

作品描述

瑞士艺术家诺特·维塔尔热衷于在世界各地建造奇异的房子，他的艺术表达了全球文化的精神，用具有功能的艺术作品模糊了现实与幻想，以及历史、现在与未来之间始终暗淡的界线。在他的世界里，这些建筑也是具有艺术性的功能物品，不仅有着典型的形式主义和极少主义的风格，还借鉴了大地艺术的手法。2008年，诺特还曾在北京制作装置及巨型雕塑作品，正如他自己所言，他把整个世界视为自己的"工作室"。

维塔尔于2000年在尼日尔的阿加德兹开始建造雕塑房屋。他建了一个复杂的泥屋群，每幢建筑都有一个明确的主题，如"抵御热浪和沙尘暴之屋""望月之屋""观日落之屋"等。其中一个是儿童学校，450名儿童上课时就坐在这个阶梯上而不是在常规的室内课堂，因此这所学校不仅是一座雕塑，而且还具有了社会功能；另一所房子作为献给月亮的颂歌，还有一所13米高"能看见日落的房子"，观众从一个楼层移到另一楼层，日落的高度也随之变化，这种状态诗意而超然。

维塔尔运用被当地有些人视为废料的牛角等材料建造了极具现代感的建筑结构，使作品既与当地的传统建筑物完全脱离开来，又能融入这座城市的审美习惯中。他在阿加德兹沙漠持续工作到 2007 年，其目的是为了唤起 18 世纪那种完全融入大自然的浪漫的崇高概念。

解读

艺术源于生活并为生活服务，而建筑是最早的艺术形式之一。早在原始社会末期，就有了"作为艺术的建筑的萌芽"，在那个万物的产生和创造都是以服务人类自身为基础的时代中，建筑的功能性和历史性展露无疑。

公共艺术建筑应先拥有建筑物的基本功能再谈公共性与艺术性。维塔尔的每件建筑作品都有明确的主题和实用功能，虽只为某个地区的居民服务，但其灵感却来自于全球各地，他的作品淡化了国家意识与民族意识，在他的理解中，世界是个地球村，所有的艺术形式都应该被共享。没有使用丰富的建筑材料，也没有花哨的表现形式，简单的建筑形式却更加的适合这片荒漠的场所。一旦成型，便会迅速成为人群聚集地，至此，才算开启了公共建筑的功能即宣扬和传播思想和理念。设计本就应为社会居民服务，设计师倘若是这么想的，则其作品也必秉持着其构想，承载着他的期望。（高浅）

Artwork Description

Switzerland Artist Not Vital has a passion for building outlandish houses in all parts of the world. These functional artworks arise out of a global culture. They dim the boundary line between reality and fantasy and between past, present and future. In his world, these buildings are artistic realizations that take a cue from formalism and minimalism, as well as from techniques of earth art. In 2008 Not was in Beijing, where he created an installation and several massive sculptures. As he said himself at the time, all the world is truly his studio.

Vital began building a complex mud-house community in Niger's Agadez desert in 2000. For each building he chose a special theme, for example The House Against Heat And Sandstorms, The Full Moon House, The House To Watch The Sunset. One of the buildings is a school that has no indoor classrooms, but instead a flight of outdoor steps that can accommodate 450 children. A distinctive sculptural form was thus wedded to the school's social function. The Full Moon House was conceived as a kind of ode to the moon, while in the 13-meter-high House to Watch the Sunset visitors can move from one level to the next in counterpoint to the sinking sun, an experience that is poetic and transcendental.

Vital incorporated materials that locals regard as refuse, such as the horns of cattle, to construct buildings very distinct from the vernacular architecture, yet with an aesthetically localized feel. His work in the Agadez desert, which continued until 2007, achieves a harmony with nature which at the same time arouses sensations of being in another century.

Artwork Excellence

Not Vital built the architectural sculptures in Niger without any commissioner or sponsor. Working with the tradition of Land Art, he went to a place and translated the geographical conditions into a formalist gesture. His sensitivity towards nature and local materials meets with a western utopian vision architecture.

无题
No title

艺术家：埃里亚斯・斯美（埃塞尔比亚）
地点：埃塞俄比亚，亚的斯亚贝巴
推荐人：维拉・托尔曼
Artist: Elias Simé (Ethiopia)
Location: Addis Ababa , Ethiopia
Researcher: Vera Tollmann

作品描述

祖玛当代艺术中心（ZCAC）位于亚的斯亚贝巴，中心所在的房屋由埃塞俄比亚艺术家埃里亚斯・斯美设计，艺术家有机地使用自己的双手建造，房屋的材料使用了泥土、稻草、石头、木材和马赛克。

最初斯美只是想建造一座供居住的房屋。建造了 8 年的房屋完工之后，公众对这座独特的房屋产生了兴趣纷纷来访，于是斯美意识到他不能在房屋里居住和工作了，它已经变成了一处公共场所。这幢复合式房屋包括至少能容纳六位宾客的三间客房、两间厨房、两间浴室、一间工作室，在建筑主体亭子里有一处大面积的公共空间。

斯美的创作艺术手法独特，清晰地扎根于他国家的传统中。自从 2001 年，他与一位人类学家兼策展人麦斯克兰・阿苏古德一起出去旅游，阿苏古德现在担任亚的斯亚贝巴 ZCAC 中心的主管，他们穿越了很多埃塞俄比亚的偏远村庄，研究如今依然存在的古老宗教仪式。斯美使用泥土和稻草创作雕塑，这种手法扎根于埃塞俄比亚的文化。他想要复兴古代使用泥土建造房屋的知识，这种方法很环保，而且造房效率更高，但如今波纹铁却是当地建房使用的标准材料。

艺术家小时候起就自己学会了缝纫、绣花、修理家具。他收集废弃的物品和材料，比如扁平的锡罐，将它们进行重新创作，使其成为他自己的艺术品。如今他继续充满热情地收集不同颜色的塑料袋、塑料鞋子、牛角等废弃物如，同时他也在位于亚的斯亚贝巴广阔的中央市场的梅尔卡托购买一些创作的原材料比如纽扣。说到纽扣举个例子，他在三间浴室里使用了纽扣装饰天花板，而没有打草图。斯美有机地建造着房屋。他的绘画都是当场发挥。例如，他使用泥土建造第一面墙，他用水泥覆盖墙面，在水泥上贴陶瓷板。这是非常有趣的墙面，但斯美改变了主意，他移除了全部的水泥部分，将它改造成如今的泥土墙。他还建造又拆毁了另一些建筑结构。天花板低了很多，然后艺术家把它推倒又重新扩建了。

正如同他从市场里和街道上收集材料用以创作艺术，斯美也在亚迪斯街道发现了自己作品的主题。他记录了亚的斯亚贝巴周边切克斯·格波雅、莱嘎哈尔和切德·特拉瓦地区埃塞俄比亚人们的生活和肖像，并在那里发现了自己的大多数创作素材。斯美和他记录的那些社区尤其是社区的儿童有着深厚的联系，那些儿童们把在街道上收集到的物品交给他。

ZCAC 是一座对生态敏感、艺术家驻场、具有教育功能的村子。它以祖玛·施弗劳的名字命名，他是一位年轻的埃塞俄比亚艺术家，1979 年死于癌症。2002 年 Giziawi #1（当地语言，意思是"临时"，是一个三天的公共艺术活动）开展时，这个概念被第一次引入公众，该中心首次举办艺术活动。阿苏古德说，ZCAC 的运营就像一个家庭，周边的社区正是中心的一种延展。尼基基金会、美国大使馆和山姆家鞋店在亚的斯亚贝巴的 ZCAC 开展与艺术相关的工作坊。那里也举办非正式的诗歌和标志设计工作坊。

在亚的斯亚贝巴的这座房屋转变成祖玛当代艺术中心以后，斯美和阿苏古德开始了后续的项目。斯美与其他艺术家一起在德雷达瓦市收购了一幢空屋并将其转变成了 ZCAC 的分部，后来又设立了第三个分部：位于哈拉的 ZCAC 哈拉分部，这是德雷达瓦东部的一座历史悠久的小村庄。

解读

根植于埃塞俄比亚传统文化的挖掘，艺术家斯美独具创意地设计与建造了如今的祖玛当代艺术中心。传统材料的运用和独特的艺术创作手法使这座房屋得到公众的关注，使其现在成为了一座对生态敏感、艺术家驻场、具有教育功能的公共空间。斯美的环保意识也值得人们赞赏与学习，艺术中心的大多数艺术作品都是她收集的废弃物品制作而成的，作品独特富有创意，让公众不可思议，在一定程度上，也引起人们对生活环境的关注与思考。

"无意"中的房屋却变成了现在的公共场所、公共空间，不可思议却又在意料之中，意义之丰只有"无题"可以承载，值得回味。（冯正龙）

Artwork Description

The Zoma Contemporary Art Center (ZCAC) in Addis Ababa resides in a house that the Ethiopian artist Elias Simé designed and constructed organically with his own hands. The house is made from mud, straw, stones, wood, and mosaic.

In the beginning, Simé's intention was to build a house to live in. When the house was completed after an eight-year construction period, the public became interested in visiting his unique house. Simé then realized he could not live and work there since it was turning more and more into a public place.

The compound includes three guestrooms for at least six guests, two kitchens, two bathrooms, a workshop, and a big communal space in the main building-pavillion.

Simé has a very unique approach to art, clearly grounded in the traditions of his country. Since 2001, he has been traveling with Meskerem Assegued—an anthropologist and curator, now director of ZCAC Addis Ababa—through many rural villages of Ethiopia to research ancient rituals that are still in practice. He works with mud and straw to create sculptures, something rooted in the Ethiopian culture. He wants to recover the ancient knowledge of using mud for construction that is more environmentally friendly and more efficient than the corrugated iron that is now the norm for local housing.

As a child, Simé taught himself to sew, embroider, and repair furniture. He collected cast-off objects and materials, such as flattened tin cans, and fashioned them into his own creations. Today he still collects energetically—plastic shopping bags in different colors, plastic shoes, horns from slaughtered cattle—and buys some of the raw material for his work, such as buttons, at the Mercato, Addis Ababa's sprawling central market. With the buttons, for example, he decorated the ceilings in the three bathrooms. There was no drawing. Simé built organically. His drawings happened on the spot. For instance, he build the first wall with mud, plastered the outside with cement and laid ceramic plates into the cement. It was an interesting wall but Simé changed his mind and removed the entire cement section and changed it to the current mud wall. He also built and knocked down other structures. The ceiling was a lot shorter then he knocked it down and extended it.

Just as he gathers materials for making art from the markets and streets, Simé finds the themes for his works in the streets of Addis. He documents the lives and portraits of Ethiopians living in the Cherkos Gebeya, Legaehar and Chide Terra areas surrounding Addis Ababa, where he finds most of his working materials. Simé has deep connections with the communities he documents, in particular with the neighborhood children who bring him objects they collect from the street.

ZCAC is an eco-sensitive and educational artist-in-residence village. It was named after Zoma Shifferaw, a young Ethiopian artist who died of cancer in 1979. The concept was first introduced to the public in 2002 during Giziawi #1, its first art happening. Assegued says ZCAC is run like a family, in which the surrounding community is an extension of the center. The Niki Foundation, U.S. Embassy and Sam's Shoes have had art-related workshops at ZCAC Addis Ababa. It has also held informal poetry and logo designing workshops.

After the house in Addis Ababa turned into Zoma Contemporary Art Center, Simé and Assegued started a follow-up project. Along with other artists, Simé adapted an existing house in the city of Dire Dawa into another ZCAC. A third location has been established afterwards: ZCAC Harla is in Harla, a small historic village east of Dire Dawa.

Artwork Excellence

The house designed and built by Elias Simé is like a meta-artwork because it is closely connected with his artistic practice. Indeed, the artistic practice and the architectural design cannot be separated. The house carries many references to his artwork and even includes pieces he made. The house has turned into his biggest artwork so far. As the house was built and decorated with only local materials—natural or from the market—it is a unique portrait of everyday Ethiopian culture. At the same time, the architecture combines different traditional Ethiopian cultural techniques and therefore naturally addresses issues of sustainability. Both the building of the house and the founding of ZCAC and additional ZCAC sites were initiated by local people, not from the outside.

Today the house is an artists' residency and center for contemporary art, the first of its kind in Addis Ababa. It's a place for international cultural exchange, artistic production and debate.

希尔布鲁、波利亚和耶尔维尔区的公共艺术项目
Public Artworks Programme for Hillbrow/
Berea/Yeoville

艺术家：约有 30 名不同的艺术家
放置地点：南非约翰内斯堡
推荐人：薇拉•托尔曼
Artist: about 30 diverse artists
Location: Johannesburg, South Africa
Researcher: Vera Tollmann

作品描述

南非约翰内斯堡在种族隔离后期，以及刚结束了种族隔离制度后的那段时期里百废待兴。为了重建、升级和改造这座城市，约翰内斯堡发展局 (JDA) 发起了一项大型的公共艺术项目：大约有 30 名艺术家受邀在城市的希尔布鲁、波利亚和耶尔维尔区创作公共艺术作品。HYB 公共艺术是那些区域内首批升级的项目，在这个意义上来讲是对这座城市的一次实验。

为了有助于管理这个项目，约翰内斯堡发展局与"三位一体会期"签订了合约，这是一个艺术家斯蒂芬•霍布斯和马库斯•诺伊斯泰特主管的当代艺术的创作团队，他们被请来协调为附近这些街区完成概念化和安装合适的公共艺术品而发出的一系列公开邀请。（"三位一体会期"调查了艺术与商业、协作实践和网络发展之间的关系。霍布斯个人的艺术兴趣在于城市环境和公共艺术之间的干预，诺伊斯泰特的关注点是电子艺术和扩大虚拟社区。"三位一体会期"通过与约翰内斯堡进行交流而得到了很强的定义。）

在地址的选择上，"三位一体会期"从开始起就把注意力放到那些交通要道以及由市政府和约翰内斯堡发展局划定要改建的地方。那些只要有公路、公园和公共广场可以容纳的道路、交通连接和可进行干预的可能地点都被添加到安置名单上。

而且艺术品的选择也要承担一些些风险，"三位一体会期"也很清楚这些空间内的很多地方由外国移民占据着，而且在这种居民成分很复杂的地区进行都市改建目标选定会引起紧张气氛也是难以避免的。大多数被挑选来开展这一项目的艺术家都是当地人。

安德鲁·林赛，1990 年他在希尔布鲁区的温德布露剧院创作了瀑布壁画，如今他在剧院对面也创作了马赛克瀑布壁画，几年前，林赛和他的工作室的工匠一起创作了一系列玻璃和蓝色瓷砖瀑布壁画面板，竖立在瀑布悬崖的崖壁上。如今，被升级的、装饰性更强的马赛克壁画被安置在同一地点。此外，通往希尔布鲁的通道侧翼堤防装饰着一系列小型的马赛克蜘蛛、蝎子和蜻蜓，目的是要让看到的人得一惊喜。

莫弗·莫利肯和布兰登·格雷在公园里开辟儿童工作坊，为公园里的每个玩耍区域搞设计。这些设计然后被转移成一种主纤维，这是一种用回收的汽车轮胎制成的柔软材料。每座公园都使用独特的主纤维为儿童游乐区域做设计。

"书雕塑"由诺克萨那·诺高贝斯创作，其作用如同图书馆的欢迎致词以及对当地社区和周边地区的友好态度。

艺术品归城市所有，但是版权归艺术家所有。这些艺术品主要的维护和修理由约翰内斯堡市和艺术文化和遗产部门投保。至于小型的维护，城市部门比如城市公园维护艺术品的周边空间，约翰内斯堡发展局积极协助维护活动。

解读

这是一个公共艺术介入或干预公共空间的典型案例。

公共艺术与环境艺术、雕塑、装置艺术的差异在于，除强调公共性和与公众的交流和互动以外，在价值观上，尊重不同的文化形态，倾听不同的声音，关注不同的方面，体现出多元化的特点。公共艺术的多元化使其远远超过了城市雕塑等单纯强调审美的单一性，呈现出新的开放的图景。虽然大多时候公共艺术会以城市雕塑、装置艺术等形式出现，在完成其美化环境基本功能的同时，却为它们赋予了新的文化内涵和社会意义，这也是公共艺术发展的一种重要倾向。

"HBY 公共艺术"的公共艺术项目不仅体现了公共艺术介入公共空间的重要作用，而且还隐含了公共艺术干预城市空间美化和改造的成功与否，关键在于公共艺术运行机制的支撑，其实这也是公共艺术发展过程中遇到的难题。（冯正龙）

Artwork Description

Johannesburg, South Africa became run down in the late apartheid and post-apartheid eras. In an effort to rebuild, upgrade and regenerate the city, the Johannesburg Development Agency (JDA) launched on a large public art project: About 30 artists were asked to produce public artworks in the city's Hillbrow, Berea and Yeoville districts (HBY Public Art). HYB Public Art was the first major upgrade in the area and in that sense was an experiment for the city.

To help manage the project, the JDA contracted The Trinity Session, a contemporary art production team directed by artists Stephen Hobbs and Marcus Neustetter, to coordinate a series of public invitations for the conceptualization and installation of suitable public artworks in these neighborhoods. (The Trinity Session investigates the relationships between art and business, collaborative practice and network development. Hobbs' personal artistic interest lies in the urban environment and public art interventions, and Neustetter's in the electronic arts and expanding virtual communities. The Trinity Session is strongly defined by its exchanges with Johannesburg.)

Regarding site selection, The Trinity Session started with a focus on gateway routes and entered into areas needing upgrading as demarcated by the city and the JDA. Routes, links and possible sites for intervention—where the road, parks and public squares could accommodate them—added to placement decisions.

While some risks were taken with selection of artworks, The Trinity Session was very conscious of both the foreign immigrant audience that occupies many of these spaces and the tensions inherent in the urban upgrade objectives in such a set of complex neighborhoods. The majority of artists selected to develop projects were local.

Andrew Lindsay, for example, worked on a mosaic Water Fall at the same site as his waterfall mural of 1990, opposite the Windybrow Theatre in Hillbrow. Years ago, Lindsay and his studio of artisans had produced a series of mirror and blue ceramic tile water fall panels, erected onto the face of the waterfall cliff. Now, the updated, more elaborate mosaic version is at the same site. In addition, the flanking embankments of this gateway into Hillbrow were treated with a series of small-scale mosaic spiders, scorpions and dragonflies, there to catch one by surprise.

Mpho Molikeng and Brenden Gray held workshops with children in the parks to create a design for each park's play area. These designs were then translated into Masterfibre, a soft material made from recycled car tires. Each park has a unique Masterfibre design for the children's playground area.

The Book Sculpture by Nkosana Ngobese was created to serve as a welcoming statement of the library and its friendliness to its community and surroundings.

The artworks are owned by the city, but the copyright resides with the artists. For major maintenance and repairs, the artworks are insured by City of Johannesburg and the Department of Arts Culture and Heritage. For smaller maintenance activities, city departments such as City Parks maintain some of the spaces around the art, and the Johannesburg Development Agency has been active in assisting in the maintenance.

Artwork Excellence

HBY Public Art was the first major upgrade in the area and in that sense was a an experiment for the city. Local artists were heavily involved in order to ensure that they would be sensitive to the complexity of the neighborhoods.

The Hillbrow, Berea and Yeoville districts are complex neighbourhoods, due to diverse cultural settings, immigrant communities, poverty, safety and security challenges, and urban decay. These aspects had an impact not only on the design, planning and implementation of a public art program, but also the questioning of relevance, target audiences, participatory processes and accessibility of the work. HBY Public Art was a very site-specific project.

It could be read as proof of success that The Trinity Session was afterwards appointed as curator-coordinator for all City Public Art Commissions from 2011 to 2013. Public Environment Upgrades Artworks Programmes currently in production include; Orlando East, Kliptown, Braamfontein, Newtown, and Diepsloot.

卢旺达治疗项目
The Rwanda Healing Project

艺术家：叶蕾蕾（美国）
地点：卢旺达西部鲁巴弗区鲁格雷洛分部，靠近吉塞尼市
推荐人：薇拉·托尔曼
Artist：Lily Yeh（USA）
Location：Rugerero Sector of Rubavu District in West Rwanda,
close to the city Gisenyi
Researcher：Vera Tollmann

作品描述

1994 年，发生在卢旺达的种族灭绝事件导致吉塞尼附近在 100 多天内有 100 万人被杀害。2005 年，美籍华裔艺术家叶蕾蕾和由她创立的费城非营利性组织"赤足艺术家"通过两个关注点表达对种族大屠杀的哀思：一处是万人坑，另一处是名为鲁格雷洛的幸存者居住的村庄。这是一个持续进行的项目，既营造了缅怀的场所，又提供教育、发展的机会和对未来生活的看法。

她和三名志愿者一道，在万人坑建造种族灭绝纪念馆，纪念馆由白色和蓝色的雕塑组成，高大的柱子上端涂绘着卢旺达国旗的颜色：天蓝色、黄色和墨绿色，凄凉的大墓地被转换成色彩缤纷的院落；在幸存者村庄，志愿者与幸存者一起工作，开展艺术、健康、社区和经济活动和儿童教育项目，面向年轻女性的支持小组和基础健康教育项目，改善环卫基础设施，推出小额贷款计划，提供教育、发展的机会，创造一个可持续的基础设施。

她还在当地学校和美国学校之间发起笔友联谊活动，教育美国儿童了解 1994 年的种族屠杀和卢旺达的现状，并得到其他机构的支持和参与，如卢旺达红十字会、无国界工程师（费城小组）、托马斯·杰弗逊医科大学、佛罗里达大学（医疗保健艺术中心）、天空热度（一个缅因州的太阳能小组）以及许多志愿者。项目的意义不仅涉及死亡和破坏，也蕴含着生命、未来和建设。

解读

我们必须依靠艺术哀悼人类在历史中犯下的错，警醒着恶魔的再次出现；也深恶痛绝地咒骂着施害人总是如此轻易创建悲伤，让善良无辜的人们如没有生命的砂石一般任人践踏、不堪一击。但政权的斗争和兵器的角逐却一而再、再而三地让我们在受伤害时手足无措。一切过后，唯一能做的就是帮助自己和受害人收拾伤痛、抚平创伤。"卢旺达治疗"项目代替了本该政府行使的职责，志愿者们和幸存者们一起努力重建家园。而当此行为作为公共行为公之于众，他们希望达到的目的却只是为了唤醒恶人们的良知，满怀希望祈祷着噩梦不会重现。（高浅）

Artwork Description

In 1994 genocide in the Rwandan city of Gisenyi led to the deaths of one million people over the course of a little more than 100 days. In 2005, Chinese American artist Lily Yeh founded the Philadelphia nonprofit organization "Barefoot Artists" with the intent of establishing two genocide memorial projects: one, a structure containing a mass grave, and two, a survivors' village called Rugerero. This is an ongoing project to create a memorial site and provide opportunities for the flourishing of education and an outlook toward the future.

She and three volunteers built a genocide memorial at a mass grave site consisting of a white and blue sculpture surrounded by tall columns painted in blue, yellow and green (the colors of the Rwandan flag). The overall effect is of a necropolis that is turned into a colorful courtyard. In the survivors village, volunteers and the survivors are working together to develop art, health, community and economic programs, including: a children's education project; a support group for young women; a basic health education program; a sanitation infrastructure improvement project; and the launch of a micro-lending program to provide education and development opportunities and create a sustainable economic infrastructure.

She also initiated pen pal activities between local schools and schools in the U.S., to teach American children about the 1994 genocide in Rwanda and local childrens' current situation. Other agencies supporting and participating in this initiative include the Red Cross Rwanda, Engineers Without Borders (Philadelphia group), Thomas Jefferson Medical University, University of Florida (Center for the Arts in Healthcare), Skyheat (a Maine-based solar group), and many volunteers. The project is not concerned with death and destruction alone, but contains a broader message about life, and building a future.

Artwork Excellence

The socially engaged Rwanda Healing Project deals in two ways with the grief of the 1994 genocide. It created a place for memory, but also offers education, development, and a perspective for life in the future. Yeh and her team of volunteers worked with genocide survivors and their families. During the project process, participants learned the Barefoot Artists' methodology on community building and economic development through art with the aim of creating a sustainable infrastructure. Yeh initiated pen friendships between the local schools and schools in the U.S. to educate children in the U.S. about the 1994 genocide and today's situation in Rwanda. The Barefoot Artists initiative gathered material support from other institutions and companies. For example, MIT brought renewable solar-powered flashlights and hand-crank dynamo electric LED lanterns to the village. Those are prototypes which ideally would go to other villages in Rwanda, too.

住在一个小城镇的 2010 个理由
Two Thousand and Ten Reasons to Live in a Small Town

艺术家：南非视觉艺术网络 (VANSA)
地点：南非多个小镇
推荐人：吉乌希·切克拉
Artist: The Visual Arts Network of South Africa (VANSA)
Location: various small towns in South Africa
Researcher: Giusy Checola

作品描述

"住在一个小城镇的 2010 个理由"是由 VANSA 筹措，由南非国家彩票发行信托基金资助的公共艺术项目。该项目的宗旨是在新的背景和环境下，以新的方式创造机会介入到当代艺术的实践中并接触新的观众群及思想。该项目使艺术家能在主要中心城市以外的地区发展公共艺术实践。

VANSA 明确远离两个常规立场：一者是（明显）不介入的观察员或评论员；二者总是将创造性启示带入黑暗和贫困地方的传教士。让艺术家们更有兴趣的是与通常概念中的小镇生活那种慢吞吞、落后于时代以及物质缺乏相反的特质，这些可以被看做资源的特质，它们是：历史感、编造神话、充满渴望以及对未来的憧憬。有了这些艺术家就可能成为"参与的过客"。

作品自然而然地强调过程而非结果的重要性，并大力鼓励艺术家的实践要着重于新策略，而不必顾及不朽性和永恒。然而，这并不排除创作永久性的作品。项目涉及到各种当代艺术实践的组合研究。参与的艺术形式包括行为艺术（表演）、特殊地点的装置作品、工作坊和雕塑作品。

其中包括：

"Noli Procrastinaire" 由 Kathryn Smith 和 Verna Jooste 在 Laingsburg 指导的测试。组织各类公共艺术实践的测试。受试者包括 Laingsburg 的斯泰伦博斯（Stellenbosch）大学的视觉艺术专业的学生和 Karoo 社区的居民以及受邀艺术家 Roderick Sauls、storytellers Edgar Whitley、Linda Fortune、Robert Weinek 和 AfrikaBurns。

"Dlala Indima" 由 Buntu Fihla，Kwanele Mboso 和 Mak1One 创作，旨在重塑一个 Phakamisa 小乡镇的公共空间。第一步是再利用和调整被遗弃的老建筑。

"穆西纳制造"由 Thenjiwe Nkosi，Ra Hlasane 和 Raymond Marlowe 在穆西纳创作。对正在进行的项目，MIM 行动委员会开发网络和服务。这些服务包括诸如边境农场，组建 Dulibadzimu 剧院团体，筹备穆西纳艺术节——有超过 300 人出席的艺术节，搭建所有艺术家 / 艺术团体在穆西纳数据库等一系列内容。该数据库正在 madeinmusina.blogspot.com 不断增长。

"黑暗和寂静" 由 Bronwyn Lace 和 Marcus Neustetter 在 Sutherland 创作。项目发展了和南非天文观测台（SAAO）的合作。特别邀请到物理学家 Kevin Govender 和天体物理学家 Carolina Ödman，Karoo Hoogland 自治区和当地社区成员并邀请其他科学家和艺术家，用以拉近看似遥远的不发达的 Sutherland 社区与国际周边天文望远镜观测活动之间的距离。

"多米诺效应" 由 Chris Murphy 和 Tracey Derrick 在 Hermon 创作。这个项目创造了一个和环境融合在一起的游戏活动场地。Hermon 地区周边不同的社区居民都可以加入游戏行列。虽然这个地区的人通常不直接接触，但在这个游戏场地里他们可以一起组队合作或比赛。

"生活在历史中" 由 Neil Coppen 和 Vaughn Sadie 在 Dundee 创作。他们从被称之为 Dundee Die Hards 历史剧演出团的访谈和文件资料中整理内容来组织表演。Dundee Die Hards 是一群中年男人，他们（穿古装）重新演绎了了英国人、布尔人和祖鲁族之间的一些历史性战役。Coppena 和 Sadie 主要着重从较小的不浮夸的个人叙述和流行的历史想像中找到历史事件。

"Thor 和 Zena" 由 Guy du Toit 在里士满创作，与 Harry Siertsema（艺术收藏家和慈善家，比勒陀利亚大学）的当代艺术项目合作。它提供了一个可能性，以探索建立一种新型小镇的公共雕塑，这种小镇涉及复杂的社区划分。

解读

都市就像一块磁石，满腔热血的青年人趋势若鹜。那么，小城镇为何而存在？谁又会因为什么缘由而在小城镇驻足？生活在都市，我们可以冠之以为了理想为了未来，为了自我实现。那在小城市停留却是为了什么？追根溯源的安定感？还是寻找一片远离世俗喧扰的桃花源？人生在世短短数十年，闲暇时间是否有人想过，究竟为了什么而生活？又为了谁而奔波？艺术家们似乎也想要探求这个答案。从社会学的角度看，无人不愿意让悠闲的生活节奏取代疲于奔命，如果小城镇也有良好的生活文化设施，便利的生活条件，那么市民们一定会停下匆匆的脚步，回归到小城市定居，生活于一片悠闲。希望艺术家们的努力真的可以给这个社会一个答案，告诉人们在小城市居住的 2010 个理由，从而让辛苦的人们停下奔向都市的脚步，静静驻留小城镇，品味生活，享受生活。（高浅）

Artwork Description

Two Thousand and Ten Reasons to Live in a Small Town is a public art project facilitated by VANSA and funded by South Africa's National Lottery Distribution Trust Fund. The project aims to create opportunities for the insertion of contemporary art practice and thinking into new contexts and environments, reaching new audiences in new ways. The project enables artists to develop public art practices outside of major urban centers.

VANSA explicitly steers away from two common positions: that of the artist as (apparently) disengaged observer or commentator, and that of the artist-as-missionary bringing creative enlightenment into dark and deprived places. They were more interested in the ways in which small towns—contrary to their common association with slowness, past-ness and lack—might be understood as a resource: places of history, mythmaking, aspiration, and futurity, in which artists might participate as "engaged strangers."

The nature of the works emphasized process over outcome and strongly encouraged artists to experiment around newer strategies away from the monumental and permanent. This however did not exclude permanent work, and the project evolved through research towards a combination and diversity of contemporary practice that included performance, site specific temporary installations, workshops, and sculptural works.

Among them:

Noli Procrastinaire led by Kathryn Smith and Verna Jooste in Laingsburg, which tested various practice approaches to public art between students of Stellenbosch University's Department of Visual Arts in Laingsburg and a Karoo community, with guest artist Roderick Sauls, storytellers Edgar Whitley and Linda Fortune, Robert Weinek and AfrikaBurns

Dlala Indima by Buntu Fihla, Kwanele Mboso, and Mak1One which aimed to regenerate the public space in a small rural township of Phakamisa, starting from the re-use and the adaption of old and abandoned buildings

Made in Musina by Thenjiwe Nkosi, Ra Hlasane and Raymond Marlowe in Musina, an on going project which through the work of the MIM Action Committee that developed a network and several projects like the Border Farm, the formation of the Dulibadzimu Theatre group, the

Musina Arts Festival attended by over 300 people, and a database of all the artists/arts organisations in Musina. The database is currently growing online at madeinmusina.blogspot.com

Dark and Silent by Bronwyn Lace and Marcus Neustetter in Sutherland, was developed—in collaboration with SAAO, in particular physicist Kevin Govender and astrophysicist Carolina Ödman, the Karoo Hoogland municipality, local community members and invited scientists and artists—in order to address the current relationship between the seeming "distance" of the disadvantaged communities in Sutherland and the international neighboring telescopic observatory

The Domino Effect by Chris Murphy and Tracey Derrick in Hermon, which created an environment for games and activities that emerged out of the place itself and would bring together the different communities that live in and around Hermon, where people who normally do not engage directly, are brought together in a joint effort – or competition

Living in History by Neil Coppen and Vaughn Sadie in Dundee, which evolved from interviews and documentation of a group of historical re-enactors known as the Dundee Die Hards—a group of mainly middle-aged men who re-enact (in period costume) a variety of historic battles between British, Boer and Zulu. Coppena nd Sadie focused on how history is founded on smaller, less grandiose personal narratives and on popular historical imagination

Thor and Zena by Guy du Toit in Richmond, which involved partnership with the Modern Art Projects of Harry Siertsema, an art collector and philanthropist, and the University of Pretoria. It presented an opportunity to explore the possibilities of creating a new kind public sculpture in a town involving the participation of a complex and divided community.

Artwork Excellence

VANSA sought to explore public art practice outside of the conventional frameworks and agendas of public or corporate commissions, with an emphasis on process over outcome, by positioning small towns—contrary to their common association with slowness, past-ness and lack—as places of history, mythmaking, aspiration, and futurity. The project brings together core concerns of the the organization in a single, ambitious intervention and creates new opportunities for more than 150 contemporary artists to develop their practice in new contexts for new audiences. The project has generated diverse networks that extend beyond the normal metropolitan confines of contemporary arts and which function as a dynamic and flexible public art infrastructure predicated on community involvement.

获奖案例报告
Top6 Case Research

提乌纳的堡垒文化公园
Tiuna el Fuerte Cultural Park

委内瑞拉建筑师 Alejandro Haiek Coll 做案例报告

提乌纳的堡垒文化公园

2013 年 4 月 12 日 12:30，来自委内瑞拉的建筑师 Alejandro Haiek Coll 就"提乌纳的堡垒文化公园"发言。

Coll 介绍说，该项目选择了一座被废弃的停车场用来开发公园，这是一场由建筑师和艺术家主导的新兴集体文化运动。建筑师、艺术家以及当地的人们为了节约经费和能源，基础设施均以非主流施工技术建造，人们把这里打造成了一个城市文化公园，不仅建立起了文化中心、信息中心、办公室、教室、绿色和体育区域，还计划通过举办工作坊和开展活动来促进艺术和科学的发展。如今，每天都有超过 500 名儿童和青少年前来参加文化和艺术活动。

如 Coll 所说："我们想看一看，没有这种学术性的、治理的规划，这个城市如何按照人们的需求，在人们的参与下能够发展，能够和社区更好地融合。因此我们就有了这样一个整体规划，充分利用社区的地形，我们和社区密切合作。我们在法律框架之内，利用法律的保护，能够使我们现在的这个社区朝更好的方向发展。我们致力于打造一种所谓的微城市主义，也就是说它在一个比较小的规模上进行社会化，并且重新打造或者是影响法律，能够让目前的局面实现再平衡。"在这个项目中，艺术家们没有将改变社区面貌的希望寄托于政府体系，而是依靠民间的力量来推动公共艺术，体现出创造性、连贯性以及改善当前社会和人类生存环境的能力。

尼日尔建筑
Niger Buildings

瑞士艺术家 Not Vital 做案例报告

尼日尔建筑

2013 年 4 月 12 日上午 9:30，瑞士艺术家诺特·维塔尔发言，阐述了"尼日尔建筑"的设计过程和理念。

他介绍了建造该项目的经过："我在尼日尔的阿加德兹买下了一块地。我用我的脚在沙子上画了这个房子的草图，然后我用牛粪做了一个房子的模型。房子造得很快，我们基本上是用赤脚来搅拌沙土，最后我们就有了非常坚固的石材。"

就这样，诺特·维塔尔在尼日尔的阿加德兹逐步建起一个复杂的泥屋群，每幢建筑都有一个明确的主题，如"抵御热浪和沙尘暴之屋""望月之屋""观日落之屋"等。

其中，维塔尔先生还特别提到了一个儿童学校的建造："尼日尔是世界上最贫穷的一个国家，于是我决定在我的房子旁边再造一个学校。我给大家看一下这个学校的形状，它就是一个金字塔形的，这样就可以坐下所有的学生。大家看到，这儿有 150 个学生，但是只占到这个建筑的四分之一。学校是建在一个小山上，风大凉快，景观不错，每年可以吸引很多学生。"

诺特·维塔尔的这种艺术表达了全球文化的精神，用具有功能的艺术作品模糊了现实与幻想，历史、现在和未来之间始终暗淡的界限。

在发言中，维塔尔先生还始终表露出作为一个艺术家的社会使命感："在世界上，有许多地方给我们艺术家创造出很多机会，仅给非洲人民给予经济支援是不长久的，我们必须和他们一起做梦，并且帮助他们实现梦想，创造挑战，这样才能给他们创造出很大的不同。"

纽约空中步道公园
NYC High Line Park & Art

美国《公共艺术评论》杂志主编、发行人 Jack Becker 做案例报告

纽约空中步道公园

2013 年 4 月 12 日上午 10：00，美国《公共艺术评论》杂志发行人杰克·贝克主编代替"纽约空中步道公园"项目的设计者发言。

贝克先生首先介绍了该项目的背景："纽约空中步道之友与纽约市政府合作保护将被拆除的空中轨道，并邀请艺术家参与其中，将其改造成一个升高的空中花园，当中既有功能性的设计，也有其他的一些艺术形式，包括告示牌、视频、艺术表演等。"

接着，贝克先生选择性地介绍了纽约空中步道公园之中一些别具特色的公共艺术形式："这个名为'发展当中的托盘'，是一个告示牌作品，用同一块告示牌的位置，这个告示牌是贴近空中步道的，能够被空中步道下司机看到；而所谓空中步道频道，就是在靠近空中步道的一个墙上，打上幻灯，也就是视频作品；这是艺术表演，这是大家一起做色拉，参与者把蔬菜搅在一起，在一个塑料布上洒上色拉酱，然后分成盘，分给大家；同时还有一些歌剧的表演；这是一个舞蹈表演，艺术家探索步道各个特征，然后进行各种舞美编辑，然后进行表演。"

贝克先生总结道："'纽约空中步道公园'项目吸引公共艺术的多样性，公共艺术家利用纽约这样一个新的公共空间，创造了很多艺术作品，非常受人欢迎。"

四川美术学院虎溪校区
Sichuan Academy of Art, Huxi Campus

四川美术学院副院长郝大鹏教授做案例报告

四川美术学院虎溪校区

2013 年 4 月 12 日上午 10:30，四川美术学院副院长郝大鹏教授在论坛上发言，从学校、大学城与重庆的关系、可持续框架、山地建筑、和谐的天然美景以及理想与诗意的坚守等五方面介绍了四川美术学院虎溪校区。

郝教授首先谈到了"建筑隐藏"理念："拥有这样一块地，我们能做的就是保护性原则。因此在规划中，我们做了一个'十面埋伏'的策略，就是怎样将建筑放到环境中去。我们的建筑尽量利用校区里的谷底、山丘之间的环境。我们追求的理念就是藏在一个山谷地里。"就是在这样构思的指引下，四川美院完整保留了 11 个山头，建筑群散落其间，以丰富的形态和朴实的材质体现合乎原地形地貌的生长状态。

接着，郝教授提到四川美院重视"地方语汇"，使当地元素再生于校园中。在郝教授的介绍中，我们可以感受到整个校区充满静谧、清幽的自然情怀，并将地域文化与校园个性联系在一起。校区保留了部分原有的天井、农舍、水渠、农田，具有地方特色的乡村长廊构成了连接校园的网络，农家生活在安静的校园中悉如从前，这对于当代城市建设无疑具有导向性的意义。

最后，郝教授表明了这个公共艺术项目的设计理念："首先，在建设当中我们要兼顾很多东西，艺术、文化、地域、政策、技术、经济等。第二，无为。就是一开始建筑时没有想过它会怎么样，就是把它保护好、控制好，最后再生好。第三，大学的责任。我们认为大学在学科和学术上应该有社会责任，因此，它应该去探索一些学术上、艺术上的思想和观念，因此在如何做地域再造上，也为社会做出探索。"

厨师、农民、他的妻子和他们的邻居
The Cook, the Farmer, His Wife and Their Neighbor

荷兰艺术家、公共空间设计师 Henriette Johanna Waal 做案例报告

厨师、农民、他的妻子和他们的邻居

2013 年 4 月 12 日 12:00，荷兰艺术家、公共空间设计师 Henriett Johanna Waal 向与会嘉宾介绍了别具一格的公共艺术项目"厨师、农民、他的妻子和他们的邻居"。

Waal 介绍说，在 1980 年代，该地区是无人之地。2004 年，城市把所有权转让给了房屋公司，但该公司在这块敞开的空间上看不到价值。最终导致一群年轻的艺术家、设计师、建筑师和社区居民合作，回归粮食生产。

Waal 说："居民们真的希望能够在自己的社区当中利用这些绿色空间种菜、种植物。后来他们出钱在我们这个项目当中能够弄到一块自己的菜园。我们发现，我们不需要做很多的组织工作，实际上我们只要打开栅栏就行。我们具体的做法就是在这个绿地上面，比如说确定一些小的空间，人们进来之后可以很好地知道菜园的分割，我们和居民进行合作，利用那些空置的小店作为公共厨房。这个项目的特殊之处在于这个社区的居民来自 22 个民族，因此，他们习惯于不同的食物。所以他们现在自己造的菜园充分反映了他们自己民族的食物，最后还可以在集体厨房里面大家进行集体交流。"

这个项目证明，居民不仅渴望参与设计他们的城市，而且这一渴望完全是可以实现的，他们有能力推进项目的进展，使公共空间永久地从企业控制下半自主的空间转换成一个充满活力的公共社区。

一如 Waal 所说："这样我们可以看到社会激活社区的方式。这些居民，他们以前彼此之间缺乏联系，他们上街都觉得不安全，但是现在这个项目完全改变了他们的居住环境。"

21 海滩单元
21Beach cells

德国艺术理论家和策展人 Natalia Schmidt 做案例报告

21 海滩单元

2013 年 4 月 12 日 11:30，德国艺术理论家和策展人 Natalia Schmidt 发言，阐述格雷戈尔·施耐德先生在城市再造方面的想法和理念。

Schmidt 介绍了"21 海滩单元"的大致情况。2007 年，施耐德先生在邦迪海滩装置了 21 个网状笼子，在这些笼子里，除了蓝色气垫、海滩遮阳伞之外，还有令人不安的黑色塑料垃圾袋。参与者在貌似舒适的环境中，却感到被这个网状结构束缚住，由此产生某种被囚禁的心理暗示。此外，装置通过人为构筑空间与个人心理空间的冲突，使参观者游移于自由和被监控、隐私和曝光、内部和外部的错觉之间。

Schmidt 补充说："施耐德先生曾说，2005 年 12 月 11 日的暴动影响到了这个作品，它们和海滩共同反映了澳大利亚的国家形象，反映了在恐怖主义的大背景之中，社会里的一些紧张局势。它体现了施耐德的一贯理念，具有强烈的批判现实的寓意。"

为了更详尽地介绍施耐德先生的作品风格，Schmidt 还讲述了施耐德先生 1988 年的"卧室"、1993 年的"咖啡屋"、1995 年的"客房"、德国馆的房间重构、"黑盒子"项目、"白色酷刑"等空间改造作品。这些作品无不反映出施耐德先生擅长驾驭不同形态的空间表现力，并认识到建筑空间能够影响人的深层意识并产生异化感，有着透彻的政治见解。

据 Schmidt 介绍说，施耐德先生也曾表明："我不是一个政治艺术家，我是一个当代艺术家。我作为艺术家也关注世界，包括这个作品也关注澳大利亚的政治问题。"

回顾
Review

评奖历程

一、酝酿：2010 年 9 月，在讨论国际交流合作事宜时，路易斯·比格斯积极回应了汪大伟的提议，并联系广泛的国际资源，和杰克·贝克尔、约翰·麦科马克一道，共同发起了国际公共艺术奖。

二、启动：2010 年 10 月，由发起人分别推荐全球公共艺术研究员，开始在各国征集 2006 年至 2011 年之间的优秀公共艺术案例，并撰写研究报告。

三、耕耘：2011 年 3 月，由上海大学美术学院构建国内第一个公共艺术网站，由研究员提交 141 个案例参加评选。同时，在听取各方意见后，在全球范围内推举出长谷川祐子、弗尔雅·厄尔德姆奇、卡提亚·坎顿和三位发起人共同组成评奖委员会，由路易斯·比格斯担任主席。

四、收获：2012 年 5 月 4—5 日，国际公共艺术奖评委在乌镇召开评审会。在为期 2 天的评审中，经过激烈讨论，评委制定出获奖提名作品（项目）的必备标准，评选出 26 个最佳案例，最后从中评出 6 个大奖。

五、开拓：2012 年 11 月 9—10 日在上海大学美术学院举行工作会议，汪大伟、路易斯、杰克、凌敏、金江波、潘力出席。主要讨论了成立公共艺术协会、网站建设、2013 年颁奖仪式和论坛筹备。

六、展望：2013 年 4 月 12 日，国际公共艺术奖颁奖仪式暨公共艺术论坛在上海大学召开，并启动新一轮评奖活动。

Award Process

Preparation: In September of 2010, when discussing the international exchange programs, Lewis gave his positive response to Wang Dawei's proposal and contacted a wide range of international resources. With the help of Mr. Jack Becker and Mr. John McCormack, we co-sponsored this groundbreaking award.

Start: In October 2010, the promoters recommended global public art researchers, who collected outstanding cases between the year of 2006—2011 in countries and wrote research reports.

Cultivation: In March 2011, the first public art website, by the College of Fine Arts, Shanghai University, began to be constructed.The researchers submitted 141 cases to participate in the selection. At the same time Yuko Hasegawa, Fulya Erdemci, Katia Canton togetherwith the three promoters composed the jury by Lewis Biggs presidency.

Achievement: 4-5 May, 2012, the International Public Art Award jury held assessing meetings in Wuzhen. After a heated discussion in two days, the jury worked out essential criteria of the award-winning nominated works (projects), and selected 26 tops, from which 6 finalists came out last.

Development:9-10 November 2012 meetings were held at the College of Fine Arts, Shanghai University, which Wang Dawei, Louis Biggs, Jack Becker, Ling Min, Jin Jiangbo, Pan Li attend in. They mainly discussed the establishment of a public art association, website construction, and the preparations of 2013 International Award for Public Art Presentation Ceremony and Public Art Forum.

Prospect: 12 April 2013, the International Public Art Awards Presentation Ceremony and Public Art Forum is held at Shanghai University, and a new round of awards event is starting.

评奖花絮

经过一年的征集，截至 2011 年 10 月 31 日，共收到由全球专家组成的研究团队推荐的 141 个优秀公共艺术案例，涵盖装置、建筑、雕塑、壁画、行为、活动等多种类型，分永久性和暂时性两大类。值得一提的是，其中许多作品在当地具有一定影响力，曾获得过奖项；也有相当一部分作品来自较为贫穷、不发达的国家和地区，公共艺术的开展在很大程度上影响当地人们的观念，进而改变人们的行为，并为他们带来新的希望。

2012 年 5 月 10 日至 11 日，《公共艺术》杂志组织召开了评审会，评选入围案例。邀请前英国利物浦双年展行政长官路易斯·比格斯担任评审委员会主席。评审团成员由具有多年从事、策划公共艺术项目经验的专家学者组成，包括美国《公共艺术评论》（Public Art Review）杂志主编杰克·贝克，荷兰阿姆斯特丹艺术和公共空间基金会主管弗尔雅·厄尔德姆奇，日本东京当代艺术馆（MOT）总馆长长谷川祐子，巴西圣保罗大学当代艺术教授卡提亚·坎顿，中国上海大学美术学院院长、《公共艺术》主编汪大伟教授。在为期 2 天的评审中，经过激烈讨论，评委制定出入围作品（项目）的必备标准，同时评选出 26 个最佳案例，其中 6 个案例入围最后大奖的评选。

2013 年 4 月 12 日至 15 日，国际公共艺术奖颁奖仪式暨公共艺术论坛将在上海大学举行。届时将揭晓 6 个大奖的获奖者名单，获奖艺术家或机构将获得由《公共艺术》和《公共艺术评论》共同颁发的荣誉奖项，26 个最佳案例将会编辑印刷成册。公共艺术论坛除了对获奖案例进行研究和评论外，还将对"上海地铁公共艺术"、"上海世博园区改造"和"浙江玉环县美丽乡村建设"等三个实际项目进行论证，提出建设性方案。通过结合实际项目来实现"地方营造"的理念，使当代艺术与地域特色相契合并产生公共效应，成为社会发展的推动力量。

国际公共艺术奖评选活动今后将每两年举办一次，任何国家都可以申请成为主办方。"国际公共艺术奖"的评选活动不仅能够拓宽公众视野，建立一个强大、广泛的全球公共艺术网络，也能促进公众更好地理解公共艺术概念，并积极参与公共艺术的建设。

评选国际公共艺术奖标志
（上海大学美术学院院长、《公共艺术》主编汪大伟教授在做评述）

国际公共艺术奖评委会评选现场

评委们观看参评国际公共艺术奖的作品

评委们在查阅案例和评分

评委们在查阅案例和评分

上海大学校长罗宏杰等专家学者出席相关活动并和评委们合影

评审们在世博园区考察参观

评委们相聚在古朴恬静的江南古镇——乌镇

评审们和工作人员的合照

主题演讲
沙迦双年展与公共艺术

Keynote Speech

长谷川祐子发表主题演讲"沙迦双年展与公共艺术"

沙迦双年展与公共艺术

2013 年 4 月 12 日下午 14:00，日本东京都现代美术馆总策展人长谷川祐子教授发表题为 "沙迦双年展与公共艺术" 的演说。

在讲演中，长谷川教授介绍了她对于公共艺术的展望和看法： "从传统来说，公共艺术会包括环境设计、建筑，往往是在现有景观的基础上补充一些文化价值。出于一些象征性意义或者是出于增加本地的旅游价值的考虑，人们希望使用一些公共艺术。但我认为，今后，公共艺术会更多的增强文化和公众之间的关系。它不应该依靠单一的艺术形式，应该增强地域性，而且能够从本地居民那儿汲取精华，能够增强本地的文化和历史背景。很重要的是，公共艺术能够帮助本地发现更多的可能性，能够释放本地居民更多的潜力。我想这是对于公共艺术的展望。"

"当然在这个过程当中，会给本地社区带来很多经济利益。比如说，丰富他们的文化，增加本地的旅游价值，我想除此之外会有更多的价值，比如说让人们发挥更多的想象力。策展人也好，这些项目的协调者也好，他们发挥着非常重要的作用。我觉得我们这次公共艺术论坛，应该能够触发我们去思考，艺术怎么样能够充分利用人际资源，能够使我们的公共艺术成为人和社区之间以及文化之间的桥梁。"

接着，长谷川祐子教授介绍了她参与设计的公共艺术案例，并借此指出公共艺术面临的一些问题。

其中，长谷川教授着重介绍了沙迦双年展项目，她说："我想摆脱欧洲的视角，希望让本地人能够描绘自己的文化蓝图。因此我们可以看到，它就会为本地赋予更多的文化氛围。" 在这样的思维的指引下，沙迦双年展上出现了诸如 Emesto Neto, Sanaa, SuperFlex 等既融合了伊斯兰风格，又契合公共艺术理念的优秀作品。

长谷川教授总结说： "我觉得我们要关注当地人，也要关注一些本地人，他们会有很多的潜力，他们可以用自己的想法，用自己的想象力去创造公共空间的各种可能性，这是我自己的一种体会，也是我在沙迦这个双年展工作当中的一些体会。"

主题论坛
公共艺术与社会发展

Keynote Forum

清华大学美术学院《装饰》杂志主编方晓风教授发言

艺术的姿态：实现公共性的路径探索

清华大学美术学院《装饰》杂志主编方晓风教授发表题为"艺术的姿态：实现公共性的路径探索"的发言，他指出："我关注的核心点是公共性。我觉得因为公共性，尤其在中国还是讨论得不够充分的一个概念。"

随后，方教授讲述了萧伯纳的故事并展示了一组"十大最丑建筑"的照片来进一步阐明公共性缺失这个命题。他说："公共性意味着不同人群对公共事务的平等介入，由这一认识出发，艺术的目的、手段以及艺术所应作出的姿态都必须有相应的调整。对于普通公众而言，姿态可能是最易感知的内容，姿态本身也应被视为艺术家创作的一部分并得到充分的讨论。而艺术究竟应该以何种姿态介入生活，是个正在被逐步认识和探索的问题。"

"现代艺术应更进一步去精英化。艺术家，如果能够去掉这种权利影子，艺术才能够得到更彻底的尊重。另外一个问题，就是参与，一方面是有利益引导的参与，另一方面我们是想激发对公共事务或者公共话题讨论的参与。"

"所以，我个人对公共艺术的理解，我觉得艺术在这个地方只是手段，而不是目的，目的反而是公共性，所以公共艺术我的理解是，以艺术为手段的、对公共性的一种追求。"

北京大学艺术学院翁剑青教授发言

日常生活与地方关怀

北京大学艺术学院翁剑青教授发表了题为"日常生活与地方关怀"的讲话，从理论角度阐述了他对公共艺术的看法。

在他看来，20世纪中期以来，在中国，当代艺术与公众社会和个人的日常生活及实践发生了更为密切的关系。尽管其中可能包含一些具有普世价值的含义，但是它不应该或者说也不可能建立一种完全适合所有地域和所有地方社会的一种艺术表现形式。

"艺术与生活之间最密切的关系，就恰在于实践性，包括公共艺术在内的当代艺术实践，需要对于现实问题和需求做出积极的介入。"

"对于21世纪的中国公共艺术而言，其社会和文化的使命，主要在于积极参与城镇一体化进程中的公共空间、公共生活场所的建构与改造。创造性地介入和干预城镇社区的日常生活形态和观念形态的传承与变革。而不仅仅是政治文化和商业资本的一种传播工具，或仅仅是纯粹的视觉美学的一种张扬。"

"所以我说，公共艺术的创作和表现方式，显然应该立足于其地域的文脉、地域的特性和地域性的价值的揭示与发挥。从而，使之成为振兴和再造地方社会的文化方式和富有生机的艺术途径。"

中国美术学院杭间教授发言

美学城市中的公共性

中国美术学院杭间教授发表题为"美学城市中的公共性"发言，他说："荷兰哲学家海因茨说，美学来源于城市生活，后者是前者的土壤。然而，这其中的内容，有的是会变成一个形式的躯壳。有的虽然有思想，但是有可能这个思想可能是一个假象。另外不可否认，在所谓的一些城市的公共艺术的过程当中，有很多是由商业策略推动的。所以我个人认为，尤其要警惕创意产业这四个字，在创意产业的口号下，公共艺术的娱乐化生产会变得令人值得怀疑，因为这里面我很想知道，它的内在的资本，它的文化的价值，以及这个文化的消费之间，究竟是一个什么样的关系。

"其次，随着新一代的成长，公共性的本质蜕变为娱乐知识的服务，公共生活的民主精神被不断地肢解，这样的状况在东亚已经变得比较明显。那么在这样的一个过程当中，我想公共艺术如何在保持，刚才有的先生发言当中提到的，公共性的本质，也是值得思考的。

"并且，我认为这种在美学城市的发展过程当中，公共艺术所出现的城市化、躯壳性以及商业性影响这样一些因素，如果在专制制度下出现就更有危险性。专制制度一个非常重要的特点就是不是经过充分吸收大多数的意见而产生的决定，而往往是通过权力交换、利益交换而来实现某一个所谓的艺术样式或者是公共行为的建成。在这样的一种情形下，可能目前会遭受到非常大的困难。

"有一种解决办法是回到公共艺术家的状态。所谓的公共艺术家，我指的是那些城市的游荡者，对于他们来说，城市如果变成这些艺术家的最好的隐居地，能够让他们自由的游荡或者闲逛。这样的艺术可能产生某一种超越一般价值之上的结果。当然，同时这种艺术原本状态的产生必须跟民众、社区，跟民众的意愿、跟社区需求产生一种非常密切的结合。"

上海大学社会学院教授顾骏发言

此岸即彼岸：中国公共艺术的"现世"性质

上海大学社会学系顾骏教授做了题为"此岸即彼岸：中国公共艺术的'现世'性质"的发言。他指出："公共性不是艺术家所认为的公共性，而是民众实际生活中体现的公共性。所以在这一点上来说，艺术家是我们许多城市公共行动的发起者和动员者，我非常欣赏这句话，但是我又反过来觉得，作为发起者和动员者，公共艺术家能否成功可能同他的技术水平、艺术表现手法和他个人的能力有关。但是根本上可能还要取决于公共艺术家的理念与公民实际需求的吻合程度，以及两者之间的良性互动。"

顾教授表明，在中国，公共艺术理念可能更需要公共艺术评论家去解读、发现，所以他愿意来解读四川美术学院虎溪校区公共艺术中所蕴含的理念。

他说："我们在四川美院的设计里看到了对农业的尊重，其实这也是对环境的尊重，对我们曾经生活状态的尊重。并且，四川美院保留的农村、农业和农民的原生态，恰恰要同今天农民的公共需求联系起来考虑。如今，中国农村已不再是人人渴望进城的阶段了，相反，现在城里人越来越希望到农村来玩玩，特别是最近一段时间，中国城市里面遭遇大规模的雾霾，所以去农村洗洗肺也成为一种时尚，更重要的是，越来越多的农民经过城市生活和农村生活的比较之后，他们开始发现农村生活是更人性的生活。因而，农民对于农村和城市两种生活的不同方式的选择是今天四川美院保留农村、保留农民、保留农业的原生态的更深刻的理念。所以我想有许多时候，艺术家的理念很重要，没有理念，在某种程度上就没有公共艺术，但是仅仅只有艺术家个人的理念，却不能得到公众的认可，那么我可以承认他是艺术家，很难承认他是公共艺术家。"

讨论会

上海大学美术学院院长汪大伟教授主持主题论坛

新西兰 Starkwhite 艺术机构总监 John McCormack 担任主持

"公共艺术与社会发展"主题论坛讨论会纪实

2013年4月12日下午14:30,"公共艺术与社会发展"主题论坛召开。在四位专家学者发言之后,主持人汪大伟教授对以上发言做了简单的总结:"方晓风教授提到关注公共性,特别是对艺术在公共性表现过程当中,只是一种手段,而不是全部。翁剑青教授特别提到了如何关注社区,关注区域的价值,关注区域的文化,同时对地区关怀提出自己的见解。杭间教授提到了警惕公共艺术的娱乐化、商业化,特别是在我们城市化进程当中如何再一次来思考我们自己的公共性。最后顾骏老师提到公共性的分享,如果说刚才方晓风说的是关注公共艺术的话,顾骏老师提到的就是去分享公共性,这些观点都非常的有新意。"

在汪大伟院长总结之后,论坛进入讨论阶段,讨论会由新西兰 Starkwhite 艺术机构总监 John 主持,上海大学美术学院潘耀昌教授、广州美术学院李公明教授作为评论员,对之前四位学者的发言进行点评。

潘耀昌教授认为,李公明先生早就预言了艺术会向社会方向发展。公共艺术具有现代性和社会性两个特性。公共艺术需要权利的平衡,但公众权利总是相对较弱。"我们更关注运作过程中的权利的平衡,方晓风的发言重视公共性,翁剑青教授关注公共艺术与社会生活的关系,杭间教授过度开发利用公共空间的观点,表明了表面繁荣背后的副作用。顾骏教授则关注城乡问题。四位的发言是从社会的角度看中国的公共艺术,透过繁华的表象,看到了本质的问题,代表了中国知识分子的思考。一方面"有为",一方面"反省"。

李公明教授也和与会众人分享了自己的看法:"以上发言,提出了一些新的命题,焦虑和企盼。方晓峰教授提到公共性,强调了平民的参与。但是去纪念性,在转折的关头,会不会有某种误解?当然,我也同意去教化功能。而翁剑青教授并不排斥对普世观念的价值关怀,但这一点表现得并不突出。另外,我非常同意杭间教授的观点,文化创意园地之内的权力纵横,利益勾结,就是当下最需要解决的问题。顾骏教授提到农民工问题太好了,我相信他提出一个最重要的问题是,需求,谁的需求?这些需求是不是真正的得到了尊重?需求背后就是权力。所以他提出来的,农民的真实的诉求怎么样能够得到公共的这种尊重,那么在所有这些问题里面,其实真的是包含中国当下公共艺术的一个趋向,而这些问题我认为是一个民主宪政的社会制度框架底下,才能回答诸位的问题。"

主题论坛
公共艺术与城市发展

Keynote Forum

上海天祥华侨城投资有限公司公共关系总监黄英彦先生发言

公共艺术深度展示"华侨城模式"魅力

上海天祥华侨城投资有限公司公共关系总监黄英彦先生做了题为"公共艺术深度展示"华侨城模式"魅力"的发言。

他说，华侨城作为文化央企，经过14年持续经营，文化艺术产业布局低调从容，对当代艺术的支持和投入有着自己的特色，彰显了致力于中国城市公共艺术发展的责任和使命。华侨城在文化艺术领域体现出专业化、公益性和公共性的重要特征。

"我们坚持赋予艺术以专业独立性，通过长期对艺术机构和大型艺术活动提供无偿资助和营运支持的方式，为中国当代艺术在本土的良性发展机制和华侨城自身的文化品质与公共艺术环境建设，树立了受到海内外称誉的'华侨城模式'。公共性就是说，我们认为艺术应该跟公众发生关系。让公众对艺术能够触手可及，成为我们所在城市的高品质的文化生活的窗口，这是我们对于华侨城模式的一个特色的理解。"

"此外，华侨城旗下各艺术馆收藏吴作人、张仃、张晓刚、方力均等艺术家作品五百余件，藏品价值上亿元。部分藏品作为美术馆的终极收藏，在华侨城的公共区域常年展出，为普通市民提供了近距离接触欣赏大师作品的机会，也促进了当代艺术在公共领域的发展。"

上海城市规划设计院副院长孙珊教授发言

关于提升上海公共空间艺术品质的思考

上海城市规划设计院副院长孙珊教授发表了题为"关于提升上海公共空间艺术品质的思考"的讲话。孙教授说:"作为上海的城市规划来讲,我们是对土地和空间这一项公共资源进行一个市级层面的配置,并且在规划实施当中进行管控,可能相对是比较宏观和理性的。"

孙教授在讲话中指出,上海公共空间主要存在四个问题,第一是公共空间的数量和面积不足,服务半径过大。第二是尺度不够宜人,有广场并不代表有人愿意在里面停留。设施的配置和活动以及人性化的设施,都是考虑不足的。第三是上海有很多历史建筑,但历史建筑的元素挖掘不够,风貌展现不够,很多历史建筑还在隐藏,而且被保护和利用的也不是非常好。第四是上海空间中的艺术设计品质还有待提升。

"我们也学习了一些国际上好的做法,譬如巴塞罗那、纽约、里昂等城市的做法。目前,我们在做的第一件事情,就是充分利用空间。第二件事情,我们觉得可以做的就是对历史古建和古树名目进行一个全方位梳理,纳入到前面讲的一些公共活动和串联的体系里面,并且把它纳入到整个的管理机制里面。第三是以街区为单位,全面推进整个城市的更新。"

"我们认为就是在"十八大"提出的新型城镇化的概念里面,其实是在不同的城市是有不同的路径的,我们上海的新型城镇化更加应该注重内涵式的发展,进一步聚焦发展质量,全面提升城市品质,在空间上大有作为,最后通过空间和艺术的发展,建设魅力上海,提升整个城市的硬实力和软实力。"

汕头大学长江艺术与设计学院院长王受之教授发言

公共艺术在城市发展中的角色

汕头大学长江艺术与设计学院院长王受之教授做了题为"公共艺术在城市发展中的角色"的发言。

王教授认为，中国的城市是不给公共空间的。自改革开放年以来，中国的城市发展很快，民众开始有了公民意识，有了公民意识，公共艺术开始走向一条路，就会说话了。中国的公共艺术和城市的关系，即当中国城市里面的居民逐步产生了比较强烈的公众意识的情况下，中国的公共艺术会走向一条应该走的道路。

他认为："目前，中国的城市化越来越快。这一届的大会，提出一个口号叫'城镇化'，我觉得这是中国一个伟大的进步，不说'大都会化'。有了这个发展，中国的公共艺术其实就可以扮演一个更加活跃的角色，在很多地方创造很多人能够生存的空间。另外，公共艺术在中国城市里面扮演的角色，不完全是一个艺术欣赏的对象，应该是给城市居民带来一个欣赏艺术的空间。"

同时，王教授指出，我们所讲的公共艺术，有几种，其中一种是公共拥有的艺术。"现在可以说所有艺术品都是公共拥有，因为国家就是我们的代表，我们假设这么说。"但是另外一种是公共参与的艺术品。第三个就是让政府对于艺术品有一定的重视，对于公共艺术能够有一定的投入，重视这个是一个城市必须的东西。最后一个就是提高市民的审美能力，让市民终于有一天能够把公共艺术变成公众参与的艺术、公众创作的艺术。

中国美术学院公共艺术学院院长杨奇瑞教授发言

关于公共艺术创作和教学的思考

中国美术学院公共艺术学院院长杨奇瑞教授做了题为"关于公共艺术创作和教学的思考"的发言。他指出："关于公共艺术的文化探讨与学术研究，在中国已经历时多年，观点多元而丰富。而城市化进程，在中国呈现出'剧变'的特征。"

杨教授依据他多年的创作和教学经验，向与会嘉宾重点介绍了"杭城九墙"和成都"宽窄巷子"两个公共艺术项目。杨教授指出："城市化进程都具有城市更新、历史街区改造这样一个特点。所以我的作品都是关于历史街区改造。在这个作品的思考当中，我觉得作为我们这代人做艺术，总是想在作品当中有一点内涵和思想性。"

在这些项目的设计创造过程中，杨教授充分尊重艺术与城市空间、艺术与大众、艺术与本土文化的关系以及艺术家的独立精神，并且始终反映"源于生活，取之于生活，还给生活，还给百姓"的公共艺术理念。

讨论会

清华大学美术学院院长鲁晓波教授主持主题论坛

长谷川祐子教授做点评

深圳雕塑院院长孙振华教授做点评

新西兰奥克兰国立美术馆馆长 Rhana Devenport 做点评

"公共艺术与城市发展" 主题论坛讨论会纪实

2013 年 4 月 12 日下午 16:00，在清华大学美术学院院长鲁晓波教授的主持下，会议进入 "公共艺术与城市发展" 主题论坛。在四位专家学者就相关话题进行发言后，深圳公共艺术中心主任孙振华教授、新西兰奥克兰国立美术馆馆长 Rhana Devenport 对以上发言进行了点评。

孙振华教授认为，四位学者的发言恰好构成了当今中国公共艺术的四大推动力量。"黄英彦先生认为公共艺术要为一个城市公共空间提供一个好的艺术品质，然后让公众能够更多接触到艺术。孙珊女士指出，公共艺术就是一个提升城市公共空间文化内涵和艺术品质的艺术。王受之教授的发言跟他们有所不同，他比较强调从政府和公众关系这个角度来定义公共艺术。所以他谈到公共艺术肯定是一种具有公民意识，能够体现公众意志和公众可以参与的艺术。杨奇瑞更多的是一个实践者，有一条我觉得很有意思，就是中国城市化的道路不一样，所以中国的公共艺术的特点肯定也不一样。所以我觉得不同的社会角色在谈公共艺术的时候有他们自己相对不同的理解和侧重点，这个挺好。"

在谈及公共艺术的本体问题时，孙教授认为，"实际上，公共艺术的本体问题有三种可能。一是一个公共艺术的发起者或者一个组织者或者一个创作者他给公众创造了一个公共艺术作品，还有一种方式是和公众一起完成了一件公共艺术作品。最后还有一种方式就是让公众自己完成了一个公共艺术的作品，或者说实践的一个过程，我觉得这个实际上是不一样的。"

"我们说到参与的问题，其实参与也有不同的参与方式。包括欣赏，能够得到或者欣赏到公共艺术这是一种参与，能够创造公共艺术，这也是一种参与；最后还有一种，你是否能够参与到公共艺术的决策，赋予公众民主权利。"

最后，孙教授认为，很多时候人们不是先想好了什么是公共艺术以后才去做的，而是做的过程中将是一个长期的定义公共艺术的过程，这个过程以后再来回顾肯定是充满趣味。

新西兰奥克兰美术馆馆长 Rhana Devenport 对上述发言一一做出了点评，她说："我觉得这一节的发言者都是非常有意思的，黄英彦谈到了公共的特性，我觉得公共性这个词是非常有意思，就是说，到底公共艺术它的本性是什么，在黄英彦的发言当中，这个所谓的性质就是一个特点。孙珊教授谈到了上海的一个雄心，也讲到了上海是可能成为一个具有巨大影响力的城市。而且我很高兴听你谈到了两点，一个是历史，就像我们最开始就涉及到的，就是历史、人文和空间的关系。我想再补充一个就是生态，这是非常重要的，因为这是代表着未来的一个趋势。"

"王受之谈到了城市主义，他直截了当地深入到这个议题。他又谈到了公民的意识，我觉得这个所谓的平民和公民，在世界不同的东西也有不同的含义。此外他也讲到了公共舆论，以及公共艺术在公共舆论方面所负起的责任。在杨奇瑞教授的发言当中，谈到了中国翻天覆地的变化，而且触及到每个人的生活，而且也谈到了自己的作品，我觉得这就像是一个未来的历史学家一样。"

在两个主题论坛结束后，路易斯·比格斯先生对主题论坛做出了总结："我们在讨论当中，一直在集中的探讨公共艺术对空间的重塑，我们今年这个奖的主题也正是如此，地方重塑。"

"我们今天上午看到了很多的范例，显示出我们的艺术是如何对地方重塑做出贡献。而且我也想从形式和内容上来看艺术的地方重塑。我觉得这也是我们公共艺术的两个维度，它都会影响到品质。形式也就是说一个艺术作品的形式，是我们给人们一种视觉上的刺激。内容则是公共艺术作品所讲述的故事、它的理念。"

"然而，公共艺术的流程在不断的变化，我们时刻都处于转型期，没有一个转轨的终期期，我们看到艺术被赋予新的含义，因为我们看到艺术形式也会不断的改变。这就是我讲的内容与形式。因此，我认为重要的是，首先，需要将国际主义和地方主义很好地结合到一起。此外，我们要塑造的地方，我们可以看到建筑以及剩余的这些地方，怎么通过艺术作品进行更好的联动。我认为我们要塑造的地方会受到使用这个地方人的行为的影响。这就是说，我们不但是主人，我们还是访客，我们进入一个空间的时候就可以看到当地人他们是否把这个空间当做自己的空间，也就是说，他们是不是会尊重他们所在的这个地域。这也是我们看到艺术如何来塑造地方，来塑造空间。"

"最后我还是想讲一下协作，一个成功的地方就能很好地体现人与人之间的协作，人的本质实际上在于他的协作性。而我们之所以会塑造一个空间，往往决定性的行为在于协作，所以可以看到协作在塑造地方方面是非常的重要。我就讲以上这些观点，再次感谢各位的参与。"

专题论坛
后世博的公共艺术

Topic Forum

上海世博局战略发展部总监毛竹晨女士发言

后世博的空间记忆和文脉传承

上海世博局战略发展部总监毛竹晨女士做了题为"后世博的空间记忆和文脉传承"的发言，她和大家一起回顾了世博文脉与记忆，并介绍了世博园区的后续开发状况。

毛竹晨女士说："上海世博会的主题'城市，让生活更美好'，因为从海边小镇到国际化大都市，上海的发展让大家自然而然地说'城市，让生活更美好'。上海世博与北京奥运一样，是在中国国力不断强大的过程中，向世界展示自身形象的一次盛会，已经成为一个时代的集体记忆。"

毛竹晨女士介绍了日本馆、城市星球馆、英国馆、未来馆等展馆，并说："在世博会当中，我们来探讨共同的城市主题，我们更多的是通过展馆，当时我们有这样一个架构，我们期望大家在这个架构上能够从不同视角探讨对城市的理解，对城市未来的描绘。"

随后，毛竹晨女士介绍到，世博园区的后续发展已经启动，在不久的将来，浦江两岸的园区土地上就会陆续矗立起总部经济楼群、高星级酒店、商业配套设施等。

毛竹晨女士说："简而言之，我在这里只是帮助大家回顾一些历史，帮助大家梳理目前世博园区的空间，可能只是提出一些问题。后世博的公共艺术就我个人而言，更希望看到我所经历的，或者说上海人和世界所有游客共同经历的记忆，这个城市的梦想，它的文脉，以某种形式浓缩被提炼，在这个地方继续传承下去，可能若干年之后，大家回到这个地方，通过某一件作品，某一个公共艺术的形式，依然能对这个时代，这个事件感兴趣，我想我要看到的可能就是这个。"

上海视觉艺术学院美术学院副院长丁乙教授发言

汇源聚韵："世博源"公共艺术策划方案

上海视觉艺术学院美术学院副院长丁乙教授做了题为"汇源聚韵：'世博源'公共艺术策划方案"的发言。丁乙教授认为，上海"世博源"公共艺术规划方案的主题概念是汇、源、聚、韵。整体公共艺术设置方面有两个比较重要的思考，一个是固态、动态和互动的，一个是永久性和临时性和虚拟性的。同时，在整个作品布局时，从景观和整体形态都做了相应的考虑。包括从艺术品互相之间的编排、连接方面做了很多思考。对作品能够反映某种地域特征，或者中国的传统，包括视觉的角度、心理层面等都做了一定研究，对一些公共艺术的点位和人流之间的关系做出分析。此外，在整个实施方案中，围绕"六点一线"进行公共艺术和世博轴之间关系的定位。通过6个阳光谷的改造，在整个最上层的平面上进行公共艺术的概念设计。6个阳光谷的改造归纳了6个词：寻找城市源头、记录城市印迹、延续城市文脉、关注当代生活、展现文化新潮、浓缩生活体验。定位为"一个是水的谷，一个是自然谷，一个是乐谷，一个是亭谷，一个是影像谷"。另外，在二楼全长1 045米南北贯通世博园平台上面设计一条以观光车为线路的丰富的线，这个线又对应了下面的商业主题：乐活、时尚、潮流、品位。可以说，这个项目既是通过文化传播的方式增添公众购物新体验，也是对公共艺术的概念、形式、功能和意义的综合实验案例的探索。

中央美术学院城市设计学院副院长王中教授发言

艺术激活空间：从 5 米到 5 000 米的公共艺术实践

中央美术学院城市设计学院副院长王中教授做了题为"艺术激活空间：从 5 米到 5 000 米的公共艺术实践"的发言。

王中教授说："我们应该看到公共艺术在中国更强调艺术营造空间。而艺术营造空间的背后我们更应该强调艺术激活空间。公共艺术应该是植入公众土壤中的一个种子，它应该诱发文化生长，它应该让文化具有生长性，这是最重要的。"王中教授介绍了他自己参与设计的一些融入中国道家传统、阴阳理念的公共艺术作品，如"风动"等。

此外，王中教授重点介绍了他从 5 米到 5 000 米的公共艺术实践活动，包括天津直径 50 米的红色球体、郑州 500 米的城市走廊、鄂尔多斯 5 000 米的岸线以及 1 000 米的世博轴。王中教授说："城市公共艺术的建设，是一种精神投射下的社会行为，艺术和美不是唯一的目标，而是一种态度、一种眼光、一种体验，甚至是一种生活方式。让城市文化从日常生活中彰显出来，并与当下生活发生关联，将艺术植入城市肌体，激活城市公共空间，使艺术成为植入城市公共生活肥沃土壤中的'种子'，诱发文化的'生长'，延伸喜悦、激发创意，让艺术成为城市的精神佳肴，令城市焕发生机和活力，并为城市带来富有创新价值的文化积累。"

法国策展人 Ami Barack 发言

为什么目前是公共艺术？未来是什么？

法国策展人 Ami Barack 做了题为"为什么目前是公共艺术？未来是什么？"的发言，并和与会嘉宾交流了自己的看法。

Ami Barack 认为，艺术作品存在于公共场所，这会让大家有种归属感和参与感。在博物馆或美术馆这类场所中，艺术品的存在是理所当然的，公众如果公开质疑就显得不合情理了，但是在你的居住地、购物的超市或常去坐坐的广场，人们就有权发表自己的看法。

Ami Barack 指出："公共艺术越来越流行，比以往任何时候都来得重要，也将变得越来越好，艺术作品是雕塑还是设施？是用于氛围已定的现有社区，还是氛围仍会受到委员会影响的新社区？是长期的还是临时的？如果是临时的，那后续呢？社区是如何参与的？艺术家是如何参与的？任何人如果考虑到公共艺术委员会，特别是试图引领或启动环境转变的委员会，都应当慎重，不应期望太多。"

"在真正重要的层面——在我们与居住、工作、购物的场所的接触中——我们预期会见到大量的公共艺术作品。这些艺术品应该是赏心悦目的，看到它们，我们会因此心情愉快，或许为之着迷，当然也会产生好奇。这样的日常接触不应该变成无趣的例行公事，而且我们对这类物品和图像所做出的反应，背后必然有深层原因。"

美国艺术家 Bill Fitzgibbons 发言

Bill FitzGibbons 的公共艺术

美国艺术家 Bill Fitzgibbons 做了 "Bill FitzGibbons 的公共艺术" 的简要发言，他介绍了自己参与设计的一些公共艺术项目，包括阿拉斯加的 "时间封锁"、San Antonio 公园改造、"母亲地球的神秘" 等作品，谈到了近三十年来短期和长期的公共艺术、建筑干预以及高温雕塑表现，讨论了在包括冰岛、芬兰和瑞典在内的五个国家的项目中，指定场所作品所体现出来的艺术家的雕刻理念以及美学方式，讨论了包括作为创意场所的公共艺术以及对社区经济和社会的影响。

Fitzgibbons 说： "如今，指定场所艺术目前在国际上越来越流行，这种艺术使公众感受并培育公众的持续场所感。它使公众和私人场所更活跃，同时使定型的环境再次焕发生机，让看到这些艺术作品的人们得到激励，感到振奋。"

香港当代文化中心主席荣念曾研究员

台湾东海大学美术系主任林文海教授

美国韦恩州立大学 Eric Troffkin 教授

"后世博的公共艺术"专题论坛评论纪实

2013年4月13日下午13:00,在上海衡山北郊宾馆凯旋宫B馆2楼,以"后世博的公共艺术"为主题的专题论坛在香港当代文化中心主席荣念曾研究员的主持下召开。在与会嘉宾分别做出精彩发言后,台湾东海大学美术系主任林文海教授、美国韦恩州立大学Eric Troffkin教授作为评论员,对发言进行点评。

林文海教授评论说:"我接到这个题目的时候,后世博,是世博之后还是后世博。现代有现代和后现代,这两个意义完全不同。我们谈的应该不是后世博时代必须要讨论的是什么。而是在世博现象产生之后我们怎么持续做它,应该怎么看待它,这个意义是不重要的。刚才毛竹晨、丁乙老师,包括王中院长对于批判在毁灭性的城市,都做出相当好的评论。我觉得我们评论最大的价值是形成一个理论,形成一个大家共同讨论的理论,我觉得这个是比较重要的。"

此外,林文海教授还提到了关于公共艺术的评价和立法问题,他说:"要告诉人们怎么看待公共艺术这件事,可以告诉他们引起兴趣的,你不能告诉他这是美的,有的人不认为它是美的,所以我的公共艺术评估、美学评估把它分成空间面,包括它是不是具有环境艺术面、理念面。另外一个好处,评估目的并不是好与不好,而是在于它应该引起我们的关心,公共艺术像我们家人一样,不是把它做好就丢在那边。

"台湾从1992年开始做公共艺术的设置法规已经20年了,但是在整个后续维护机制里是缺乏关心的,做了之后就丢在那里,谁维护不晓得,有维护管理单位,但是没有人去做。我觉得这种事情应该在大陆也会发生。20年来,台湾每年以二三十件乃至上百件的速度在增长。台湾现在登记在案的有2 000多件,这些公共艺术作品过得好不好,这个可能需要在学校、在教育圈里面必须要维护,这样才能达到可持续性和有序性。"

美国韦恩州立大学Eric Troffkin教授评论道:"今天我听各位讲座的时候,什么会让我们觉得这个空间是我自己的空间呢?会有这样一个主人翁的感觉呢?这是我的空间,所以我可以改造这个空间。我发现非常有趣的是,你们各位介绍了改造一个空间的各种方式,你们讲的世博中心的公共艺术规划,丁教授讲到了永久的以及虚拟的和暂时的空间装置,还包括有专门给学生的作品。王中教授还介绍了自己的作品。我看到你的作品,我非常

专题讨论会

"后世博的公共艺术"为主题的专题论坛现场

关注其中个人的体验，比如说你的里面吸引很多人握着手走过铁轨，还有跳舞这两个作品给我留下非常深的印象。"

"有一位人文学家讲到理性，你们刚才一直在讲规划，规划有的时候让人们觉得非常的害怕，这位人文学家讲到理性无论存在宗教还是科学当中，都涉及个人体验的不同情形，就涉及人在不同状态之下的感受。确实，我们这些公共艺术作品在短期内能够激活人们体现这个空间的一些电流。这两天，你们在做规划的时候也想使用自己的理性，但是理性是不够的，需不需要给我们提供智慧，这个时候就要超越理性，这就是公共艺术在规划之外，人们利用公共空间，成为自己的空间，人们自己做主空间的一种方式。而且我们说公共艺术也不光是用理性来做规划，或者说我们用理性做规划的时候，公共艺术也给我们一个不破坏理性的更好方式。"

专题论坛
轨道交通的公共艺术

Topic Forum

日本商环境设计家协会国际委员广川启智教授发言

大阪地铁公共艺术

日本商环境设计家协会国际委员、上海大学美术学院客座教授广川启智教授做了题为"大阪地铁公共艺术"的发言，他介绍了大阪车站的公共艺术设计。"大阪是日本最重要的商都，地处日本中心位置。大阪车站更是大阪的门户，每天约有五六十万人出入，在不影响日常交通购物和生活功能的情况下，历时 6 年建成。"

据广川教授介绍，大阪车站公共空间有地上 20 层大楼的酒店及购物商场——阪神、阪急、伊势丹、三越、JR 等五大百货公司，公共艺术空间有日本庭院景观绿化、最新环保太阳能、风力发电、利用雨水浇灌天空农园和卫生设施，尤其雨水和杂排水的再利用均是世界最新的构想设计。

在车站内共有地下 5 层的 5 条地铁以及 3 条私铁，以及 JR 共约 10 个交通电车网交织。这里不仅有最现代化的设备和艺术空间广场，还有鸟语花香的休闲活动设施，可供饮茶喝咖啡、朋友聚会、家族购物、情人约会或外出旅行。

在这有限的空间中，艺术家和设计师们将公共艺术由高空或地面充分潜入地下，卓越地完成了超空间的立体公共艺术工程，不愧是日本公共艺术最杰出的代表。

上海大学美术学院章莉莉副教授发言

上海轨道交通公共艺术总体规划

上海大学美术学院章莉莉副教授发表了题为"上海轨道交通公共艺术总体规划"的讲话，她从三个方面进行阐述：第一是介绍目前上海地铁公共艺术的现状；第二是在此基础上提出上海地铁公共艺术总体规划以及策划；第三是公共艺术运作机制的探讨。

章莉莉副教授介绍，上海地铁里早期广告泛滥的情况已经得到了控制，目前上海地铁有意识地推动公共艺术的全面发展。但目前在上海地铁的280多个车站中，公共艺术覆盖率仅为20%。面对现状，她提出了地铁公共艺术总体规划以及策划方案。这个方案共分为三个层次：第一个层次，首先在三线换乘车站建立一个大型的公共艺术馆；第二个层次，选择一些有特色的车站建立若干中型公共艺术馆；第三个层次，计划建立遍布上海各个车站的小型地铁公共艺术馆。同时，选择采用四种模数化展示单元的方式，为上海地铁空间提供一个选择。在展示主题和内容等七方面，与上海重要的艺术展览、文化节庆活动形成合作，为社会主旋律文化提供窗口，为上海艺术家、艺术院校、艺术机构和企业文化提供展示窗口。

此外，章莉莉副教授指出，针对地铁公共艺术，还需要一套稳定的组织机构和运作模式，目前上海地铁在这方面做着紧张的筹备工作。首先，在组织机构上建立上海地铁公共艺术委员会；其次，在运作模式上采用以策展人为核心的项目运作机制；第三，确立经费来源模式；最后，要通过一定流程保证规范操作。

中国社会科学院哲学研究所刘悦笛副研究员发言

从"生活美学"定位公共艺术

中国社会科学院哲学研究所刘悦笛副研究员做了题为"从'生活美学'定位公共艺术"的发言,从接受者的角度对"轨道交通的公共艺术"进行剖析。

刘悦笛先生认为,其实公共艺术的问题不仅仅是艺术的问题,公共艺术的本质和公众性是相关的,公共艺术可以回到生活美学加以看待。一方面公共艺术是按照现代艺术原则与美学理论构建而来的矛盾修饰法,并逐渐拓展到当代艺术的疆域之中,而今是否要走出这种现代主义模式,回到生活美学,来重新思考公共艺术的公共定位。另一方面,恰恰由于走出了圈子,出现公共空间的时候与公众趣味相联系,这是不是要回到中国美学重思公共艺术的公共定位?

刘悦笛先生借"永久的梦露""倾斜的弧线""越战纪念碑"等著名公共艺术作品来探讨"到底是公众趣味自我决定艺术,还是通过艺术来儒化公众?""公共艺术作品在广大社群那里被拒绝,从美学角度看,是否意味着这件作品的失败?""公共艺术的决定权在谁手中?"等问题。

此外,刘悦笛先生还提出,公共艺术强调参与者参与到公共艺术所营造的情境当中,这样可以把公共艺术分为三种:第一种是前卫艺术,第二种是环境空间艺术,第三种就是民众生活艺术。

最后,刘悦笛先生谈到另一个重要的方向即生态美学,他以"无声的进化"为例来展示艺术如何融入自然,如何塑造环境,如何保持生态。刘悦笛先生说:"其实这个作品本身也是和生活有关的,我们要知道公共艺术怎么和生活相关,艺术家本人怎样生活。这也和艺术家本人经历有关,这个表述了艺术生活的理念。"

英国伦敦地铁管理局总监 Tamsin Dillon 女士发言

伦敦地铁项目

英国伦敦地铁管理局总监 Tamsin Dillon 女士就伦敦地铁项目与嘉宾交流，她说："如何在伦敦地铁这样复杂的环境中做出有意义的尝试，以及怎样来做，我们伦敦地铁项目就是为了回答这些主要的问题。"

Dillon 女士据此介绍了他们所做的一些项目，包括 Mark Wallinge 的 270 件不同的艺术作品、Sarah Morris 的地铁站展台、海报艺术品、数字艺术作品、实验歌曲、现场互动等一系列形式多样、内容丰富的公共艺术作品。

Dillon 女士总结说："这是一个很简短的介绍，总结了我们最近的一些项目，我们要为世界级的地铁创造世界级的艺术，这是我们地铁本身的使命，而世界级的城市要有世界级的地铁，我们要创造世界级的艺术，我们决定要提高地铁乘客的体验环境，而且要提高我们伦敦地铁的声誉。"

美国皮特工作室主任 Peter Kaufmann 先生发言

都市交通艺术中关于玻璃艺术的思考

美国皮特工作室主任 Peter Kaufmann 先生做了题为"都市交通艺术中关于玻璃艺术的思考"的发言。

Kaufmann 先生介绍说："这是我们刚刚完成的一个项目，两个星期之前安装完毕对外开放，是在一个中转车站，把铁路、地铁、公交车站结合在一起的转乘车站，那里面有很多现代设计，我们也有电梯可以送大家到这个月台，另外还有电梯、自动扶梯把大家送到楼上，也有屋顶。而且我们这里用的很多都是全玻璃的设计。这里面我们还和灯光照明设计师进行合作。"此外，他还提到了一个全玻璃的电视塔的设计、与 MTA 合作的公共艺术项目等，简要概括都市轨道中当代建筑艺术的玻璃设计，介绍了艺术品引入公共交通建筑的众多机会。

最后，Kaufmann 先生还提及当代玻璃艺术与现代交通建筑的无缝连接。

中国美术学院艺术人文学院院长曹意强教授主持论坛

奥克兰大学美术学院院长 Derrick Cherries 教授

上海书画出版社副总编辑徐明松先生

"轨道交通的公共艺术"专题论坛评论纪实

2013 年 4 月 13 日下午 13：00，"轨道交通的公共艺术"分论坛在上海衡山北郊宾馆凯旋宫 C 馆 2 楼召开，国务院艺术学科评议组召集人、中国美术学院艺术人文学院院长曹意强教授主持论坛。在五位嘉宾发言后，上海书画出版社副总编辑徐明松先生、奥克兰大学美术学院院长 Derrick Cherries 教授作为评论员，对嘉宾发言进行点评。

徐明松副总编表示认同刘悦笛先生对于生活美学的定义以及通过案例演绎的对于生活美学所给大家带来的启示。同时，他提出两个问题：一个是公共艺术本身是艺术问题，又不完全是艺术问题，公共艺术是由公共性决定的；另一个则是希望大家走出现代主义的个性原则。这对公共艺术是一个非常好的启示。

此外，徐明松先生就广川启智发言的大阪地铁案例作出评价，他认为这个案例给我们提供了很好的线索，呈现了关于环保、太阳能、风力发电、水的再利用等诸多启示。另外，日本庭园景观在车站的呈现，也反映出大家普遍运用本土文化资源、吸收本土文化资源的心态。这一点实际上在章莉莉老师所阐述的上海地铁公共艺术的规划中也可以看到。此外，还有人与自然、新的科技发展问题，我们应当如何面对自然，面对社会发展等问题在刘悦笛先生讲的"无声的进化"中得到演示。

最后，徐明松先生指出："本世纪初，已经有大量公共艺术的内容进入伦敦地铁。比如说艺术家进行海报设计以及地铁的导向、标志系统等。今天，更是表现出了它的多样性和丰富性，比我们上海地铁可能走得更具开放性，艺术家在地铁里用了多种艺术创造的样式进行创作。今天，公共艺术进入地铁空间、进入轨道交通的空间中，实际上在整个创作方法多元化和创作形态多元化的过程中，呈现出越来越丰富的状态，有很好启示。"

奥克兰大学美术学院院长 Derrick Cherries 教授也发表了自己的见解："我从广川启智先生介绍的大阪地铁想到，实际上整个地铁站周围像一个小城镇，里面有各种各样的功能，规模非常大。这要求有关非常复杂的技术管理能力。我们也听到了上海地铁公共艺术总体规划报告，应该说这样一些实用的问题、功能的问题都在上海地铁里得到了解决。但下一轮，如何在人的体验上使之更加人性化。我也看到了上海在这方面所做的总体方案，考虑到了将艺术融入空间的表现。这里面有一些永久性的项目，还有一些临时性的艺术项目，这种体验的变化也非常重要。"

专题讨论会

"轨道交通的公共艺术"专题论坛现场

对于刘悦笛先生讲到人在地铁系统中的体验，讲到要建立一种生态美学，Derrick Cherries 教授表示认同，他认为上海地铁系统里的环境应该具有独特性。"我认为公共艺术要发挥大的作用，其实也是可以发挥各个方面的作用，有各个方面的功效。这也是公共艺术现在所遇到的挑战。所以我认同公共艺术一定要和它所在的环境相呼应，而且要能够体现、同时也要体现我们自己所要表达的意思。"

针对 Peter Kaufmann 的发言，Derrick Cherries 教授认为这个发言所展示的作品其实做了各方面的工作，其中一个就是通过这些作品能够让我们看到作品的人性化，而另一方面，包括诸如人行道等设施也是通过艺术作品能够让它更加人性化。

专题论坛
美丽乡村中的公共艺术

Topic Forum

上海大学美术学院潘力教授发言

"地方重塑"的文化内涵

上海大学美术学院潘力教授做了题为"'地方重塑'的文化内涵"的发言。

潘力教授说:"我们这次论坛的主题是'地方重塑',这是公共艺术的核心要义,从不同角度关注和诠释城市生活与地域文化,关注空间环境,关注人文、历史脉络和日常生活,提升人的生活质量和幸福指数,为当代艺术进入大众生活提供了契机。围绕这个要义,能够体现出公共艺术对重塑城市文明、农村文明和居民文化生态的意义。"

潘力教授重点介绍了上海大学美术学院在浙江玉环县进行的乡村公共艺术建设项目。他谈了依托美术学院的力量,利用当地的材料和废弃的木料等模拟卡通人物,以"动漫花谷"为主题来提升文化氛围的公共艺术实践,还特别提到了借鉴日本"一乡一品"的理念,以及改造玉环干江、坎门、沙门镇的具体情况。

在谈及"地方重塑"观念时,潘力教授说:"以'地方重塑'为主题的公共艺术的概念,绝对不是外来的艺术家对当地的一种施舍,一种外来力量的符号性的添加,而是把它作为一种激励机制,去激活当地自身的动力,只有这样,以'地方重塑'为核心的公共艺术才能真正有一个持续的发展空间,也才能真正能够造福当地……美丽乡村建设是美丽中国建设一个最重要的环节,其核心就是激发人们对生活的追求,而不是在于艺术家个人对公共艺术本体的表现。受中国国情影响,它决定了中国的公共艺术面貌和国外的公共艺术面貌是必然不同的。"

浙江省玉环县龙溪乡党委书记施明强发言

公共艺术让乡村更美丽

浙江省玉环县龙溪乡党委书记施明强做了题为"公共艺术让乡村更美丽"的发言，就玉环县龙溪乡美丽乡村公共艺术实践做出清晰介绍。

施明强书记首先讲述了美丽乡村和公共艺术结缘的过程，他说："自和上大美院联合调研以后，我们一致认为发展关键是重塑农村的文化生态，让城市居民到乡村寻找精神家园，让农民激活文化自觉和发展信心，为新农村置入内生动力，进而催生由文化意义、经济产业带动农民致富。由此看来美丽乡村建设本身就是一个大的公共艺术课题。于是，我们把公共艺术和地域文化、现代农业、生活体验、课外素质教育、民间文化、生态文明、社团组织、地方特色节会等结合起来，使得乡村重塑成为一种模式。"

施明强书记还提到了公共艺术在美丽乡村发展中起到的积极作用，包括乡村文化再造、文化福利普惠、解决公共问题、唤醒文化自觉、形成创意景观、提高审美意识、转变价值观念、实现城乡交流、寻找文化认同等。

最后，施明强书记谈到了美丽乡村和公共艺术互动发展的几个问题，他说："首先是体制层面的问题。公共艺术介入美丽乡村建设还缺乏支持体系。现在主要是来自行政体制内部的支持，在规划编制、发展策划、项目实施等方面还缺乏外部支持和专业指导。公共艺术组织介入美丽乡村建设，需要加强与社会科学研究机构、政府部门的双向沟通，以期达成统一思想和合作的架构，形成互惠互利、互动发展的良性的体制；第二是制度层面问题，如何通过跨界研讨以及政府官员培训等方式，扩大公共艺术的普及宣传和对社会的影响力，进而增加与政府间的交流合作，使公共艺术的社会介入和解决公共问题职能纳入政府的社会制度层面。第三是农民主体意识和文化自觉的问题。农民的素质如何提升，还有文化自觉的培养，是农村持续健康发展的一个重要问题。如何加强对美丽乡村的跟踪考察，着力培养智慧农民，完成农民教育培训体系，值得特别关注。"

新西兰奥克兰国立美术馆馆长 Rhana Devenport 女士发言

美丽乡村中的公共艺术："澳新体验"

新西兰奥克兰国立美术馆馆长 Rhana Devenport 女士从澳新体验的角度谈了她所理解的"美丽乡村的公共艺术"。

Rhana Devenport 女士指出："在这里有三个问题我们要讨论。第一，为什么要做公共艺术？第二，谁创造了公共艺术以及为什么？第三，谁会关心公共艺术？这三个问题好像太生硬了，但是的确值得我们思考。我们必须要考虑整个公共艺术创立的动机是什么以及跟观众是如何联系起来的。今天我想以新西兰为例，谈一下景观和雕塑实践之间的一种特殊和紧密的关系。"

Devenport 女士介绍了位于北岛乡村地区景致优美的雕塑公园 Gibbs Farm，这里随处可见艺术家依地形、景观所雕塑的作品；位于怀赫科岛海湾处的海角雕塑，该雕塑在短时间内转变了人们探索自然的经历；奥克兰的 Britomart 公司发展，及其进行的互动艺术品工作；针对那个灾难性的地震进行的克莱斯特彻奇美术馆外太空项目作品；近期新西兰范围内的一天雕塑项目以及先锋电影制片人和活跃艺术家 Len Lye 在公共艺术领域的雕塑作品等。

借此，Devenport 女士向与会嘉宾表达了对"公共艺术到底有多公共？""这些作品怎样影响我们对景观的看法以及我们对自然环境的理念和感官联系？"等问题的看法。

美国《公共艺术评论》杂志主编杰克·贝克先生发言

美丽乡村中的公共艺术

美国《公共艺术评论》杂志主编杰克·贝克先生谈了他对"美丽乡村的公共艺术"的思考。杰克认为，从长期讲，要获得什么样的成功，就必须要知道通过推出公共艺术要达到什么样的目标。比如说，我们希望公共艺术能够增加游客量，或者希望能够迎来各个年龄层次的人士，或者说我们希望能加强荣誉感等。公共艺术能让每个乡村都变得不同，就必须要侧重于这个地区或这个乡村的独特历史和文化，对这个地区或者城镇的品牌起到促进作用。因此，让来自外部的艺术家能融入到日常的工作中，随着他们进一步了解这个乡村的独特之处，才能够真正有独特的体验。

针对以上思考，杰克谈论了几个他认为在公共艺术的创作过程中尤为重要的问题。首先要突破常规和传统，也就是对公共艺术定义的传统；第二需要不同媒介创造公共艺术，不仅仅是视觉艺术。"我们需要有不同的资源来创造这个公共艺术。同时也必须有专业人士和协调员在创作过程中发挥作用。另外，还要有一些教育性的项目，让本地的居民对这种艺术形式有所了解。"

此外，杰克举了一些他所在的明尼苏达州和全球的优秀公共艺术案例，包括爱华达小镇的家庭系列雕塑作品、"马赛克瓷砖"、"艺术之旅"、史派罗作品、盖普斯农场等。

贝克先生总结说："谈到地方重塑和公共艺术之间的关系，我们必须看到，我们要创造的是有意义的社会空间，这是核心。我觉得公共艺术就是关爱，我们要关爱我们所生活的这个世界的健康，我相信我们需要把艺术家的才能跟社区真正的需要结合起来。"

美国艺术家 Ta-coumba Aiken 先生发言

公共艺术是众人合作的结果

美国艺术家 Ta-coumba Aiken 先生在发言中系统地介绍了自己的创作经历。 Ta-coumba Aiken 谈到了他从事艺术工作的原因："因为我父母的原因，我的父亲和母亲都是农民，他们离开了农田到了城市，我还是小孩的时候，跟他们回老家，农场上已经什么都没有了，但是在那边度过的时间总是很愉快。"

在非洲，Aiken 先生意识到艺术可以是三维的，此后，他做了许多公共艺术作品。包括位于非洲的"儿童世界"、向当地文化致敬的爪哇餐厅、连接天和地的过道、改造当地停车场、充满历史感的出版物封面、儿童游乐场"手拉手"墙画、伏特加海报、101 个花生雕塑、会移动的公共艺术项目等。这些作品根据不同的环境、不同文脉和空间而有不同的效果，风格多样。

Aiken 先生用他自身的经历告诉大家，公共艺术是众人合作的结果，他说："艺术家和学生都要会和各方合作，公共艺术更是如此，一定要跟相关的比如政府机构合作，要知道他们有什么要求，他们有什么想法。而且合作的一个好处，就是合作总是能够让我们碰撞出很多的新想法，有很多新主意。"

日本东京艺术大学美术学部学部长保科丰巳教授发言

"艺术工程"推进造乡运动

日本东京艺术大学美术学部学部长保科丰巳教授做了题为"'艺术工程'推进造乡运动"的发言，向大家介绍了日本的公共艺术情况和东京艺术大学最近正在进行的一些公共艺术活动。

保科丰巳教授表示，如今日本面临的最严峻的社会问题是人口出生率下降，老龄化趋势日益严重。此外，两年前日本遇到的灾难性大地震给人们的精神带来很大冲击。因此，"公共艺术如何能够对社会发生作用？这是我们要考虑的首要问题。"

保科丰巳教授介绍了日本的一些公共艺术作品，包括 Art Line 以及从日本民间形象中脱胎而来的艺术作品。保科教授介绍说："今天，日本的当代艺术与社会以及日常生活的关系愈加密切，尤其是 2011 年 3 月的日本东部大地震以来，以自然环境和人类自身为中心的'社区设计'被作为公共艺术的新主题，以艺术的力量作为推进'造乡运动''地域振兴'等事业的重要环节的策划越来越多。相对于以东京为中心的当代艺术，这样的'艺术工程'正在日渐显示出向地方展开的可能性。"

专题讨论会

中国国家画院副院长张晓凌教授担任主持

新西兰 Starkwhite 艺术机构总监 John McCormack 作点评

"美丽乡村的公共艺术"专题论坛评论纪实

"美丽乡村的公共艺术"专题论坛现场

2013年4月13日下午13：00，"美丽乡村的公共艺术"专题论坛在上海衡山北郊宾馆贵宾楼会议室召开，中国国家画院副院长张晓凌教授担任主持。在嘉宾发言之后，新西兰 Starkwhite 艺术机构总监 John McCormack 先生对本次"美丽乡村的公共艺术"做总结。

John McCormack 先生说："第一个讲者提到中国浙江省的政策说要打造美丽乡村，当地相关方也很积极的推动这种美丽乡村的工作。这是自下而上和自上而下很好的结合。我们也听到了龙溪乡代表介绍的关于项目的一些实际情况，之后我们听到了 Rhana Devenport 介绍了盖普斯农场一些公共艺术项目，这个地方不光是展示艺术，确实在我们人的生活中起到了很重要的作用。再就是《公共艺术评论》的演讲者给我们介绍了一个非常重要的概念，就是关于艺术家在项目中的重要性。有很多的公共艺术项目跟当地合作，对当地做了很多贡献。我觉得这个是非常有意义的。最后一个演讲，给我们介绍了日本的情况。总的来说，前面六个演讲都非常有意思，让我学到了很多东西。"

最后，John McCormack 先生认为，在中国，大家对公共艺术关注的是公共的部分，这一点是需要大家去思考的。公共艺术对公共空间有很深厚的社会影响，如果在未来想继续推动这种所谓的地方重塑，他觉得很重要的就是大家需要进一步在这方面进行讨论、思考。

"大会论坛 " 总结会

Final Report

"大会论坛" 总结会

总结会现场及嘉宾出席相关活动

国外学者和艺术家在论坛上

总结会现场及嘉宾出席相关活动

国内外学者考察上海世博园区

新西兰奥克兰大学美术学院院长 Derrick Cherries 在讲话中说："公共艺术能够鼓励公众的参与和体验，而且能丰富他们的体验。那我们就要讨论究竟什么是参与，对你来说或对我来说，什么是参与？什么是体验的丰富呢？是不是一种社会学的体验？从这几天接触到的公共艺术作品中，我们可以看到那些作品都要体现某种身份和某种目的，有时候本身就是一个非常简单明了的流程，但有时候也是非常复杂的流程。因此，在这个过程中我们需要能够识别在这些作品背后的驱动力和愿望。

此外，我们看到了全面的知识的重要性，而且要应用于不同的情形。在一个环境中的知识，在另外一个环境中可能就是缺乏的。所以说，重要的是我们需要有一个了解这个特定环境和情形中具有特别知识的人，而且我们也需要其他知识的补充,比如在一个新的环境中我们可能需要某一种新的知识。

这些项目的关键品质在于作品用户的参与，他们不仅仅是观察者，并不仅仅是被动的观察，而是一种积极的参与，他们就是积极的观众。这种参与，并不是只有一种方法。

我们也需要帮助培养我们受众的专业知识。公共艺术的观众，他们艺术的专业知识可能没有像在座的这么多，所以说我们怎么知道他们对公共艺术有怎样的期望值，这就需要我们去帮助他们，让他们知道公共艺术是什么样的内容，他们会看到什么样的公共艺术作品。我这里没有一个一成不变的模式，但是确实我们需要去积极的做起来。"

美国韦恩州立大学艺术设计系助理教授 Eric Troffkin 发表讲话时说："我也很认同终端用户或观众的看法。最开始也谈到了什么是成功的公共艺术，什么是好的公共艺术，过去几天我一直在想这个问题。公共艺术在现在这样一个不断变化的社会文化环境下，应该如何发展？前面一些学者谈到大家会为未来做规划，说我们未来的情境应该是怎样的，但是未来你不可能光通过逻辑或者是通过推导能推导出来。对我来说，成功的公共艺术它将能够让你留下深刻的印象，此外能够让你对未来保持乐观。

我们都知道，我们确实会为未来做规划，但事实上未来真正的情况很可能和我们的规划是事与愿违。但是由于有公共艺术，让我们有一种归属感，并且对未来始终保持一种信心，对未来保持一种信念。

Derrick 讲到的用户的重要性，我觉得这确实也是判断公共艺术是否成功的一个指标。在公共艺术界，如果我觉得一个项目不成功，那我至少会比较宽容地说，他好歹做过尝试了，而且通过尝试他可能做得越来越好，只要他更好地和终端用户互动，考虑到终端用户的想法。"

新西兰 Starkwhite 艺术机构总监 John McCormack 先生为"美丽乡村的公共艺术论坛"做总结。他说："公共艺术有什么样的好处，谁获得这样的好处以及我们如何能够实现这些好处。我们分论坛的演讲者讲了各种各样的话题，最先的两位，潘力教授和施明强先生主要讲了中国的乡村公共艺术，在中国的小山村，地理位置比较偏远的渔村做的公共艺术项目，并且向我们介绍了这些公共项目对当地的社会、经济、文化产生的积极影响。也就是说，其实这些公共艺术项目不光是能够创造很好的艺术价值或者社会价值，而且也能使乡村实现重生。我们也讨论了自下而上或者自上而下的不同方法。

第三位 Rhana Devenport 介绍了艺术对于新西兰当地风景的影响，体现了公共艺术能够承担的社会作用。

第四位是 Jack Becker 先生，他讲了如何通过公共艺术解决生态的、挑战经济的问题，更好改善大家的生活。给我们介绍了很多充满创意的项目，以及如何利用公共艺术项目为当地居民服务。

下一个演讲者是美国艺术家 Ta-coumba Aiken 先生，也讲了他的公共艺术项目如何和当地互动，如何和当地居民合作，让他们能从艺术作品中获得更多的收益。

最后一位是来自日本的保科丰巳教授，他向我们介绍了日本的公共艺术。我们都知道，日本出现了老龄化和年轻人数量不断减少的挑战，所以公共艺术也是为了回应这样一个现象，去做了很多的努力。

最后我想说的是，当时我们主持说到，他昨天和很多讲者沟通之后，发现谈公共艺术方面，中国艺术家和研究者比较关心的是公共艺术中公共的一部分，而西方的研究者和艺术家可能关注的是公共艺术中艺术的一方面。我也很同意，我们一定要求同存异。实际上，我们有很多相同点，很多项目现在已经成为了实验室，也就是可以让我们尝试不同想法、不同做法，而且我们也可以去把不同的相关方纳入到我们的公共艺术的项目合作里面。所以，如果我们要问公共艺术的未来走向是怎么样的，那我觉得最好的答案就是我们要很好地预测未来，最好的方法就是成为创造未来的一部分。也就是说，我们要努力共同合作，为我们打造一个自己的公共艺术的未来。"

中国美术学院人文艺术学院院长曹意强教授主持"轨道交通的公共艺术"分论坛，他总结说："我们主要从三个角度讨论公共交通的艺术，一个是谈艺术怎样融入整体的地铁建设，而不是作为一种装饰，如何使艺术成为地铁建造包括功能各方面的一个有机组成部分，其他三个案例比如上海、伦敦、皮特的玻璃艺术。一个是在受限的建筑空间里面，怎样利用艺术，让艺术再重新融入到建筑里面去，使它构成一个整体。这是两种不同角度来探讨公共艺术。还有一个是从理论方面，就是刘悦笛研究员，主要谈生活美学，里面涉及公共艺术的价值问题。非常有意思。

对我个人有一个很大的启发，因为我一直在想，我们考虑艺术的时候，往往把艺术当成一种丰富公共环境的东西，其实这种考虑在某种程度上是没有真正体现艺术更深层的价值。我认为艺术在给我们提供美的过程中，其实在不断塑造我们的一种敏感性，而且对于我们的思想、思维以及对于我们的科学创造包括我们对于生活问题的处理，都提供的不光是一种想象力，而是一种非常高级的技术手段。这一点从大阪的案例中，刚才广川教授以及 Tamsin 这两位演讲者没有用公共艺术的概念，但是从他们的谈论中，我就感觉到公共艺术不是一种添加物，而是解决我们的技术问题，解决我们的能源、环保、安全以及生活问题的一种智慧，同时也是一种创造性手段。这给我非常深的体会。也就是说，我们说的公共艺术，其实是艺术与科学、与我们生活的一个有机结合体。我认为，公共艺术用中文解释，就是人类整体环境的有机艺术。"

"后世博的公共艺术"分论坛的主持人、香港当代文化中心主席荣念曾研究员发言总结，他说："第一个讲者是世博局战略发展部的毛竹晨，她给我们一个回顾，也讨论了世博计划的前瞻；第二个讲者是丁乙老师，讨论的是现代中国公共艺术的概念和西方公共艺术的概念的差异切入点在哪里。丁老师也谈到关于从商业角度去看文化还是从文化角度去看商业；王中老师谈的是他自己的创作，但他提了很多关心的问题，其中一个是现在中国的城市面对一个危机，毁坏的程度超越大家的想象；Ami是世博会公共艺术品的重要策划人，回顾了世博会艺术品的情况，详细解说了这些艺术品产生的过程，也提出了一些关于公共艺术的论述，探讨关于外与内、与公共、与私人的问题；美国艺术家Bill向我们展示了他过去几十年的创作，我的感受就是他的创作反映了高科技的发展。这就不只是在公共地做做创作，而是在这个公共地域里与社会、与文化如何真正开始互动。之后是两位评论员，一位是来自台湾的林文海老师，谈的是后世博还是世博后的区别，还有怎么由评论慢慢发展到舆论。我们也谈到领导与群众、公共与非公共。最后是Eric Troffkin，他给了一些比较哲学性的说法，最后谈的是公共空间怎样让每个公民都拥有，这是非常重要的一点。"

此外，荣先生也就"公共艺术"谈了自己的看法和理解："艺术的公共性，艺术在不同时代与环境的关系，只要引发到公共的讨论与关注的话，它就显示公共性。今天我们将公共放在艺术前面，可能就是对艺术的一种批评，对艺术现在在经济、政治强势下的一个批评，同时也是对艺术缺乏教育的一个批评。艺术留到今天，必然有它的公共性，今天的艺术能流传到明天，也一定有它的公共性。我们关注的是'流传'这两个字，关注这两个字本身的机制和运营，以及背后的政策与价值观，它是不是真的能走在社会公共理念发展的前端，能否发展成为拓展、维护社会公共空间的力量。我想这里关心的都是公共艺术的机制、营运、政策以及价值观，了解这些客观的情况，再去审视艺术创作与当下环境公共的关系，自然就会发展出相对的评估准则，这样才能协助我们更深入地审视论述艺术创作本身。我想大家不会反对说公共艺术应该是可以影响公共空间的艺术，同时也是应该用来探讨目前跟未来空间的艺术。

当然，公共是一个非常政治的名词。公共空间在发展中国家能否独立于第一部门和第二部门就是政府和企业，还是有挑战的。它的严肃性、辩证性，都是可以拓展和强化公共空间。我个人觉得这个论谈本身就是一个公共空间最好的案例。我们其实是靠这里每个朋友，怎么让这个论坛有严肃性、辩证性，然后让公共空间跟公共艺术开拓新的论述。更实际一点，是通过公共和艺术的讨论，让我们的文化发展有更健康、更前瞻的机制和营运。"

上海书画出版社副总编辑徐明松先生发言说："我特别想提到的就是下午轨道交通的公共艺术论坛给我两点启示。一点是对公共艺术的认知中不断触及人与自然、人与环境的关系，广川先生对大阪地铁站这样一个复合的公共空间的描述，让我们看到了公共艺术像盐溶于水的有机机理和艺术表征，让我们知道人与自然的关系中，存在着很多需要解决的城市问题，这也是公共艺术所要面对的一个重要发展方向。

另外一点是我们对公共艺术在地铁公共空间的发展，实际看到了它的历史文脉，或者每个国家的地铁发展过程中都有自身城市发展的特点。

因此，以伦敦地铁为师，今天上海地铁的发展，也可以从我们的历史文化资源中获得很多启示，包括上海很多地铁站已经通过形象塑造使历史文化资源得到了很好的呈现。这个过程中，我们可以看到伦敦地铁的整个公共艺术的艺术家委托机制，给上海的公共艺术的发展，艺术管理机制、运行机制也提供了很好的参照。

上海地铁公共艺术的规划中，也提到了运输委员会的设立和策展人机制的设立，但是整个具体的运行中如何真正体现它的公共性，体现出很多嘉宾所提到的业主方、政府方和艺术家和公众之间的关系。我想这个随着中国城市化的发展，公民的公共艺术的意识会越来越强，从这个角度讲，实际上上海地铁公共艺术在规划过程中所做的很多民调，基本反映出非常强烈的发展公共艺术的热情和意志。

还有一点，我们现在很多地铁公共艺术的创作手段和创作方法是多元化的，具有发展的前景。实际上，我们在伦敦地铁近些年的项目实施过程中，可以看到概念艺术、当代艺术在地铁空间的呈现。为我们地铁艺术的多元化或者是国际化、多样性，提供了一个很好的参照。"

上海大学美术学院院长汪大伟教授做总结陈词。他说："我听下来以后，觉得有三个不必担心和一个要担心。

第一个不必担心，是我们公共艺术的定义不清。我们大家都在讨论公共艺术到底是什么，什么是好的公共艺术。我想它的定义不清，才带给了我们魅力，才带给了我们无限的想象力。所以我们不要求定义它到底是什么，这个不必担心。

第二个不必担心，我们一直讲民主意识、公民意识缺乏，这不必担心。为什么呢？信息化时代到来，它带给我们人类的文明是什么？就是民主的文明。工业化社会带给我们是物质文明享受，而信息化时代带来的是民主文明享受。所以这个不必担心。

第三个不必担心，是我们现在和政府、艺术家、企业等跨境合作的难题，这个不必担心。因为这个难题的出现，是因为需求不明确，等到政府需求明确，等到企业需求明确的时候，他们自然就会走到一起。

所以这三个，都是我们纠结的事情，但是我觉得都不必担心。

我担心的是什么呢？我们大家都在谈国际化、多样性。我们今天坐在一起，就是要求一个'一'。这个'一'，实际是没有的。我们公共艺术的推广发展，不希望在最后走向又是一个单一的统一，去追求统一。这是我担心的。

下面我以主办者的身份讲几句。

今天我做一个自我批评，我们还做得不够的地方。只有批评了以后，下次才会更好的进步。

第一个，实际刚才荣念曾先生已经提到了，问在座哪位是政府界的，在座哪位是企业界的，没有人举手。因此，这还是我们小圈子，高校、学术界的小圈子，这是这次论坛做得很失败的地方。

第二，我们讲的是公共艺术的讨论，但是我们的会议组织形式本身就缺乏公共意识。比方说今天在座的就分台上和台下，这样很难能够有一种交融。但是我觉得这样的形式，是我们的一种惯性，是惯性带给了我们没有公共性，缺乏这所有的一切。也值得今后我们在工作中改进。

最后一点，我想讲一些希望。上海大学美术学院承办这次论坛，更寄希望于能持续下去。这个持续，实际上我还是希望能够成为一种机制，不是成为一种固定化的东西。作为这一届的主办方，我向 Lewis 主席提出了三个期待。一个就是希望下一届有一个好的议题或主题，第二要有一个好的工作机制，最后希望还是要给地方确实实解决问题的一个方式和方法。

玉环东沙·美丽渔村
国际公共艺术论坛

Forum

活动纪实

玉环东沙国际公共艺术基地授牌仪式

苏泊尔集团董事长苏增福向嘉宾介绍大鹿岛的建设情况

嘉宾讨论

嘉宾沿途品尝村民自制小食

嘉宾观看坎门东沙舞龙灯表演

坎门东沙舞龙灯表演

坎门东沙舞龙灯表演

坎门东沙舞龙灯表演

玉环坎门村民在看美国艺术家 Ta-coumba Aiken 的作品

2013 年 4 月 14 日上午，国际公共艺术论坛国内外嘉宾一行从上海出发赴玉环，下午抵达坎门街道东沙社区。嘉宾下车时，当地村民进行了隆重的传统民间非遗艺术项目表演：喊海、坎门花龙表演、鳌龙鱼灯舞等。在阵阵微咸的海风中，嘉宾们感受了一场动态的视觉盛宴，这也是渔民们自己独特的公共艺术。

大家随后沿途步行参观渔村风貌。放眼望去，浓郁的渔区文化气息扑面而来，石阶、石墙、石屋，一幅幅石头演绎的景致随处可见。村里留存着道教信徒用以供天、地、水三官神的庙——三官堂。渔民祖祖辈辈出海捕鱼，人们通过烧香、拜祭的方式向天神、地神、水神祈求平安和丰收。

嘉宾们还赴东沙营房参观玉环东沙建设规划图板展，漫步石阶，感受海风，这是在现代都市里难以体验的惬意。

在玉环县副县长胡载彬的主持下，坎门街道与上海大学美术学院签订战略合作协议，并举行了玉环东沙国际公共艺术基地授牌仪式和上海大学美术学院艺术创作基地授牌仪式。

玉环县委副书记朱立国首先致辞：随着玉环东沙国际公共艺术基地、上海大学美术学院艺术创作基地的建立，以及玉环东沙国际公共艺术家驻地计划的实施，将为玉环打造一个渔村文化与公共艺术相融相生的美丽渔村，为玉环提供一个渔村文化传播、交流的新平台。此次论坛正是一个契机，让玉环不断引入国内外先进的公共艺术建设理念，进一步整合自然生态资源和民俗文化资源，努力探索美丽渔村和公共艺术融合发展的新路子。

Lewis Biggs 主席在为玉环东沙国际公共艺术基地授牌后说：他感受到了当地生活的美好状态，看到了玉环坎门的美好前景。他承诺，在国际艺术家的引领下，会尽力帮大家达成理想的结果，希望在公共艺术的影响下，国际艺术家给当地带来更美好的未来。

汪大伟院长也在授牌仪式上表示，34 年前他曾到过坎门，并在这里住过一个月。现在重返故地，带着上海大学美术学院与坎门的合作项目，旧地重游，难以掩饰激动心情。他说："当地的喜庆气氛让大家感受到民间艺术带来的幸福，体验到当地渔民火热的生活，更点燃艺术家们的创作热情。坎门将成为中国国际公共艺术的窗口。"他代表美院承诺，上大美院会将坎门视为自己的后方基地，这是对接社会、为社会服务的基地。国际艺术家们、上大美院师生、玉环人民在共同享受生活的前提下，将精诚合作，共建美丽乡村和美好未来。

傍晚，嘉宾一行乘船赴大鹿岛。次日（4 月 15 日）上午，"玉环东沙•美丽渔村国际公共艺术论坛"召开，论坛由坎门街道主任曾志斌主持。

苏泊尔集团董事长苏增福首先在会上致辞，他追溯了大鹿岛的历史，感谢洪世清教授自 1985 年以来为继承中国岩雕艺术，在大鹿岛辛勤创作 14 年留下百余件岩雕艺术瑰宝。同时对苏泊尔集团为大鹿岛的建设所作的努力做了陈述。

论坛共分为三个阶段，第一阶段分别由坎门街道党工委书记陈海鸣、上海大学美术学院潘力教授发言。陈海鸣书记表示，以公共艺术的"地方重塑"为载体，通过社区改造、空间转换、艺术交流等方式诠释社区生活、乡风民俗和渔区文化，将促进坎门及玉环滨海景观和人文景观的交融。潘力教授主要围绕玉环东沙坎门街道历史文化村落规划报告展开阐述，展示了坎门街道总体风貌改造前后的对比效果图和局部民居外装修的效果图，对风貌修复的规划内容进行了重点讲述；第二阶段则由专家 Lewis Biggs、汪大伟院长和 Jack Becker 发言；第三阶段，由上海大学美术学院院长助理金江波主持，数位专家学者就建设美丽渔村展开讨论交流。

最后，玉环县委副书记朱立国做总结发言，他希望美丽乡村建设能够有利于当地历史文化的传承与发展，希望东沙美丽渔村成为可持续发展的生态文明，也希望由于公共艺术和国际艺术家们的介入，美丽渔村建设最终实现村美、人和、民富。

傍晚，嘉宾一行返回上海，大家表示，在国际公共艺术专家和艺术家的支持下，东沙美丽渔村建设一定会盛开出绚烂之花。

Lewis Biggs（国际公共艺术奖评委会主席）

沙渔村具有很强的地方文化特征，阳光、沙滩与海水是当地的特色。通过一系列考察，也看到了当地文化与环境的呼应，并提出小小疑问：东沙规划报告是确定要发生的还是处在讨论的阶段？每个规划报告应该是需要长期讨论的。建筑师领域有句名言："如果要建造一个好建筑，需要20年时间的讨论，然后用1年的时间来建造它"。所以，讨论的环节非常重要。所有优秀的公共艺术作品都有一个共同点：艺术家都在当地生活过、接触过当地风物人情，这样的合作计划才是双赢的项目。将那些远景设计图变为现实也许需要很长的时间。但要建设一个有质量的驻地项目，需要艺术家深入观察总结。花费时间越长，项目存续时间也会越长。希望这里的艺术设计是一个不断延续、有吸引力的过程，希望更广大的艺术家聚集此地，为国际公共艺术与玉环建设做出成绩。

曾志斌（坎门街道党委主任）

感谢与会专家、嘉宾来玉环东沙指点江山。坎门背靠海洋渔区乡土文化，面对现代生活方式及外来文化的交汇。坎门及东沙的美丽渔村建设，应基于保护优先、彰显特色、以人为本的原则，以公共艺术的"地方塑造"为载体，通过社区改造、空间转换、艺术交流等方式，来诠释社区生活、乡风民俗和渔区文化，促进坎门及玉环滨海景观和人文景观的交融并茂。

我就坎门东沙的美丽渔村建设提出以下几点看法：

一、打造展示渔区文化的窗口。渔区多彩的风土人情孕育了绚丽的特色文化，坎门拥有丰富的民俗、饮食、非遗等渔区文化，包括家喻户晓的坎门花龙、鳌龙鱼灯舞、渔工号子等，这些原汁原味的本土文化深深植根于坎门渔区，也是本地文化的精粹。希望在公共艺术塑造中有机融合渔区文化的元素。兼具运用现代造型与民间美术手法，点线结合东沙及坎门沿海景观带，将坎门及玉环塑造成为东海渔区的亮丽窗口。

二、打造保护海洋文化的典范。坎门以海为邻，千百年来经受海洋文化的浸润和洗礼。其海洋文化丰富而具有多样性，涵盖了移民文化、妈祖文化及"三色文化"等，兼具开放包容、兼收并蓄的秉性，与坎门渔区的生产、生活、乡风、民俗密切相关，有极强凝聚力与向心力。顺应海洋文化主体的审美情趣，以"保护塔台，文化唱戏"，在坎门渔区形成一种稳固的文化态势，从而为为海洋文化的长足繁荣打下坚实基础。

三、打造对外文化交流的平台。借助国际公共艺术基地、上大美院创作基地和国际公共艺术家驻地计划的东风，立足地方塑造，以论坛对话、艺术采风、节庆活动等为依托，深化跨文化的撞击与合作，真正实现文化交流的辐射与叠加效应。

四、打造研究区域文化的基地。文化保护，特别是非遗文化保护需要渐进式的挖掘保护。如针对本地的建筑风格与历史遗迹以科学的修复调整来替代简单的破坏性拆建，以保留岁月的记忆，展现出区域文化的本真性与活态性。这就需要借助学术机构的专业指导，凝聚艺术家们的智慧和力量，得益于实施团队的建造技艺。以地域文化为先导，以互动实践为通道，将海洋渔区提炼得更加醇厚。艺术无国界，文化和而不同，真诚寄望今天人文与历史脉络之间获得最佳平衡点，为坎门的美丽渔村建设和文化生态重塑指引导航。

汪大伟（上海大学美术学院院长）

从两个感受谈起：一是从外国艺术家的感受和表情知道，这个合作的项目是有意思的；二是昨晚与国内外艺术家挥毫交流、沟通，激发了无限创造力。今天的论坛名称是"美丽渔村国际公共艺术论坛"，把几个字拆开：渔村、国际、美丽、公共艺术，这些字如何组合交融呢？

先说"渔村"和"国际"，艺术家们融合交流，正是渔村走向国际、国际介入渔村的很好案例。想让国际人士关注此地，就需要保留、挖掘本土文化，进行原汁原味的展现，这是吸引国际人士到来的根本。我们的设计目的不是把渔村变成欧洲村落，而如何保留渔村的本色和风貌，是我们需要关注的问题。

再说"美丽"和"公共艺术"，这些天谈了很多好的案例，重要的是如何落地和操作，让我们艺术家一到这里就有创作的热情。如何让艺术家的这种创作冲动持久，如何建立一种运作机制，值得思考。我们希望达成三方合作的模式，例如通过民间成立一个国际艺术家驻地基金会，希望政府能给出相关政策来推动整个过程，使项目发酵生长。希望通过这样的创作活动激活当地文化的发展，引发世界的关注，吸引更多的人气。后续能否做好，还要看当地是否能留住和利用这种人气，持续有效地发展。

Jack Becker（美国《公共艺术评论》杂志发行人、主编

玉环东沙已有良好的公共艺术意识与基础。国际艺术家们认为这里的自然
环境、生态状况以及历史建筑等都很有特色，但需要善加保护和利用。有
时人的活动往往会干扰当地的生活和现状。所以在面对这些其实非常脆弱
的历史文化遗产时，需要人们格外的关注和维护。在生态环境艺术与大地
艺术的概念中，强调在不破坏当地环境和生活方式的基础上，创建美好的
生态艺术作品，保护和提升当地生态环境资源，以增添生活的艺术美感。

整个公共艺术计划应该是一个漫长的过程，需要基金会与当地政府的有力
支持和一个管理系统的支持。此外，还需要给当地政府领导、大众普及生
态知识。当地政府背后有很多潜在资源，比如教育、媒体、企业方面的资源。

公共艺术和艺术家驻地计划是长远发展的计划，整个公共艺术计划涉及广
泛，在实施过程中，需要涉及历史、经济、生态等各个方面问题。在实施
整个地区规划的同时，还需引入"公共艺术品规划"的系统，更离不开当
地政府与合作单位的支持。公共艺术吸引人的特点在于能够影响人们对某
一地区的关注度。可以先请艺术家来实地考察，然后根据当地的实际情况，
建立长远的合作计划，才会行之有效。在定计划之前，要成立一个公共评
论团，一同参与规划的整个流程。诸如成本控制、时间等各种客观因素，
都需要一一考虑。当然，不管现在这里有怎样的公共艺术即将发生，我们
都会尽力传播整个艺术计划，直到大洋彼岸。

嘉宾讨论

上海大学美术学院院长助理金江波主持嘉宾讨论

嘉宾讨论

村干部代表发言

玉环县委副书记朱立国做总结发言

论坛实录

金江波（上海大学美术学院院长助理）

我以虔诚和感恩的心主持这个活动，因为我是地地道道的玉环人。这里发生的一切都与我有关。我也希望这个活动，能带给我们更美好的未来。我先请上海大学社会学院教授顾骏先生回答公共艺术在社会管理中的作用和地位问题，以及关于公共艺术如何介入生活的问题。公共艺术是可持续发展的机制，公共艺术的介入为当地产业提供契机，并在调和各种冲突上有所贡献。

顾骏（上海大学社会学院教授）

如今，中国经济发展正遭遇瓶颈。一个地方要发展，经济发展不可缺少。玉环原来的经济发展相当不错，但是现在也遇到瓶颈。比如人口太多、资源不足等问题。昨天在渔村看到渔民家的门前都有一个水柜，用来储存饮用水，因为现在全县用水都很紧张。玉环经济发展面临转型，我们就在这个背景上，观察玉环引入公共艺术的意义到底在哪里。第一个作用是继续提升玉环当地人民的生活质量。所有的发展都是为了人的发展，物质生活到一定水平之后就需要精神发展。

昨天看到的渔民表演，如果在不能维持生活的基础上进行，就没有意义。反过来，物质生活有了改善，但目的一直在捕获更多的鱼上面，这样的生活也是单调乏味的。所以，应该在改善当地生活的基础上，引入公共艺术，而引入公共艺术的目的就是为了美化当地的生活。第二点，引入公共艺术，有可能就是引入一种产业。就好像捕到的鱼越来越少以后，我们需要其他的生活方式，需要引入其他产业。如果公共艺术的开展，使渔民可以在原来的生活方式上，找到新的谋生途径与生活方式，那么这个地方就能得到更好发展。例如小吃，在家庭上叫饭食，在旅游上，叫饮食文化。公共艺术的引入，能否为当地带来新的产业发展契机，这是非常重要的。如果有更多渔民可以通过鱼龙舞而不是捕鱼获得生机，这就是产业升级。第三，在发展产业时，可能会面临各种利益调整。有近期的、远期的利益冲突，也有各个群体之间的利益冲突。我们昨天看到的部分设施用途的改变，涉及当地人是否愿意这样做的问题。所有的问题都是在一定体制框架内进行的。但这些体制框架不是很容易改变的。公共艺术就代表一种解决问题的方法。其最大优势在于我们是用智慧解决问题，而不是用其他简单的方法解决问题。公共艺术可以给我们带来讨论的机会，也可以给我们带来相互妥协的平台。

总体来说，发展公共艺术，在玉环不仅符合发展的目标，也是发展的手段。所以如果能很好运用公共艺术这个载体，玉环的发展前景是远大的。玉环是因为一个传说而得名的，若干年后，玉环不一定以这个传说，可能以公共艺术著称于世。浙江有个小地方原来默默无闻，现在全世界都知道，那就是义乌，全世界过节的小商品，都从那里批发。如果说哪天世界上许多公共艺术都与玉环东沙有联系，那么公共艺术有多大影响，玉环也就会有多大影响。

金江波：公共艺术是可持续发展的机制，公共艺术的介入为当地产业提供契机，并在调和各种冲突上有所贡献。同样也提出一个问题，当地的村民是否需要公共艺术介入他们的生活？他们是否意识到公共艺术的介入能为他们的当地生态与生活品质带来提升呢？请清华大学美术学院《装饰》杂志主编方晓风先生回应这个问题。

方晓风（清华大学美术学院《装饰》杂志主编）

玉环东沙要做这个项目，我心目中最理想的模式叫做"微循环模式"，就是看公共艺术在改变当地人生活的过程中，到底起什么作用。打个比方，如果把公共艺术介入后，东沙的整个改变理解成一个化学反应的话，我希望在这个过程中，艺术家和政府官员应起到催化剂的作用，而不仅仅是变成化学反应物的一部分。其中更强调的或更理想的场景，是让村民直接参与到整个环境的改变、生机和产业链之中。今天的东沙村看上去有个非常美丽的基础，但这个基础是自然形成的，并非规划的产物。

20世纪60年代，在美国现代博物馆中，有个很有名的展览，叫"没有建筑师的建筑"，讲的是世界各地民居的状况。因为他自己也是建筑师，建筑师很容易把自己当上帝。希望在这个项目中，艺术家和设计师会是当地人的朋友，是顾问的角色，而不是上帝。昨晚的聚会场景，让他看到任何人对美、对艺术，都有参与的热情，也相信此地的村民和干部中，必有藏龙卧虎者，有很多人才，他们可以依据自己的生活，产生许多想象不到的创作热情和能量。他希望将来变得更美好的东沙村，其主人仍然是东沙村的村民，而不是外来人员。

金江波：由此想到此次获大奖的荷兰艺术家的作品："农民、农场与他的妻子"。希望这位艺术家谈谈如何让村民、艺术家、当地政府相互协调、联手打造更好的生活方式。

Henerrite Wall （首届国际公共艺术奖获奖艺术家）

昨天大家看完表演后，有机会来到一个农民家庭，看到桌上放置一个神龛。我好奇地问当地人这个物件的功用。当地人回答：这个在当地很常见，当丈夫出去捕鱼的时候，家里供着神龛，作为平安和祈福的寄托。我就想到当地人和艺术家的关系问题，比如艺术家也可以写祝愿在神龛上，这也可以成为一种互动的艺术。就像大家沿路走来，一路挑了些村民现做的小吃来品尝，这也是一种公共艺术。关于艺术家驻地计划，需要给艺术家一种充分的信任。每个国家的人都有不同特点，需要建立一种平衡。其中艺术家与当地相互交融、平衡的过程是最重要的。

丁乙（上海视觉艺术学院美术学院副院长）

我自己虽然是一个艺术家，但所在学校有公共艺术专业，在整个教学理念中，更关注的是区域公共艺术的概念。我希望建立一个客观的思想者的概念，让学生学会更多地倾听当地的声音。公共艺术存在项目的差异性，每个地区的项目也完全不同。如何客观面对要研究的对象是公共艺术在教学理念上最基本的特点。同时应该注重社会调研这个课程，可以让学生深入了解创作的对象，以及整体与局部的信息。这些调研来自于所研究对象的各种角度，如经济、建筑特点、色彩系统等，并且都有相应的课程让学生去深入学习探讨。在公共艺术创作过程中，应采用以渗透型的方式而不是强硬的介入性方式，不应该改变区域的现状。这就如同用一种微创手术，来对今天的城市古建筑进行保护，使它更有存在和延续的意义。

金江波： 赞同丁乙先生的发言。此外，台湾与坎门，在地域上有一定的相似性与差异性，接下来请台湾东海大学美术系主任林文海谈谈台湾在乡村实践中的历史经验，以及他看到两地存在着哪些差异性？

林文海（台湾东海大学美术系教授）

由于今天全球化、国际化的发展趋势，原以为西方很强大的文化介入会消灭小地方的文化。因为人们在思维中喜欢大、新、异己的事物。像今天，这么多来自不同领域的人相遇，是一件奇异的事。那么我们要融合，还是混合，还是同化呢？我来自台中，台中有一种著名的饮料——珍珠奶茶，珍珠是台湾的、地方的，奶茶是西方的，通过人把它们混合在一起，彼此并没有被消灭掉，而是形成了新的品种。就像手机里面还有其他随身听、相机等功能，成为了一个很好使用的工具。而台湾的木瓜牛奶，里面既看不到木瓜，也看不到牛奶，形成一个新的口味，这是一种融合。玉环东沙如今正在积极进行公共艺术项目计划，我比较感兴趣的是，这个项目是以居民为主角，还是艺术家或者政府为主角？谁才是真正的主角？在台湾，此类艺术进程大致分为三个阶段：一是社区营造；二是艺术进入空间；三是公共艺术。以艺术家为主导，艺术家进驻村落后，所呈现的影响力，能为地方重新诠释区域的特性。在台湾，这三个方面有一个时间性、阶段性的问题，应该循序渐进，有计划地进行，当然也包括立法的完善。

金江波： 林教授提到的关键点，也是我们面临的问题。谁是主导？谁将受益？建设历史文化村落，如何在保护文化遗产的同时又能提升村民居住品质？政府有什么作为？如何保持这之间的平衡？请东道主曾志斌先生做出回应。

曾志斌（坎门街道主任）

人民群众是最大的历史创造者，政府是政策的制定和引导者。在规划的实施过程中，政府将不遗余力地推动政策的实施和落实。在规划实施中，组织好当地居民，参与共同保护开发，共享成果，更好地提升当地群众的生活和生产。作为当地政府，在文化保护和环境建设上责无旁贷，但要做到"有所为而有所不为"。

一要树立"保护优先"的传承意识。文化是一个地方的特色和魅力所在，保护文化遗产，既传承传统文脉，又维护精神家园，意义深远。对此我们始终坚持将"保存修护"放在首要位置，保留历史的真实性，凸显风貌的完整性，呈现生活的延续性，力求由内而外地拓展文化张力。

二要明确"因地制宜"的建设思路，在保护优先的前提下，注重规划的全面性和整体性，统筹兼顾建筑单体、地理环境、传统格局、风土人情、文化遗产等的保护、发掘和修复，在建筑、环境与景观营造上突出地域特色，尊重和延续村落特有的习俗，塑造村落个性，彰显具有独特吸引力的乡土风貌。

三要健全"以人为本"的管理机制。在推进做好保护和建设两大基础性工作的同时，我们也要着力抓好历史文化村落的配套管理，切实履行好各项中心工作的岗位职能（包括美丽渔村建设、清洁家园、非遗传承、环保监察等），做到"软硬兼施"，促进文化与人居、传统与现代、人与自然达到有机融合。

John McCormack 新西兰 Starkwhite 艺术机构总监

想用一种与众不同的方式来解说公共艺术的一些问题。先从自己的经历说起。首先，昨天的欢迎仪式和沿途看到的风景，令我倍感惊喜；其次，欢迎仪式所创造的别出心裁的空间，也令我久久难忘。昨晚一起画画时，汪院长平易近人，大家欢声笑语，其乐融融，这种场合对我而言是非常美好的时光。这也让我深思，到底要组织什么，要让什么在渔村发生？

沿路无论是表演还是当地的饮食，都是一道风景线。此外，沿海的海岸线，在新西兰也有，如何在海岸线上做文章，怎样搭建一条能够观赏到美丽风景的海岸路线，非常重要。除了沿海线，还可以设计一条去森林的路线。我想可以运用现代先进的技术，外加相应的介绍和地方音乐的设计，让人们更多感受地域文化，从而对乡村有更全面的认识。

昨天在签字的时候，就知道有一排房子，可能要建成美术馆。看到很多人在参与设计，这过程远比建造房子的结果更有意思。东沙到底要建成什么样？是旅游中心还是艺术中心？政府有什么主意？我觉得应该让大家享受渔民生活，让大家感受乡村文化，这是这个项目的乐趣所在。所有观景的海岸线设计、取名、驻地计划等，都是让当地民众一起参与且相互影响的过程。还可以在森林中创造一个空间，让人们有一个特别的角度和地点，停下脚步观看风景、享受生活，这也是我们所要关注的。

朱立国（玉环县委副书记）总结发言

听完关于东沙渔村的建设及有关玉环的整个建设理念，很受启发。针对东沙美丽渔村建设，主要谈两点，第一，东沙美丽渔村建设前景美好。主要基于三方面原因，一是当地居民的基础。因为玉环、坎门原来是多种文化的交融地，有海洋文化、农耕文化和移民文化，其包容性、开放性特别强。玉环在改革开放以后，最先发起的也是坎门，在改革开放初期，坎门最先走出去，引领整个玉环乃至台州的时尚。坎门东沙美丽渔村建设，有公共艺术介入，也有很好的群众文化基础与底蕴。二是国际艺术家们介入坎门东沙的美丽渔村建设，大家利用自己的知名度和理念创造了很大的平台。三、当地政府官员对于这个问题的看法。这个计划需要城乡统筹发展，城镇化是我们今后的必走之路，需要通过美丽乡村建设来实现。希望通过这个计划，让农村享受现代文明。

第二，美丽渔村建设的路径。一是公共艺术介入渔村建设，最关键的在于前期的理念策划。渔村各有差异，需要差异性的发展和建设，公共艺术的介入就是理念的创新。二是有了良好的理念和策划后，还需要规划设计，需要落地可操作。三是建设的主体是东沙村的村民，村干部应当积极发动老百姓自愿自主的参与。具体实施以村集体为单位，体现社会主义制度的优越性。政府给予帮助和资金的扶持，带动整个社会资金的投入，这是整个工作的理念和路径。美丽乡村建设，有利于当地历史文化的传承与发展。希望东沙美丽渔村成为可持续发展的生态文明。也希望由于公共艺术和国际艺术家的介入，美丽渔村建设最终实现村美、人和、民富。

访谈

Interview

用公共艺术打破"博物馆泡沫"
——"2012 国际公共艺术奖"评委会主席路易斯·比格斯访谈

路易斯·比格斯（Lewis Biggs）是"2012 国际公共艺术奖"评委会主席，曾任英国泰特美术馆馆长和利物浦双年展艺术总监。他积极推动英国当代艺术走向公共空间，促进艺术与城市之间的联系，为英国公共艺术的发展起到了重要的作用。在访谈中，路易斯对"2012 国际公共艺术奖"的评奖过程做出了总结，并对中国的获奖作品进行了点评。此外，他还从自身经验出发，介绍了英国公共艺术的发展现状与成功经验，并阐述了对中国公共艺术的独特见解。

我们知道，您长期关注公共艺术，从作为利物浦双年展的策展人开始，我们想知道公共艺术在您的职业生涯中起到了什么样的作用。

路易斯：我的职业生涯最早是以艺术史家开始的，然后是策展人。我对公共艺术的感情源于我在博物馆工作的一些挫折经历。从博物馆的工作经历中，我发现艺术在博物馆的展出方式总是受艺术史的制约。那么，对于那些对艺术史并不感兴趣的公众而言，这就是一个跟艺术有关的难题。因为我的第一个学位是历史学，所以我比较了解历史，也乐于见到历史和艺术的结合。对很多人来说，历史并不重要，所以，当他们有机会欣赏当代艺术时，历史就会成为一种障碍。我认识许多艺术家，他们很多人都喜欢博物馆，因为他们通常希望将作品卖给来自博物馆环境的人们。但这种环境对艺术家的创作而言也是一种很大的局限性，其实艺术家也期望摆脱在历史环境中工作的束缚，历史对艺术家而言也是一种巨大的负担。所以，最后我对那些希望反映现实生活的艺术家更感兴趣，而失去了对那些仅仅创造"博物馆泡沫"的艺术家的兴趣。

现在回到你的问题，我之所以对公共艺术越来越感兴趣，是因为我意识到博物馆对我自己以及我感兴趣的艺术家存在很大束缚。所以，我一直在用公共艺术来打破"博物馆泡沫"。

您什么时候开始介入公共艺术？是否有一些成功例子？

路易斯：我在担任利物浦泰特美术馆馆长时，首次尝试做的公共艺术项目，就是博物馆的一种外延形式。我的同事罗伯特·霍珀，他在利兹的亨利·摩尔基金会工作，资助我筹备了"Artranspennine98"，这是个大型的展览，一半作品展出在博物馆和画廊里，另外则展出在画廊和博物馆外面。展览遍及英格兰全境，去过 30 个地方，走过 150 英里。从某个方面来说，作为展览这次活动是失败的，因为参观者并不愿意长途跋涉去参观所有的展品。但从另一方面来看，它又是个前所未有的展览，也是个成功的展览。因为艺术家们继承了这个展览。自我们 1998 年举办以来，每五年就举办一次类似的展览。2003 年和 2008 年，艺术家们自发地举办了展览，如今，2013 年又举办了一次。他们问我是否可以使用相同的名称，因为他们十分推崇这个展览，并希望再次举办。

这个展览的名称是什么？

路易斯：Artranspennine 98。

从参观人数来说，那次展览是不成功的。对于游客而言，因为展览的展出地点很多，很难参观完所有展览。然而展览背后的意义非常深远，这也是展览多次被重复的原因。展览还影响了政府的政策。因为我们把英格兰视角概念化了，当时这种视角是不被政府官员们所接受的。但是在我们展出 3 年以后，政府也开始采用英格兰视角来发展旅游业，政府也认为这是一个很好的想法，2002 年，这个有关英格兰地区的展览被命名为"北方之路"（the Northern Way）。

对我来说，这种艺术主动性在社会中能起到引领作用。艺术家们擅长想象与现实生活不一样的生活，所以像我这样的人（也可能与文化工作有关）希望充当艺术家与政府之间的桥梁，让政府从艺术家的想法里得到启发，从而把它变为现实。这就是为什么我对公共艺术感兴趣的原因。当然，艺术家的一个作用是丰富博物馆的艺术史，但对我而言，我更感兴趣的是艺术家促进社会发展的作用。

在英国的公共艺术发展中有什么比较成功的案例？无论您本人参加与否。

路易斯：有个案例其实是政府倡议的结果。有个沿海的小镇，当时它的经济优势还没有被开发，该镇负责经济发展的部门找到了我，问我是否有艺术作品能吸引更多的人到该镇的海边旅游。当我们接受这个工作时，镇里的居民支持和反对的各占一半。其实引起争论也是好事，因为人们开始思考艺术能做什么。

我们准备先做短期的，只把作品放在那里 18 个月，18 个月后再决定是否要长期放置。在 18 个月内，这个展览成功了，它带来了巨大的观光游客量，从而对当地经济产生了积极的影响。因此，希望作品保留的人们获胜了——因为它刺激了旅游业和商业，经济也得到了改善。

这件事的成功有三个方面：首先是经济方面，因为参观该地区艺术作品的人数增加，所以当地居民挣到钱了；其次是效应评估方面，作品对该地区的外在影响力产生了效果，过去这个海边小镇籍籍无名，但现在借助于艺术作品，海滩的照片无处不在，它变得众所周知。在英格兰的任何地方，都可以看到艺术品在海滩上的照片。它推动了英格兰北部的发展，这个地区已经成为一个非常著名的地方；第三是"社会资本"的增益，这是一个很难评估的概念，它关注的是群体内部的交际方式。任一群体中不同个体的认识程度和关联程度是可以估量的。如果我们把一个有争议的艺术品放入社区，就会使相互从不说话的人坐在一起讨论。因此，"社会资本"的增加，对社区是好事。

再讲一个例子，这个例子提出了公共艺术应该是永久性的还是临时性的问题。我认为这对公共艺术来说是一个非常有趣的问题。我想谈论的是维多利亚别墅（Villa Victoria，Tazro Niscino 在 2002 年建的只有一个房间的酒店）。这个作品引起人们对公共空间私有化的关注。其次，它解决历史的传承问题，如何为现代创作带来灵感；第三，因为是暂时的，它更像一个演出（存在于人们的记忆中），而不像是一件物品。这是一个经历而不是一个展品。总的来说，我认为，公共艺术应被看作是一种体验，而不是物品。

中国有较多报道，英国在公共艺术方面取得很多成就。英国虽然是老牌的工业国家，但保留了文化魅力。中国也开始现代化转型，英国的这些经验对中国有什么启示吗？

路易斯：这个问题的有趣之处在于，在更广泛的文化层面上，公共艺术体验究竟是一种导向还是一种现象。在某种意义上，文化创造在英国是一个历史悠久的成功产业。文化是不可能立竿见影的。利物浦就是一个很好的例子，2008 年，利物浦是如何成为欧洲的文化都市，是因为，200 年前，利物浦就已经是一个富裕的城镇，100 年前，它就成为一座文化城市。利物浦文化产业体系的发展是许多年前私人和公共资金投入的结果。因此中国现在巨资投入文化产业是件好事，但成果可能要二三十年之后才能显现。

这次举办国际公共艺术奖和学术论坛，您为什么会这么热心地参与其中？

路易斯：我一直生活在利物浦，我的职业都与艺术有关，我所感兴趣的是创造发展的机会。我希望我能为利物浦打造更好的艺术基础，这正是我所尝试做的。我希望我也能为中国的艺术基础做出贡献，这也就是我目前在香港和日本所做的尝试。我应香港和日本的要求关注、考察和分析当地艺术发展的需要，并努力促成这些需求。

此次评奖是国际性的，论坛也是国际性的，这和您以前的职业经历有什么不一样的地方？

路易斯：当然对于我来说，这是一个很好的学习过程。虽然多年以来，我一直在英国担任比赛评委，但我喜欢了解不同背景下的公共艺术状况。国际评奖仍然是个挑战，因为存在着许多有趣的文化差异。我不知道在中国公共空间的概念是什么，因为这里的环境同英国不一样。如何定义公共艺术非常重要，因为公共艺术必须同公共空间相关联。这种文化差异深深地吸引了我。当然，如果我能帮助其他人更好地了解各自的实际情况是件好事。我坚持利物浦双年展邀请许多国际艺术家参与，因为比起根植于某种文化的艺术家，来自不同文化背景的艺术家对同一事件会有更多有趣的见解。

与过去相比，这些新的经历对您来说有什么新的挑战，给您什么新的认识？

路易斯：我不认为这是一个全新的体验，当然也有新的人和新的内容，但谈不上真正的新经验。有一点我要说明，虽然我的动力源自我对艺术的热爱，以及艺术家为社会做出贡献的使命感，但有时我的动力也来自于愤怒。比如当我走在利物浦的街道上，看到整条街都是空置的房子或是废弃的房子，当孩子们途径这样的街道去上学时，我意识到这样的社会会毁了孩子们，他们在充满社会问题的环境中长大，而这些问题正是我们造成的。我去过很多亚洲城市，这些城市正在或者已经破坏住在那里的人们的生活，这也使我很愤怒。

在利物浦的愤怒来自哪个方面？

路易斯：我的愤怒是一种反应，来自中央和地方政府对人民的一种虐待。街道之所以成为这个样子是由于政府政策，人们被迫生活在给他们带来困难的环境中。我们都受到周遭物理环境的影响，我是说，我们的生活受我们居住地的物理环境的影响，就像我们的家庭对我们的影响一样。如果物理条件和自然条件都很差，我们都会成长为残缺之人，之后又将危害整个社会。

在中国，事事都可能有变化，这对孩子的成长有利吗？

路易斯：有变化当然是好事，特别是在环境不好或不明了的时候。在城市里推倒老房子，又取代之地盖起高楼大厦。这个现代化的过程同样发生在 1960 年的英国。当然，这些新房子都是全新的，又有很多改进，肯定要比老房子好。但是现在在英国，我们不得不将 20 世纪 60 年代所建的高层建筑都推倒，因为人们发现居住在高楼大厦里有许多问题，甚至会引发社会问题。目前我们正在为他们建造平房或类似平房的房子，并做一些不同的安排。

在世界各地我们可以看到，所建造的房子都是为建筑商谋利，从来不会考虑人们住进公寓里会遇上什么困难。我的意思是，每个人都应该享受高质量。质量不是商标，更不是金钱，而应该是种关怀，给予人们做出决定的权利，并为他们考虑。

您会愤怒吗？

路易斯：是的，我很愤怒。因为这件事带来这么多麻烦。在英国、法国郊外、圣保罗、里约热内卢等地都能看到社会住房引发的矛盾，甚至世界各地都存在同样的问题。在规划和发展住房的商业进程中，人们被剥夺了他们居住地的文化。

您认为公共协会会做些什么吗？您对此有何期待？

路易斯：我们想知道不同国家的公共艺术信息，所以我们就设立国际公共艺术奖。这些消息的传播并不是很容易。我们的愿望是分享信息，因为公共艺术的本质是建立在本土文化上的，而分享本土文化信息是非常困难的，所以我们在许多国家开展研究，并把这些研究成果集中到一个地方，以此推进公共艺术的发展。虽然公共艺术协会是个比较小的机构，但它致力成为信息交流中心，联络其他相关机构。虽然现在许多国家都有公共艺术组织，但相互之间并没有像预期那样保持联系，公共艺术协会可以尝试鼓励他们相互联络。

您觉得在当前中国公共艺术热的情况下，公共艺术协会会不会对中国的公

共艺术发挥一些专家顾问的作用？或是其他作用？

路易斯：我不认为公共艺术协会会做一些咨询的事务，因为我认为公共艺术咨询最好是在当地，或本国，不可能是国际咨询。我觉得协会是教学研究和信息研究的交流机构，而不是行政机构。世界各地的人们可以保持联系，同时，承担起实际的咨询工作，并使用协会提供的有用信息。

您对这次获得大奖的六个案例有什么评论吗？为什么它们会在众多案例中脱颖而出？

路易斯：我想最重要的是这第一次大奖表明我们的研究是全球性的。我们希望确保六大洲都各自有一个备受赞誉的项目，所以设定六个奖项。我们也应该认识到，在非洲做公共艺术和在北美做公共艺术的方式是完全不一样的，所以我们从北美洲和非洲寻找不同的作品。但我们也希望来自世界各地的项目存在着共同点，也就是艺术家在项目中起到的作用。艺术家的作用应该是很重要的，同时，工艺、材料和技术的质量都应该是不错的，参与的程度应该也是不错的。但是，以非洲大部分地区为例，那里没有私人或公共赞助体系，所有项目都是艺术家自发创作的。而在北美，就有非常完善的赞助体系。

对中国的四川美院虎溪校区您的印象如何？有什么样的评价？

路易斯：对我而言，这是一种全新的经历。让我印象最深刻的是，校园没有把所有农民都赶走，而是让他们继续生活在这里，他们不但在那里从事园艺，而且也参与校园环境建设。来校的学生和教授们能在这充满活力的公共艺术中同农民交流，了解校园所在地方的特点。这是公共艺术的一次成功运用，住在那里的人们都感到自豪。

我们知道您正在筹备今夏在日本爱知举办的三年展，你能谈谈公共艺术在中国和在日本有哪些区别吗？有哪些经验值得我们学习？

路易斯：我认为在不同文化和不同环境里可以很好地学习公共艺术。但由于文化不同，国情不同，所以不一定要学习别人的经验。然而，日本当代公共艺术很有意思，在 20 世纪 80 年代的房地产热潮中，有许多企业，甚至地方政府买入和卖出放置在户外、街道或公园的艺术品。那不是真正的(与公共有关)公共艺术，这只是在画廊和博物馆内外举行的交易活动。这种行为一直到房地产繁荣结束时才停止，然后，艺术市场发展放缓。在 20 世纪 90 年代和 21 世纪初，政府资助当地社区创作艺术，这是一种振奋民心的方式。在这些受到资助的艺术家中，有些可能不是优秀的艺术家，有些也可能不擅长公共艺术。第一、第二期都没出现优秀的公共艺术作品。但也有一些好的例子，在地方政府或企业的赞助下，一些海外艺术家同日本艺术家合作创作出一些杰出的作品。越后妻有 (Echigo Tsumari) 和贝尼塞基金会 (Benesse Foundation) 就是两个很好的例子。

用公共艺术打破"博物馆泡沫"
——"2012 国际公共艺术奖"评委会主席路易斯·比格斯访谈

You have a history of engagement with public art, starting from your leadership of Liverpool Biennial. We want to know what kind of role public art plays in your career.

Lewis: Well, my career really started as an art historian and then as an exhibition organizer in museums. And so my feeling about public art grew out of some frustration I experienced while working in museums. I understood from my work in museums that the way art is experienced in the museum is always conditioned by the history of art. And so for people who are not interested in the history of art, it is a big problem in relation to art. Although I know the history of art myself, and my first degree was also in history, so I am very happy with history as a context for art. But for a lot of people history is not important, and this is barrier when they have the opportunity to enjoy contemporary art. I know a lot of artists, and some of them love the museum context-often because they want to sell their works to people who are also from the museum context. But this context is actually also a great limitation on what artists can do and many artists want to be free of the restriction of working 'within history' – art history is a huge burden for artists to carry. So I found in the end that I was more interested in the artists who want to react directly to life round them rather than making art for the 'museum bubble'.

So, to come back to your question, I became most interested in public art because I realised that the limitations of the museum were too restrictive for myself, and also too restrictive for the artists I am interested in. So public art for me has been about breaking out of the museum bubble.

When did you start involving in public art, and can you mention any successful cases?

Lewis: My first sustained attempt to engage in public art was a form of outreach from the museum, when I was director of Tate Liverpool. With my colleague Robert Hopper of the Henry Moore Foundation in Leeds, I organized Artranspennine98, a very big exhibition that took place about half inside museums and galleries, and half outside the gallery and museum context. The sites for the exhibition stretched from coast to coast of England in 30 different locations, over 150 miles. In some ways it failed as an exhibition – visitors were quite unwilling to travel to see it all. But you could just say it was 'before its time'. It was also a success, because artists have chosen to repeat the same concept each five years since we did it (1998). In 2003 some artists organized themselves to repeat it. And 2008 they repeated it again. And now again in 2013. And they did this without my support. They even asked me if they can use the same name, because they honour the original exhibition and want to repeat it.

What's the name of this exhibition?

Lewis: Artranspennine98.

The exhibition was not successful from the point of view of the numbers of visitors; for the visitors, the exhibition was very hard to negotiate, because it was so spread out. However the idea behind it was very, very strong, and that's why it has been repeated by artists. And also actually it influenced government policy after the event. Because we conceptualized the view of England from coast to coast which was not accepted generally at that time by politicians. But three years after we had done it, there was a government initiative to make use of this view of England to encourage tourism. Even government realized that actually it was a good idea, and in 2002 the area of England defined by the exhibition was named 'the Northern Way'.

So for me, this is an example of how artistic initiatives can play a leadership role in society. What artists are good at is to imagine how life can be different from the way it is. So the job of people like me (involved with culture) is to try to act as a bridge between artists and politicians, to allow politicians to steal the ideas from artists in order to make them happen. The reality of this situation is why I am interested in public art. Of course there is a role for artists in adding another step into the history of art in the museum. But for me, this is not as interesting as artists that are adding another step into the development of society.

What do you think are the successful public art cases in England, whether you have participated in or not?

Lewis: The example I am going to give you now was in fact the result of a government initiative. There was government agency responsible for economic development of a small town where there was a beach that was also an unrealised economic asset. The agency approached me to find an art work that would increase the number of visitors to this beach. So we did this job, but half of the population was very against it, half of the population was for it. This was good thing that created a lot of discussion and debate and people started to think about what art can do.

We decided to act on a short term basis, that we would put the artwork there for just 18 months. After 18 months, there had to be a decision about whether the work would stay or the work would be taken away again. But the success during the 18 months was that it brought so many people to see the beach, and it had a positive affect on the local economy. So the people who wanted it stay won over the people who didn' t want it to stay - because it had stimulated tourism and business, and the economy had been improved.

We were able to say it had been a success for three reasons. One economics-local people are making money because many people come to see the art in this area. The second way is to evaluate the effect it has on the external imagine of the area. In this case, the beach was not known to anybody before, but now-thanks to the artwork - photographs of the beach are everywhere and it is very, very well known. Anywhere in England you can see photographs of this artwork on the beach, because it is used to promote the North of England, so it's become a very well known place. A third way to evaluate success is the gain in 'social capital', which is a hard concept to evaluate, but it concerns the way which the community communicates with itself. You can measure the degree to which different people in any community are connected with each other, know about each other. If you put a controversial artwork into a community, it very often makes people get around the table who normally would not speak to each other. So there is an increase in 'social capital'. General speaking, it is good thing for community.

And I'd like to mention one more work, and this one raises the question as to whether it is better for pulic art to be permanent or not permanent. I think this is a very interesting question with public art. I want to talk about the Villa Victoria (one room hotel, by Tazro Niscino, 2002). That work directed peoples' attention to the privatisation of public space. Secondly, it addressed the question of the heritage - how to take contemporary energy from heritage. Thirdly, because it was temporary, it is more like a performance (which lives in peoples' memories) rather than being an object. It's an experience not an object. In general I do think that public art is best thought of as an experience, not an object.

Many reports in China suggest that the UK has achieved a lot in public art. Although the UK is a tradional industrial country, it still retains the cultural charm. Since China has also begun to become a modern society, what helpful experience can be brought to China?

Lewis: I think the interesting thing about this question is the degree to which the experience of public art is a driver or a symptom of the wider culture. Cultural creativity in the sense in which it is now considered a successful industry in the UK has come from a very long history. It's very hard to do cultural things quickly. Liverpool is a good example - why it was able to be European Capital of Culture in 2008. Because it was already a wealthy town 200 years ago it also became a very cultural place 100 years ago. The institutions of the culture industry in the city resulted from an investment made by private and public money a long, long time ago. So China is now investing a great deal in culture, which is great, but the good results may take 20 or 30 years to come through even so.

Why are you so actively involved in the International Public Art Award and the Public Art Forum here in Shanghai?

Lewis: Over the entire time that I was working in Liverpool, although art was and is my subject, what came to interest me most was creating the opportunity for development. I hoped I could enable Liverpool to develop a better art infra-structure -hat was what I was trying to do. And I hope that in China, I can also contribute to developing the infra-structure for art. That is also what I'm trying to do in Hong Kong and in Japan too. This is why I am also invited to Japan and Hong Kong, to try to look at the situation, analyze and see what needs to be developed further, and try to make that happen.

The public art award is international, and the forum is also international. As regard to your career, any fresh experience do you think you can get from your participation?

Lewis: Of course, it is a learning experience for me. And I love trying to understand unfamiliar situations, although I have been involved as a judge of competitions in the UK for many years. The international stage is still a challenge because of the many fascinating cultural differences. I am not sure what the concept of public space means in China, the situation is so different. So this is very important to discuss what public art can mean because public art must be related to public space, so this cultural difference is fascinating for me. And of course if I can contribute to other people understanding more about their own situation, that is wonderful too. For Liverpool Biennial, I used to insist on bringing many international artists to Liverpool, because people from a different culture always have so much more interesting things to say about a particular situation than the people who are already inside that culture.

Are there any new experience for you, while participating in the activities?

Lewis: I don't think it's a completely new experience: of course there are new people and new content, but not really new experience. Something I want to say, although I am motivated by love of art and what artists can contribute to the society, I am also motivated by anger. When I walk down a street in Liverpool and see whole streets of houses that are empty, derelict and boarded up, and there are children walking to school along these streets, then I know that these children are being damaged by society, and that they will grow up with problems, of course, by what we do to them. And I have visited many cities of Asia, which have been or are being built in such a way as to damage the people who live there. This makes me very angry.

What is the origin of your anger in Liverpool?

Lewis: My anger is a reaction to the mistreatment of people by central and local government. Because these streets are the way they are because of government policies: people are forced to live their lives in conditions that give them problems. We are all influenced by our physical surroundings. I mean we are much influenced by the quality of the physical environment where we live, as we are by the quality of our family. And if the physical and environmental conditions are bad, then we grow up as damaged people – who later do damage to society.

In china, everything is changing dramatically, is it good for the way children grow up?

Lewis: Change can of course be good, particularly if conditions are bad, but actually I don't think the situation just now is clear. You are knocking down the old houses in towns and building high-rise apartment blocks instead. This is the same process of modernisation that happened in the UK in the 1960s. Of course, the new buildings are new and they have many good things about them that are better than the old houses. But in the UK we now have to knock down all the high-rise buildings from the 1960s because people find them too difficult to live in, and many people who do live in these high-rise apartments have been creating social problems. Now we are trying to help them live on the ground level or near the ground level and arrange differently.

And you can see everywhere in the world, the way that housing is built is to make a profit for the developer without consideration as to whether the kind of building is difficult for people to live their lives in. What I mean is that everybody deserves quality. Quality isn't about brands, and it isn't about money. It is about thoughtfulness, it is about giving people some power to make decisions, and about giving them consideration.

One reason I' m interested in place-making is because the people who are planning and building the new towns, for the most part, are not proposing to live there themselves. Everywhere in the world you can see land is treated as money, it is not treated as social capital. Where people live should be about the people who live there, it should not be reduced to a question of how much money some people can make from the situation. Because housing development is treated as a business, rather than as a cultural enterprise. If governments and businesses treated the development of housing as a cultural enterprise, then we would have new 'places for people' being built; instead, we just have buildings, with no soul and no culture attached to them.

Are you angry with that?

Lewis: I am angry, yes, because so many troubles come from this. You see this trouble in social housing projects, in England, in the outskirts of Paris, in San Paolo and Rio de Janeiro, actually all over the world, the same problem. People are deprived of culture where they live, usually through the process of planning and developing housing as a business.

What do you think the IPA will do, and what are you looking forward to?

Lewis: When we had the idea to make the International Award for Public Art, we were motivated by seeking information about public art made in many different countries. This information doesn' t travel very easy. So we were motivated by the desire to share this information, because the nature of public art is based on local knowledge, and to shareinformation about local knowledge is really difficult, so we had the idea to research in many countries and to bring all this research together in one place, and to use this research as an advocacy tool. So the idea of an Institute of Public Art is for a very small organization, which will try to keep other organizations in touch with each other, by acting as a central clearing-house for information. Because existing public art associations in many countries don't speak to each other as much as they might, an IPA could try to encourage them to speak with each other.

You see that the public art is quite popular here in China, do you think that IPA will play a consulting role or what kind of role IPA can play?

Lewis: I don't think IPA will have a consulting role because I think consultation is better done on the local—or national - level, not the international level. So I think we shall just have an institution involved with the exchange of education/information research, rather than an executive body. So people around the world can keep in touch with each other, but undertake the actual consulting themselves making use of the research information when it is useful to them.

About the six winning cases: why could they stand out of many entry works? Can you comment on these successful cases?

Lewis: We thought the most important thing for the first time we made this award was to demonstrate that the research being undertaken really was global. And so we wanted to make sure that we had one highly commended project from each of six continents. So that is why it was six.

It was also important to recognise that the way public art is done in Africa is very different from the way in which the public art is done in North America. So we looked for different things from the projects from North America and from Africa. But of course there was something that we wanted the projects from across the world to have in common, and that was that we were looking for the role played by the artists. The artist' s role should be important, and the quality of workmanship, material and technology should be good, and the quality of engagement should be good. But in much of Africa, for instance, there is no system of private or public patronage, so all the projects were very much were self-initiated. While in North America there is a very established system of patronage.

What is your comment on the HUXI campus of Chinese Sichuan Fine Arts Academy?

Lewis: I found it an extraordinary experience when I visited. I was particularly impressed by the way in which the farmers who previously lived on this area of land have continued to live on the university campus and have been involved, not only ingardening, but also in constructing the physical environment. The incoming students and professors could meet the farmers in this activity of making public art, and learn about the particular qualities of the place on which the campus was built. It's a very brilliant use of public art. They are proud of the place where they live.

We know that you are planning the Aichi Triennial in the summer of 2013 in Japan. Would you tell us the difference between public art in Japan and public art in China? What experience we can learn from?

Lewis: I think it is good to study public art in different cultures and circumstances, but it is not necessarily possible to learn from others' experiences, because each culture is different and each country has different circumstances.

However, I do think that the recent history of public art in Japan is very interesting: during the property boom of the 1980s there were many businesses and even city governments buying and selling artworks that were placed out of doors, in streets and parks. This was not really about public art (related to the public), it was just carrying on the same art dealing activity inside and outside the galleries and museums. This activity stopped when the property boom ended and the art market slowed down.

Then in the 1990s and 2000s government gave money to local communities to make art, as a way to help the morale of those neighbourhoods. The money went to local artists, but maybe some of them were not very good artists or not very good at public art. Neither the first nor the second period produced much good public art. But there are some examples where artists from outside Japan worked alongside Japanese artists under regional government patronage or the patronage of business to produce extraordinary work. I will just mention Echigo Tsumari and the Benesse Foundation as two high quality examples.

感悟

Personal Understading

不是任意一件放置在户外的艺术作品都可以被称作公共艺术作品。公共艺术必须兼顾精神和实体。它是特定区域内人们思维方式、社会文化背景的外在表现形式。而公共性与参与性是评价艺术是否应被称作公共艺术的最主要部分，其次是否敢于对现实问题和社会需求做出迅速和积极的回应。中国的公共艺术更需要对生活方式和思考方式进行主动干预。而地方重塑作为第一届国际公共艺术奖评选的主题，提倡的并不是简单的还原建筑物，而更像对人类本根的寻根溯源，为拥有相同情怀的人们寻找情感依托。（高浅）

人类之于世界，渺小、短暂、孤立。与外界建立某种关联，是自知与知人的机会，走向天地的同时照见内心。公共艺术的出现，正是这种谋求的直呈。公共对应私有，当不满足或不确定小我时，个体会渴求到更广阔的空间中寻求归属感。不同于普通公共设施的服务功能，仅满足人们的实体需求，艺术化的公共环境，借由艺术，期待更难能可贵的精神共鸣。都说人生如寄，沧海又一粟，把微观的人投入宏观的境，感受艺术与环境缔造的温暖和恒久，公共艺术是人类亘古自然观的现代进化。这个演进过程中，人，人与人，人与环境，人与美好的环境，结成同盟，当环境与人达成默契，人与外界不再对立，夏虫，亦可语以冰。（吴昉）

如今伴随着社会转型、城市化进程的背景，公共艺术已成为一种崭新的艺术观念，它不仅涵盖公共空间中，民众对于美的追求，同时强调文化认同、地方意识等形式在内的多元化艺术，成为民众参与城市发展，增进互信的纽带。当今中国正以阔步发展的形式迈入现代化国家的行列，艺术不可避免地成为引导城市生活、关注地域文化、探寻未来发展的动因，公共空间下的艺术作品也越来越多地反映出公众的文化诉求，更以重塑地域文明和文化生态的理念去探寻人类文明发展的最高形态。（马熙逵）

第一届国际公共艺术的众多优秀案例，让我们对地方重塑的主题更加明确，一件典型的公共艺术作品，是代表艺术家个人的观点还是代表公众和表现社会问题对于最终定性这个作品的意义很重要，虽然从形式、范畴、场所等可以把它归于公共艺术，但最重要评判标准还在于这个作品是不是与公众发生了关系，是不是对地区的再造和复兴起到作用，是不是具有公共性。对于公共艺术的判断，从艺术学和社会学角度来看，标准有明显的差异，艺术学侧重于视觉审美，社会学侧重于公共性。不管评价标准怎么变化，最核心的还是公众的参与和互动。（冯正龙）

撰稿名单
Editor Group

编撰者名单：高浅　吴昉　周娴　马熙逵　冯正龙

校对：陈锴洋　徐谦　秦勉　谢靖　洪靓　郭博文

后记

Afterword

本书是上海大学美术学院在 2013 年组织举办的首届国际公共艺术奖评选活动的梳理，也是对国际公共艺术评奖活动阶段性成果的总结。从前期活动开展到后期案例的评论、包括本书的出版、印制与发行，都与大家的努力密不可分。非常感谢上海大学美术学院院长汪大伟教授发起与组织该活动，为国际公共艺术领域的研究填补了空白；感谢学校领导和相关部门的大力支持；感谢国际公共艺术协会（IPA）的 Lewis Biggs、Jack Becker 先生不远万里，千里迢迢参与组织与评选，给予来自国际资源与学术力量的支持与指导，分享他们的专业评论和国际经验；感谢我的同事潘力、周娴、章莉莉、胡建君、姚舰等老师们在各个方面给予的帮助；感谢我院的博、硕研究生们在此次活动和书稿编辑过程中付出的一切努力；感谢上海大学出版社与柯国富老师为本书的出版、发行给予的积极的帮助与支持。正因为有了大家的合作和帮助，才有此书的顺利诞生和问世。

感谢所有为本书编辑、出版、发行而付出努力并给予过帮助的朋友们。

金江波
2014 年 11 月

图书在版编目（CIP）数据

地方重塑：国际公共艺术奖案例解读（全二册）/ 潘力，金江波
主编 . —上海：上海大学出版社，2014.11
ISBN 978-7-5671-1493-7
I. ①设⋯ II. ①潘⋯ ②金⋯ III. ①城市景观—公共艺术—
研究 IV. ① TU-856

中国版本图书馆 CIP 数据核字（2014）第 251077 号

责任编辑：柯国富

美术编辑：谷　夫

技术编辑：章　斐

书籍设计：章莉莉

排版助理：周姣姣　杨　巍　房兆骅

书名字体：周姣姣　柯孝琴

地方重塑案例解读 1
金江波　主编

出版发行　上海大学出版社
社　　址　上海市上大路99号
邮政编码　200444
网　　址　www.shangdapress.com
发行热线　021-66135112
出 版 人　郭纯生

印　　刷　上海上大印刷有限公司
经　　销　各地新华书店
开　　本　787×1092　1/24
印　　张　27.25
字　　数　600千字
版　　次　2014年11月第1版
印　　次　2014年11月第1次
书　　号　ISBN 978-7-5671-1493-7/TU・002
定　　价　260.00元（全二册）